ISBN 978-0-266-90283-6
PIBN 10907181

Historic, archived document

Do not assume content reflects current
scientific knowledge, policies, or practices.

List of Chemical Compounds

Authorized for Use Under USDA Inspection and Grading Programs

Listings Effective as of March 1, 1980

IITED STATES	FOOD SAFETY	MISCELLANEOUS	ISSUED
:PARTMENT OF	AND	PUBLICATION	APRIL 1980
iRICULTURE	QUALITY SERVICE	NUMBER 1373	

The LIST OF CHEMICAL COMPOUNDS is revised annually, with an update issued once between the annual editions. If you wish to be notified when new editions and supplements are issued, write to the following address and ask to be put on mailing N-504:

Superintendent of Documents
U.S. Government Printing Office
Attn: Mail List, Stop SSOM
Washington, D C. 20401

CONTENTS

LIST OF CHEMICAL COMPOUNDS

AUTHORIZED FOR USE UNDER USDA

INSPECTION AND GRADING PROGRAMS

Listings Effective as of March 1, 1980

INTRODUCTION

This publication lists nonfood compounds authorized by Food Ingredient Assessment Division, Science, Food Safety and Quality Service for use in slaughtering and processing plants operating under the U. S. Department of Agriculture poultry, meat, rabbit, shell egg grading and egg products inspection programs.

The U. S. Department of Commerce, National Marine Fisheries Service, recognizes this publication as an official list of nonfood compounds accepted for use in plants operating under the U. S. Department of Commerce, Fishery Products Inspection Program.

THIS LIST IS EFFECTIVE AS OF MARCH 1, 1980

Compounds authorized by letters from the U. S. Department of Agriculture after issue of this publication are also acceptable.

A nonfood compound is defined as any agent proposed for use in a federally inspected plant but not intended as an ingredient of a meat, poultry, rabbit, or egg product.

Compounds Which Require Evaluation

Listed Compounds

Compounds classified in the categories described on pages 6 through 12 of this publication require evaluation by USDA prior to use in plants operating under the USDA poultry, meat, rabbit, shell egg grading and egg products inspection programs. Letters indicating acceptability of the compounds are issued to suppliers by the Compounds Evaluation Unit. Copies of the letters must be supplied to Federal inspectors as proof of authorization until the compounds appear in the subsequent revision of this publication. Once a compound appears in the publication, the letters are no longer valid as proof of authorization.

Unlisted Compounds

Paints and some other types of nonfood compounds also require chemical acceptance by USDA, but are not categorized in this publication. Letters indicating continuing chemical acceptability are issued to suppliers by the Food Ingredient Assessment Division. Copies of such letters must be supplied to Federal inspectors as proof of authorization at all times, since the compounds cannot be listed in this publication. Such letters remain valid as proof of authorization until revoked by USDA. The final granting of authorization to use such compounds which do not come in direct or indirect contact with edible products or packaging materials is the responsibility of the Inspector in Charge at the official plant. Technical assistance will be provided by the Food Ingredient Assessment Division upon request.

Procedures For Obtaining Compound Authorization

1.   Maintenance and cleaning chemicals; sanitizing and pesticide chemicals; food processing chemicals such as scald media, tripe processing compounds; and fruit and vegetable washing compounds; and structural or equipment coatings proposed for application on site in federally inspected plants.

Send requests for such evaluation to:

> Compounds Evaluation Unit
> Food Ingredient Assessment Division, Science, FSQS
> U.S. Department of Agriculture
> Building 306, BARC-East
> Beltsville, MD 20705
> Telephone (301) 344-2566

All preparations to be evaluated by the Compounds Evaluation Unit must be submitted using the appropriate application form. A description of the forms and their use is provided in Section 4.1 of the "Guidelines For Obtaining Authorization of Compounds To Be Used in Meat and Poultry Plants." Obtain copies of the "Guidelines" and blank forms from the Compound Evaluation Unit.

DO NOT USE THOSE FORMS FOR ANY PREPARATIONS OTHER THAN THOSE SUBMITTED TO THE COMPOUNDS EVALUATION UNIT.

2. Shell egg cleaning, defoaming, destaining, and sanitizing compounds (categories Q1,Q2,Q3,Q4,Q5,Q6)

Although such compounds are included in the "List of Chemical Compounds," their evaluation is not the responsibility of the Compounds Evaluation Unit.

Send all requests and inquiries concerning such compounds to:

> Poultry Grading Branch
> Poultry And Dairy Quality Division, FSQS
> U. S. Department of Agriculture
> Washington, DC 20250
> Telephone (202) 447-7410

Copies of authorization issued by the Poultry Grading Branch are sent by them to the Compounds Evaluation Unit, and the compounds are subsequently inserted in the "List of Chemical Compounds."

3. Packaging materials, marking and branding inks, coatings applied to equipment or structural members prior to installation in the plant, and all materials such as metal alloys, plastics, belting, hose, etc., proposed for use in association with processing facilities and equipment.

Send requests for such evaluation to:

> Compounds & Packaging Unit
> Food Ingredient Assessment Division, Science, FSQS
> U.S. Department of Agriculture
> Washington, DC 20250
> Telephone (202) 447-7680

4. Other chemicals proposed for use in meat and poultry plants.

    Send requests or inquiries to:

> Director
> Food Ingredient Assessment Division, Science, FSQS
> U.S. Department of Agriculture
> Washington, DC 20250
> Telephone (202) 447-7623

5. Chemicals or materials proposed for uses other than those previously specified.

The evaluation of such preparations is not the responsibility of Food Safety and Quality Service, nor can it be performed by that Agency. Requests or inquiries concerning such evaluation should not be directed to the addresses given above.

Compounds Which Do Not Require Evaluation

The compounds referred to in this part may be used under the conditions described in the following paragraphs without being evaluated or authorized by USDA. Any such compounds which are classified as economic poisons must be registered with the Registration Division, Office of Pesticides Programs, Environmental Protection Agency, Washington, DC 20460, and must be used according to label directions. Any compounds governed by FDA regulations must be used in accordance with those regulations. All compounds must be used in a manner which, in the opinion of the Inspector in Charge, will not result in the direct or indirect contamination of food products.

1. **Denaturants.** Denaturants formulated and labeled according to the following subparagraphs require no authorization by USDA prior to their designated use.

Preparations consisting only of the substances listed below may be used for denaturing carcasses, parts thereof, meat and meat food products (other than rendered animal fat) condemned for pathology and intended for disposal only as tankage. The denaturant must be deposited in all portions of the carcass or product to the extent necessary to prevent its use for food purposes. The container label must identify all ingredients present.

Crude carbolic acid	A formula consisting of:
Cresylic disinfectant	1 part FD&C Green No. 3,
	40 parts oil of citronella,
	40 parts liquid detergent,
	and 40 parts water.

Preparations consisting only of the substances listed below may be used for denaturing carcasses, parts thereof, meat and meat food products intended for disposal for purposes other than human food. The denaturant must be deposited in all portions of the carcass or product to the extent necessary to prevent its use as human food. The container label must identify all ingredients present.

Charcoal (finely powdered)	FD&C Blue No. 2
FD&C Blue No. 1	FD&C Green No. 3

Preparations consisting only of substances listed below may be used for denaturing poultry and poultry products condemned for pathology and intended for disposal only as tankage. The denaturant must be deposited in all portions of the carcass or product to the extent necessary to prevent its use for food purposes. The container label must identify all ingredients present.

Crude carbolic acid	Phenolic disinfectants conforming
Fuel oil	to commercial standards CS
Kerosene	70-41 or CS 71-41
Used crankcase oil	

Preparations consisting only of substances listed below may be used for denaturing poultry and poultry products intended for disposal for purposes other than human food. The denaturant must be deposited in all portions of the carcass or product to the extent necessary to prevent its use as human food. The container label must identify all ingredients present.

FD&C Blue No. 1	FD&C Red No. 3
FD&C Blue No. 2	FD&C Red No. 40
FD&C Green No. 3	Ultramarine Blue

Denaturants including ingredients other than those listed in the preceding paragraph, or those identified only by proprietary name, must be submitted for evaluation according to Section 2.1 of the "Guidelines For Obtaining Authorization of Compounds To Be Used in Meat and Poultry Plants," prior to use.

2. Outdoor pesticides.

Compounds such as herbicides, bird control compounds, and other pesticides labeled for outdoor use only may be used around the premises of federally inspected plants without authorization by USDA, provided that they are registered by the Registration Division, Office of Pesticides Programs, Environmental Protection Agency, Washington, DC 20460; and they are used in accordance with label directions and in a manner which prevents the direct or indirect contamination of food products.

3. Compounds used in offices and other similar nonprocessing areas.

Compounds such as dusting aids, furniture waxes or polishes, wall or floor finishes or waxes, rug or upholstery shampoos or cleaners, rug antistatic treatments, window or glass cleaners, and other similar compounds specifically designated by the Compounds Evaluation Unit require no authorization by USDA prior to use in such areas. They may not be used in any processing area of a federally inspected plant.

4. Compounds used in cafeterias or other retail food service areas.

Compounds such as rinse additives to prevent spotting and streaking of dishes and utensils, utensil tarnish removers, some hand dishwashing soaps or detergents, and other similar compounds specifically designated by the Compounds Evaluation Unit require no authorization by USDA prior to use in those areas. They may not be used in any processing area of a federally inspected plant.

5. Compounds used in heating systems.

Compounds such as fuel additives, firebox or flue treatments or cleaners, and other similar compounds specifically designated by the Compounds Evaluation Unit require no authorization by USDA prior to their proposed use.

6. Compounds used in holding pens, trucks, transporting cages, etc.

Compounds such as pesticides, disinfectants or cleaners used in transporting or holding facilities for animals prior to slaughter, or on animals being transported or held prior to slaughter; chemicals for deodorizing or dissolving animal waste; or other similar compounds specifically designated by the Compounds Evaluation Unit require no evaluation or authorization by USDA prior to their intended use.

Compounds Which Are Not Applicable For Evaluation

None of the compounds referred to in this part may be used in a processing area of the plant. However, they may be used under the conditions described in the following paragraphs without being evaluated or authorized by USDA. They may be stored within the plant provided that such storage is in a designated nonprocessing area acceptable to the Inspector in Charge. Any such compounds which are classified as economic poisons must be registered with the Registration Division, Office of Pesticides Programs, Environmental Protection Agency, Washington, DC 20460, and must be used according to label directions. Any compounds governed by Food and Drug Administration regulations must be used in accordance with those regulations. All compounds must be used in a manner which, in the opinion of the Inspector in Charge, will not result in the direct or indirect contamination of products.

1. Compounds used in sewage or waste water systems outside of the plant.

Compounds such as grease solvents for traps or lines, sewage odor control compounds to maintain biological balance in sewage lagoons or holding ponds, or other similar compounds specifically designated by the Compounds Evaluation Unit.

2. Compounds used in cooling towers or evaporative condensers. Compounds such as corrosion inhibitors, algaecides, compounds for cleaning or maintaining the tower or condenser, or other similar compounds specifically designated by the Compounds Evaluation Unit.

3. Compounds used on the exterior of buildings or immediate surrounding areas.

Compounds such as exterior coatings; mortar cleaning or etching compounds; tar, asphalt, or mortar removers; driveway patching or finishing compounds; or other similar compounds specifically designated by the Compounds Evaluation Unit.

4. Compounds used for cleaning or maintenance of the exterior of vehicles.

Compounds such as car washes or shampoos, tire cleaners, truck body brighteners, or other similar compounds specifically designated by the Compounds Evaluation Unit.

Category Code Letters And Their Meanings

The permissible use for each authorized compound is designated by the following code letters, and conditions for use are restricted by the category in which it is placed.

NOTE:

In several categories, reference is made to the need for rinsing with potable water. USDA accepts water used in federally inspected plants as being potable when such certification is made by state health agency authorities.

A. CLEANING COMPOUNDS

A1. Compounds for use as general cleaning agents on all surfaces, or for use with steam or mechanical cleaning devices in all departments.

Before using these compounds, food products and packaging materials must be removed from the room or carefully protected. After using these compounds, surfaces must be thoroughly rinsed with potable water.

A2. Compounds for use only in soak tanks or with steam or mechanical cleaning devices in all departments.

Before using these compounds, food products and packaging materials must be removed from the room or carefully protected. After using these compounds, all surfaces in the area must be thoroughly rinsed with potable water.

A3. Acid cleaners for use in all departments.

Before using these compounds, food products and packaging materials must be removed from the room or carefully protected. After using these compounds, all surfaces in the area must be thoroughly rinsed with potable water.

A4. Floor and wall cleaners for use in all departments.

Before using these compounds, food products and packaging materials must be removed from the room or carefully protected. After using these compounds, all surfaces in the area must be thoroughly rinsed with potable water.

A5. Floor and wall cleaners for subfreezing temperatures.

When used in areas with subfreezing temperatures, potable water rinsing is not required following use provided that the solution and solubilized soil are effectively removed by wiping or wet vacuuming.

A6. Scouring cleaners.

Residues resulting from the use of scouring cleaners must be carefully removed from surfaces by thorough rinsing with potable water.

A7. Metal cleaners and polishes or nonfood contact surfaces.

A8. Degreasers or carbon removers for food cooking or smoking equipment, utensils, or other associated surfaces.

Before using these compounds, food products and packaging materials must be removed from the area or carefully protected. After using these compounds, all surfaces must be thoroughly rinsed with potable water. The compounds must be used in a manner so that all odors associated with the compounds are dissipated before food products or packaging materials are reexposed in the area.

B. COMPOUNDS FOR LAUNDRY USE

B1. Laundry compounds.

Laundry detergents, bleaches, and sours may be used on fabric which contacts meat or poultry products, directly or indirectly, provided that the fabric is thoroughly rinsed with potable water at the end of the laundering operation.

B2. Laundry compounds for uniforms or other fabric which does not come in direct contact with food products.

C. COMPOUNDS USED IN INEDIBLE AND NONPROCESSING AREAS

C1. Compounds for use on all surfaces in inedible product processing areas, nonprocessing areas, and/or exterior areas. These compounds must not be used to mask odors resulting from insanitary conditions. They must be used in a manner which prevents penetration of any characteristic odor or fragrance into edible product areas. Compounds containing isomers of dichlorobenzene, or other substances toxic by inhalation, may be used only in areas where there is adequate ventilation to prevent accumulation of hazardous vapors.

Permission for the use of these compounds on loading docks and other similar areas is left to the discretion of the Inspectors in Charge of the plants.

C2. Compounds for use in toilets and/or dressing rooms.

C3. Paint removers for use in nonprocessing areas.

Equipment and utensils which directly contact edible products must be thoroughly cleaned and rinsed with potable water after treatment with such products before being returned to a processing area.

D. SANITIZERS

D1. Sanitizers for all surfaces always requiring a rinse.

Before using these compounds, food products and packaging materials must be removed from the room or carefully protected. After using these compounds, surfaces must be thoroughly rinsed with potable water before operations are resumed. The compounds must always be used at dilutions and according to applicable directions provided on the EPA registered label.

D2. Sanitizers for all surfaces not always requiring a rinse.

Before using these compounds, food products and packaging materials must be removed from the room or carefully protected. A potable water rinse is not required following use of these compounds on previously cleaned hard surfaces provided that the surfaces are adequately drained before contact with food so that little or no residue remains which can adulterate or have a deleterious effect on edible products. A potable water rinse is required following use of these compounds under conditions other than those stated above. The compounds must always be used at dilutions and according to applicable directions provided on the EPA registered label.

E. EMPLOYEE HAND CARE

E1. Handwashing compounds for use in all departments.

The compounds must be dispensed from adequate dispensers located a sufficient distance from the processing line to prevent accidental product contamination. After the use of the compounds, the hands must be thoroughly rinsed with potable water. Under conditions of use, there can be no odor or fragrance left on the hands.

7

E2. Handwashing and sanitizing compounds.

The compounds must be dispensed from adequate dispensers located a sufficient
distance from the processing line to prevent accidental product contamination. The
hands need not be washed prior to the use of the compounds. After the use of the
compounds, the hands must be thoroughly rinsed with potable water.

E3. Hand sanitizing compounds.

The hands must be washed and thoroughly rinsed prior to sanitizing with the
compound. The hands need not be rinsed following the use of the compound.

E4. Hand creams, lotions, and cleaners.

The use of such compounds is limited to toilets and dressing rooms. Employees who
handle edible products may use the compounds only when leaving the plant.

F. PESTICIDES

F1. Pesticides for use in all areas.

These compounds must be used in accordance with the conditions for nonresidual
insecticides set forth in Section 8.48 of the Meat and Poultry Inspection Manual.

F2. Pesticides for use in nonprocessing and inedible areas only.

These compounds must be used in accordance with the conditions for residual
insecticides set forth in Section 8.48 of the Meat and Poultry Inspection Manual.
They must be used in a manner that prevents their entry into edible product areas
through open windows, ventilating systems, etc.

F3. Rodenticides for controlled use only.

Compounds listed in this category are to be used in accordance with the conditions
for rodenticides set forth in Section 8.49 of the Meat and Poultry Inspection
Manual.

F4. Fumigants for controlled use only.

Compounds listed in this category are to be used in accordance with the conditions
for fumigants set forth in Section 8.48 of the meat and Poultry Inspection Manual.

F5. Fumigants for controlled use only.

Before using these compounds, all edible products and packaging materials must be
removed from the room to be fumigated. After fumigation, the treated equipment and
space must be thoroughly aerated to remove all vapors before inspectors or
employees reenter the area. Food contact surfaces must be rinsed with potable
water before edible products are returned to the room. Use of such compounds is
not covered in the Meat and Poultry Inspection Manual.

G. WATER TREATMENT COMPOUNDS

NOTE: Compounds used in such treatment should not remain in the water in
concentrations greater than required by good practice. Compounds containing
substances which may subsequenly result in the adulteration or contamination of

meat or poultry products may not be introduced into the system.

G1. General potable water treatment compounds.

G2. Phosphate potable water treatment compounds.

The concentration of phosphates may not exceed 10 parts per million calculated as phosphate ion.

G3. Silicate potable water treatment compounds.

The concentration of silicates may not exceed 10 parts per million calculated as silicon dioxide.

G4. Chlorine potable water treatment compounds.

Chlorine may be present in process water of meat plants at concentrations up to 5 parts per million calculated as available chlorine. Chlorine may be used in process water of poultry plants at levels acceptable to plant management, recognizing the self-limiting factors of effect on product, corrosion of equipment, and acceptability by plant personnel. Plant management must notify the Inspector in Charge when the chlorine level is increased above 20 parts per million. Chlorine must be dispensed at a constant and uniform level and the method or system must be such that a controlled rate is maintained.

G5. Cooling and retort water treatment compounds.

The amount used should be the minimum sufficient for the purpose.

G6. Compounds for treating boilers, steam lines, where the steam produced may contact edible products and/or cooling systems where the treated water may not contact edible products.

In compounds containing volatile amines, the amine in the steam may not exceed the indicated concentration.

Cyclohexylamine	10 ppm	Morpholine	10 ppm
Octadecylamine	3 ppm	Hydrazine	0 ppm
Diethylaminoethanol	15 ppm		

G7. Compounds for treating boilers, steam lines, and/or cooling systems where neither the treated water nor the steam produced may contact edible products.

H. LUBRICANTS

H1. Lubricants with incidental contact.

These compounds may be used as a lubricant or antirust film on equipment and machine parts in locations in which there is exposure of the lubricated part to edible products. They may also be used as a release agent on gaskets or seals of tank closures. The amount used should be the minimum required to accomplish the desired technical effect on the equipment. If used as an antirust film, the compound must be removed from the equipment surface by washing or wiping as required to leave the surface effectively free of any substance which could be transferred to food being processed.

H2. Lubricants with no contact.

These compounds may be used as a lubricant, release agent, or antirust film on equipment and machine parts in locations in which there is no possibility of the lubricant or lubricated part contacting edible products.

H3. Soluble oils.

These products are chemically acceptable for application to hooks, trolleys, and similar equipment to clean and prevent rust. Those portions of the equipment that contact edible products must be made clean and free of the mixture before reuse.

J. ABSORBENTS

J1. Absorbents or antislip agents for spot application to floors.

Such compounds may be used in all areas provided that use is limited to the portion of the floor area where the hazard exists, and that such use does not result in dusting, tracking, or other objectionable conditions. Compounds may not be used as a substitute for good sanitation. They must be removed as a part of the routine floor cleaning operation.

K. SOLVENT CLEANERS

K1. Solvents and solvent degreasers for use in nonprocessing areas.

Following the the use of these compounds, equipment and utensils must be thoroughly washed and rinsed with potable water before being returned to a processing area.

K2. Solvents for cleaning electronic instruments.

These compounds are chemically acceptable for cleaning electronic instruments and devices which will not tolerate aqueous cleaning solutions. Before using these compounds, food products and packaging materials must be removed from the area or carefully protected. These compounds must be used in a manner so that all odors associated with the compound are dissipated before food products or packaging materials are reexposed in the area.

K3. Adhesives or glue removers.

Before using these compounds, food products and packaging materials must be removed from the area or carefully protected. After using these compounds, all surfaces, must be thoroughly washed and rinsed with potable water. These compounds must be used in a manner so that all odors associated with the compound are dissipated before food products or packaging materials are reexposed in the area.

L. SEWER AND DRAIN CLEANERS

L1. Compounds for use in sewage and/or drain lines.

L2. Enzymatic compounds for use in sewage and/or drain lines.

Manufacturers must provide USDA with a record of salmonellae analysis for each lot of the finished enzymatic treatment proposed for sale to federally inspected meat and poultry plants. Testing must be performed by a qualified microbiologist.

TURANTS

aturants for meat.

nts for carcasses, parts thereof, meat and meat food products (other than
 animal fat) condemned for pathology and intended for disposal only as
  The denaturant must be deposited in all portions of the carcass or
to the extent necessary to prevent its use for food purposes.

aturants for meat.

nts for carcasses, parts thereof, meat and meat food products intended for
 for purposes other than human food. The denaturant must be deposited in
ions of the carcass or product to the extent necessary to prevent its use
n food.

aturants for poultry.

nts for poultry carcasses, parts thereof, or poultry products intended for
 for purposes other than human food. The denaturant must be deposited in
ions of the carcass or products to the extent necessary to prevent its use
 purposes.

aturants for poultry.

nts for poultry carcasses, parts thereof, or poultry products condemned for
y and intended for disposal only as tankage. The denaturants must be
d in all portions of the carcass or product to the extent necessary to
its use for food purposes.

) PROCESSING COMPOUNDS

tting agents for use in poultry scald vats.

is used for this purpose may not alter the water retentive properties of
 products being processed.

3 scalding compounds.

is must be labeled in accordance with Section 318.1 (d) of the Meat
ion Regulations.

ipe denuding, bleaching, and neutralizing agents.

it must be followed by thorough rinsing with potable water to remove added
:es. Compounds must be labeled in accordance with Section 318.1 (d) of the
spection Regulations.

npounds for washing fruits and vegetables that are to become ingredients of
, meat, and rabbit products.

3ing these compounds, the products must be thoroughly rinsed with potable

npounds for cleaning or treating feet or other edible parts.

11

Treatment must be followed by thorough rinsing with potable water to remove added substances.

P. MISCELLANEOUS

P1. Compounds requiring letter indicating authorized use.

Compounds listed in this category are to be used in accordance with the conditions set forth in the letter of acceptance. Copies of the acceptance letter must be supplied to Federal inspectors as proof of the acceptable use(s) of the compound. These letters remain valid as long as the compound is continuously listed in this publication.

Q. COMPOUNDS FOR USE ON SHELL EGGS.

NOTE: Although such compounds are included in the "List of Chemical Compounds," their evaluation is not the responsibility of the Compounds Evaluation Unit. Send all requests and inquiries concerning such compounds to:

> Poultry Grading Branch
> Poultry And Dairy Quality Division, FSQS
> U. S. Department of Agriculture
> Washington, DC   20250
> Telephone (202) 447-7410

Q1.   Shell egg cleaning compounds.

Eggs that have been washed with these compounds shall be immediately subjected to a thorough potable water rinse.

Q2.   Shell egg destaining compounds.

Eggs that have been destained with these compounds are to be rewashed and spray rinsed with potable water.

Q3.   Shell egg sanitizing compounds.

Eggs that have been sanitized with these compounds shall be subjected to a thorough potable water rinse only if they are to be immediately broken for use in the manufacture of egg products.

Q4.   Shell egg sanitizing compounds.

Eggs that have been sanitized with these compounds may be broken for use in the manufacture of egg products without a prior potable water rinse.

Q5.   Foam control compounds.

These compounds may be used to control foam in egg washing machines. Eggs washed in water containing these compounds shall be immediately subject to a thorough potable water rinse.

Q6.   Shell egg sanitizing compounds.

Freshly washed eggs may be sanitized with these compounds only if the eggs are rinsed prior to application of the compound. A subsequent rinse is not required.

12

How To Use This List

The sequence of listing companies, and compounds within companies, as printed by the computer is as follows:

Compounds or companies, having a special character (e.g., #, %, ") as the first character are printed before any compound, or company, having an alphabetic character (e.g., a, b, c) as the first character. For instance, under the heading of Abbott Supply Company, #777 Bottle Washing Compound is listed before Abbott Concentrated Lymee.

Compounds, or companies, having a numeric character (e.g., 1, 2, 3) as the first character are printed after all compounds, or companies, having an alphabetic character as the first character. For instance, under the heading of Alpha Chemical Services, Inc., 1017 is listed after Trolley Cleaner for that company. As another example, the listing for the company named 2-Cleen Chemical Laboratories, immediately follows the listing for Zurn Industries, Inc.

Because of the sorting sequence utilized by the computer, series of compounds which have numeric characters may not be printed in numeric order according to magnitude. The left-most digit of such compounds will primarily determine the sequence in which they are printed. For instance, under the heading of Alex C. Fergusson Company, AFCO 4324 is printed before AFCO 52.

Please examine all listings carefully before concluding that a compound, or company, is not listed.

LIST OF AUTHORIZED COMPOUNDS

(In the following list, compounds are arranged alphabetically
by trade name under the heading of the firm by which they are marketed)

A

& A Industrial Supply Diversified
Purpose II Concentrate A4
er Plus Concentrate A1

& L Laboratories, Inc.
Clor D2
Dine..................................... D1
ıvy Duty Floor Cleaner A4
L Special Washing Powder .. A1,Q1
kstone Remover
oncentrate A3
-Clor.............................. A2,Q1
Dairy Detergent A1,Q1

& M Distributing Company
nd Dishwashing Compound .. A1
? Grill & Oven Cleaner A4,A8
uid Hand Dishwash A1
juid Hand Soap 40% E1
chine Dishwashing
Compound A1

& M Maintenance Supply, Inc.
& M-3 Cleaner...................... A4
eed Cleaner.......................... A4

& R Products, Inc.
erry Bowl Cleaner C2
ynamo................................. A1
ncore A1
xtra Heavy Duty Concrete
Cleaner............................... A4
rime-Scat A4
dustrial Cleaner A4
ulti Blue A1
-150 Chlorinated Machine
Dishwash A1
)0 Plus Drain Pipe Opener ... L1

. A A Chemical/ A Division Of Drummac Co., Inc.
AA GB-1000......................... A1
AA 100-C Concentrate D1
AA-306 Concentrate A1
'/S Degreaser Concentrate A1
)30 Cocoanut Oil Hand Soap... E1

. A L Kleen Products Inc.
AL-Kleen Liquid Drain
Opener L1
eluxe Liquid Hand Soap.......... E1
ormula 300 Floor & Wall
Cleaner.............................. C1
eavy Duty Concentrate A4
ulti-Purpose Concentrate...... A1
erfumed Beads..................... A1

. B C Chemical Corporation
BC #20 Hand Soap................. E1

ABC Chlorinated Pdrd Mach
Dish Wash Cmpd A1
ABC Powdered Bleach............. B1
ABC Powdered Machine
Dishwashing Cmpd. A1
Deterge............................... B1
No. 201 Aluminum & Ferrous
Metal Cleaner..................... A3
No. 202 Lime And Soap
Remover A3
No. 203 Window Cleaner C1
No. 205 Bowl and Basin
Cleaner C2
No. 207 Lime And Rust
Remover A3
No. 209 Detergent For Dishes... A1
No. 213 Liquid Paint and
Varnish Remover................. C3
No. 214 Silicone Lub. &
Release Agent H1
No. 215 Liquid Steam Cleaner... A1
No. 216 Multi-Purpose Cleaner/
Dgrsr A4
No. 219 Heavy Duty Degreaser C1
No. 220 Non-Conductive Safety
Solvent K1
No. 223 Pine Scent Cleaner/
Deodz. C1
No. 226 Rust And Tarnish
Remover A3
No. 228 Sludge Dissolver.......... C1
No. 229 All Purpose Cleaner A1
No. 302 Drain De-Clogger........ L1
No. 304 Powdered Floor Clnr/
Dgrsr A4
No. 407 Contact Insect Killer ... F2
No. 408 Residual Insect Killer ... F2
No. 409 Flying Insect Killer F1
No. 418 Lemon Scent
Disinfectant C1

A C F Industries, Inc./ W-K-M Valve Division
Key Graphite Paste.................. P1
Key Tite P1

A G A Corporation
AGA Food Machinery Gr. No.
AA H1
AGA Food Machinery Gr. No.
CA..................................... H1

A I M International Chemicals Corp.
Assault A4
ADS Automatic Dishwashing
Soap A1
Boiler Guard G6
Bug Shot F1
Coca Wash E1
Coil Clean A3
Crossfire F1
Derma-Clean E4
Derma-Shield E4

Di-Gestor L2
Double Action A4
DE HY L2
Electro-Kleen......................... K2
Final..................................... D2
Foam Power A1
Germa-Clense II D1
Glide.................................... H1
Grime Buster......................... A4
Hot Stuff L1
Irish Mint C1
Kleen Up C1
Laundra Bright B1
Liquid Steam A1
Od-Out................................. C1
Penelube 4 .... H2
Phospho-Clean A3
Porcelain And Bowl Cleaner C2
Pure Lube 500......................... H1
Quik Wash A1
Radiant C1
Rust Away C1
Rust Away II A3
Sila-Cote G1
Sparkle A4
Wash Up E1
Whirlpool L1

A M C E C O/ American Chemicals Corporation
Oil Lave A4

A R C O/Chemical Company/ Division of Atlantic Richfield Co.
AcroHib BT-850 Boiler
Treatment G6
ArcoHib BT-840..................... G6
ArcoHib BT-845 Boiler
Treatment G6
ArcoHib BT-851...................... G6
ArcoHib BT-857...................... G6

A.B.C. Chemical Products
CDD Disinfectant & Sanitizer ... D1
Neutra-L-Cleaner Concentrate... A1
Powdered T.E.N. A1
Ready Clean........................... A1
TEN All Purpose Cleaner.......... A1

A.B.C. Compounding Company, Inc.
"Easy-Kleen" Inhibited Bowl
Cleaner............................... C2
A-145 Disinfectant Cleaner A4
A-150 Superior Concentrate...... A1
A-151 Pink Concentrate............ A1
A-152 Syntheso Neutral
Cleaner............................... A1
A-153 Atomic Action................ A1
A-161 Sof-T-Suds A1
A-162 Beaded Hand Dishwash
Comp. A1
A-225 Blue Sea Beads A1
A-225 Pink Sky Beads............... A1

A-225 White Foam Beads........... A1
A-230 Super Chlrd. Machine
  Dishwash Dtrg..................... A1
A-231 Prize Machine Dishwash
  Dtrgnt A1
A-235 Deluxe Machine
  Dishwash Dtrgnt A1
A-270 All Purpose Cleaner........ A1
A-280 Egg Wash A1
A-290 Po-Pac Ho Cleaner A1
A-300 B-Gone HD Cleaner........ A2
A-300 Foam-Up A4
A-310 Labor Saver Chloro
  Cleaner.............................. A1
AA106 Chain Lub & Cleaner..... H1
B-165 Liquid Machine
  Dishwash Dtrgnt A2
B-190 Jet White Odless. St. Clg.
  Comp. A2
B-192 Hvy Duty Odless St. Clg.
  Comp. A2
B-193 Ex. Hvy.Duty Odless St.
  Clg. Comp A2
B-194 White Odorless Steam
  Clng. Compd....................... A2
B-195 Steam Cleaner "Heavy
  Duty" A2
B-196 Medium Duty Steam
  Cleaner.............................. A2
B-197 Extra Heavy Duty Steam
  Clnr. A2
B-250 Restaurant King DFF
  Cleaner.............................. A2
Betty's Pink Lotionized Hand
  Soap E1
Bionic Emulsifier Cleaner........ A1
C-700 Milkstone Remover.........A3
Cherry-O-Dis 19-30 .... A4
Cherry-O-Dis 27-42 .... A4
Concrete Cleaner C1
D-156 Odorless Concrete
  Cleaner.............................. A4
D-160 ADR Sudsing Cleanser ... A4
Ecomomy Electro Kleen
  Solvent.............................. K2
F-186 Chlorinated Krystals A1
F-187 Super Germicide............. D1
F-188 Germ 1 San:.................. D2
Fire Proof Electro Kleen
  Solvent.............................. K2
G-100 Cocoanut Oil Liq Hand
  Soap E1
G-105 Econo Cocont Oil Liq
  Hand Soap.......................... E1
General All-Purpose Cleaner .. A4,A8
H-280 Egg Wash Q1
Hound Dog Concrete Cleaner C1
Knock Down Defoamer A1
KA-GO C-601 Acid Cleaner...... A3
L-120 Scal-Away Inhibited
  Bowl Cl. C2
L-125 Blue Inhibited Bowl
  Cleaner.............................. C2
L-130 Emulsion Bowl Cleaner ... C2
L-135 Sparkle Paste Cleaner C2
L-180 Pine Oil Disinfectant....... C1
L-210 Hog Scald Compound...... N2
Lem-O-Dis 19-30 .... A4
Lem-O-Dis 27-42 .... A4
Lemon Disinfectant Coef. 5 ...... C1
Lif-Out Tile & Grout Cleaner.... C2
Liquid Deep Fat Fry Cleaner A8
M-157 Heavy Duty Magic
  Concrete Cl. C1
M-157 Magic Concrete Cleaner.. C1
M-159 Econo Concrete Cleaner. C1
M-181 Pine Odor Disinfectant ... C1
M-185 General Disinfectant C1
Margie's Pearl Lotionzd Hd
  Dshwshng Cmpd A1

Mint Disinfectant................... C1
Mint-O-Dis 19-30................... A4
Mint-O-Dis 27-42 .... A4
N-211 Insecticide No. 111 .... F1
N-212 Space Spray No. 3 .... F1
Neutronic Powerhouse Cleaner
  L-255 .... A1
New Improved Pride A8
O-215 Malathion-50 Spray F2
O-880 Sewer SolVent L1
Pine-O-Dis 19-30.................... C1
Pine-O-Dis 27-42.................... C1
R-140 Metals & Stnless.Steel
  Cln. & Poli C1
Sil-Kon H2
Super Sewage Enzymes.............. L1

A.C. Chemical Industries, Inc.
A.C. Super Lemon Disinfectant. C2
A.C. Super Mint Disinfectant C2
A.C. Super Pine Odor
  Disinfectant C1
AC Air Sanitizer Bouquet
  Fragrance C1
AC Bee & Wasp Killer F2
AC Disinfectant Foam Cleaner.. C1
AC Dry Lubricant H2
AC Glass Cleaner C1
AC Hospital Disinfectant
  Deodorant C1
AC Lemon-Lime Air Sanitizer .. C1
AC Neutra Clean.................... A1
AC Nut Bolt Loosner H2
AC Residual Insecticide F2
AC Safety Solvent................... K1
AC Time Dispensing
  Insecticide.......................... F2
Mr. Gets-It Extra Hvy Duty
  Clnr-Degrsr A1

A.C. Towel Supply Co., The
Sid's Tripe Wash N3

A.F.I. Incorporated
Chain & Cable H2

A.J. Sales & Supply Company
Concentrate Many Purpose
  Cleaner.............................. A1
Fireproof Cleaner A4
Glass Cleaner C1
Sani-Kleen Bowl Clnr &
  Disinfect. C2
Stainless Steel Cleaner A7
Steam Cleaner A4
Swinger A4

A.J.'S Equipment Sales and Service
Steam Cleaner....................... A1

A.R.C.Y. Corporation
A-231 Delux Machine
  Dishwash Detergent............... A1
Atomic Action A-153............... A1
ABC A-300 B-Gone Heavy
  Duty Cleaner....................... A2
ABC A-310 Labor Saver
  Chloro Cleaner A1
ABC-150 Superior Concentrate . A1
ARC-10 .... A8
B-165 Liquid Machine
  Dishwash Detergent............... A2
B-192 Heavy Duty Odorless
  Stm Clng Cmpd A2
Betty's Pink Lotionized Hand
  Soap E1
Big "G" A4
Blue Inhibited Bowl Cleaner L-
  125.................................... C2

C-700 Milkstone Remover A3
Concrete Cleaner C1
Disinfectant Cleaner A-145 .... A4
Econo Coconut Oil Liquid
  Hand Soap G-105 .... E1
Eze Kleen............................. E1
Foam-Up A-300 .... A4
Fresh Air Orange C1
General Disinfectant M-185 .... C1
Hound Dog Concrete Cleaner ... C1
Insecticide No. 111 N-211 .... F1
Ka-Go C-601 Acid Cleaner....... A3
Klean Oven A8
Klear Glass........................... C1
L-210 Hog Scald Compound...... N2
Lem-O-Dis 19-30 .... A4
Life Ice Melt......................... C1
Malathion-50 Spray O-215......... F2
Mozel Germicidal Multi
  Purpose Cleaner................... C1
Odorless Concrete Cleaner D-
  156.................................... A4
Pine Odor Disinfectant M-181 ... C1
Pine Oil Disinfectant L-180...... C1
Quall F2
Rust Buster H2
Rust-Ban H2
Saf-T-100 .... K1
Sani-Lube Food Equipment
  Lubricant H1
Slide Silicone Lubricant........... H1
Smooth & Clean Waterless
  Hand Cleaner E4
Space Spray No. 3 N-212.......... F1
Super Germicide F-187............. D1
White Odorless Steam Cleaning
  Cmpd B-194 .... A2

A-Chem Corporation
A-Chem C-20......................... G6
A-Chem LPA G6
A-Chem 150 .... G6
A-Chem 319 .... G6,G2
A-Chem 354 .... G6
A-Chem 63 .... G6
A-Chem 661 .... G6
A-Chem 722 .... G7
A-Chem 760 .... G7
A-Chem 900 .... G7
A-Chem 920 .... G7

A-Jon Products LTD.
AJ 140 Multi Purpose Boiler
  Treatment G6
AJ 149 Water Tower Cleaner ... G7
AJ 164 Boiler Water Adjunct ... G6
AJ 210 Foaming Degreaser A1
AJ 211 Freeze Clean................ A5
AJ-212 Tire Mark & Adhesive
  Remover K1

A-K Chemical Supply
871 .... A2

A-OK Products, Inc.
Brite Glass Cleaner................. C1
Bug Out Roach N'Ant Killer F2
Clinger............................... C2
Knuckles N'Nails.................... E4
No. 15 Liquid Hand Soap.......... E1
No. 310 Stripper and Degreaser A4
Porter's #210 .... A4
Sani-Fresh Disinfectant
  Deodorant C2
Sparkling Glass Cleaner C1
Squeeks N'Leeks.................... H2
T.S. Dry Spray Lubricant H2

A-Z Supply Company
Active................................. A1

14

Toro D C1
Twirl B1
Ultimate C1
Utensi Clean A1
Utili Clean A4
Val Cleen A4
Victor F1
Vit C2
Work Horse A1
Zime...L2
40% Costile Liquid Hand Soap .E1
40% Skin Cleanse.......................E1

Aborn Chemical Industries, Inc.
Ark Arkomatic A4
Ark Automaid A1
Ark Flash C1
Ark Freezer & Low Temp.
  Hard Surface Cln..................A5
Ark Hy-Dis D1
Ark Industrial Spray F1
Ark Light A4
Ark Light-A A1
Ark Oven Cleaner A8
Ark Prilled Drain Opener.............L1
Ark Residual Insecticide.............F2
Ark San 200 ....D2
Ark Sanidet D2
Ark Soft Anti.Liq. Hand
  Soap(15%) E1
Ark Soft Anti.Liq.Hand
  Soap(40%)Conc. E1
Ark Soft Liquid Hand Soap
  (15%) E1
Ark-Lube...................................H1
Ark-San D2
Bac-Gard A4

Abso-Clean Chemical Company
"Quik" S-5500" A4
A.S.C. No. 100 Automatic
  Scrubber Conc. A4
A-C Concentrate #AC 600 ....A4
Abso-Clean #1025-N Liq Tile
  & Grout Clnr A3
AC 906 ....A1
De-Grease Liquid Steam Cl
  Conc. Form 334.......................A1
Flake Caustic Soda:....................A2
Formula CS-100..........................D1
Odor-Nil #200.............................D1
PF-127 Phosphate-Free All-
  Purp Cl Conc. A4
S-5000 Liquid Emulsifying
  Cleaner...................................A1
803 Heavy Duty Degreaser/
  Cleaner Conc. A4

Academy Chemical Co.
Blue Diamond Liquid Scouring
  Power A6
Lotionized Hand Soap E1

Academy Supply Company, Inc.
Combo A4
Concentrated Heavy Duty A4
Pink General Purpose Cleaner ...A4
Terra-Clean A4

Accent Paper Co., Inc.
B-C Concentrate A4

Accommodation Sanitary Supply Co., Inc.
Sanco Degreaser Cl. Wax
  Stripper A4,A8
Sanco Liquid Hand Soap...........E1
Sanco Spray On Wipe Off..........C1
Sanco Strike-Out D1

Sanco Super Detergent A4

Ace Chemical Co.
Ace Slip Silicone Lubricant H1
Hospital Disinfectant
  Deodorant C2
4-Way Nut & Bolt Loosener
  Penetrant H2

Ace Chemical Products
Ace Special A1
Acid Detergent A3
Agre-Sol A4
AG Detergent A1
AG-505 ....A1
AGP Detergent A1
AGP-509.....................................A1
AGP-510.....................................A1
AMC...A1
AMC H.D. A1
Cir Alk HD A2
Cir-Chlor....................................A2,Q1
Concentrate LC-300 ....A1
DS 540.......................................A4,A8
Floor Klean A4
Floor Klean H.D. A4
Foam Add A1
Formula 73 ....A1,Q1
Formula 83 ....A1
FM-301 ....A1
G.P. Chlor..................................A1
H-55 Boiler Water Additive.......G6
H-56 Boiler Water Additive.......G6
H-57 Boiler Water Additive.......G6
H-60 Boiler Water Additive.......G6
Heavy Duty Liquid Detergent...A2
HTP 545.....................................A2
L.F.C. Low Foam Caustic A2
Liquid Clear...............................A1
Liquid-Cir-Chlor........................A1
LHT..A1
Milkstone Remover.....................A3
Milkstone Remover Conc...........A3
P-81 Boiler Water Additive........G6
Strip Away..................................A2
Super Bottle Wash A2
Super Chlor Egg Wash...............Q1
Super-AGP A1
SC-501..A1
Wash Brite Laundry Detergent .B1
Wash Brite WB-200....................B1

Ace Chemical Products, Inc.
Ace Can Lube A1,H2
Ace Lube 8 ....A1,H2
Ace-M-All #403 Liquid Parts
  Degreaser C1
Ace-M-All Acid Cooperage
  Cleaner...................................A3
Ace-M-All Acid Descale
  Copper Inhibited....................A3
Ace-M-All Acid Star A6
Ace-M-All All Purpose
  Detergent...............................A1
Ace-M-All Antis. Liq. Hand
  Soap E1
Ace-M-All AB A1
Ace-M-All AY.............................A1
Ace-M-All Bleach Cleanser........A6,D1
Ace-M-All BD.............................A1
Ace-M-All Can Wash STP...........G5
Ace-M-All Canners Cleaner A1
Ace-M-All Chlorinated Cl.#1-P A1
Ace-M-All Chlorinated Cl.#1-S A1
Ace-M-All Chlorinated
  Cleanser.................................A6
Ace-M-All Concentrated Bowl
  Cleaner...................................C2
Ace-M-All CG.............................A2

Ace-M-All Detergent-Sntzr
  Iodophor Type.......................D2
Ace-M-All Foamax A1
Ace-M-All FG:.............................A3
Ace-M-All FH.............................A3
Ace-M-All FS..............................A3
Ace-M-All Galva Cleaner #20 ..A3
Ace-M-All GA A3
Ace-M-All Heavy Duty Gel #1 A2
Ace-M-All Hog Scald & Tripe
  Cl...N2,N3
Ace-M-All J 305 ....A1
Ace-M-All Laundry Detergent..B1
Ace-M-All Liqui-Clean...............A1
Ace-M-All Liquid Bactericide ...D2
Ace-M-All Liquid Detergent A1
Ace-M-All Low Foam Additive A1
Ace-M-All Low Foam
  Iodophor Det. Sntzr...............D2,E3
Ace-M-All LC-60 ....A2
Ace-M-All No. TK-77 ....A2
Ace-M-All No. 110 ....A2
Ace-M-All No. 110 Stripper........A2
Ace-M-All No. 110 Stripper
  Liq...A2
Ace-M-All No. 110-B...................A2
Ace-M-All No. 110-K...................A2
Ace-M-All No. 110F....................A2
Ace-M-All No. 17 ....A1
Ace-M-All No. 17A
  Chlorinated Cleaner A1
Ace-M-All No. 18 ....A1
Ace-M-All No. 220 Steam
  Cleaner...................................A1
Ace-M-All No. 26 ....A1
Ace-M-All No. 28 ....A1
Ace-M-All No. 280 ....A1
Ace-M-All No. 3 Utensil
  Cleaner...................................A1
Ace-M-All No. 36 ....A2
Ace-M-All No. 43 ....A1
Ace-M-All No. 440 Chelated
  Cleaner...................................A1
Ace-M-All No. 45 ....A1
Ace-M-All No. 47 ....A2
Ace-M-All No. 50 ....A1
Ace-M-All No. 5020.....................A2
Ace-M-All No. 53 ....A2
Ace-M-All No. 54 ....A1
Ace-M-All No. 56-1 ....A2
Ace-M-All No. 6 Bake Pan
  Cleaner...................................A1
Ace-M-All No. 62 Retort
  Additive.................................A1
Ace-M-All No. 65 ....A2
Ace-M-All No. 660 ....A2
Ace-M-All No. 73 ....A2
Ace-M-All No. 75 ....A2
Ace-M-All No. 76 ....A2
Ace-M-All No. 761-P...................A2
Ace-M-All No. 770 ....A2
Ace-M-All No. 78 ....A2
Ace-M-All No. 8 ....A1
Ace-M-All No. 87 ....A2
Ace-M-All No. 88 ....A2
Ace-M-All No. 90-10 ....A2
Ace-M-All No. 98 Low Foam
  Wetting Agent A1
Ace-M-All Poultry Spray...........A1
Ace-M-All Powdered Hand
  Cleaner...................................C2
Ace-M-All Quat Rinse................D2,E3
Ace-M-All Quat 42.....................D1
Ace-M-All Sanitizer Pink D1
Ace-M-All Sanitizer White.........D1
Ace-M-All Sanna Gel A1
Ace-M-All Sanni-Rinse...............D1
Ace-M-All Shroud Cleaner A1
Ace-M-All Strip 750 ....A2
Ace-M-All Super Heat Drn
  Pipe Opnr. L1

pany,

17

Rustoscale PB 200 .... G6
Rustoscale PB 300 .... G6
Rustoscale PC-66 .... G1
Rustoscale Regular #2 .... G6
Rustoscale Regular #5 .... G6
Rustoscale Regular #6 .... G6
Rustoscale Special #11 .... G6
Rustoscale Special #12 .... G6
Rustoscale Special #13K G6
Rustoscale Special #19 .... G6
Saniterge A4
Solvin AAA C2
Swype C2
Ukazone Blocs C2
Ukazone Disks C2
Vitra-Kleen C2
Voltreet G7
Voltreet FP 100: G6
Voltreet FP-20 .... G6
Voltreet FP-40 .... G6

Acme Industrial Compounds Corp.
Acme Food Equipment
  Lubricant H1
Acme Silicone Spray.................... H1
Enzyme Drain Cleaner L2
Ortholene................................... A1
RLT G6
Safety Solvent Degreaser K1
Sanit-Aire Sewage & Garb Odr
  Contl&Dgrsr C1
Scale-Out MIXS1........................ G6
Scale-Out NPT-N G6

Acme Janitor Equipment Co.
A-J Concentrate A4

Acme Soap Company
Acme Mild A1
General Purpose Cleaner A1
Hog Scalding Compound N2
Meat Hook & Trolley Cl. &
  Deruster.................................. A2
Shroud Cleaner B1
Shroud Cleaner No. 2 .... B1
Smokleen A2
Smokleen No. 10 .... A2
Spray On Car & Truck Wash A1
Tripe Cleaner N3

Acorn Chemical Company
Product No. 1 Boiler Water
  Treat. G6
Product No. 10 Hi-Phos G1
Product No. 16 Indust. Odor
  Control.................................... C1
Product No. 2 Boiler Water
  Treat. G6

Acro Chem
Eggwhite..................................... Q1

Acrocem, Incorporated
Thrift Grease Trap & Drain
  Cleaner................................... L1
Thrift Sewer & Septic Tank
  Cleaner.................................... L1

Action Chemical Company
Shroud Cleaner #11..................... B1

Action Chemical Company
Action Chlorinated Cleanser A6
Action Eze A1
PHC-70 .... A1
PHC-71 .... A1
PHC-72 .... A1
PHX.. E1
Twenty Percent Hand Soap E1

Action Chemical Company
Action Cowboy Heavy Duty
  Clnr. Concnrt........................... A4
Action Lemon-Dis Disinfectant . C2
Action Plus(Ammoniated) Clnr. C1
Action Thrifty Mint
  Disinfectant C2
Action Thrifty Pine Pine Odor
  Disf. Five................................. C1
Action Winda Shine Glass
  Cleaner.................................... C1
Big Daddy 69.............................. A4
Cardinal All Purpose
  Concentrated Clnr. A4
Pink Lotion Hand Soap E1
Pink Squeeky Lotion
  Dishwashing Detrgt A1
Silicone Spray............................. H1

Action Chemical Supply
Chem-Power Non Butyl
  Degreaser A1

Action, Incorporated Division of Nobles Industries, Inc.
FAC 41..................................... A1
FAC 42..................................... A1
FAC 43..................................... A1
FAC 44..................................... A1
FAC 46..................................... A2
FAC 61..................................... A1
FAC 83..................................... A1

Active Chemical Corporation
Acet Chem-Ox Boiler Water
  Treat. G6
Acet 282 Steam & Return Line
  Treat. G6
Acet 525 Boiler Water
  Treatment G6
Acet 550 AF Boiler Water
  Treatment G6
Acet 550 Boiler Water
  Treatment G6

Active Organic Mfg. Corp.
Ak-Tiv-X Concentrate G6

Ad-Rem, Inc.
Clean Ease.................................. A1

Adams Best Company, Inc.
Bright Eyes A4
Bright Eyes Plus......................... A4,A8
BCS-290..................................... A4
Causti Jell.................................. A1
Fly'n Mosquito Insecticide F1
Foam-Plus A1
Glass Cleaner C1
Instant Oven & Grill Cleaner A1
Liquid Steam Cleaner A1
Pink Panther E1
Pipe-Wrap.................................. P1
Roach N'Ant Killer..................... F2
Smooth E1
Sno-Kleen................................... C2
Stripclene A1
Wax Strip C1

Adams Best Corporation See Adams Best Company, Inc.
Adams Company, The
Adams Cleaner P-1...................... A4
Adams Cleaner T3....................... A4
Adams Drain Pipe Cleaner......... L1
Adams L-Tox Spray F1
Adams Lectra Solv K2
Adams Liquid Hand Soap E1

Adams Lo-Sudz A4
Adams Mint Disinfectant............ C1
Adams Odo-Rout Concentrate... C1
Adams Super Pine(Phenol ·
  Coef.5)..................................... C1
Adams Window Cleaner.............. C1
Adsol Emulsion Bowl Cleaner... C2
Cleaner CPC................................ A4
Pine Oil Disinf. (Phenol Coef.5) C1
Super Cleaner A1
Super Cleaner 075 .... A1

Adchem, Incorporated
Radiant Red A1

Adco Chemical Corporation
Bio-D7 .... A1
Drain Cleaner L1
Sanizyme L1

Adco Chemicals, Incorporated
Adco Fog Insecticide................... F1
AC-62 .... A1
Cleaner 7-11................................ A1
Deodorant No. 98........................ C1
HS-11 .... N2
Liquid CPM A1
Mist Spray Insecticide................ F1
MP-20 .... A1
Sewer Solvent L1

Adco, Incorporated
Concentrated Liquid Hand
  Soap E1
Neutral Cleaner A4

Adcoa, Inc.
Pink Bacteriostatic...................... E1
Vac-Ez.. C2

Admiral Chemical Corp.
Super Solv #2.............................. A4,A8

Advance Chemical Company
Advanced Industrial Cleaner...... A1
Colena 24.................................... L1
First For Floors........................... A4

Advance Chemical Company, Inc.
AC-105 .... G1
AC-106 .... G1
AC-107 (L).................................. G1
AC-114 Boiler Water Treatment G6
AC-116 .... G6
AC-200 .... A1
AC-201 .... A1
AC-202 .... A1
AC-211 .... A1
AC-220 .... A1
AC-242 .... A4
AC-275 .... A3
Formula 88 General Purpose
  Clnr... A1
700 Liq. Det. Concentrated........ A1

Advance Industries
Advance #230 .... G6
Boiler Water Treatment
  Product #261 .... G6

Advance Machine Company
Advance Formula 60 Detrgnt
  Sanitizer.................................. D1
Formula 10.................................. A1
Formula 10 Plus.......................... A1
Formula 20.................................. A1
Formula 20 Plus.......................... A1
Formula 40.................................. A1
Formula 60 Plus.......................... D2

Aero Chemical Corporation

Aero Distributing

Aero Mist, Incorporated

Misty Stainless Steel Cleaner A7
Misty Surf. Dis.and Air
  Deodorizer C1
Misty Textile Grade Dry
  Silicone Spray H1
Misty Utility Cleaner C1
Misty Vandalism Mark
  Remover K1
Misty Waterless Hand Cleaner
  Cream E4
Misty White Lithium Grease...... H2

Aero-Master, Inc.
Aero-Master Air Sanitizing
  Compound C1
Fogging Insecticide with 0.25%
  SEP-1382 .... F1
Formula B-20-200 .... F1
Mill Fogging Formula F1
Spraying And Fogging
  Insecticide F1
Super Fogging Insect. w/0.50%
  SEP-1382 .... F1

Aerochem Industries
Kitchen Chembrite A1

Aetna Chemical Corporation
Actajet A1
Actana 200 .... G6
Actana 25L G6
Actana 400 .... G6
Actana 477 .... G6
Actana 489 .... G7
Actana 600 .... G7
Acterge S A4
Acterge 100 .... A1
Acterge 101 .... A1
Liquid Hand Soap E1

Agra-Labs
Stainless Steel Cleaner A7

Agri International, Inc.
Agri C Farm Implement
  Cleaner A4

Agri-Bio Corporation
En-Zap...................................... A1,L2

Agricultural Laboratories/ Division of
  A G Chemicals, Inc.
Bio-Gest.................................... L2

Agway, Incorporated
Acid Cleanser 2 .... A3
Agway Space Spray.................... F1
Bulk Tank & Utensil Cleaner A1
Cleanser A1
Egg Washing & Sanitizing
  Compound Q1
Egg Washing Compound............ Q1
Food Processing Machinery
  Grease H1
Liquid Bactericide...................... D2
Livestock And Farm Spray........ F1
Sanitizer And Teat Dip.............. D2

Aidex Corporation
Aidex Malathion Premium-5E.... F2
Aidex Mist No. 3644 .... C1
Di-V-Ex.................................... F2
Dimex-267................................. F2
Fogicide P.................................. F1
Mirasect.................................... F2
Pyrenex...................................... F1
Vap-O-Ate................................. F2

Aiken Chemical Company
All Purpose Cleaner................... A4

Aim-Chem Enterprise, Inc.
Chemi Pro Away......................... C2
Chemi Pro Bug-B-Gone............. F2
Chemi Pro Fly-Away F2
Chemi Pro Foam Dis.................. C1
Chemi Pro Loosen-Up H2
Chemi Pro Mintone.................... C1
Chemi Pro Sani-Bowl C2
Chemi Pro Sur Dis..................... C2
Chemi Pro Un-Clog L1
CS-100 Cleaner/Degreaser A4,A8

Air City Janitor Supply Co.
Silicone Spray H1
Vandal Mark Remover C1

Air Distribution Associates, Inc.
Deep Foam A1

Air Master Services, Inc.
Super Clean B-150...................... A4
Super Concentrate #20.............. A4

Air Technology
AT-19B...................................... A2

Air-Tite Products Company
Air-Tite 772................................ C1

Aire-Mate Division Carmel Chemical
  Corporation
Banish Bits High Level Odor
  Counteract............................. C1
Disinfecting Bowl Cleaner C2
Medisan C2
Steri Kleen Germicidal Spray &
  Wipe Clnr C1
Sterikleen.................................. D1

Airkem Central Indiana
Enzymatic L2

Airkem Sales
Formula 10-10............................ A1

Airkem Upper Midwest Sales Co.
Cling-N-Clean A4,A8

Airkem/ A Div. of Airwick Industries,
  Inc.
A-125 Dry A4
A-125 Dry Phosphate Free C1
A-33 .... C1
Airkem A-125 .... A4
Airkem A-3................................ A4
Airkem CBC Redi-Mix C2
Airkem Deoterge........................ A4
Airkem F/H............................... D2
Airkem Multipurpose Degreaser A1
Airkem Odor Controlled Insect.
  Vap. Spy............................... F1
Aquatoc C1
Aquinoc C1
ABC Accelerated Bowl Cleaner C2
Brawn A3
CBC... C2
CBC Plus Bowl Cleaner C2
Empire Bowl Cleaner C2
Entacide................................... F1
Entacide Insecticide Vaporizer
  Spray.................................... F1
Entacide-R F2
Escape....................................... C1
Extra-Point A7

Face-Off Vandal Mark
  Remover K1
Finish-Line General Purpose
  Cleaner C1
First-Place Stainless Steel Clnr
  & Plsh A7
Grid-Iron A8
Multi Purpose Insecticide Spray F1
Multi-Purpose Solvent Cleaner .. P1
New Citrus Scented Solidaire C1
New Floral Scented Solidaire C1
Odor Counteractant Osmix No.
  21 .... P1
Odor Counteractant Spec. 6521. C1
Omega Concentrate.................... C1
Omnipak C1
Premeasured A-33 Dry............... A4
Solidaire Blue Label................... C1
Solidaire Gold Label.................. C1
Solidaire Green Label................. C1
Solidaire Red Label.................... C1
Specification 105....................... P1
Specification 198....................... P1
Spritz A4
Touch-Down.............................. C1

Airwick Institutional Products A Div.
  of Airwick Industries, Inc.
Breaker-Ph C2
Extra-Point A7
Finish Line (Aerosol)................. C1
Finish-Line (Bulk) A1
Finish-Line Dry A1
First-Place A7
Half-Time A3
Mr. Brilliance C1
Power-Play A8
Sharp-Shooter A4
Touch-Down Detergent-
  Disinfectant D1
Touch-Down Disinfectant
  Foam Cleaner C1

Ajax Sanitary Supply Company
Concentrate............................... A4
Detergent Jel Floor Cleaner A4
Liquid Hand Soap E1
Lo-Suds Floor Cleaner A4
Swifty Cleaner A4

Akers of Supplies
Al-Phil I A1
Al-Phil II A1
Citron Hand Shampoo E1

Akro Chem
A-229 .... H1

Akwell Industries, Inc.
SK-80.. E4

Al Margo, Incorporated
Lift-Off A1
Penetrate................................... A4
Penetrate-1 .... A4
Penetrate-10 .... A1
Penetrate-11 .... A1
Penetrate-2 .... A2
Penetrate-3 .... A1
Penetrate-7 .... A1
Penetrate-8 .... A1
Penetrate-9 .... A1

Al Meyer Co.
Green Coco................................ E1

Al Nelson Company
Al's Hi-P Contact Cleaner.......... K2
Al's No Sweat............................ P1

Alamo Biochemical
Liquid Live Micro Organism L2

Albany Chemical & Pool Co.
Cherry Bomb Insect Spray F2
CP-45 All Purpose Cleaner and
    Degreaser A1
Super Pink Concentrate A1
White Grease H2

Albemarle Supply Co.
Activ-Plus A1

Albert G. Maas Company
Bowl & Porcelain Cleaner C2
High Concentrate Bowl Cleaner C2
Sanni Rinse D2

Albert Verley & Company
Malabate R-90-W C1
Malabate R-93 .... C1

Alco Chemical Company
FPI Spray F2
Rat & Mouse Killer Bait F3

Alco Products, Inc.
F-E-L H1
Grip Dressing for V-Flat-
    Round Belts P1
GGC Grease Gun In A Can H2
Kleen Waterless Hand Cleaner .. E4
Litho-Lub H2
No Sweat P1
OP-20 .... P1
SS 32 .... K2
SS-40 .... K1
SSC Stainless Steel Metal
    Polish A7
TEF Release Fluorocarbon
    Lubricant H2

Alconox, Incorporated
Alcojet A1
Alconox A1
Alcotabs A1
Liqui-Nox A1
Terg-A-Zyme A1

Alcott-Haig And Co.
Amaze A1
Argonaut A4
Cold War A5
Seize D1

Aldac, Inc.
MX 237 The Master Oil H2

Alde Chemicals, Inc.
Aquagard 244 .... G6
Catalyzed Sulfite Organic G6
Formula AAF G6
Formula Polymer EMG-6 .... G6
Formula 1024 .... G6
Formula 680-L G6
Formula 753 .... G6
Formula 802-AF G6
Formula 807-OS G6
Formula 903-AF G6
One-Shot G6
Polymer EMG G6
Polymer 802-CAF G6

Aldran Chemical Company
Aldran A1
Dynamo A1
El-Zid A4

Alex C. Fergusson Company
Afco Chlorilizer D1,Q4
Afco Low Foam In-Place Acid
    Cl. A3
Afco Low Foam Liquid
    Detergent A1
Afco Poultry Scald N1
Afco S# 5529 .... H2
Afco San-I-Two D2
Afco Spark D2
Afco Spray-All A1
Afco Tops D2
Afco TR-100 .... G6
Afco TR-101 .... G6
Afco TR-102 .... G6
Afco TR-103 .... G6
Afco TR-104 .... G6
Afco TR-109 .... G5
Afco TR-110 .... G5
Afco TR-111 .... G6
Afco TR-113 .... G6
Afco TR-114 .... G6
Afco TR-115 .... G6
Afco TR-119 .... G6
Afco TR-121 .... G6
Afco TR-123 .... G6
Afco TR-124 .... G6
Afco TR-125 .... G1
Afco TR-169 .... G5
Afco 1056 .... A1
Afco 1072 Super Defoamer Q5
Afco 2626 .... Q3
Afco 311 .... A1
Afco 3137 .... G6
Afco 3140 .... G6
Afco 3141 .... G6
Afco 3142 .... G6
Afco 3143 .... G6
Afco 3144 .... G6
Afco 3145 .... G6
Afco 3146 .... G6
Afco 3147 .... G6
Afco 3148 .... G6
Afco 3149 .... G6
Afco 3151 .... G6
Afco 3160 .... G6
Afco 3161 .... G6
Afco 3162 .... G6
Afco 3163 .... G6
Afco 3164 .... G6
Afco 3170 .... A1
Afco 3180 .... A1
Afco 3185 .... A1
Afco 3196 .... A3
Afco 3202 .... A4
Afco 3203 .... A1
Afco 3204 .... A1
Afco 3210 .... A1
Afco 3212 Campbell Blend A1
Afco 3214 .... A2
Afco 3218 .... L1
Afco 3222 .... A4
Afco 3224 .... A2
Afco 3226 .... A2
Afco 3229 .... A2
Afco 3233 .... A1
Afco 3234 .... A1
Afco 3235 .... A1,Q1
Afco 3237 .... A1
Afco 3238 .... A1
Afco 3241 .... A1
Afco 3243 .... B1
Afco 3249 .... N2
Afco 3257 .... N3
Afco 3259 .... N2
Afco 3269 .... A1
Afco 3271 .... A1
Afco 3272 .... A1
Afco 3274 .... A2
Afco 3280 .... A1

Afco 3284 .... A1
Afco 3285 .... A1
Afco 3295 .... A2
Afco 3297 .... A1
Afco 3298 .... A2
Afco 3302 .... A1
Afco 3304 .... A1
Afco 3310 .... A1
Afco 3312 .... A1
Afco 3314 .... A1
Afco 3317 .... A1
Afco 3318 .... A1
Afco 3319 .... A1
Afco 3321 .... A1
Afco 3322 .... A1
Afco 3324 .... A1
Afco 3325 .... A2,Q1
Afco 3326 .... A1,Q1
Afco 3328 .... A1
Afco 3329 .... A1
Afco 3330 .... A2
Afco 3331 .... A2
Afco 3332 .... A2
Afco 3333 .... A2
Afco 3334 .... A2
Afco 3335 .... A2
Afco 3336 .... A2
Afco 3337 .... A2
Afco 3338 .... A8
Afco 3339 .... A2
Afco 3340 .... A1
Afco 3341 .... A1
Afco 3343 .... D1
Afco 3347 .... A1
Afco 3348 .... D1,Q3
Afco 3349 .... A3
Afco 3354 .... A2
Afco 3357 .... Q1
Afco 3358 .... A1,Q1
Afco 3359 .... D1,Q4
Afco 3360 .... Q1
Afco 3362 .... A1
Afco 3363 .... Q1
Afco 3364 .... A1
Afco 3369 .... A1
Afco 3370 .... A1
Afco 3372 .... A2
Afco 3373 .... A2
Afco 3375 .... A1
Afco 3376 .... A1
Afco 3377 .... A1
Afco 3378 .... A1
Afco 3380 .... A2
Afco 3381 .... A2
Afco 3384 .... A1,Q1
Afco 3386 .... A2
Afco 3388 .... A1
Afco 3391 .... A1
Afco 3394 .... A1
Afco 3395 .... A1
Afco 3396 .... A1
Afco 3397 .... A2
Afco 3401 .... A1
Afco 3402 .... A1
Afco 3405 .... A4
Afco 3406 .... A1
Afco 4317 .... C1
Afco 4319 .... D1,Q3
Afco 4320 .... C2
Afco 4324 .... D1
Afco 52 .... A3
Afco 5213 .... A4,H2
Afco 5219 .... H2
Afco 5222 .... N1
Afco 5228 .... H2
Afco 5229 .... A1
Afco 5231 .... A4
Afco 5232 .... A2
Afco 5233 .... G5
Afco 5234 .... A1

Afco 5235 .... A1
Afco 5236 .... A1,Q1
Afco 5237 .... A2
Afco 5238 .... A1
Afco 5244 .... A1
Afco 5245 .... A1
Afco 5246 .... N1,N2
Afco 5248 .... A1
Afco 5252 .... A1
Afco 5254 .... A1
Afco 5255 .... A1
Afco 5256 .... A1
Afco 5257 .... A1
Afco 5259 .... A1
Afco 5261 .... A1
Afco 5262 .... A1
Afco 5263 .... A1
Afco 5264 .... A2
Afco 5265 .... A2
Afco 5266 .... A2
Afco 5269 .... A2
Afco 5271 .... A1
Afco 5272 .... A1
Afco 5281 .... A3
Afco 5282 .... A3
Afco 5283 .... A3
Afco 5284 .... A3
Afco 5289 .... A3,Q2
Afco 5291 .... A3
Afco 5293 .... A3
Afco 5294 .... A3
Afco 5296 .... A3,Q2
Afco 5320 Fergusson 2700 ...... A1
Afco 538 .... G6
Afco 549 .... G6
Afco 5501 .... E1
Afco 5503 .... H1
Afco 5505 .... A1,H2
Afco 5508 .... A1,H2
Afco 5520 .... H2
Afco 5524 .... H2
Afco 5525 Chain Slick A1,H2
Afco 5526 .... H2
Afco 5527 .... A1,H2
Afco 5528 .... H2
Afco 5536 .... H2
Afco 5538 .... H2
Afco 5640 .... H2
Afco 5641 62 Conveyor
   Lubricant H2
Afco 5642 Conveyor Lubricant . H2
Afco 5643 62 Conveyor
   Lubricant H2
Afco 5645 .... H2
Afco 565 .... G6
Afco 5660 Conveyor Lubricant
   Liq. H2
Afco 5661 .... H2
Afco 567 .... D1,Q4
Afco 568 .... G6
Afco 899 .... F1
Afco 9501 .... G6
Afco 9502 .... G6
Afco 9504 .... G7
Afco 9505 .... G7
Afco 9513 .... G6
Afco 9514 .... G6
Afco 9520 .... G6
Afco 9522 .... G6
Afco 9523 .... G6
Afco 9525 .... G6
Afco 9564 .... G6
Afco 9566 .... G6
Afco 9576 .... G6
Afco 9580 .... G6
Afco 9586 .... G6
Afco 9587 .... G6
Afco 9588 .... G6
Afco 9592 .... G7
Afco 9611 Synsolv K1

Afco 9624 .... A2
Blast Off A1
Campbell Cleaner 'A' A2
Chloro-Plus H Q1
Con-Clean A3
Quik Klene Chlor Non Foam
   Egg Cl. A1,Q1
Super Slash D1
Tetra San D2
Utopia Active C1
Utopia Dexterous A1
5511 Anti-Septic Liquid Hand
   Soap E1

Alexander Chemical Co.
Anti-Clog L1

Alexander Chemicals Division
D-Solve G1

Alfa Chemical Company
Alfa Kleen Hand Cleaner E1
Alfa Kleen Heavy Duty
   Industrial Cleaner A4,E1,A8
Alfa Kleen Hood & Grill
   Cleaner A8

Alfa-Kem
Spotlite A4
Steam E-Z A1

Alfred Chemical Corporation
A.P. 59 .... A1
APC Concentrate A4,A8
Chlorinated APC A1
Egg Wash 326 .... Q1
Formula EO A8
Formula HS Liquid Hand Soap . E1
Formula MDW-H Machine
   Dishwashing Cmpd. A2
Formula MDW-H Machine
   Dishwashing Compou A2
Formula OG Cleaner &
   Degreaser C1
Formula PP Pot & Pan Cleaner. A1
Formula PTW Hot High
   Pressure Truck Wash A1
Formula RC Concrete Floor
   Cleaner C1
Formula 211 All Purp. Cl with
   Anti-Stat A4
Formula 307 Hand Cleaner C2
Formula 671 All Purpose Steam
   Clng Compd A1
Formula 771 Heavy Duty
   Steam Clng Cmpd A4
Freezer Clean A5
Hot Hi Pressure Truck Wash
   Powder A4
Liquid MDW-Machine
   Dishwashing Compound A2
Neutral APC A4
Pan Cleaner 'S/S' A1
Pan Cleaner 'TM' A1
Pan Cleaner AB A1
Pan Cleaner LI A1
Pan Cleaner MP A1
Smokehouse Cleaner 887 .... A1
Stone Remover 'TM' Acid A3
TMT-P Tile Marble & Terazzo
   Cleaner A1
102 All Purpose Steam
   Cleaning Compound A1
200 Heavy Duty Steam
   Cleaning Compound A1

Alger Oil Company
C-I-P 358 .... A1
Chlorinated Super Cleaner 322 .. A1

Liquid C-I-P 34.............................A1
Liquid General Purpose
  Detergent HD62.....................A1

Alkem Labs, Inc.
Silicone Lubricant H1

Alken-Murray Corporation
Alken Catalyzer-878...................G6
Alken Disperse-300 ....G6
Alken KP-200 ....G6
Alken KP-225 ....G6
Alken Liquid Treatment C.........G6
Alken Scaloff-461 ....G6
Alken Scaloff-462 ....G6
Alken Scaloff-463 ....G6
Alken Scaloff-464 ....G6
Alken Scaloff-466 ....G6
Alken Scaloff-466A G6
Alken Scaloff-521 ....G6
Alken Scaloff-522 ....G6
Alken Scaloff-523 ....G6
Alken Scaloff-524 ....G6
Alken Scaloff-525 ....G6
Alken Scaloff-526 ....G6
Alken Scaloff-528 ....G6
Alken Scaloff-529 ....G6
Alken Scaloff-531 ....G6
Alken Scaloff-535 ....G6
Alken Scaloff-536 ....G6
Alken Scaloff-542 ....G6
Alken Scaloff-543 ....G6
Alken Scaloff-544 ....G6
Alken Scaloff-545 ....G6
Alken Total A-467 D G7
Alken Total-468 D G6
Alken Treatment "C" G6
Alken Treatment "J"-660...........G6
Alken Treatment "J"-661...........G6
Alken Treatment "J"-662...........G7
Alken Treatment "J"-663...........G7
Alken Treatment "J"-665...........G6
Alken Treatment "J"-666...........G6
Alken Treatment "J"-668...........G7
Alken Treatment J-670 ....G6
Alken Treatment J-672 ....G6
Alken Treatment 478 ....G6
Alken Treatment 50 ....G6
Alken Treatment 61 ....G6
Alken Treatment 62 ....G6

All American Chemicals Co.
CP 00 ....A1
Foamanol...................................A1
GS 37 ....A2
GS 41 ....A1
GS-22..A1
GS-32..A4
GS-38..A1
Just-Rite.....................................A1
Liquid Sanitizer Q4
SS 112...A1
SS 112...A2
SS 151...A2
SS 28 ....C1
SS 3..A1
SS 30..A1
SS 43 ....A1
SS 88..A1
SS-111...A1
SS-134...A2
SS-20..A3
SS-29..C1
SS-5..A1
SS-61..A1
SS-74..A2
SS-75..A1
123 Spray Cleaner A4
20% Hand Soap...........................E1
20% Liquid Hand Soap E1

All American Janitorial Supply Co., Inc.
Non Butyl Cleaner Degreaser:...A1

All Chemical Systems Company
JFS Acid....................................A3
JFS Alkaline Cleaner A2
JFS Oil H1

All Imports, Inc.
All Belts P1
Big "G" A1
Brandex Plus..............................A4
Cocoanut Oil Hand Soap
  Concentrate E1
Disinfectant Plus........................A4
Fogging Spray Concentrate F1
Handy E4
L-Tox Spray F1
Ovec A8
R.I.S.-15 Residual Insect. Cont.
  Baygon...................................F2
Remov-Ox Plus..........................A3
Safety Solvent 88.......................K1
Slide H1
Super Big "G" A1
Thrift 100 ....A4
Vita-Pine Coef. 5 ....C1
Wyte-Ease H2

All Maintenance Supply
Free-N-Clean A1

All Right Products, Inc.
Super Solv #2.............................A4,A8

All State Chemical
#115 Organic Drain Line
  Opener&Maintnr....................L1
Safety Solvent Clnr. &
  Degreaser K1
Super Enzymes...........................L2
Super Neutral Cleaner A1

All-Brite Sales Co.
Grease Cutter..............................A1

All-Chem Corporation
A.P.C. A4
All-Brite Laundry Compound.....B1
Bottlewash #100.........................A2
Bottlewash 99.............................A2
Bowl & Porcelain Cleaner #508 C2
Cement Floor Cleaner #528.......C1
Chlorinated Dishwash #603.......A1
Chlorinated Egg Wash................Q1
Cleaner Degreaser A4
Dish Suds #606 ....A1
Fome-All A1
Formula No. 700 ....C1
Formula 503 Auto-Scrub
  Concentrate A1
Formula 504 General Purpose
  Cl...A1
Formula 507 Acid Maint. Clnr. ..A3
Formula 553 Powdered Bleach..B1
Formula 578 Pressure Wash
  Det. A1
Formula 580 Steam Cleaner A1
Formula 581 H.D. Steam Clnr....A1
Formula 583 H.D. Liq. Stm.
  Clnr. A1
Formula 584 Med. Duty Liq.
  Stm. Cl...................................A1
Formula 586 Liq. Conveyor
  Lubrcnt. H2
Formula 612 Liq. Acid Cleaner..A3
Formula 626 Drain Cleaner.......L1

Formula 629 Gen. Purp. Clnr. ...A2
Formula 701 Gen. Purp. Clnr. ...A1
Formula 704 Hvy. Dty. Non-
  Caustic Cl...............................A1
Formula 707 Gen. Purp. Clnr. ...A1
Formula 709 All Purpose
  Cleaner...................................A1
Formula 710 Pwdrd. Acid Det...A3
Hog Scald...................................N2
Jell-It..A1
Liquid Castile Hand Soap #511..E1
Liquid Hand Soap 20%..............E1
Liquid Hand Soap 40%..............E1
Liquid Pressure Wash A1
Lotionized Dishwash #607 ....A1
Mint 7 ....C1
Non-Butyl Cleaner-Degreaser
  #502-GX A4,A8
Oven & Grill Cleaner #614........A8
Pine Odor 6................................C1
Pressure Terg #595 ....A4

Alladdin Laboratories, Inc.
Boiler Water Conditioner G6
Chelate Condenser Treatment....G7
Duall Boiler System Treatment .G6
Hot and Cold Water Treatment.G1
Slime Fungus Treatment.............G7
Steam and Condensate
  Treatment C...........................G6
Steam and Condensate
  Treatment M G6
Steam Condensate Treatment.....G7

Allanco Supply
Degrease-All A1

Allchem Products Company
Lusteel A7

Allegheny Water Services Corp.
BWT-20......................................G6
BWT-23......................................G6
BWT-24......................................G6
BWT-27......................................G6
BWT-28......................................G6
CST-5 ....G6
DWT-1..G1
F.G.-10.......................................G7
SR-50 ....G6
SR-51 ....G6

Allen Chemical Company
Hi-P Contact Cleaner.................K2
Liquid Live Micro Organisms....L2
Merit A4

Allen Food, Inc.
Lasco #88 Insect Fogger F1

Allen Maintenance Supply Co.
No. 25 Insect Spray....................F1
Sani-Lube H1
Stainless Steel and Metal Polish.A7

Alliance Industrial Products, Inc.
G-Cide Super..............................D1

Alliance Maintenance Supply Co.
Al Alliance Odrless
  Disinf.&Sanitizer....................D1
Al Concentrate A1
Al 500 ....A4
Anti-Bact E1
Drive...F1
Fast-Solv A1
Liquid Soap................................E1

Allied Block Chemical Company
Jumbo 2B Deodorant Block C2,L1
Jumbo 3B Deodorant Block C2,L1
Neat Dry J1

Allied Chemical Corporation
Caustic Soda Liquid 50%
  Regular A2
Caustic Soda Liquid 50%-Food
  Grade G6,N2,N3
Soda Ash A1,G1
Solvay Snowfine A1
Solvay Snowflake Crystals A1

Allied Enterprises, Inc.
All-Lube H2
Allied Deo-Gran C1
AE Formula 200 .... C1
AE Formula 2500 .... A8
Formula 100 .... K1
Formula 100 .... A4
Formula 1038 .... H2
Formula 1309 Stainless Steel
  Clnr & Plsh A7
Formula 2100 .... C1
Formula 250 .... A4
Formula 420 .... A4
Formula 780 .... D1
Hi-Count Bacteria Enzyme
  Complex L2
Hydrolytic Enzyme Bacteria
  Complex L2
Power 'us............................... C2
Smoke Scrub 46 SC................. A1
Smoke Scrub 547 OC A8
Syntho Scrub 146L A1

Allied Mills, Incorporated
Wayne D-Odor #II L2
Wayne Super Cleaner A1
Wayne-O-Dine D2

Allied Products & Services Inc.
APS-90 Plus E1
CL-400 .... A1
Dissolve A1
Foam Kleen A1
Formula 3000 .... A1
Quatra-Kleen D1

Allied Products Company
Corodex Rust Remover A2

Allied Service & Supply Company
Poultry House Cleaner Formula
  II. A1

Allied-Kelite Products Division
Acid Cleaner #65 .... A3
Acid Cleaner #75 .... A3
Alklor D2,Q4
Antifoam 10 .... Q5
C.I.P. Cleaner #305 .... A1
C.I.P. Cleaner #305 Special A1
C.I.P. Cleaner #325 .... A2
Foam Cleaner #45 .... A1
Formula #24 .... A2
Formula #24-L A2
Formula #27-A Q1
Formula #29 .... A2
Formula #48 Deluxe................... A1
Hook and Trolley Cleaner #410 A2
Hook and Trolley Cleaner
  #410-L A2
Hook and Trolley Cleaner #420 A2
Hook and Trolley Cleaner #430 A2
HTST A2
I P Descaler L A2
Isoprep #136.............................. A1

Isoprep #7 .... A2
Ke-Klor L Q1
Ke-Scrub-2 .... C1
Kelite #22-Klor Q1
Kelite Disinfectant Cleaner
  #100... D1
Kelite Keystar A1
Kelite KeKleen E A1
Kelite Mech-Chem #402 .... H2
Kelite Mech-Chem #300 .... A1
Kelite Mech-Chem #301 .... A3
Kelite Mech-Chem #305 .... A1
Kelite Mech-Chem #325 .... A1
Kelite Mech-Chem #400 .... A1
Kelite Mech-Chem #401 .... A3
Kelite Mech-Chem #402 .... A3
Kelite Mech-Chem #403 .... N3
Kelite Mech-Chem #404 .... A3
Kelite Mech-Chem #405 .... N3
Kelite Mech-Chem #406 .... A3
Kelite Mech-Chem #407 .... A2
Kelite Shroud Cleaner #20 .... B1
Kelite Tripe Bleach #640 .... N3
Kelite Tripe Bleach #650 .... N3
Kelite Tripe Cleaner #100......... N3
Kelite Tripe Cleaner #615......... N3
Kelite Vari-Kleen A1
Kelite Vari-Kleen-S A1
Kelite 102 .... A4
Kool Spray 141 .... A2
M.P. Cleaner 200 .... A1
Pressure Cleaner #12 .... A2
Process 235 .... A2
Process 236................................ A2
Process 242 .... A2
RR-520 .... A4
Soaker 35 .... A2
Socolite H A1
Socolite O A1
Spray White A1

Allion Chemical Co.
Non-Solvent Degreaser/Cleaner A1

Allstate Chemical Service, Inc.
AC-SLV G6
AC-10 .... G6
AC-30-M G6
AC-36 .... G6
AC-45 .... G6
AC-61 .... G6
AC-80 .... G6
AC-83 .... G6
AC-87 .... G6

Allstate Chemicals, Inc.
Grease Trap Formula.................. L2
Kneet Floor Kleen A1
Linger C1
Somethin E4
Super-Super Power
  Concentrate A4

Allstate Industrial Supply Co.
E-Z Dri J1

Allstates Chemical Corporation
Action IV C2
All Grease H2
Automatic A4
Badger A3
Boiler-Treat G6
Conquer F2
Cream Clean C2
De-Dis C2
De-Mise Insecticide F1
Disinfectant-Sanitizer D2
Dri-Lube..................................... H2
E-Z Clean................................... E1

Foam Away A4
Go Grain Solvent L1
Golden Mist C1
GGT Steam Cleaner A4
HD Steam Cleaner A1
Job.. E1
Lemon Scent Disinfectant-
  Deodorant C1
Lemon Scent 15 Disinfectant-
  Deodorant C1
Lubé All H2
Mint Fragrance Disinfectant-
  Deodorant C1
Mint Fragrance 15...................... C1
Mr Power K1
Offense F1
Onward Disinfectant Detergent. A4
Onward XL A1
Pepcotex A1
Pine Scent Disinfectant-
  Deodorant C1
Pine Scent 13 Disinfectant-
  Deodorant C1
Ren-O-Vate A3
Residual Roach Liquid F2
Silicone Spray H1
Sta-Kill Fogger Insecticide....... F2
Super Go Hand Cleaner E4
Super Handy Clean Cleansing
  Cream E4
Super Sewer Solvent.................. L1
SC-100 Porcelain Cleaner C2
SS-345 .... A1
SS-400 Insecticide..................... F1
SS-400 Insecticide With SBP-
  1382 .... F2
SS-400 Space Spray F1
Whilcat A4
Wipe-A-Way C2

Alltech Corporation
Silicone Spray H1

Allube Division Far Best Corporation
Allube Aqua-Shield Polyfilm
  No. 3 .... H1
Allube Aqua-Shield Polyfilm
  No. 5 .... H1
Allube Aqua-Shield 1059 .... H1
Allube Aqua-Shield 1059 2T H1
Allube Aqua-Shield 109C.......... H2
Lubri-Shield CPM 90 HD H2
Moly-Shield Pelgear 90.............. H2
Moly-Shield Pelmill 2011 HVY. H2
Watershield 1009 F. G............... H1

Almo Laboratories Co., Inc.
Alclor A1
Almo Scale Remover A3
Almochlor D2
Almosan...................................... D2
Blast A1
Bon Argent Liquid A1
Chlorace A1
Degreaser C1
Erase A1
Formula-10 .... A1
N.F.D. A1
Oven-Grill Cleaner A8
Tight A1

Alpha Chemical Co.
1013 Detergent A4
1018 Cleaner A4
1028 Detergent A1
1187 Steam Cleaner................... A1

Alpha Chemical Services, Inc.
Acid Cleaner A3

24

25

Amalie Refining Company Division of Witco Chemical Corp.

All Purpose Grease H2

Amatron/ Div. of Partsmaster, Inc.

Ari-Slick H2
Bayonet H2
Belt-Buddy P1
Cushion H1
Dis-Place P1
Fall Off C3
Hand-Plus E1
Perma-Gear Open Gear Lube H2
Sil-Lube H1
Spray Loose H2
Sta-Brite C1
Tac-Guard K2

Amaza Laboratories, Inc.

E-Z Go Waterless Hand
  Cleaner E4

Amazing, Inc.

Amaz-Clean A4

Ambassador Industries LTD.

Elbow Drain-Flo L1
Hot Shot Truck Wash Stm Clnr
  Press. A1
Pro-Clean Hvy Dty Wtr Sol
  Indust Cl Dgrs A1

Ambeco Products Co See DuChemCo Products Company
Amchem Corp.

Acid Quat Dairy & Food
  Industry Cleaner D2
Amer Safe K1
Amer Safe Extra K1
Amerkleen 20 .... E1
Belt Dressing P1
Bio-Kleen A1
CD Cleaner D2
Dandy E4
Fluoro Carbon H2
Formula 550 .... H2
Frigid Cut H2
Graphite Fast Dry Lube H2
Grease Thief K1
Jell C1
Mint Deodorant C1
Non-Acid Lime and Scale
  Remover A1
NP D2
Paint Zoff C3
Penetrating Oil H2
Polysiloxane Release Agent H1
Precision Instrument &
  Electronic Pts Cl K2
QAC #10 .... D1
Rust and Corrosion
  Preventative H2
Sewer Sol-Granular L1
Sewer Sweet-Lemon L1
Silicone Lube H1
Space Spray F1
Sulfa Bac C1
Sulfa Bac VHC C1
Super Power A4
Super Solv #2 .... A4,A8
Veto E4
X-Ox A3

Amchem Products, Incorporated

Fumarin-22 .... F3
Fumasol A F3
Fumasol C F3
Fumasol-S (Green) F3

Amclean Corporation

"X-OX" A3
Acid Spray Detergent A3
Alkaline Foam Additive A2
Amer Joy A1
Amer Safe K1
Extra Heavy Duty Steam
  Cleaner A1
Fast Acting Sewer&Drain
  Opener L1
Floor Cleaner A4
General Purpose Manual
  Cleaner A1
Heavy Duty Acid Detergent
  Cleaner A3
Liquid Detergent A1
Pressure Wash A1
Thrifty Kleen C2
3-A-K Cleaner A4

Amco Chemical Company

Chicken Little Meat Room
  Degreaser A1

Amco Laboratories Corporation

Eee Zee Clean Super
  Concentrate A1

Amerchem Corporation

Amer-San 100 .... A4
Castor Oil Grade C H1
Compound AP-10 .... A2
Compound AP-15 .... A2
Compound AP-30 .... A2
Compound AP-35 .... A2
Compound AP-57 .... A2
Compound AP-60 .... A2
Compound C-75 .... A1
Compound CH-312 .... A1
Compound Deruster AP A2
Compound N-125 .... A1
Concentrate #50 .... A1
Concentrate 75 .... A4
Hog Scald N2
Liqua-Clean A1,E1
Liquid-Sope E1
Sewer Cleaner L1
Stainless Steel Cleaner A3

American Andesite/ A Division of Ca-Ja, Inc.

Butcher's Helper J1

American Building Equipment Corp.

Heavy Duty Degreaser Non
  Butyl Form. 305 .... A1

American Chemical

Dynamic A1

American Chemical Co., Inc.

A-Plus L1
American Medicide C2
Amersol C2
Coconut Oil Base Soap
  Concentrate E1
Formula 1808 Industrial
  Cleaner & Dgrsr. A4,A8
Mr. Brite C2

American Chemical Laboratories

American #100 .... A4
American #200 .... A4
American #300 .... A4,A8
American #400 .... A5
Oven & Grill Cleaner A8

American Chemical Sanitary Supplies Inc.

A-4 Ammoniated Stripper C1
AC-5 Cleaner A4
Germicidal Bowl Cleaner C2
Hvy Dty Steam Cleaning
  Compound A2
Pink Magic A1

American Chemical Works

Answer A4

American Cleaning Equipment Corp.

CSD 909 Degreasol A1
Formula 300 Quick Clean A4
Formula 75 Germidyne D1

American Cleaning Equipment Corp.

Heavy Duty Degreaser Non
  Butyl Form. A1
Non Butyl Degreaser A1

American Custom Chemicals

Accu-Zymes L2
Big "G" A4
Big "3" C1
Fresh Pine C1
Koil Klean P1
Mean Klean A3
Mighty Ox A3
Min-Tee Disinfectant C1
Quat-27 .... D1

American Cyanamid Co.

Cyanamer P-35 .... G6
Cyanamer P-35 Liquid G6
Cyanamer P-70 Liquid G6
Cyanamer P-70-D G6
Cygon 2-E F2
Cythion F2

American Farm Bureau Service Company

Maxon Egg Oil H1

American Feedwater Engineering Co.

Kemloid G6

American Fluoride Corp. Pest Control Equipment Div.

AFC Baygon Residual Spray F2
AFC Fogging Concentrate F2
AFC Fogging Concentrate
  Type PD F1
AFC General Purpose Spray
  Type II F1
AFC General Purpose Spray
  Type 5 .... F1
AFC Silicone Release Spray H1
AFC Synergized Pyrethrins
  Spray F1
AFC 1-10 Emulsifiable
  Concentrate F1
AFC 1-2-3 Fogging
  Concentrate F1
Bait Tickets F3
Formulation 120 .... F2
Gold Crest C-100 Emlsfbl.
  Concntrt. F2
Insecticide Liquid, Diazinon,
  1% F2
PCE Fog Oil F1
PCE Vestibule Spray F1

American Heritage Products of Texas

Lube-Eze H1
Oven Cleaner A8

Top Choice A4
Tuf Stuf E4
Tuff-N-Tender E1
Two-Twelve L1
Vaporize A1

American Inn Foods, Inc.

Concentrated Bowl Cleaner C2
Degreaser & Cleaner C1
Deluxe All Purpose Cleaner A4
Drain Pipe Opener L1
Germicidal Cleaner D1
Liq. Dish/Glasswash Pot & Pan
  Clnr. A4
Low Suds Laundry Detergent ... B1
Machine Dishwash A1
Porcelain Tile and Fixture Clnr. C1
Stripper A4

American Kleaner Mfg. Co., Inc.

Compound HTR A1
Compound K-5 .... A1
Compound K90C A1
Compound K90P A1
Compound LHS-400 .... A1
Compound P-45-HD C1
Compound U-45-S C1
Compound X-101 .... A1

American Lincoln Division

Amerblitz A1
Concentrated All Purpose
  Detergent A1
Concentrated
  H.D.Indstrl.Cl.&Wax Strppr. . A4
Concentrated Hvy Dty
  Autoscrub Cl. A4
Concentrated Speed Demon ·
  Autoscrub Cl. A4
Concentrated Supermarket
  Special Det. A1
Dollar Saver Concentrated
  Detergent A1
GD No. 8 Cleaner and
  Degreaser A4
Heavyweight C1
Heavyweight Plus A4
Speed Demon A4
Versatile Spray & Wipe C1

American Linen Supply

Clean All Non Butyl Degreaser. A1

American Lubricants Co., The

All Temperature Oil #1 .... H2
All Temperature Oil #2 .... H2
Alubco All-Spex Motor Oil 30 .. H2
Alubco Bison #454 .... H1
Alubco Food Machinery
  Grease H1
Alubco Grpht Pressure Grease
  #1 .... H2
Alubco Grpht Pressure Grease
  #3 .... H2
Alubco Improved Molyshield H2
Alubco L-K Chain Lube H2
Alubco Moly Motor Oil H2
Alubco Non Drip Oil H2
Alubco Solvent Degreaser K1
Alubco Ultra-Tec Gear Lube H2
Bison M-X2 .... H2
Bison 1900 .... A2
Chain Oil H2
Gear Stix H2
Magic Shield H2
Moly All Temp Oil #1 .... H2
Moly All Temp Oil #2 .... H2
Moly Chain Oil H2
Moly Non Drip Oil H2

Moly Ultra-Tec Gear Lube
  SAE 90/SAE 140 .... H2
PMT-SAE 140 .... H2
PMT-SAE 80 .... H2
PMT-SAE 90 .... H2
Spindle Oil Heavy H2
Spindle Oil Light H2
Spindle Oil Medium H2
Super-Clean #2851 .... A4
Tap A4
Tiffany T-A-P A1

American Municipal Chemical Co., Inc.

Acid Quat Dairy & Food
  Industry Cleaner D2
Aerosol Disinfectant C1
All Purpose Non Flammable
  Safety Solvent K1
AC 68 Compound L1
Belt Dressing P1
Bio-Kleen A1
CD D2
De-Grease K1
Digestant 250 .... L2
Disinfectant Sanitizer
  Deodorizer D2
Drain-Away L1
Fluoro Carbon H2
Frigid Cut H2
Graphite Fast Dry Lube H2
Heavy Duty Liquid Drain
  Cleaner L1
Jell C1
Lemon Sewer Sweetener L1
Liquid Hand Soap E1
Mint Deodorant C1
Multi-Purpose Penetrant H2
NP D2
Odor-Ban Odor Control C1
Odor-Ban Odor Control with
  Mask C1
Paint And Varnish Remover C3
Penetrating Oil H2
Polysiloxane Release Agent H1
Precision Instrument & Elect.
  Pts. Cln. K2
Rust Preventative H2
Smooth Silicone Lube H1
Spray-Fog F1
Stainless Steel Cleaner A7
Stericide D1
Super Power A4
Super Solv #2 .... A4,A8
Waterless Hand Cleaner E4

American National Chemical Company

A.N.C. Hand Soap E1
A.N.C. Sewer Compound L1
ANC Disinfectant & Deodorant C1
ANC Disinfectant Foam
  Cleaner C1
ANC Residual Insecticide F2
Super-Steam A1

American Oil And Supply Co.

PQ C-105 .... H2
PQ C-109 .... H2
PQ C-300 .... H2
PQ C-303 .... H2
PQ C-304 .... H2
PQ C40AA-1 .... H1
PQ C4001-0 .... H2
PQ C4001-1 .... H2
PQ C4001-2 .... H2
PQ Gear Oil AGMA 2 EP H2
PQ Gear Oil AGMA 3 EP H2
PQ Gear Oil AGMA 4 EP H2
PQ Gear Oil AGMA 5 EP H2

PQ Gear Oil AGMA 6 EP........ H2
PQ Gear Oil AGMA 7 EP....... H2
PQ Gear Oil AGMA 8 EP........ H2
PQ L-50 Way Lube.................... H2
PQ L-80 Way Lube.................... H2
PQ L-90 Way Lube.................... H2
PQ 100 Hydraulic Oil H2
PQ 32 Hydraulic Oil H2
PQ 46 Hydraulic Oil H2
PQ 68 Hydraulic Oil H2

American Pacemaker Products, Inc.
All Away....................................... A1
Gentle Giant A1
Pace 72 .... A4
Royal Flush................................. L1
Steam-It A1

American Paper & Twine Co.
Hysil Anti-Stick Multi-Purp
Silicone Spy H2

American Paper Towel Company
Aptco #17 Bowl Cleaner C2
Aptco Liquid Hand Soap
(Plain) E1

American Petrofina Co. of Texas
Fina FMG-2................................ H1

American Power Ind. LTD
Odor Goe C1
Power Quat II.............................. D1
Quat Disinfectant And Cleaner-
II... D1

American Process Equipment Corp.
Apte No. GP-3............................ A1

American Products
Americo Motor Degreaser &
Equipment Clnr K1
Americo Silicone Spray............. H1

American Refining & Manufacturing Inc.
Heavy Duty All Purpose
Cleaner................................ A4,A8

American Research Corporation
Arc Saf-T-100 .... K1
ARC Concentrate......................... A4
ARC Protector E4
ARC Thermo-Solve..................... A1
ARC 10 .... A4
ARC 299 .... L1
ARC 40 to 1................................ A4
ARC 598 .... L1
B.O.G. C1
D-I-N-G F2
Dyno-Mite................................... L1
Enzymes...................................... L2
Foggcide...................................... F1
Go Liquid Drain Opener........... L1
Surgent Liquid Hand Soap......... E1
TF7 Outdoor Fogicide F2

American Salt
American Granulated Kiln
Dried Salt............................... P1
American Industrial Compacted
Salt... P1
American Medium Rock Salt..... C1,P1
American Mini-Cube Cmpt Wtr
Soft... P1
American Royal Fine Rock Salt C1
American Water Softener Gems G1
4-In-1 Salt.................................... B1

American Sanitary Specialties, Inc.
Hand Soap.................................... E1
Lab 80... A4
Stubborn Mike A1
ST-27 .... A4

American Sanitary Supply Co.
Magic Shine Stainless Steel A7
Pink Cream Antiseptic Hd. Clr..E1
Pink Creme Bacteriostatic Hnd
Lotn. Soap....................... E1
SS-215 Silicone Spray................. H1

American Shortening & Oil Co.
Lubrikol...................................... H2

American Water Treatment, Inc.
Formula No. 10-M G6
Formula No. 1524-A G7
Formula No. 33 .... G6
Formula No. 452 .... G6
Formula No. 55 .... G5
Formula No. 56 .... G5
Formula No. 77 .... G5
No. 840 Industrial Bactercide..... G7

American Wax Company
Degresol A1
Detergent F-11 .... A4

Americhem
Action IV.................................... C2
Blitz.. A4
Blue Chem Concentrate.............. A1
Boot-Hill..................................... F2
Crawl-Kil F2
Draineze L1
Green Lightning.......................... A4
Kleen Quik C2
Kleer-Vue.................................... C1
Mag-I-Bol.................................... C2
Mig-I-San C2
Saf T Kleen A4
San-I-Sept.................................... D1
Scent-Gard C2
Sewerola...................................... L1
Sili-Cote...................................... H2
Tri-Action A4
Z-Ox A3

Amerolite Industries, Inc.
Activated Alkaline Powdered
Drain Opener.......................... L1
Amerozyme.................................. L2
B-Sect Residual Insect Spray F2
Hercules...................................... A1
Odorless F-Sect Insect Spray..... F1
Prolube F H1
Samson.. A1

Ames, E. O. And Son
Gold Bond No-Slip Rub.&Can.
Preserv Belt............................ P1

Amitron Chemical Corporation
Rins A Mint A1
Rins A Mint Plus....................... A4,A8
Ultra-Solv................................... K2
Uni-Solv K1
XL-201 Strip (Liquid)................ A2
XL-70... C1
XL-95... C1

Amoco Oil Company
Amdex Grease No. 1EP H2
Amdex Grease No. 2EP H2
American Industrial Oil No. 100 H2
American Industrial Oil No. 150 H2

American Industrial Oil No. 220 H2
American Industrial Oil No. 260 H2
American Industrial Oil No. 32..H2
American Industrial Oil No. 320 H2
American Industrial Oil No. 46..H2
American Industrial Oil No. 460 H2
American Industrial Oil No. 500 H2
American Industrial Oil No. 68..H2
Amoco Amolube Motor Oil
SAE 10W H2
Amoco Amolube Motor Oil
SAE 20-20W H2
Amoco Amolube Motor Oil
SAE 30 .... H2
Amoco Amolube Motor Oil
SAE 40 .... H2
Amoco Amolube Motor Oil
SAE 50 .... H2
Amoco C-3 Fluid H2
Amoco Compressor Oil No. 110 H2
Amoco Compressor Oil No. 260 H2
Amoco Cylinder Oil No. 1000...H2
Amoco Cylinder Oil No. 460......H2
Amoco Cylinder Oil No. 680-T. H2
Amoco Dexron II ATF.............. H2
Amoco Food Machinery
Grease.................................... H1
Amoco Hydraulic Oil All-
Weather.................................. H2
Amoco Lithium Multi-Purp.
Grease.................................... H2
Amoco Multi-Prps Gear Lbrcnt
#140.. H2
Amoco Multi-Prps Gear Lbrcnt
#90.. H2
Amoco Multi-Purp Gear
Lubrant SAE 80W-90 .... H2
Amoco Multi-Purp Gear
Lubrant SAE85W-140............ H2
Amoco Multi-Purp Gear
Lubricant SAE 80W H2
Amoco Packers Technical Oil ... H1
Amoco Supermill ASU Grease
No. 31052................................ H2
Amoco Syntholube H-5 Oil........ H2
Amoco Syntholube HG-5 Oil ...H2
Amoco Syntholube Oil No. 25-
HA H2
Amoco Trough Grease............... H1
Amoco Trough Grease Soft....... H1
Amoco White Mineral Oil No.
10-NF...................................... H1
Amoco White Mineral Oil No.
13-NF...................................... H1
Amoco White Mineral Oil No.
18-USP.................................... H1
Amoco White Mineral Oil No.
21-USP.................................... H1
Amoco White Mineral Oil No.
31-USP.................................... H1
Amoco White Mineral Oil No.
35-USP.................................... H1
Amoco White Mineral Oil No.
5-NF.. H1
Amoco White Mineral Oil No.
7-NF.. H1
Amoco White Mineral Oil No.
9-NF.. H1
Amoco White Oil AFP.............. H1
Amoco White Oil No. 8-T......... H1
Amoco White Oil No. 9-T......... H1
Amoco Worm Gear Oil.............. H2
Amoco 100 Motor Oil SAE
10W H2
Amoco 100 Motor Oil SAE 20-
20W H2
Amoco 100 Motor Oil SAE 30..H2
Amoco 100 Motor Oil SAE 40..H2
Amoco 1000 Fluid...................... H2

Amine Return Line Treatment #6D G6
Amine Return Line Treatment #91D G6
Chelate "C" Boiler Water Treatment G6
Chelaté Boiler Treatment E G6
City Formula B.W.T. G6
Formula CA G6
Formula S-4 .... G6
Hot & Cold Water System Treat. G6
Liquid Drain Cleaner L1
Liquid Hand Soap E1
Liquid Oxygen Scavenger G6
New Colloid Boiler Water Treat P-1 .... G6
Ox-Scav G6
Qxygen Scavenger "S" G6
Phosphate Control G6
Polymeric Adjunct "A" G6
PCA Boiler Water Treatment G6
Retort and Cooling Water Conditioner G5
Return Line Treatment # 12-6... G6
T.S.P. G6
Velox All Purpose Cleaner A1
Velox S17 All Purpose Cleaner . A1

Amsco Wholesalers, Inc.

Boiler Treatment 300 .... G7
Breaks A1
Bug Getter Space Spray F1
Bug Getter-Residual F2
Descaler A3
Hand Soap (40% Concentrate) .. E1
Heavy Duty All Purpose Cleaner A1
Heavy Duty Pine Cleaner (Pine Odor) C1
Liquid Degreaser A1
Odor-Rid Concentrate C1

Amtek Supply Company

I.C.C. Concentrate Degreaser-Cleaner A1
I.S.C. Concentrate Steam-Clnr. & Dgrsr. A1

Amway Corporation

Allano E4
Auto. Dshwshr Comp-Soft Water Formula....................... A1
Auto. Dshwshr Compound-8.7% P Formula A1
Automatic Dishwasher Compound A1
Chrome & Glass Cleaner A6,A7
Concntrted Fabric Softener & Brightener.................................. B2
Dish Drops................................. A1,E1
Drain Mate................................. C1
Engine Degreaser K1
Germicidal Concentrate D1,Q3,C2
Germicidal Concentrate & Sanitizer D1
Hard Water Film Remover A3
Improved 3-D Form Conert F1 Cl Comp. A4
Industroclean A4,C1
Industroclean No-Phosphate Formula A4,C1
Kool Wash Liquid...................... B1
L.O.C. High Suds...................... A4,E1
L.O.C. Regular.......................... A4,E1,C1
Paste Metal Cleaner A6
Pursue....................................... C2
Quick-Killing Bug Spray F1
Smashing White.......................... B1

Super Concntrtd Dry Chlorine Bleach B1
SA8 Limited Phosphate A1,B1
SA8 No Phosphate A1,B1
SA8 Plus.................................... A1,B1
Tri-Zyme B1
Water Softening Comp. Ltd. Phosphate G2,B1
Waterless Hand Cleaner E4
Wonder Mist H1
Zoom Instant Spr. Cl. No-Phosphate Form...................... A4,C1
Zoom Instant Spray Cleaner A4,C1
Zoom Spray Concentrate A4,C1
3-D Formula Concrete Floor Clng. Comp. C1

AmGard Division of American Can Company

AmGard Alkaterg A1
AmGard All Purpose Cleaner.... A1
AmGard All-Purpose Liquid Cleaner..................................... A1
AmGard Allude........................... A1
AmGard Applause E1
AmGard AC-1 .... A3
AmGard Bactozyme L2
AmGard Bottle-Kare A2
AmGard Brite A3
AmGard BG-26 .... A2
AmGard C-1 .... A2
AmGard Clean-Belt A1
AmGard CC-649 .... A1
AmGard CL-47 .... A1
AmGard D'Germ C2
AmGard D'Smoke A2
AmGard Deterg-"K" A1
AmGard DE-87 .... K1
AmGard DG-66 .... K1
AmGard DL-90............................ C1
AmGard DV-68............................ C1
AmGard Ease-All........................ H2
AmGard Equipmate A1
AmGard Exit A3
AmGard Fast-Flo A3
AmGard Ferogel A3
AmGard Foam N' Kleen............. A1
AmGard Foam-Add A1
AmGard Foam-All....................... A1
AmGard Foamzit A1
AmGard Formula "C" D2
AmGard G-42.............................. A1
AmGard Germ-Go....................... C2
AmGard Hi-Foam A1
AmGard Hydro-Clean C1
AmGard Hygenol........................ E1
AmGard HD-88............................ A2
AmGard I.P.C. A1
AmGard Insto-Clean.................... A4
AmGard Kilz-M F2
AmGard Laundri-Kleen B1
AmGard Lift................................ A1
AmGard Liquid GM-27................ A1
AmGard Liquid HD Cleaner....... A1
AmGard Liquid X-240................. A2
AmGard Mello E1
AmGard Micro-Phene A4
AmGard Mo-Pow A2
AmGard MR-56 .... A3
AmGard Nox K1
AmGard OP-34 .... A4
AmGard Perma Kill Brand F2
AmGard Powdered HD Cleaner.................................... A2
AmGard Pro-Med E1
AmGard Rid-O............................ A3
AmGard RM-457 .... A3
AmGard Saf-Kleen A1
AmGard San-O-Sep E1
AmGard San-O-Sep Odorless E1

AmGard Steamex A2
AmGard Super Brand
  Insecticide Conc. F1
AmGard Super-Fog F1
AmGard Swash A1
AmGard SHS D2
AmGard SR-Remover A3
AmGard Tough-Clean A2
AmGard Vey-Eze H2
AmGard X-38 .... A2
AmGard 222 .... L1
AmGard 342 .... A3
AmGard 514 Insecticide F1
AmGard 522 .... D1
AmGard 580 .... F2
AmGard 583 .... F2
AmGard 722 .... B1
AmGard 77 .... E1
AmGard 919 Iodophor
  Concentrate D1

An-Fo Manufacturing Company
Acid-Du A3
An-Fo Insecticide F1
Bottle Wash-Du A2
Chlorinated Dish-A-Du A1
Dairy-Du Double Action
  Cleaner A1
Dairy-Du Milkstone Remover ... A3
Double Strength Liquid Dairy-
  Du A1
Du-Suds A1
Egg-Du #7 .... Q1
Floor-Du A4
Fly-Du Aerosol F1
Iodu D2
Milk-Master A1
Pipe Line-Du D1
Pipe Line-Du A D1
Sani-Du D2
Sterl-Aid D1
Super Wash-Du B1
Super-Du A3
Thrifty-Chlor D1
Trouble Shooter D1
Vac-Line-Du A2

Analab Laboratories, Inc.
A-N Concentrate A1
Anamine D1
Citron C1
LaFleur Liquid Soap E1
Superfoam Liquid Hand Soap ... E1
Wintergreen Drip Fluid C2

Analytics, Inc.
"Scrubb-Up" E2

Anar Sanitary Supply Co., Inc.
Anar Oil Concentrate #3610 ..... F1
Anar Pan Wash A1
Anar Pride Powder A1
Scrub-It A1

Anco Refining Co.
Adjunct P-4 .... G6
Adjunct P-4L G6
Adjunct T G6
Al-Con G6
Al-Con Crystals G6
Alk-X G6
Alpha PH 1300 .... A3
Alpha PH 1305 .... G7
Alpha PH 1310 .... G7
Amine-X #600 .... G6
Anco Polymer Treatment Series
  800 .... G6
Anco-Cide G7
Ancor 1070 .... G7

Aqua-Solv G6,G3
Boiler Compound NS G6
Colloid Adjunct G6
Corrax-605 .... G6
CT-1020 Cooling Water Treat. ... G7
Dalsperse 1021 .... G7
Dynazine G6
Filmex G6
Keylate G6
Keylate 715 .... G6
Keylate-P G6
Ox-Out G6
Ox-Out Catalyzed G6
Phosphex G6
Poly-Alk G6
Polymer Treatment 810 .... G7
Saf-T Cleaner A3
Tower Cleaner P G7

Andersen Water Engineering Co.
Boiler Water Treatment BB-3 .... G6
Chemitrol Clear G6
Chemitrol SS G6

Anderson Box Company
Egg-Brite SG-5 .... Q1
Egg-Brite SG-7 .... Q1

Anderson Chemical Company
Alkasteem G6
Anco Algaecide G7
Anco Ice Machine Cleaner A3
Anco Liquid Descaler A3
Anco Water Treatment G1
Anco Water Treatment-S G3
Braxon #1 .... G6
Braxon #2 .... G6
Flako G6
Korromeen G6
Ox-Gem Mp G6
SR-2 .... G7

Anderson Chemical Company
"Attack" A4
"Insect Minus" F1
Acco Acid A3
Aerol X A1
Aerol X Chlorinated A1
Aerol X-100 .... A1
Aerol XX A1
Alkali #1 .... A1
An Lube 10 .... H2
An-Care 601 .... A1
An-Chlor A1
An-Foam A1
An-Glo Liquid Detergent A1
An-Klor Plus A2
An-Matic 675 .... G2
Analox De-lime A3
Analox T-75 .... A3
Anderson's An-O-Dine D1
Anderson's Defoamer E A1
Anderson's Liquid Anchlor A1
Aqua Treat G2
ADC MACH LSR A3
ALA A1
AN GLO 85 .... A1
Beck Tower Algaecide G7
Bex #2 .... G6
Bex Regal G6
Bex Regal 40 .... G6
Bex-Amine No. 190 .... G6
Chloro Anox B1
Chloro-Anite A1
Clor-Minus A1
CLR Compound A1
Egg Cleaner-Chlorinated A1,Q1
Egg Wash-X Q1
Fomite A1,Q1
Food Plant Spray F1

Formula REG-13 Liq. Chlorine. D2,Q'
Geronite B-25 .... A2
Geronite L-10 .... A3,Q?
Geronite M-10 .... A1
Halverson Defome Q1
Halverson Egg Wash Q1
Heavy Duty An-Chlor 701 .... A1
Liquid An-Chlor Egg Cleaner ... Q1
Litalox A3
Litalox C5 .... A3
Litebo A3
Liteho X A3
Microlox Special 20 .... A3
Microlox Special 50 .... A3
Microlox Special 70 .... A3
Red Tile Floor Cleaner A4
Scanalite X A1
Scandol No. 30 .... A1
Sequestering Agent N1
Tec-An-20 .... Q1
Wallup B1
Z 85 Alkali Defoamer Q5
Z-92 Egg Defoamer Q5

Anderson Industrial Chemical,
Shur-Kleen A4
Sure Clecn A4
Z.I.P. C1

Anderson Industrial Sales
AmGard Laundri-Kleen B1
Anderson Germicidal Cleaner ... D1
Sewer Solvent L1
Shure-Kleen A4

Anderson Oil & Chemical Co.,
Lusol Cleaner Degreaser A1

Anderson Supply Company
G.L.D. A4

Anderson-Stolz Corporation
Sol-Vet 01 Acid Cleaner A3
Sol-Vet 02 Acid Cleaner A3
Sol-Vet 03 .... A3
Sol-Vet 104 Boiler Treatment G6
Sol-Vet 108 Boiler Treatment G6
Sol-Vet 120-A Boiler Treatment G6
Sol-Vet 146 .... G6
Sol-Vet 164 Boiler Treatment G6
Sol-Vet 181 Boiler Treatment G6
Sol-Vet 185 Boiler Treatment G6
Sol-Vet 33 .... G6
Sol-Vet 334 Potable Water
  Treatment G1
Sol-Vet 341 PC G6
Sol-Vet 341 PCN Boiler Water
  Treat. G6
Sol-Vet 345 Boiler Treatment G6
Sol-Vet 348 .... G6
Sol-Vet 348-X Boiler Treatment G6
Sol-Vet 352 .... G6
Sol-Vet 399 .... G6
Sol-Vet 421 .... G7
Sol-Vet 421 .... G7
Sol-Vet 599 .... G7
Sol-Vet 599D Circulating
  Water Treat. G7
Sol-Vet 599DX Circulating
  Water Treat. G7
Sol-Vet 599Z Circulating Water
  Treat. G6
Sol-Vet 600 Boiler Treatment G6
Sol-Vet 613 Boiler Treatment G6
Sol-Vet 617 Boiler Treatment G6
Sol-Vet 617-A G6
Sol-Vet 658 Boiler Water
  Treatment G6
Sol-Vet 810 .... G6

Sol-Vet 84 Potable Water
Trtmnt.................................... G1
Sol-Vet 88 .... G7
Sol-Vet 900L.............................. P1
Sol-Vet 99 .... G6

Anderson, J. F. Distributors, Inc.
AC 200 Low Foam Degreaser..: A4
AC 210 All Purpose Cleaner..... A4
AC 500 Grease Off...................... A1

Andesite Of California, Inc.
Friction...................................... J1

Andrak Chemical Corporation
APC-700.................................... A4
BW-80....................................... A2
Calorite-225.............................. A3
Clean-Up-702 .... A1
CC-90.. A2
Dignify 240 .... A3
Foam-Up 480 .... A1
Foamac-250................................ A3
LAC-610.................................... A1
LAC-620.................................... A1
MC-650..................................... A1
PAC-615.................................... A1
PAC-625.................................... A1
Sani-Foam 470 .... A1
SA-201 .... A3

Andrews Chemical Products Co.
Butyl-K..................................... A4

Angelina Chemical Company
Heavy Duty Cleaner Degreaser. A1
Lemon Disinfectant.................... C2
Mint Disinfectant....................... C2
Pine Odor Disinfectant C1

Angus Chemical Corporation
Alko 90............,,,...................... A2
Alko 90-BQ................................ A2
Alko 90-T Tree Wash A2
Alko 90-400 X........................... A2
Anga Blue A1
Anga Fog Insecticide.................. F1
Anga Super Strength
Insecticide................................ F1
Anga Tex Disinfectant Cleaner.. A4
Anga Tex 128 .... A4
Angel Blue A1
Angus Fly Spray F1
Angus Fog Insecticide F1
Angus Fog Plus Insecticide........ F1
Angus No. 2 Insecticide.............. F1
Angus QT-164............................ D1
Angus Super Fog Insecticide F1
Angus Super Strength
Insecticide................................ F1
Can Washing Compound A1
Chlorinated Cleaner A1
Clean-Me................................... A1
Defoamer PF-50 .... N1
DO-49 .... C1
DO-58 .... C1
Egg Polish.................................. Q1
Egg Polish #2............................. Q1
Galvanized Cleaner A3
Glissen...................................... A3
High Yield Plus 1 .... N3
High Yield Plus 2 .... N3
Hot Rod L1
Hot Rod Liquid Drain Opener .. L1
Laundry Bleach B1
Pic-Kwik N1
Polt 45 .... A1
Prime Time A1
Residual Insecticide.................... F2

Smokey.. A2
Smokey Foam.............................. A1
Trolley Polish A2
XS-6 Chlor.................................. A1
3% Malathion Insecticide F2
727 L... B1

Anichem Division
Trolley Lubricant W-2................ H1

Antical Chemicals, Incorporated
Antical Acid Cleanser A3
Antical Acid Cleanser
Concentrate A3
Antical Acid Sanitizer................. D1
Antical Alkali.............................. A2
Antical Alkali No. 5.................... A2
Antical Bottle Brite A2
Antical Bottle Wash A1
Antical C-I-P Acid Cleanser A3
Antical Can Brite........................ A1
Antical Chelated Alkali #5 .... A2
Antical Cleanitizer...................... D1
Antical Cleanser No. 20.............. A1
Antical Cleanser No. 35.............. A1
Antical Cleanser No. 35-X A1
Antical Cleanser No. 60.............. A1
Antical Cleanser No. 60
Chlorinated............................... A1
Antical Cleanser No. 81-X A1
Antical Escort Lube.................... H2
Antical Flo A2
Antical Flo Brite Liquid
Cleaner...................................... A2
Antical Flo-Brite Liquid
Cleanser 303 .... A2
Antical Flo-X.............................. A2
Antical Foam Add XXX A1
Antical Fryer Cleaner................. A2
Antical FC-34 .... A1
Antical FC-68 .... A1
Antical H-T Cleanser.................. A1
Antical Hard Floor
Cleaner(Scntd) C1
Antical Hypo-Chlor D2,Q4
Antical HCX-322........................ A1
Antical I-Odo-Cal Cleaner-
Sanitizer................................... D2
Antical I-Odo-For D2,Q6
Antical I-P-C Acid A3
Antical I-P-C Cleaner (E)........... Q1
Antical I-P-C Cleaner (F).......... A2
Antical I-P-C-X Cleaner A1
Antical IPC Cleanser A1,Q1
Antical J-100.............................. A3
Antical Jell-Add A1
Antical Laundry Detergent......... B1
Antical Liquid Cleanser A1
Antical Liquid Kee-Add A3
Antical Liquid Phos-Add A1
Antical Low-Foam I-Odo-For
Sanitizer................................... D1
Antical LI-Quat D1
Antical Machine Dishwashing
Cmpd. No. 1............................. A1
Antical Rustripper...................... A2
Antical Sani-Chlor...................... D1
Antical Single O......................... A2
Antical Special A1
Antical Special Floor Cleaner.... A4
Antical Speedy Lube................... H2
Antical Super Cleanser A1
Antical Super Solvent................. A1
Antical Super Solvent
Chlorinated............................... A1
Antical Swish.............................. A1
Antical 1404 .... A1
Antical 1409 .... A1
Antical-3500 .... G7

Antisoil A1
Antisoil 301-A............................ A4,A8
Antisoil 500................................ A4
Antisoil-HD A1
Antisoil-301............................... A1
Antisoil-305............................... A1
Cleanser No. 81 .... A1
Flo-XXX A2
IPC-E Egg Wash.............E...... Q1
IPC-F Liquid Cleanser............... A1

Antiseptol Chemical Corp.
Degreaser K-30 Concentrate...... A4
Mellofoam Antiseptic Creme E1
30 Below Deep Freeze Clnr
Solution A A5
30 Below Deep Freeze Clnr
Solution B................................ A5

Apache Chemical & Janitor Supply Co.
Apache Egg Wash...................... Q1

Apex Commodities
Jell-Lube................................... H2

Apex Distributing
All Purpose Cleaner Degreaser . A1

Apex Services
A.S. 47....................................... A1
Big Con A1
Clean Concentrate A4
Destroy...................................... B1
Flo-Way L1
Flow-Way Drain Opener............ L1
Formula 74 Steam Cleaner
Liquid A1
Germ-Side D1
K-Kleen C1
Odo-Rout Concentrate............... C1
Sparkle...................................... C2

Apex Soap & Sanitary Corp.
Chem Plus Degreaser Cleaner ... A1

Apollo Chemical & Supply
Extra Heavy Duty Lift-Off Cl-
Degreaser A1

Apollo Chemical Company, Inc.
Energy Plus 680 .... A1,E1

Apollo Chemical Corporation
Pentron Deoxo........................... G6
Pentron E-3................................ G7
Pentron E-5................................ G7

Apollo Home-Care Products, Inc.
Powerbrite.................................. B1
TLC... A4
2001.. A4

Apollo Industries, Inc.
Ant & Roach Residual Spray..... F2
Dual Synergist Insect Spray....... F1
Flying Insect Killer No. 11........ F2
Germicidal Spray Cleaner C1
Industrial Strength Insect Killer F2
Moisture And Corrosion
Preventative Aer...................... H2
Total Release Fogger.................. F2

Apollo-Sol Distributors
Apollo-Sol................................. A4

Apperson Chemicals, Inc.
Ex-Skale A3

31

Applied Biochemists, Inc.
Septictrine L2

Applied Chemicals Corp.
Nature's Award A1

Applied Products Corporation
3060 Scale Remover.................... A3
9000 All Purp. Cleaner & Degr. A1
9030 Medium Duty Steam
  Cleaner.................................. A2
9050 Smoke House & Hot Vat
  Degrsr.................................. A2
9070 Hot Vat Car Rust
  Grease&Paint Remvr A2
9090 Rust & Scale Remover....... A2
9399 Thickening Agent.............. A1

Apter Industries, Incorporated
Appro Killer Fog Spray F1
Appro Powerclean A1
Easy A4,A8
Hans-Down E1
Macro Concentrate Super D...... A1
Phenz A4

Aqua Laboratories
Aqua-Chem-8............................ G7
Aqua-Coll................................ G6
Aqua-Coll-P............................. G6
Aqua-Gard 303 .... G7
Aqua-Phos............................... G6
Aqua-Treet-450........................ G7
Aquamine G6
Deoxit-D G6

Aqua Laboratories and Chemical Co., Inc.
Alcco Tower Kure..................... A3
Alco-Cide #100 C G7
Alco-Cide #200 P G7
Alco-Cide #300 S..................... G7
Alco-Floc #3........................... G1
Alco-Floc #5........................... G1
Alco-Floc #8........................... G1
Aqua-Ayd................................ G2
Boiler Kure #C G7
Boiler-de Scaler G7
Boiler-Kure #A G6
Boiler-Kure-Reg. G7
Boiler-Treat 20........................ G6

Aqua Products, Incorporated
Aqua Clor A1
Matic.................................... A1
Pan...................................... A1
Soak A1

Aqua Treat Chemicals, Inc.
B-200.................................... G6
B-201.................................... G6
B-202.................................... G6
B-204.................................... G6
B-205.................................... G6
B-206.................................... G6
B-207.................................... G6
B-209.................................... G6,G2
B-211.................................... G6,G2
B-212.................................... G6
B-212X................................... G6
B-213.................................... G7
B-214.................................... G6
B-215.................................... G6
B-217.................................... G6
B-242.................................... G6
B-244.................................... G6
B-246.................................... G6
B-250.................................... G6

B-252.................................... G7
B-254.................................... G6
B-256.................................... G6
B-260.................................... G6
B-261.................................... G7
B-262.................................... G6
B-264.................................... G7
B-265.................................... G6
B-266.................................... G6
B-268.................................... G6
C-510.................................... G3
C-512.................................... G7,G2

Aqua-Day Corp.
Aqua-Remove-All 1020.............. A1

Aqua-Lyte, Inc.
Aqua Lyte B1000L..................... G6
Aqua Lyte B102L...................... G6
B 205 Boiler Water Treatment... G6
B 235 Condensate Line
  Treatment G6
C 305.................................... G7

Aqua-Serv Engineers, Inc.
A-101 Algaecide Briquettes........ G7
A-102 Algacide G7
A-103 Liquid Algaecide............. G7
A-104 Algaecide G7
A-105 Liquid Algaecide............. G7
A-106 Algaecide G7
A-108 .... G7
A-109 .... G7
A-109 Microbiocide.................. G7
B-201-B................................. G6
B-201-P................................. G6
B-202 S G6
B-202-B................................. G6
B-202-P................................. G6
B-206.................................... G6
B-208.................................... G6
B-211.................................... G6
B-212.................................... G6
B-212 U G7
B-212-X................................. G6
B-213.................................... G7
B-214.................................... G6
B-215.................................... G6
B-217.................................... G6
B-242.................................... G6
B-246.................................... G6
B-250.................................... G6
B-253.................................... G6
B-261.................................... G7
B-702.................................... G6
B-703.................................... G6
B-704.................................... G6
B-706.................................... G6
B-743.................................... G6
B-794.................................... G6
B216-30................................. G6
C-315.................................... G1
C-318.................................... G1
C-320.................................... G7
C-322.................................... G7
C-323.................................... G7
C-324.................................... G7
C-325.................................... G7
C-416.................................... G7
C-716.................................... G6
C-720.................................... G7
C-720X.................................. G7
DSR...................................... A3

Aquafine Chemical & Equipment Co.
NAL-310 .... G6

Aqualock Manufacturing Co.
Aqualock Compound G2

Aquaness Chemical Company Div. of Magna Corporation
Calnox 214.............................. G6
Calnox 214 DN........................ G6

Aquaphase Laboratories, Inc.
Aquaphase R-Tec-4D-USDA..... G7
Aquaphase RTEC-4 Adjunct-A-
  USDA.................................. G7
Aquaphase RTEC-4C-USDA G7
Aquaphase S-Tec-10.................. G7
Aquaphase S-Tec-11.................. G7
Aquaphase S-Tec-11A G7
Aquaphase S-Tec-2.................... G7
Aquaphase S-Tec-3C.................. G7
Aquaphase S-Tec-6.................... G7
Aquaphase S-Tec-6A G7
Aquaphase 100-S-USDA G7
Aquaphase 100-S2-USDA........... G7
Aquaphase 100-S5-USDA........... G7
Aquaphase 101-P-USDA G7
Aquaphase 101-PQ-USDA G7
Aquaphase 101-Q-USDA............ G7
Aquaphase 101A-USDA.............. G6
Aquaphase 110-S-USDA G7
Aquaphase 110-S2-USDA........... G7
Aquaphase 110-S5-USDA........... G7
Aquaphase 111A-USDA.............. G6
Aquaphase 1110-LA-LB-USDA G7
Aquaphase 1110-LA-SP-USDA. G7
Aquaphase 1130-SP................... G2
Aquaphase 120-S-USDA G7
Aquaphase 120-S2-USDA........... G7
Aquaphase 120-S5-USDA........... G7
Aquaphase 1201-USDA P1
Aquaphase 121-P-USDA G7
Aquaphase 121-PQ-USDA G7
Aquaphase 121-Q-USDA............ G7
Aquaphase 121A-USDA.............. G6
Aquaphase 130-S-USDA G7
Aquaphase 130-S2-USDA........... G7
Aquaphase 130-S5-USDA........... G7
Aquaphase 131A-USDA.............. G6
Aquaphase 2001 USDA A3
Aquaphase 300-P-USDA G6
Aquaphase 300-USDA-SA G6
Aquaphase 310-P-USDA G6
Aquaphase 320-P-USDA G6
Aquaphase 410-USDA-L............. G6
Aquaphase 410-USDA-SA G6
Aquaphase 510-A-USDA............. G6
Aquaphase 520-A-USDA............. G6
Aquaphase 520-D-USDA............. G6
Aquaphase 721-BX-USDA G7
Aquaphase 721-USDA................ G7
Aquaphase 722-BX-USDA G7
Aquaphase 723-BX-USDA G7
Aquaphase 724-BX-USDA G7
Aquaphase 724-USDA................ G7
Aquaphase 731-L-2-USDA G7
Aquaphase 733-L-2-USDA G7
Aquaphase 741-BX-USDA G7
Aquaphase 741-USDA................ G7
Aquaphase 742-BX-USDA G7
Aquaphase 743-BX-USDA G7
Aquaphase 744-BX-USDA G7
Aquaphase 744-USDA................ G7
Aquaphase 751-L-2-USDA G7
Aquaphase 753-L-2-USDA G7

Aquaterra Biochemicals Corp. of America
Alive Bacteria/Enzyme LS-
  1471.................................. L2
Bioprime-FP............................ L2
Exod..................................... C1
FPCC.................................... A1
X-Foam.................................. L2

32

Aquatherm, Inc.
Boiler Water Treatment SA-2 .... G6
Boiler Water Treatment TA G6
Boiler Water Treatment 705 ...... G6
Boiler Water Treatment 707 .... G6
Boiler Water Treatment 750 .... G6
Boiler Water Treatment 752 .... G6
Boiler Water Trtmnt Formula
AF-2 .... G6
Boiler Water Trtmnt Formula
AOH G6
Boiler Water Trtmnt Formula
DA G6
Boiler Water Trtmnt Formula
610 .... G6
Boiler Water Trtmnt Formula
708 .... G6
Boiler Water Trtmnt Formula
711 .... G6
Boiler Water Trtmnt Formula
711-H G6
Boiler Water Trtmnt Formula
715 .... G6
Boiler Water Trtmnt Formula
715-N G6
Boiler Water Trtmnt Formula
720 .... G6
Formula KA G6

Aquatrol Laboratories, Inc.
Balancer BWT FM-1 .... G6
Balancer BWT FM-2 .... G6
Balancer BWT FM-3 .... G6
Balancer BWT FM-4 .... G6
Balancer BWT FM-5 .... G6
Balancer FM. No. 6 .... G6
Basic BWT FM-1 .... G6
Basic BWT FM-2 .... G6
Basic No. 3 .... G6

Ar Chem Corporation, The
Guardian Rat Bait F3
Rodere F3

Arcal Chemicals, Inc.
#20 Stainless Steel....................... A7
"D-Scent" C1
"MCR"-12 .... A1
"Ravage" Chemical Liq.
Concentrate C1
"Tracer" A1
"Uni-Clean" A1
A-118 Concrete Cleaner A1
Crust Strip A2
D' Scale #24 .... A3
Formula 20 (Aerosol).................. A7
Kwell Q1
Mintar A1
PC Cleaner A1
Rout-Out ...:.................................. A2
163-A A4

Archem, Inc.
AC Concentrate............................ A4
AC Industrial Odor Killer C1
AC-III.. A4,A8
HI-P Contact Cleaner K2
Insect Spray F2
Kleen Waterless Hand Cleaner ..E4
OVEC Ammoniated Oven
Cleaner... A8
Sani-Lube Food Equipment
Lubricant H1
Silicone Spray H1
Touch-Up All Purpose Cleaner . C1

Archibald & Kendall, Inc.
Ster-Bac #9003 .... A1

Arden Sales
Arden Foam-Ban.......................... Q5
Ardochlor #15.............................. D2
Extra Heavy Duty........................ A1
Hand Dishwashing Compound ..A1
Kwik Acid Cleaner A3
Zip Chlor D1,Q4

Ardmor Chemical Company
Ardbrite A1
Hot Tank 70................................. A2
Meat Hook Soaker A2
Steam Kleen #60.......................... A2
201 Laundry Detergent............... B1

Arizona Quality Chemicals
Lit-Ning Lube............................... H2

Arkansas Globe Chemical Company, Inc.
Monarch A1

Armor Research Company
Armor Action All Purpose
Cleaner C1
Armor Action Foam
Disinfectant Cleaner C1
Armor Armor 55.......................... H2
Armor Break It Parts & Brake
Cleaner K1
Armor Brilliance.......................... A7
Armor Brite Stainless Steel
Cleaner A7
Armor Cold Kleen A5
Armor Dry-27 Dry Lubricant.... H2
Armor Hoodlum Vandalism
Mark Remover............................. K1
Armor Oasis Hand Cleaner E4
Armor Protector Barrier Cream E4
Armor R.D. 30 Hospital
Disinfectant Deo...................... C1
Armor Refresh Mint Air
Sanitizer C1
Armor Refresh Pine Air
Freshener C1
Armor Renue Tile & Grout
Cleaner C2
Armor S.A.M. Wasp & Hornet
Killer F2
Armor Slide Silicone Lubricant. H1
Armor Spot-XS K1
Armor Static Safety Solvent &
Contact Cl.................................... K1
Armor Unloc Penetrating
Lubricant H2
Armor Vision Glass Cleaner C1
Armor 10....................................... A1
Armor 27....................................... A8
Armoros A3
Bomber F1
Digest L1
L.H.S. E1
Neutral A1
No-Rinse P1
Power A1
Sparkle A1
Staff-X C2
T.N.T. E4
Unique A1
Zapper A1
Zymex L2

Armour and Company
Armour Standard Detergent
110.. A2

Armour-Dial, Incorporated
Flint Chips Stock No. 818 .... A1
Flint Powder Stock No. 757 ...... A1
Flint Powder Stock No. 823 ...... A1
Flint Powder Stock No. 824 ...... A1
Giant Powder Stock No. 749 ..... A1
Hospital Green Soap No. 856 ... E1
Lustro Soap No. 741 .... E1
Parsons' Sudsy Deter.
Ammonia C1
Perfumed Green Liquid 524....... E1
40% Liquid Soap No. 545 .... E1

Armstrong Chemical Company
Chem-Clean A1
Grease-Off..................................... K1
Hook and Trolley A2
Lektri-Safety-Sol K1
Liquid DSO L1
Multi-Purpose A1
Smokehouse Cleaner A2
Steem Kleen A1
SSD-Sewer Solvent &
Degreaser L1
WD-40 .... D1

Arol Chemical Products Company
Arodet AA-350.............................. A1
Arodet DYN.................................. B1
Arodet R-100 .... A1
Arodet SC Special........................ A1
Arodet ST A1
Arodet 25 S................................... A1
Arodet 60 T Soft A1
Arofos 326..................................... A1
Arolterge RAB A1
Arolterge 100 M A1
Arolube LB-300 .... H2
Bonansa A4
Coconut Soap 40% A1,E1
Crystal Bright Green Label........ A1
Crystal Bright Red Label A1
Detergent CR A1
General Cleaning Fluid EMC-
175 .... K2
Hi Foam Powder A1
Laundry Compound MG............. B1
Liquid Hand Soap E1
Loradet I-201 .... A4
Loragleam BW A1
Scour Number 5 .... A1
Siturge A1

Arrow Chemical Corporation
Alltreat-300 .... G7
Aquacide F1
Arrow Kleen A1
Arrow Safety Solvent And
Super Degreaser K1
Arrow Solv K1
Belt Dressing................................ P1
Black Disinfectant, Coef. 5......... C1
Blast Solv A1
Blue.. L1
Bole Aire C2
Bowl Blox C2
Bulldozer A4
Citosene 10.................................... C1
Cocoanut Oil Hand Soap
(Concentrated) E1
Coil-Ite P1
Contact Cleaner K2
Deodorant Blox C2
Deodorant Cakes C2
Dri-M-Lube................................... H2
Dual - 27....................................... D1
Flush Off....................................... K1
Fogging Spray Concentrate F1
Food Processor's D'Grees........... A1
Grime Off...................................... E4
Heavy Duty Degreaser K1
Heavy Duty Degreaser K1
Hi-Slip H1

33

Hospital Disinfectant...................D1
Hydrolytic Enzyme Bacteria
  Cmplx L2
Lectra Solv K2
Lemonene..............................C1
Luball H2
Marks Off..............................K1
Mr. Brite Porcelain Cleaner C2
No Lock................................H2
No-Con.................................P1
Odo-Rout...............................C1
Orange Deodorant.......................C1
Oven Protector.........................P1
Penilube...............................H2
Pine Oil Disinfectant, Coef. 5 ....C1
Power Concentrate......................C2
Power Foam.............................C1
Power Kleen............................A8
Power Kleen E4
Protection.............................E4
Quick Kill.............................F1
R.I.S.-15..............................F2
Red Hot................................L1
Remuv-Ox...............................A3
Safety Solve...........................K1
Sani-Lube..............................H1
Seal-Ease..............................H2
Siltrol................................H2
Steam Solv.............................A1
Steel Bright...........................A7
Super-Solv.............................A4
Tef-Lube...............................H2
Trak-Lube..............................H2
Vita-Pine..............................C1
Whirlpool Drain Opener.................L1
Whomp..................................L1
Wonder Solv............................A4

Arrow Chemical Products, Inc.
Arofect Pine Odor Disinfectant.C1
Arrow Super M..........................D1
FP Special Detergent...................A1
FP-10 ....A1
FP-15 ....A1
FP-18 ....A1
FP-20 Liquid Hand Cleaner E1
FP-35 Hand Dshwsh Pots-Pans.A1
FP-38 Cement Floor Cleaner
  (Pwdr)...............................A4
FP-45 Hvy Dty Clnr Cement
  Floors...............................C1
FP-70 Hand Cleaner Powder C2
Pine-O-San.............................C1
150 Arrow Sanitizer....................D2

Arrow Kleen Products
Hefty Kleener A1

Arrow Laboratories
Acid Cleaner "A".......................A3
Arrow Clean............................A1
Arrow Food Plant Cleaner A1
Arrow Hand Dishwashing
  Liquid...............................A1
Arrow Liquid Hand Soap E1
Arrow Sink Cleaner.....................A1
Arrow T-Bowl Cleaner...................C2
Arrow Vat Cleaner......................A2
Cement Floor Cleaner A4
Hi-Speed Steam Clean A1
Machine Dishwashing Powder ..A1
Steam Clean............................A1
Steam Clean Powder.....................A2
Steam Clean-Light Duty.............A1
Super 88...............................A1

Arrow Products Co.
Quik-Clean A1

Arrowhead Chemicals, Inc.
Arrow Kleen 128........................D1

Arrow Kleen 64.........................D1
Arrowhead Sanitizer D2
Release................................A4,A8
Release 2..............................A1

Art Mattson Distributing Co.
Hand Dishwashing Compound ..A1
KP Grill And Oven Cleaner A4,K1
Special Degreaser......................A4,K1

Art Snyder Sanitary Supply Co.
Snyder's Speedy Spray A4
Snyder's Speedy Trust..............D1
Snyder's Super Speedy
  Concentrate..........................A1

Artel Chemical Co.
ZZT A4,A8

Ash Grove Cement Company
Kemilime Hydrated Lime...........N3
Snowflake Hydrated Lime N3
Veri-Fat Quicklime.....................N3

Ashco Company
Rac 1V.................................A4,A8

Ashdod Chemicals, LTD
Klordet................................A1
Pots And Pans A1

Ashland Chemical Company Div. of
Ashland Oil, Inc.
Ashland 70T White Oil..............H1
Ashland 95T White Oil..............H1
Ashland's Globrite 11..............A3
Bakery Pan Oil.........................H1
Clo-White Bleach.......................D2,B1
Globrite Boiler Treatment.......G7
Globrite Bowl & Porcelain
  Cleaner..............................C2
Globrite Crawling Insecticide ...F2
Globrite Descaler H-4000
  Liquid...............................A3
Globrite Descaler S-4000
  Powder...............................A3
Globrite Flying Insecticide........F1
Globrite Glass Cleaner.............C1
Globrite Heavy Duty Bowl
  Cleaner..............................C2
Globrite Heavy Duty Cleaner...C1
Globrite Oven Cleaner..............A8
Globrite Scouring Cream
  Cleanser.............................C2
Globrite Stainless Steel Cleaner.A7
Globrite Steam Line Treatment.G6
Globrite 14............................A3
Globrite 3001 NC.......................A8
Globrite 3002 HDC......................A4
Globrite 4256 PD.......................A4
Globrite 5001 QD.......................C1
Globrite 5025 QD.......................D1
Globrite 5025 QDC......................A4
Globrite 5064 QD.......................D1
Globrite 5064 QDC......................A4
Globrite 5256 QDC......................A4
NF White Oil No. 100..............H1
NF White Oil No. 145..............H1
NF White Oil No. 85...............H1
Tobin Cleaner No. 3 ....A1
Tobin Special Cleaner.............A1
U.S.P. White Oil No. 185.........H1
U.S.P. White Oil No. 340.........H1
U.S.P. White Oil No. 460.........H1
2056 Solvent...........................A4

Ashland Oil, Incorporated
Elix Grease #2.........................H1

Asiatic Petroleum Corporation
Shell Ondina Oil 17....................H1

Asko Chemical Supply Co.
E.M.S. 535.............................A1

Aspden Associates, Inc. Div. of
Rochester Germicide Co.
Formula 260............................A4

Associated Chemists, Inc.
#607...................................A1
All Purpose Steam Clning
  Comp. 8550...........................A1
All Purpose Steam Clning
  Comp. 8650...........................A1
AC-10..................................A1
AC-15..................................A1
AC-15 LF...............................A1
AC-17..................................A1
AC-5...................................A1
AL-20..................................A3
AL-23..................................C2
AL-34..................................A3
AL-50..................................C2
Concrete Etch & Cleaner #634.A3
Crown DSC..............................D1
Crown Microcide........................D1
DL-14..................................A4
DL-18..................................C1
DL-28..................................A1
DL-29..................................A1
DL-40..................................A1
DL-48..................................A1
DL-67..................................A1
DL-68..................................A4
DL-80..................................A1
DL-88..................................A1
DL-89..................................C1
DL-90..................................C1
DSC....................................D1
Extra Hvy Dty Steam Clning
  Comp 8670............................A2
FC-20..................................A2
Hvy Dty Steam Clning
  Compound 8560........................A1
Hvy Dty Steam Clning
  Compound 8660........................A1
Hyseptic Surgical Soap...........E1
HD-37..................................A2
HL-52..................................E1
KL-48 ....A1
Lo-Foam #680-B.........................A1
LK-20..................................A1
LK-30..................................A1
LK-40..................................A1
LK-45..................................A1
Pro-Quad...............................D1
Quill..................................D1
Ringcide...............................C2
Royal Grand Slam.......................A3
Serf...................................D1
Spray Cleaner #480...............A4
Stab...................................A1
SS-35..................................K1

Associated Grocers Of Iowa
Detergent Concentrate.............A1
Stripper...............................A4

Astor Exterminating Company
Astor-X Drione Dust
  Insecticide Spray...................F2
Astor-X Food Plant Fogging
  Insecticide..........................F1
Astor-X General Purpose Insect
  Spray................................F1
Astor-X Pro-Cide F1
Astor-X Residual Surface Spray F2

Astor-X Roach & Ant Preszd.
c/Diazinon F2
Astor-X Wasp and Hornet
Killer F2

Astor Products, Incorporated
Dixisol Hand Scrub.................... E1
White Arrow Bleach.................. D2,B1

Astor Supply Company, Inc.
Antiseptic Hand-Cleaner Lotion E1
Formula 214 .... A1
Formula 412 .... A1
Germicidal Cleaner D1
Giant Auto Scrubber Detergent A1
H-I-C Concentrate A4
Han-Sof Liquid Soap E1
Hard Surface Cleaner................. A4
Power-Kleen All Purp Clnr &
Degr. A1
Pressure Washer Detergent A1
Wint-O-Cide.............................. C1
Zap High Pressure Degrsr and
Clnr. A4

Astro Products, Inc.
Acid Clean................................. A3
Astro Clean ACP-10 .... A1
Astro Clean ACP-15 .... A4,A8
Astro Clean ACP-50 .... A1
Astro Clean ACP-55 .... A4
Astro-Ho A1
AP 1000 .... D2
Ezy-Treat 100 Boiler Water
Treatment G7
Ezy-Treat 200 Cooling Water
Treatment G7
Foam Clean EP-50 .... A1
Foam Clean EP-55 .... A4,A8
Foamit A1
Liquid Live Micro Organisms.... L2
Peachy Clean 100 .... A1
Peachy Clean 200 .... A4
Rich'n Thick Hand Cleaner E1
Super Clean 500.......................... A1

Astro-Chem, Inc.
Alltreat 300 .... G7
Astro Jet.................................... A1
Astro Kleen................................ C1
Astro-Cide.................................. F1
Astro-Fog.................................. F1
Astrox F1
Boilerite 100 .... G6
Bole Aire.................................... C2
Cloud Nine................................. A4
Cocoanut Oil Hand Soap
Concentrate E1
Concrete Cleaner........................ C1
Conveyor Lube............................ H2
Fast Klean Concentrate A3
GB Breakdown............................ L1
GPC-24...................................... A4
Hangs In There Bowl &
Porcelain Cleaner C2
Lemon Tree C1
Mercury Marvel A4
Mr. Brite Porcelain Cleaner C2
Neptune Concentrate A4
Neutrox 400................................ G7
Odo-Rout Concentrate................. C1
Perfumed Deodorant Blox........... C2
Perfumed Deodorant Cakes C2
Perfumed-Wired Bowl Blox....... C2
Pluto Pellets............................... L1
Remov-Ox................................... A3
Sky Air Industrial Deodorant C1
Solar Whirlpool L1
Solarex....................................... D1
Star Tack.................................... P1

Steam Cleaner Liquid A2
Steamrite 200 .... G6
Tank Saver................................. G7
Tower All NC Concentrate G7
Trak-Lube.................................. H2
Venus Vanquish.......................... F2
W.T.C. Algaecide and Algal
Slimicide................................. G7

Astrochem Industries
Astro-Bryte A1

Atco Manufacturing Company
"Glee" Liquid Machine
Dishwash Detergent................. A2
A.F.D.-20 .... A1
A-T Concentrate C1
A-T Plus.................................... A4
A-T Plus-K A1
A-T Plus-K (Aerosol) A8
A-T Plus-K Ammoniated Oven
Cleaner................................... A8
A-T 201 Concentrate A4
A-161 Velvet Suds A1
A-231 Jewel A1
Algex G7
Atcorex Porcelain Cleaner C2
Atox... A3
Attack Residual Insecticide F2
ASR... C2
ATA Lemon C1
ATA Mint................................... C2
ATA Pine................................... C1
Big "A" C1
Big Daddy.................................. A1
Black Disinfectant Coef. 5.......... C1
Blast F1
Blue Gran. Sewer & Drain
Solvent................................... L1
Boratex 111 .... G6
Broadside.................................... A1
Chain Lubricant.......................... H2
Con-Tec...................................... K2
Dead-eye Insecticide F1
Drain Free.................................. L1
Drain-X...................................... L1
Dynamite B-192 Odorless............ A2
DCD... D1
Econ-O-Sept................................ D1
Emulsion Bowl Cleaner C2
EG-90 Expandable White
Grease..................................... H2
Foam Out C1
Foam-It....................................... A4
Food-Sil Silicone Release H1
Food-Sil Silicone Spray............... H1
Formula No. 67 Steam Cleaner.. A2
Garb-Spray Insecticide F2
Hang-Up Jr. C2
Hang-Up Sr. C2
Ice-Go.. C1
J.D. 901 Jell Degreaser............... K1
K.O. "2" F2
Kling 'N' Kleen Bowl &
Porcelain Cleaner C2
Levelox 413................................ G7
MM445 .... A1
N-C 76 .... A4
Odor-Out C1
Pride E1
Red Hot...................................... L1
Revu C1
Sani-Lube H1
Silicone Spray............................ H1
Skalonite.................................... G7
Sparkle A-230 .... A1
Steamtrol 220.............................. G6
Steel Guard Stainless Steel &
Metal Plsh A7
SS Magic A7

Unitrol 805 .... G7
Universal E1
Vandex....................................... C1

Athena Corporation
Echols Roach Ant & Waterbug
Killer....................................... F2

Athens Janitor Supply Co., Inc.
Atomic Action A1
AA Fly Spray.............................. F1
Extra Hvy Dty Odorless Steam
Clnr. A2
Heavy Duty Concrete Cleaner .. A4
Heavy Duty Odrls Steam
Cleaner................................... A2
Jet Odorless Steam Cleaner........ A2
L. D. Concentrate A4
Liquid Hand Soap E1
Liquid Steam Cleaner A2
Malathion 50%............................ F2
Pine Oil Disinfectant Coef. 5 ..... C1
Pine Type Disinfectant Coef. 5.. C1
Quaternary Concentrate.............. D1
Sewer Solvent L1
Special Cleaner A1
Steam Cleaning Compound A2
Super Germicide D1
Super Pink.................................. A2
University Creme Hand Cleaner E4
White Magic C2
111 Insecticide F1
66 Industrial Spray F1

Atkins Chemical Co., Inc.
Break-A-Way A1
Grease Buster A1
Limesolv.................................... A3
San-A-Dish................................. A1
San-A-Dyne Detergent-
Sanitizer................................. D2
San-A-Klor.................................. A1
San-A-Lotion............................... A1
San-A-Mation.............................. A1
San-A-Pan................................... A1
San-A-Quat................................. D1
San-A-Suds................................. A1

Atlantic Chemicals, Inc.
Direx.. A3
KD-4.. A2
KM-2 Cleaner A2

Atlantic National Corp.
Anox... C1
Bingo Roach and Bug Killer....... F2
Black Jack C1
Cyclone C1
Firebird...................................... C1
Perfect A4
Rain Barrel A4
Round Up................................... F2
Sampson..................................... L1
Surprise...................................... A1
Titanic.. A4
Tornado...................................... C1
True Blue A4
Winner 1..................................... C2
Zenith C1
100 Plus A1

Atlantic Research Laboratories Corp.
Insect Spray "B" F1
Liquid Household Insect Spray
#1.. F2
No Sweat.................................... P1

Atlantic Richfield Company
Atreol 18 .... H1

Atreol 34 .... H1
Atreol 6 .... H1
Atreol 7 .... H1
Atreol 9 .... H1
Autokut Oil 250 .... H2
ARCO Gear Oil HD SAE 140.. H2
ARCO Gear Oil HD SAE 80-
  90 .... H2
ARCO Gear Oil HD SAE 90..... H2
ARCO Gear Oil SAE 140.......... H2
ARCO Gear Oil SAE 90 .... H2
ARCO Mineral Gear Oil SAE
  140 .... H2
ARCO Mineral Gear Oil SAE
  250 .... H2
ARCO Mineral Gear Oil SAE
  90 .... H2
ARCO Multipurpose Grease...... H2
ARCO Polyplex EP No. 1 .... H2
ARCO Polyplex EP No. 2 .... H2
ARCOfleet HD Motor Oil SAE
  10W H2
ARCOfleet HD Motor Oil SAE
  20-20W H2
ARCOfleet HD Motor Oil SAE
  30 .... H2
ARCOfleet HD Motor Oil SAE
  40 .... H2
ARCOfleet HD Motor Oil SAE
  50 .... H2
ARCOfleet Motor Oil SAE
  10W H2
ARCOfleet Motor Oil SAE 20-
  20W H2
ARCOfleet Motor Oil SAE 30.. H2
ARCOfleet Motor Oil SAE 40.. H2
ARCOfleet Motor Oil SAE 50.. H2
ARCOpac 70 .... H1
ARCOpac 90 .... H1
ARCOprime 100 .... H1
ARCOprime 130 .... H1
ARCOprime 180 .... H1
ARCOprime 200 .... H1
ARCOprime 350 .... H1
ARCOprime 70 .... H1
ARCOprime 90 .... H1
ARCOtech 10 .... H1
ARCOtech 6 .... H1
ARCOtech 8 .... H1
ARCOwhite HF Grease............... H1
ARCOwhite 2 Grease H1
Caldron EP 1 Grease H2
Caldron EP 2 Grease H2
Commander S 1500 .... H2
Commander S-315 .... H2
Commander S-465 .... H2
Commander S-700 .... H2
Commander 1900 .... H2
Commander 2800 .... H2
Crystex Oil H2
Diamond S-105 .... H2
Diamond S-40 .... H2
Diamond S-75 .... H2
Diamond 55 .... H2
Dominion H-2 Grease H2
Duopale 100 .... H1
Duopale 60 .... H1
Duopale 80 .... H1
Duopar 100 .... H1
Duopar 90 .... H1
Duro AW S-150 .... H2
Duro AW S-215 .... H2
Duro AW S-315 .... H2
Duro S-1000 .... H2
Duro S-105 .... H2
Duro S-150 .... H2
Duro S-150 LP H2
Duro S-215 .... H2
Duro S-315 .... H2
Duro S-315 LP H2

Duro S-465 .... H2
Duro 55 .... H2
Duro 600 .... H2
Eclipse 1050 H Grease H2
Gascon S-105 .... H2
Gascon S-150 .... H2
Gascon S-215 .... H2
Gascon S-2150 .... H2
Gascon S-315 .... H2
Gascon S-465 .... H2
Gascon S-60 .... H2
Gascon S-700 .... H2
Gascon 1200 .... H2
Insulating Oil No. 8 .... H2
Marine Engine Oil No. 10 .... H2
Marine Engine Oil No. 5 .... H2
Modoc 125 .... H2
Modoc 138 .... H2
Modoc 145 .... H2
Modoc 165 .... H2
Modoc 195 .... H2
No Drip S-1500 .... H2
No Drip S-315 .... H2
No Drip S-700 .... H2
Polar S-150 .... H2
Polar S-315 .... H2
Polar S-465 .... H2
Rubilene S-150 .... H2
Rubilene S-315 .... H2
Rubilene S-465 .... H2
Rubilene S-700 .... H2
Rubilene 1200 .... H2
Rubilene 500 .... H2
Rust-O-Lene 10 .... H2
Rust-O-Lene 20 .... H2
Rust-O-Lene 30 .... H2
Truslide S-315 .... H2
Truslide S-700 .... H2
Ultrol 5 .... H1

Atlantic Service Company

Hook-Eye Acid Cleaner A3
Hook-Eye Chlorinated Cleaner.. A1
Hook-Eye Concrete Cleaner A4
Hook-Eye Drychlorine Shroud
  Bleach B1
Hook-Eye Granulated
  Detergent A1
Hook-Eye Heavy Duty Cleaner A2
Hook-Eye Light Duty Cleaner .. A1
Hook-Eye Liquid Detergent A1
Hook-Eye Liquid Heavy Duty
  Cleaner A1
Hook-Eye Sewer Cleaner.......... L1
Hook-Eye Shroud Detergent B1
Hook-Eye Tripe Denuding
  Compound B N3
Hook-Eye Trolly Cleaner........... A2
Multi-Purpose Food Machine
  Oil H1

Atlas Manufacturing Company

Atlas 108 Sewer Pipe Cleaner.... L1
Atlas 115 Truck and Car Wash.. A1
Atlas 122 Laundrbrite B1
Atlas 304 Algi-Rid NF............... G7
Atlas 500 All Purpose Cleaner .. A4
Atlas 502....................................... A4
Atlas 503 Liquid Dishwash A1
Atlas 504 Heavy Duty Cleaner
  & Degreaser A4,A8
Atlas 506 Pure Coco. Hand
  Soap Almd.Odr. E1
Atlas 507 40 Plus....................... E1
Atlas 508 Window Cleaner........ C1
Atlas 509 Sudsng Ammn. All-
  Pur.Cl. C1
Atlas 512 Drain Free.................. L1
Atlas 517....................................... K1
Atlas 520 Concrete Truck
  Cleaner A3

Atlas 522 Solvo............................ K2
Atlas 523 Safety Solvent K2
Atlas 525 Blue Concentrate
  Cleaner A1
Atlas 539 Paint And Decal
  Remover C3
Atlas 540 Liquid Steam Cleaner A1
Atlas 542 Disinfectant A4
Atlas 548 Spray-Off Oven
  Cleaner A8
AT-110 Steam Cleaner A1
AT-111 Extra H. D. Steam
  Cleaner A2
AT-4005.. A1
AT-4007.. A1
Ferrocator 208 .... A2
Jel... A1
Lemon Odor Disinfectant........... A4,C1
Mint Odor Disinfectant.............. C1
Pine Odor Disinfectant C1
100 All Purpose Cleaner A4
1006... A1
1007... A1
1015... A1
103 Chlrm Mach. Dishwashing
  Cmpnd. A1
104 Delux Mach. Dishwashing
  Cmpnd. A1
106 Concrete Cleaner A4
1104... A1
117 Blue Spry Car Wash Det.... A1
4002... A4
5003... A4
5005... A1
543 Disinfectant Sanitizer-
  Deodorant D2
547 Off Concentrated Hvy Dty
  Clnr A1
562 Heavy Duty Cleaner A4
700 Tox... A1
7005... A3
701 Nox-Al Residual Roach &
  Insect Spry F2
702 Fog-9 Insecticide.................. F1

Atomic Steam Company

Asco Steam Cleaner A1

Atwell Chemical Corp.

Boiler Water Treatment............. G6
Concentrated All Purpose
  Detergent A1
Excel.. A1
Heavy Duty Cleaner-Degreaser. A1
Hi-Alk.. A2
Liquid Soap 15%......................... E1
LPP A4
X-Tra Clean A1

Auerbach Chemical Co., Inc.

"Flash" C1
Acemco RD-76 Rapid-Sol.......... A4
Chem-Surf..................................... A4
ODS-3 Disinfectant-Cleaner....... D1
Rust Remover A3
Super Acri-Cote Stripper A4

Augusta Janitorial Supplies & Equipment, Inc.

AJ's Powdered Steam Cleaner... A1

Aurora Manufacturing Corp.

Auro-Gel A1
Enduro Auro-Bac........................ A1
Enduro Auro-Chlor...................... A1
Enduro Auro-Move Liquid A1
Enduro Auro-Power..................... A2
Enduro Duro-Bac......................... A2
Enduro Duro-Chlor A1

565 Insect Fogger F1

B & H Chemical
Protein & Milkstone Remover ... A3

B & J Janitor Supplies, Inc.
Best Klean A4

B & L Control Service Inc.
BL 20 .... G6 ·

B & L Supply Company
All Purpose Cleaner A4
Amco A1
Amodet A4
Atco Liquid Steam Cleaner A1
Atlas III Hvy Dty Steam
  Cleaner A2
Atlas 110 Non-Caustic Steam
  Clnr. A1
B & L Atco Liquid Steam
  Cleaner A1
B & L Lemon Odor
  Disinfectant No. 20 .... A4,C1
B & L Mint Odor Disinfectant... C1
B & L Surgent Triclosan Liquid
  Soap E1
B & L 301 Pine Odor
  Disinfectant C1
B & L 542 Disinfectant
  Deodorizer Clnr. A4,C1
Blue Spray Car Wash
  Detergent A1
Concrete Cleaner A4
Deluxe Machine Dish Washing
  Compound A1
Drain Free A3
Fog 9 Concentrate F1
Fog 9 Insecticide F1
Formula 10 .... K1
Kalm A3
Laundrbrite B1
Nox-Al F2
Off Concentrated Hvy Dty
  Cleaner A1
Packing House Cleaner A1
Paint And Decal Remover C3
Pot & Pan Soap A1
Pure Coconut Hand Soap
  Almond Odor E1
Safety Solvent K1
Sewer Pipe Cleaner L1
Solvo K1
Spray-Off A8
Steam Clean A1
Super D A4,A8
The Big Blue A1
Tox F1
Unscented Pinky A1
White Hurricane C1
Winda-Kleen C1
1006 .... A1
1007 .... A1
103 Chlrntd Mach Dish Wshng
  Cmpnd A1
1104 .... A1
115 Truck Wash A1
40 Plus E1
4002 .... A1
4005 .... A1
5003 .... A4
5005 .... A1
542 Disinfectant A4
543 Sanitizer D2
562 .... A4
7005 .... A3

B & M Chemical Company
Sanitizer No. 1 .... A4

B & M International, Inc.
B-M Concentrate All-Purpose
  Cleaner A4
Bowlex C2
Chloro-9 .... A2
Kitchen Glow A4
Micro-Lemon 15 .... A4,C1
Micro-Lemon 22 .... A4,C1
Micro-Lemon 7 .... A4
Pandex All Purpose Cleaner A4
Scour Power A6
Steam-X Heavy Duty Cleaner ... C1
Steel Glo A7
Super Kill F1
Super Whiz A4
Whiz A4

B & M Supply Company
Lotion Soap E1

B & W Chemical Company
All Mighty Cleaner Degreaser... A1

B A S F Wyandotte Corporation
Accord D2,Q6,E3
Acti-Lan A1,Q1
Adjust A2
Agile B1
Agile II B1
Ahead H1
Alkali F201 .... B1
Alnate A2
Ammosene A4
Antibac B D2,Q4,E3
Apache B2
Arlac A2,B1
Aspect A1
Atane A1,Q1
Attitude A3,Q2
Avid A1,Q1
Bentec B1
Best Egg A1,Q1
Best Egg Plus A1,Q1
Blot J1
BWC A2,Q1
Caustic Soda Crystal A2,G6,L1,N2,N3
Caustic Soda Granular A2,G6,L1,N2,N3
Caustic Soda, 1/2" Flake ... A2,G6,L1,N2,N3
Caustic Soda, 1/4" Flake ... A2,G6,L1,N2,N3
Clarene A2,B1
Clarix A2,B1
Cle Chlor A1,Q1
Clenzit A1,Q1
Command A3
Conquer II .... A1·
Conveyor Lubricant Lujob L H2
Deepsol C2
Definite A1
Deltro A1,Q1
Delvak A2
Detergent A6
Digress B1
Diligent B2
Dividend D2
Doxite B1
Durzak A2
Egg Wash Formulation U-
  12495-A Q1
El-Bee C1
Elevate A3,Q2
Elmac A1
F-100 .... C1
Faspeel N4
Fernak A2
Flo Chilled Caustic Soda A2,G6,L1,N2,N3
Foamicide P Q5
Food Plant Fogging Insecticide . F1
Frigi-Lube H2
Grime-Go A4

GLX A1,Q1
Halox B1
Hazzit A2,Q1,G7,N4,G1
Impede B1
Interest A2,Q1
K.B.X. B1
Keego A1,G7,N2,G2
Kelochlor A1
Kelvar A1,Q1
Ker-Cell A1,B1
Kromet A1,Q1
Kwikpeel N4
Kwikpeel T N4
Laud B1
Light Soda Ash A1,G6,N2,N3,G5
Lo-Shun B2
Low Temperataure Emulsifier
  F901 .... B1
Low Temperature Bleach F401 . B1
Low Temperature Detergent
  F101 .... B1
Low Temperature Liquid
  Bleach L402 .... B1
Low Temperature Liquid
  Detergent L101 .... B1
Low Temperature Machine
  Detergent W100L A1
Low Temperature Machine
  Detergent W101P A1
Lujob H2
Lujob C A1,H2
Lutop H2
Machine Dishwashing
  Detergent W102P A1
Multi-Chlor D2,Q4,E3
Multi-Chlor D D2
Nuflo N3
NSQ A1,N1,G7,G2
Observe A1
One-Shot Built Detergent F102. B1
Oven Cleaner And Degreaser ... A8
Pasturite A1
Pendulum A3
Per-Vad D2
Polamine C1
Predict B2
Preferred A1
Preview A4,C1
Primary Size B2
Realm B2
Refrane B1
Render D1
Residual Spray Insecticide F2
Resource A2
Respect B1
Respect Plus B1
Rintex A1,B1
Rotate A2,Q1
Rusko G1,B1
Salute A1,Q1
Sanel N4
Sentol A3,B1
Servac A3,Q2
Servac-NP A3
Simbol A1,Q1
Skortex A1,B1
Soda Ash Dense Grade A1,G6,N2,N3,G2
Sodium Hypochlorite D2,Q4
Sodium Metasilicate A1
Spartee D2,Q3
Speedac B1
Spray Cleaner Plus C1
Stainless Steel Polish A7
Star 5X Alkali A2,N3
Starlene B2
Steri-Chlor D2,Q1,Q4
Super Deepsol C2
Super Fame A1
SR-10 .... A3,Q2,B1
Tendec Q5

BP Energol DC-1500 ....H2
BP Energol DC-600-C H2
BP Energol DC-680-C H2
BP Energol EM Motor Oil
  SAE 10W H2
BP Energol EM Motor Oil
  SAE 20W H2
BP Energol EM Motor Oil
  SAE 30 ....H2
BP Energol EM Motor Oil
  SAE 40 ....H2
BP Energol EM Motor Oil
  SAE 50 ....H2
BP Energol HL-100 ....H2
BP Energol HL-150 ....H2
BP Energol HL-220 ....H2
BP Energol HL-32 ....H2
BP Energol HL-46 ....H2
BP Energol HL-68 ....H2
BP Energol HLP-C100 ....H2
BP Energol HLP-C150 ....H2
BP Energol HLP-C22 ....H2
BP Energol HLP-C220 ....H2
BP Energol HLP-C32 ....H2
BP Energol HLP-C46 ....H2
BP Energol HLP-C460 ....H2
BP Energol HLP-C68 ....H2
BP Energol HLP-C680 ....H2
BP Energol LPT-32 ....H2
BP Energol LPT-68 ....H2
BP Energol NT-150 Oil H2
BP Energol NT-68 Oil H2
BP Energol OE-100 ....H2
BP Energol OE-150 ....H2
BP Energol OE-22 ....H2
BP Energol OE-275 ....H2
BP Energol OE-46 ....H2
BP Energol OE-68 ....H2
BP Energol Protective SAE 20 . H2
BP Extra Duty Gear Oil SAE
  140...H2
BP Extra Duty Gear Oil SAE
  80...H2
BP Extra Duty Gear Oil SAE
  90...H2
BP Mineral Gear Oil SAE 140 .. H2
BP Mineral Gear Oil SAE 80 ....H2
BP Mineral Gear Oil SAE 90 ....H2
BP Mineral Oil H-C5 ....H2
BP Mineral Oil L-10 ....H2
BP White Oil NF-70 ....H1
BP White Oil NF-90 ....H1
BP White Oil T-100 ....H1
BP White Oil T-60 ....H1
BP White Oil T-80 ....H1
BP White Oil USP-200 ....H1
BP White Oil USP-350 ....H1
Facto Motor Oil 30 ....H2
Factopet PA-NF..........................H1
Factopet SPR..............................H1
Factopet STC H1
Factopet WW-USP H1
Factopure FM-0 ....H1
Factopure FM-2 ....H1
Factopure NF-130 ....H1
Factopure NF-90 ....H1
Factopure T-70...........................H1
Factopure USP-185 ....H1
Factopure USP-350 ....H1
Factovis 43.................................H2
Factovis 52.................................H2
Factovis 65.................................H2
Gearep OG.................................H2
Gearep 140 ....H2
Gearep 85...................................H2
Gearep 90...................................H2

B Z D International, Inc.
Subdu IP...L2

B. F. Goodrich Company, The
Vynaloy V-66 White 136...........P1

B. F. Laboratories
Grease Cut A1
Par Floor Cleaner.........................A1

B-K Chemical Company
Acid Clene BK A3
Bot. Glo BK.................................A2
BK Cloral.....................................A1
BK· Cloral Ad A1
BK Concentrate A1
BK Derust 4.................................A2
BK Klener E.................................A2
Fom-it A1
Gen. Clene BK A1
Klene Vat 50................................A2
Pan Clene A2

B/L Chemical Corp.
ACF-108.......................................A1
APPC-107 ....A1
CC-124...A1
CLB-104.......................................B1
LH-400 ....E1
LHS-118.......................................E1
SL-100..H1
SL-200..H1
SLB-400..B1
SLD-100..B1
SS-111...K1
TC-1000..A2

B/L Sales Co., Inc.
LHS-117.......................................E1
SL-1000...H1
SLB-110..B1
SLD-100..B1
TC-104...A2

Babb Supply Company
Non Butyl Degreaser A1

Badger Soap Company
Liquid Detergent..........................A4

Badger State Chemical Co.
Activated Alkaline Powdered
  Drain Opener L1
All-Purpose Non Flammable
  Safety Solvent.........................K1
Bio-Kleen A1
Conc. Inhibited Muriatic Acid
  Cleaner...................................A3
Concentrated Liquid Cocoanut
  Hand Soap..............................E1
De-Grease K1
Drain-Away L1
Emulsion Engine Dgrsr
  Tar&Asph Remvr Jel..............C1
Lemon Sewer Sweetener............L1
Mint Deodorant..........................C1
Non-Acid Lime and Scale
  Remover A1
Odor-Ban Odor Control C1
Ready-To-Use Non-Flammable
  Solvent Dgrsr..........................K1
Rust And Corrosion
  Preventative H2
Waterless Hand Cleaner.............E4

Baer Supply Company, Inc.
Dyn-O-Mite.................................A4
Dyn-O-Mite II A1
Grease Police C1
Kleen-All.....................................A1
Mint Foam Surgical Hand Soap E1
Panda Gold Lemon.....................C2

Panda Grease Release Foaming
Cleaner A1
Panda Poultry Producers 128P .. A4
Panda Poultry Producers 25Q D1
Panda Poultry Producers 256P .. A4
Panda Poultry Producers 256Q.. D1
Panda Poultry Producers 64Q.... D1
Panda Poultry Producers 64Q-
Citrus A4
Panda Power-Foam A1
Panda Solv-It-Quik K1
Phylm Free A4
Polar-Baer A5
Saf-D-Solv K1
Steam-King A1
Steam-King II A1

Baer/Slade Corp.
Dyn-O-Mite A4
Grease Police C1
Panda Solv-It-Quik K1
Power-Foam A1
Steam King A1

Bailey's Kim-Ko, Incorporated
Floats #8 .... A4
Grease Gun H2
Hi-P Contact Cleaner K2
HD 7-L Steam Cleaner Heavy
Duty A1
Purj L1
Silicone Spray H1
SMC-17 .... A3

Baird & McGuire, Incorporated
BM Special Cleaner A1
BM Special Cleaner Plus A1
Champion Disinfectant Cleaner . D1
Chlordane 8 lb. F2
Chlorpyrifos-Pyrethrins Liq.
Residl. Spy F2
Coal-Tar Disinfectant Coef. 2 .... C1
Coal-Tar Disinfectant Coef. 20 .. C1
Coal-Tar Disinfectant Coef. 5 .... C1
Coal-Tar Disinfectant Coef. 6 .... C1
Concentrate No. 6 .... F1
Crestall Fluid C1
Desktop Cleaner A4
Mint-O-Phene C2
No. 15 Hand Soap E1
No. 40 Hand Soap E1
No-Roma Disinfectant D1
Pine Odor Disinfectant C1
Pine Oil Disinfectant No. 5 .... C1
Pyronyl Concentrate F1
Pyronyl ULV Concentrate F1
S-D Cleaner No. 10 .... A4
Vapona Insecticide F2

Baird Industries
Hard Hat Cleaner A1

Balco Chemical Corporation
E-Z Clean A1
Super Clean A4
Tough Clean A4

Ball Chemical Company
Ball-O Cide D1
Deo-Drane C2

Ball Industries
Auto-Scrub Detergent A4
Ball De-Limer A3
Ball-O-Zene C1
Castillian Liquid Hand Soap E1
D. O. C. D1
De-Car A8
N.B. Cleaner-Degreaser A1

Red-Eye General Purpose
Cleaner A4
Saf-T-Solv K2
Spray Clean Heavy Duty
Cleaner & Degreas A4,A8
Stainless Steel Cleaner And
Polish A7

Ballantyne Pest Control Co.
Mouse Bait F3
Rat And Mouse Bait F3

Balmar Corporation
Artic-U-Late A1
Boiler/Tower Treatment No. 2 . A3
Charex A8
Close-N-Counters A1
Controlex G7
Controlex 300 .... G7
Controlex 301 .... G6
Controlex 302 .... G7
Controlex 303 .... G7
Controlex 440 .... G7
Controlex 442 .... G7
Controlex-XL G6
CRP 111 .... C2
Exit 17 .... L1
Honey Bare Concentrated
Liquid Skin Clnr E1
Klean Tech A1
Multi Klean A4
Multi Klean II A4
NB-100 .... A1
Pine Tech C1
Safe Tech K1
Steam Klean Liquid A2
40% Pure Coconut Oil Hand
Soap E1

Bama Supply Company, Inc.
"Foam-Away" A1

Bancroft Paper Co., Inc.
Al-Pine C1
Al-Pine II Pine Odor
Disinfectant C1
Big Job A1
Big Mack C1
BDG-82 .... K1
Coconut Oil Hand Soap E1
Dawn Laundry Detergent B1
Deodorant Blocks C1
Down 'n Out L1
Heavy Duty Scatfat A1
Hey-Day Neutral Cleaner A4
Hospital Disinfectant-
Deodorant C2
Kleer Glass Cleaner C1
Lemon Fresh Disinfectant C2
Liquid Live Micro Organisms
Drain Wolves L2
Liquid Live Micro Organisms
Grs. Monkeys L2
Liquid Live Micro Organisms
Scavengers L2
Liquid Live Micro Organisms
Sewer Rats L2
Lotion Soap E1
M-P Plus D1
Nitro-Kleen A4
Open-24 .... L1
Phosphate-Free Imprvd. B. L.
C. A4
Pink Magic A4
Power Kleen A1
Professional Hand Soap E1
Ready For Use Hand Soap E1
Spring Mint C1
Ster-Dyne C2
Super Mint C2

Super Soap E1
Swash A1
Tri-Action Germ. Detergent A4
Wax Stripper A1
White Knight Toilet Bowl
Cleaner C2
Winter-Mint C2

Bancroft Paper Company Of
Beaumont, Inc.
Power Clean A1

Bankle Supply Company
Lem N Clean E1

Banks Industries, Incorporated
Action A4
B-50 Concentrate A1
Break Away A1
Cuts-It-Plus A4
Remove C1
Right-Off A4
Solar Shine A7

Banner Chemical Corporation
Cocoanut Oil Liquid Hand
Soap E1

Banner Systems, Inc.
Banner Hand Cleaner E4

Banquet Foods Corporation
Banquet Special A1
Banquet-BFC-45 .... C2
BFC-15 .... A2
BFC-17 .... A2
BFC-19 Alkaline Carbon
Cleaner A2
BFC-20 Hvy-Dty Acid
Detergent Clnr A3
BFC-23 Acid Spray Detergent .. A3
BFC-24 Hvy-Dty Acd Dtrgnt
Cncnt Cl. A3
BFC-27 C. I. P. Acid Cleaner A3
BFC-40 C.I.P. Liq. Caustic
Clnr. A2
BFC-50 Granular Laundry
Detergent B1
BFC-6 Alkaline Foam Additive . A1
BFC-6 Concentrate Foam
Additive A1
BFC-60 Feather Softening
Compound N1
Caustic Cleaner BFC-18 .... A2
Chlor Prssr Sp. Equip Dtrgnt
BFC-5 .... A1
Chlor. Alkln Vat & Ppln Cl.
BFC-15A A2
Concrete Floor Cleaner BFC-25 A4
Concrete Floor Cleaner BFC-26 C1
Fast Actng Swr & Drn Opnr
BFC-41 .... L1
General Prps Manual Clnr
BFC-1 .... A1
General Prps Manual Clnr
BFC-2 .... A1
Gr. Atmtc Mch Dshwshng
Cmp BFC-8 .... A1
Gr. Clr Atmtc Ma Dsh. Cmp
BFC-8-C A1
Heavy-Duty Steam Cleaner
BFC-3 .... A1
Hy Pressure Spray Detergent
BFC-9 .... A1
Liq. Chlorine Brng Disf. &
Germ. BFC-12 .... D2
Liq. Iodine Bearng Det/Santzr
BFC-13 .... A3,D2
Liquid Detergent BFC-4 .... A1

Powdered Toilet Bowl Clnr
BFC-46 .... C2
Scouring Cleanser BFC-30 .... A6
Sol. Chlor. Brg Pwdr Dis &
Germ BFC-14 .... D1
Spcl Alkali Vat & Insert Cl.
BFC-16 .... A2
Spel Sewer & Drain Opnr
BFC-42 .... L1
Sulfamic Acid Descaler BFC-21 A3

Barclay Chemical Company, Inc.
Air Con AX G7
Air Con BX G7
Air Con LN G7
Air Con LS G7
Air Con LS 220 .... G7
Air Con LS 330 .... G7
Air Con 414 .... G7
Algeacide CR G7
Alka Sperse G6
Bar Kem No. 4 .... G6
Bar-Kem No. 3 .... G6
Caustic Soda G6
Dispersall LT G7
Kem Coll C G6
Kem Coll M G6
Kem Coll MS G6
Kem-X G6
Kem-X-E G6
N S Powder G6
Oxogon G6
Volamine G6
Volamine RL-2 .... G6
Volamine SDP G7

Barco Chemicals Division, Inc.
Act 10 .... A4
Airofresh C1
Attack F2
B 776 .... H2
B. C. Concentrate A4
B. C. Hospital Disinfectant
Deodorizer C2
Bar-Gleem C1
Delight A4
Dooms-Day F1
Easy-Go H2
Elektrol H2
Enzact Universal Enzymes L2
Epog Steam Cleaner Compound A1
Exo-Therm L1
Germxseptic D1
Maxim A4
No. 103 Garbo-Dis C1
Perm-A-Sect Residual Spray C1
Perm-A-Sect Residual Type
Insected F2
Pink Antiseptic Lotion Hand
Soap E1
Power Pack L1
Quatergent D1
Royale Ceramic XZT A3
Sparklize C1
Steel Shine A7
Steri-Lube H1
Teflex H2
Zippo Concentrate A4
20/20 Glass Cleaner C1

Barco Supply Company
Liquid Snake L1
Super Dazzle B1
Super Speed Spray A1

Bardahl Manufacturing Corporation
Bardahl Food Machinery
Grease H1
Extra Purpose Aluminum
Complex Grease H2

Bardan Industrial Chem Corp.
Formula #BD-707 .... A4

Bardon Industries, Inc.
Bardon I Concentrate C1
Below-Zero A5
Shine 'N Glow A7
Spray-Foam A1

Barlyn Chemicals
Action-Liquid Skin Cleaner &
Conditioner E1
All Purpose Cleaner A1
Antiseptic Liquid Hand Soap E1
Antiseptic Liquid Hand Soap
Concentrate E1
Bar-lo-cide Formula PR-14 .... D1
Bar-Lo Cherry Deodorant C1
Bar-Lo Pine Oil Deodorant C1
Bar-Lo-Cide Formula C-26 .... D1
Bar-Lo-Cide Formula I-32 .... D2
Bar-Lo-Cide Formula M-28 .... D2
Bar-Lo-Cide Formula P-30 .... D1
Bar-Lo-Cide Formula X-24 .... A4
Bar-Lo-Cide Fortified Pine
Odor Disinf C1
Bar-Lo-Cide Mint Disinfectant .. C2
Bar-Lo-Cide Pine Odor
Disinfectant C1
Bar-Lo-Cide Pine Oil
Disinfectant C1
Bartrol Odor & Grease Control . C1
Bullet-Non-Butyl Degreaser A1
Chicken Scald N1
Complete Concrete Cleaner A4
Deoxidizing Chemical Cleaner .. A3
Duo Fog F1
Dynamite Cold Oven Cleaner ... A8
Formula 1000 Conveyor Lube ... H2
Formula 747 Heavy-Duty
Truck Wash C1
Greaseless Conveyor Lubricant . H2
Great Scott Butyl Cleaner A4
GC Foam Cleaner A1
Hot Shot High Pressure
Cleaner A1
Industrial Formula Skin & Body
Cleaner E1
Jell Degreaser K1
Liquid Hand Soap E1
Lo Foam-A-Vac Cleaner A4
Lotion Antiseptic Hand Cleaner E1
Miracle Floor Finish Maintainer A1
Neutral Cleaner A4
Odor and Grease Control C1
Perfect Laundry Detergent B1
Perform Foaming Cleaner A1
Pink Concentrate Cleaner A4
Pink-L-Lotion Hand Soap E1
Power Cleaner A4
Power Plus A4,A8
Quarry Tile Cleaner A4
Rapid Mechanic Hand Soap E1
Safety Solvent K1
Scum Remover A1
Secret Oven Cleaner A8
Smooth Hand Sanitizer E3
Space Age Degreaser A4
Sparkle Liquid Hand Dish
Wash A1
Special Cherry Deodorant C1
Special Lube D H2
Special Pink Cleaner A1
Steam Cleaner A1
Strike Degreaser/Cleaner A1
Strip-It A4
Supreme Stripper A4
Supreme Wax & Floor Finish
Stripper A4

Synthetic Conveyor Lube H2

Baroid Division/ NL Industries, Inc.
Barochem-AF452 .... G7
Barochem-AF454 .... G7
Barochem-BW480 .... G7
Barochem-BW482 .... G7
Barochem-BW484 .... G6
Barochem-BW485 .... G6
Barochem-BW486 .... G6
Barochem-B460 .... G7
Barochem-B464 .... G7
Barochem-CS438 .... G2
Barochem-C394 .... G7
Barochem-C448 .... G7
Barochem-C510 .... G7
Barochem-P432 .... G7
Barochem-R456 .... G7
Barochem-R475 .... G6
Barochem-S257 .... G6
Barochem-S31 .... G6
Barochem-S458 .... G6

Baron Chemical Co., Inc.
Blue Boy Hi-Foam Concentrate A1
Bright Red Thick Concentrated
Cleaner A4

Barrett Chemical Company
All Purpose Detergent A1
ACL-2 .... A1
B. O. C. A1
Barrett's B-S-S Silicone Spray .. H2
Barrett's Disinfectant Cleaner
No. 11 .... D1
Barrett's Disinfectant Cleaner
No. 12 .... D1
Barrett's Disinfectant Cleaner
No. 14 .... A4
Barrett's EX-50 .... A3
Barrett's Pink Dishwash A1
Barrett's R-3500 .... A3
Barrett's S.A.S. A3
Barrett's Solvall A4
CP-8773 .... A1
Dynamote K1
Essee A1
Koil Brite P1
Liquid Drain Opener &
Maintainer L1
Multifix H2
No. 3 Cleaner Concentrate A4
RSC-75 .... A1
Soap, Hand, Liquid E1
Stride A4,A8
SC-1 .... A1
Tile & Grout Cleaner C2

Barrier Chemicals, Incorporated
Barcrobe D1
Barrier Stainless Steel&Metal
Polish A7
Blue Pearl Synthetic Detergent . A1
Glass Kare C1
Klean-All Multi Purpose Foam
Cleaner C1
Relkem O A1

Barrier Industries Inc.
All Star Pine Cleaner &
Deodorant C1
BA-201 .... A1
Free'N'Clear A4
Fresh N'Clean A4,A8
Muscle Strip C1
Oven-N-Grill Cleaner A8
Relkem "99" A4
Slash A4,A8
Super Barcrobe D1

41

Top-Notch Cream Cleanser C1
Wildcat A4

Barrier Midwest Chemicals, Inc.
Bar-Foam All Surface Cleaner...A1

Barth, T. & Traum Co.
Cleaner/Degreaser A1

Bartok Industries, Inc.
500-SX A4

Barton Chemical Corporation
Chloro-San D1

Basic Chemicals, Incorporated
De-Greese A4
E-Z Strip A4
Glass Class C1
Kleen-Off...K1
Paladis D1
Phase-10..A4
Sanitate D1
White Bowl C2
X-1002-5 ....A1

Basso Chemicals, Incorporated
Sure Clean 121................................A2
Sure Clean 141................................A3
Sure Clean 170................................A2

Bauer, J. G., Inc./ Bauer Chemical Co.
Wash Away.......................................A1

Baver Custodial Supply & Consulting Co.
Duro Blue Label.............................A1
Duro Germicidal Cleaner D2

Baxter Manufacturing Company
Compound #22................................N1
Special Compound #10 ....A1
Super Compound #8.................A1,Q1

Bay Chemical Products Company
AC-1 ....A3
AC-4 ....A3
Bay Active A1
Bay Klor..A1
Bay SD A1
Bay 603 ....A2
Bay 607 ....A2
Bay 804 ....A2
Bay 806 ....A1
Foamer...A1
George SP-L.....................................A1
GPC...A1
Steam-Det...A1

Bay Supply & Chemical Co., Inc.
Bay C-D-Q..D1

Baylor Chemical Company
Ammo Strip C1
Hard Surface Cleaner..................A4

Bayside Chemical Labs., Inc.
Super-Sol A1

Bayside Industries, Inc.
Pink Lotion Hand Soap E1
Tincture of Green Soap N.F......E1

Beach Associates
AOK C2
APT...A1
Big Fyte...C2

Blue Glass Cleaner C1
C.I.K. Roach'n Ant Killer....F2
Decrust A3
Dirge...F1
Evenflo...H2
Gosh E4
Hy-Od-Abate C1
Hy-Tef H2
Hydene A4
Hysilan Barrier Creme E4
HP-673 ....G6
HP-88 Sewer Solvent....................L1
Liqui-Klenz C2
Mintene Disinfectant C2
Mr. W. C..C2
New Super Smooth (Aerosol)....E4
Pow..L1
S-T-Bane..F1
Slam Bang A2
Spoox Residual Roach Liquid....F2
Sprease...H2
Strip-Ex A2
Super Hy-Kil F1
Swinger Germicidal Detergent..A4
Teffy A2
Transmit C1
Wint Mint C1
15% Coco Opal Lemon Hand
  Soap E1
6209 Heavy Duty Cleaner............A1

Beacon Chemical & Supply
No Stick Silicone Spray..............H1

Beam Chemical Company, Inc.
BVC Beam Vat Cleaner A3
BVC Heavy Duty Vat Cleaner..A3
Oven, Grill & Stainless Steel
  Clnr. A1

Beam Supply, Incorporated
Antiseptic Lotion Soap...............E1
Safe-T-Walk J1

Bear Chemical, Incorporated
BC-125 ....L1
BC-126 ....L1
BC-17..A4
BC-206 ....A1
BC-50..L1
BC-60..C1
BC-65..F3
BC-66..F2

Bear Paw Company, The
Adjunct OSS.....................................G6
Boiler Chmel Liq Org Colloid
  A-00 ....G6
Boiler Chmcl Liq Org Colloid
  A-01 ....G6
Boiler Chmcl Liq Org Colloid
  A-02 ....G6
Boiler Chmel Liq Org Colloid
  A-05 ....G6
Boiler Chmel Liq Org Colloid
  Q-00...G6
Boiler Chmcl Liq Org Colloid
  Q-01...G6
Boiler Chmel Liq Org Colloid
  Q-02...G6
Boiler Chmel Liq Org Colloid
  Q-05...G6
Boiler Water Treatment WSL-
  L..G6
Boiler Water Treatment 900-G ..G6
Boiler Water Treatment 900-Q...G6
Boiler Water Treatment 900-S ..G6
Steam Line Treatment SLT-A..G6
Steam Line Treatment SLT-C..G6

Steam Line Treatment SLT-M ..G6

Beatrice Foods Company
Beat-A-Cide D2
Bottle Wash-HTST A2
General Purpose Cleaner...........A1
Liq Chloring Bearing Disinf &
  Germicide......................................D2
Liquid Hand Soap E1
Milkstone Remover No. 33 ....A3
No. 75 Liquid Cleaner A1
Rudolph Floor Cleaner...............A4
Rudolph General Purpose
  Cleaner..A1
Spray Cleaner A2
White Mountain A.P.C. A1
White Mountain All-San..............D2
White Mountain Conveyor
  Lube..H2
White Mountain EXP Cleaner...A1
White Mountain Hi Chlor A1
White Mountain Liquid C.I.P....A1
White Mountain Smokehouse
  Cleaner..A8

Beaulieu Chemical Company
B-San D2
Blu-Chlor..D1
Blu-Pearl..A1
BC-255...A2
BC-535...A3
Formula BC-280 Conveyor
  Lubricant H2
Low Count..A1
Pertek Q1
Twink A4
Vertex A1
Wiz..C1
3'N 1 ....D1

Beaver Chemical Company, Inc.
A-108 ....G6
A-108-5 ....G6
A-1082 ....G6
All Purpose Cleaner.....................A1
B C Chlor..D2
B-C Concentrate A4
Beaver Aluminum Safe Hot
  Tank Compound A1
Beaver B-100 Floor Cleaner.......C1
Beaver Chain Liquid Lube
  Soap H2
Beaver Concrete Cleaner............C1
Beaver Kleerview Glass
  Cleaner..C1
Beaver Paint Safe (Med. Dty)
  Stm, Comp C1
Beaver San D1
Beaver Sudzy All Purp.
  Detergent.......................................A1
Bottle Wash Compound...............A2
C-100 Corrosion Control............G7
Can Cleaner No. 2.........................A1
Chain Lube #1 ....H2
Chain Lube 3X H2
Chlor-Bactericide Disinfectant..D1,Q4
Con-Tac...A2
D-Scaler A3
Dry Bleach B1
DO-1 ....G6
Egg Stain Remover Q2
Egg Wash #1....................................Q1
Egg Wash #2....................................Q1
Flor-Clean C1
Foam Booster...................................A4
Fore Concentrate...........................A4
Formula-C G6
Formula-K G7
Heavy Duty Compound C1
Heavy Duty Liquid Compound .A1

42

Bee-Kim Co., Inc.
Super Power A1

Beech Chemical & Paper Co.
Grease-Gone A1

Beehive Machinery, Incorporated
Deboner Cleaner A2
Deboner Cleaner B71372 .... A2

Behnke Lubricants, Incorporated
Jax Magna Plate 36-1 .... H1
Jax Magna Plate 36-2 .... H1
Jax Magna-Plate 100 .... H2
Jax Magna-Plate 1100.................. H2
Jax Magna-Plate 140 .... H2
Jax Magna-Plate 200 .... H2
Jax Magna-Plate 33 .... H1
Jax Magna-Plate 400 .... H2
Jax Magna-Plate 42 .... H2
Jax Magna-Plate 500 .... H2
Jax Magna-Plate 6-H.................... H2
Jax Magna-Plate 600 .... H2
Jax Magna-Plate 74 .... H1
Jax Magna-Plate 78 .... H1
Jax Magna-Plate 8 .... H1
Jax Magna-Plate 8-00 .... H1
Jax Magna-Plate 90 .... H2
Jax MP Gear Oil GL-5 SAE
   140 w/Paratac.......................... H2
Jax MP Gear Oil GL-5 SAE 90
   w/Paratac................................. H2

Bel-Ray Company, Incorporated
Amber Grease............................... H2
B-R Divider Oil No. 169 .... H1
Bakerylube H2
Bakerylube Regular...................... H2
Bakerylube 200 .... H2
Bakerylube 260 .... H2
Bakerylube 400 .... H2
Bel-Ray Compuvac Oil SAE 30 H2
Bel-Ray Compuvac Oil SAE 40 H2
Bel-Ray No-Tox Oil 1600.......... H1
Bel-Ray No-Tox Oil 900............. H1
Bel-Ray SM-500 Lubricant......... H2
Bel-Ray 500 No-Tox Oil............. H1
Belco 800 Grease......................... H2
Belco 801 Grease H2
Belco 802 Grease H2
Molylube 6 in 1 Fluid H2
No-Tox Grease H1
No-Tox Grease AA-1 .... H1
No-Tox Oil................................... H1
Raylene EP Hydraulic Oil 10 .... H2
Raylene EP Hydraulic Oil 20 .... H2
Raylene EP Hydraulic Oil 30 .... H2
Raylene EP Oil No. 1.................. H2
Raylene EP Oil No. 140.............. H2
Raylene EP Oil No. 2.................. H2
Raylene EP Oil No. 3.................. H2
Raylene EP Oil No. 4.................. H2
Raylene EP Oil No. 5.................. H2
Termalene 87 Grease.................... H2

Belk Chemical & Industrial Supplies
Anti-Crawl Residual Insecticide F2
Bacti.. D1
Big Dog Concrete Cleaner.......... A4
Chemial & Maint. Supplies
   Glass Clnr................................. C1
Dynamo Detergent....................... A1
Grime Scat A1
Iodo-175 Concentrated Iodine.... D2
Liq. Steam Clnr. All Purp. Hvy
   Dty... A2
Liquid Hand Soap 15 .... E1
Multi-Blue A1
Orange Bact. Germ. Clnr.
   Hosp. Formula C1

Panacea... F1
Pine Oil Disinfectant Coef, 5 ..... C1
Pink Cleaner................................. A4
Pink Satin Hand Soap E1
Stainless Steel Cleaner A7
Steam Clng Comp. All Purp.
   Hvy Dty A2
Wax Scat A1
X-150 Chlorinated Mach.
   Dishwash A1

Bell Chemical & Fastener
Bell Heavy Duty Steam
   Cleaning Cmpd. A2
Work-Horse A4
19 All Purpose Cleaner............... A4

Bell Chemical Company
Belco Bell-Chlor.......................... D2
Belco Bell-Solve A4
Belco Black No. 5 .... C1
Belco Blast Grease & Oil
   Remover A1,L
Belco Bole Gleem C2
Belco Bug-Dize #839-38............. F2
Belco G-P Concentrate
   Insecticide................................ F1
Belco Grime-Off.......................... A2
Belco Kreme Klenzer A6
Belco Lgntng Drain & Pipe
   Opener L1
Belco Mag Rust & Scale
   Remover A3
Belco Magic Mist Silicone
   Lubricant.................................. H1
Belco Malat #839-24................... F2
Belco None Better........................ E1
Belco Nu-Life A1
Belco Quat 42 .... D1
Belco Quik A1
Belco S-A-S-S Scale Remover... G7
Belco Steri-Clene......................... C2
Belco Vapo-Cide #839-47 .... F2
Belco W-S-1 Indtrl Scale
   Remover G7
Belco 25% DDT W.E. #839-20 F2
Belco 40% Chlordane W.E.
   #839-23..................................... F2
Staph-I-Cide-20 .... C1
Staph-I-Cide-30 .... C1

Bell Chemical, Inc.
Bell Acid II.................................. A3
Bell-Sheen A1
Chloro-Sheen A1
Komplete....................................... A1
Liqui-Kleen A1
Supreme... A2

Bell Company, Incorporated
Abrasive Detergent A6
Acid Vat Clur & Descaler No.
   300.. A3
Ammoniated Oven Cleaner A8
Bell's Athletic Soap 15................ E1
Big B... A1
Big Dog Concrete Cleaner.......... A4
Blue Flame Extra Hvy Dty
   Sewer Sol. L1
Check-Mate................................... A4
Coil-X A1
Control P1
Cotton Picker Spindle Clnr &
   Lubret.. A1
Deep Chlor D2
Delux Machine Dishwashing
   Comp. A1
E-Z Foam All Purpose Cleaner. C1
E-Z Grout A3
Egg Wash...................................... Q1

43

Extra Hvy Dty Concrete
  Cleaner................................ A4
Golden Gator Concrete Cleaner C1
Hvy Dty Det & Neutrlzr & Vat
  Cl. No. 400........................ A1
Hydrolytic Enzyme Bacteria
  Complex............................ L2
Iodo-175.............................. D2
Kwik Foam.......................... A1
Lemon Fresh 20.................... C1
Liquid Hand Soap 15............ E1
Liquid Hand Soap 20............ E1
Liquid Hand Soap 40............ E1
Liquid Smoke House Clnr
  Formula II........................ A1
Metal Brite.......................... A7
Milkstone Remover.............. A3
No. 200 Smoke House Cleaner .. A2
No. 500 Tripe Cleaner.......... N3
No. 600 Hog Scald................ N2
No. 80 Space Spray.............. F1
Orange Bact. Germ. Clnr.
  Hosp. Formula.................. C1
Oven Gard Oven Protector....... H1
P.D. Concentrate.................. A4
Pine Odor Disinf. Phenol Fort.
  Coef. 5.............................. C1
Pine Oil Disinf. Form.-B Coef.
  5...................................... C1
Pink Cleaner........................ A4
Premium Fly & Roach Spray.... F1
Skin Shield Cream................ E4
Slick Silicone Spray.............. H1
Sparkle Glass Cleaner.......... C1
Steam Cleaning Compound........ A2
Super "B"............................ A1
Super Enzymes for Sewerage
  Systems............................ L2
Sure Clean Shampoo............ A1
SX-11.................................. A3
Tef-Tape.............................. H2
Treatrite 300........................ G7
TFE Fluorocarbon Lubricant H2
Wax Scat............................ A1
X-150 Chlorinated Machine
  Dishwash.......................... A1
1000 Plus............................ L1
1000 Plus Drain Pipe Opener..... L1
1000 Plus Hvy Dty Sewer
  Solvent............................ L1
360 Pot & Pan Cleaner.......... A3

Bell Laboratories, Inc.
"Raze".................................. F3
"Rodent Cake"...................... F3
Dispose................................ C1
Final Rat And Mouse Bait...... F3
P.C.Q. Rat And Mouse Bait...... F3
ZP Rodent Bait.................... F3
ZP Tracking Powder.............. F3

Bell Maintenance Products Co.
#108 Big Red........................ A1
#112 Coconut Oil Hand Soap.... E1
#124 Pink Lotion Dish
  Detergent.......................... A1
Pink Lotion Hand Soap #125.... E1
106 Blue Cleaner.................. A4

Bell Products Corporation
All Purpose Cleaner 350........ A4
Bell Insect - 100 319.............. F1
Bell Quat 335...................... A1
Food Locker Cleaner & Thaw
  372.................................. A5
H.D. Concrete Remover 308...... A3
Neutral Clean 327................ A4
Oxidation-X 332.................. A3
Power-Wash 343.................. A1
Pure Food Grease 210............ H1

Soap Scum Remover 340.......... A4
Steam Cleaner 344................ A4
Ultra Clean 309.................... A4
201 H.D. Drain Opener.......... L1

Bell Supply, Incorporated
Big Dog Concrete Cleaner........ A4

Bell Technology, Inc.
Environ-All.......................... A1

Beltraction Co.
Food Safe Lubricant.............. H1
Food Safe Silicone Lubricant.... H1

**Ben's Service Center & Sanitary
Supply**
Bensan Heavy Duty Cleaner/
  Degreaser.......................... A4
Big Shot Non-Ammoniated
  Wax Remover.................... A4
Blockbuster Spec Multiple-Ct
  Bld-Up Rmvr.................... A4
Free Spirit............................ A4
Hot Stuff.............................. A1
Kitchen Kleen Spray Cleaner A4
Onward Multi-Purpose
  Germicidal Cleaner............ D1
Renegade Ammoniated Stripper C1
Rise & Shine Heavy Duty All
  Purpose Clnr.................... A4
San-O-Side.......................... A4
Spearmint Disnf. Deo. All
  Purp. Clnr........................ C2

Benlo Chemicals, Inc.
Cleaner No. 24...................... A2
Penetrant............................ A1

Bennett Paper Company
Breakway............................ A1

Benson & Zimmerman Company
B & Z Hot Tank H D.............. A2
B & Z Medium L.................. A4
B & Z Powdered Full............ A2
B & Z Powdered Medium........ A1

Benz Oil, Incorporated
Food Machinery Grease.......... H1

Berke Products, Inc.
Heavy Duty Butyl Cleaner........ A4

Berkel, Incorporated
Berkel Food Machine Grease..... H1
Berkel Food Machine Oil.......... H1

Berman Chemical Company
No. 149 Surface Scum Remover A4
Terminate............................ D1

Bernhardt Bros., Incorporated
P. O. W............................... A4

Besco Chemical Company
Cleaner No.-100.................... A1
Heavy Duty Cleaner.............. A1

Besco Corporation, The
All Purpose Cleaner.............. A1
Glass Cleaner...................... C1
Hi-P.................................... K2
No. 25 Insect Spray.............. F1
Silicone Spray...................... H1
Skin Shield Cream................ E4
Solv #500............................ A1
Spray and Wipe All Purpose
  Clnr................................ C1

Triple-Play.......................... D1

Besst Chemical Corporation
Besst 101............................ G6
Besst 101 S.......................... G6
Besst 103............................ G6
Besst 103 S.......................... G6
Besst 105 S.......................... G6
Besst 106............................ G6
Besst 107............................ G6
Besst 108............................ G6
Besst 112............................ G6
Besst 124............................ G6
Besst 125............................ G7
Besst 126 S.......................... G6
Besst 202............................ G7

Best Chemical Products Company
Wax Stripper&Degreaser
  Product FC-61.................. A4

Best Janitorial & Chemical Compa
Best 2816 Cleaner................ A2
Best 7687 Cleaner................ A2

Best Maintenance Supply Co.
All Off Multi Purpose Cleaner... A1
Best Auto-Scrub Compound...... A1
Best Heavy Duty Insecticide...... F2
Best Liquid Hand Soap.......... E1
Bowlite................................ C2
Brite-Bak............................ A3
Bulldog Cleaner.................... A2
Deluxe #5 Dishwashing
  Compound........................ A1
DC-7.................................. D1
E-Z Concrete Floor Etch........ A3
Esteem Cleaner.................... A1
IF...................................... A1
Len-O-Lite.......................... A4
N B 711 Non Butyl.............. A1
New Besticide...................... D1
New DC-6 Disinfecting Cleaner C2
Open Ses'ame...................... L1
Qwik-Off............................ A1
Rapyd Kleen Cncntrt Indstrl
  Clnr................................ A1
Scale Solvent...................... A3
Speed Clean........................ A1
Super Bowl.......................... C2
Thoro-Kleen........................ A4

Bestgo Chemical Company
Super Concentrated Cleaner
  7153................................ A4

Bestline Products, Incorporated
B-7 Laundry Compound.......... A1,B1
B-70 Laundry Compound.......... A1,Q1,B1
Liquid Concentrate.............. A4
Zif...................................... A4

Betco Corporation
"Spray Foam"...................... A1
Acid Cleaner........................ A3
Betco Cncntrt All Purpose
  Cleaner............................ A1
Betco Mint Disinfectant........ C1
Betco Pine Odor Disinfectant C1
Betco 88.............................. D1
Betsy Mild Toilet Bowl Cleanse C1
Bol Maid............................ C2
Deep Blue Window Glass
  Cleaner............................ C1
Deep Freeze Cold Room
  Cleaner............................ A5
Kling Bowl & Porcelain
  Cleaner............................ C2
Liquid Chisel...................... A1

K-Gel X-20 .... G6
K-Gel X-40 .... G6
K-Gel X-50 .... G6
K-Gel X-67AE G6
K-Gel X-67BE G6
K-Gel X-77 .... G6
K-Gel 20 .... G6
K-Gel 77 .... P1
KI-2 .... P1
Liqui-Treat AL-1237 .... G6
Liqui-Treat AL-1300A G6
Liqui-Treat AL-1564 .... G6
Liqui-Treat AL-8583 .... G6
Liqui-Treat AL-956 .... G6
Liqui-Treat AL-981 .... G6
Liqui-Treat ASM-1237 .... G6
Liqui-Treat ASM-1591 .... G6
Liqui-Treat ASM-325 .... G6
Liqui-Treat ASM-531 .... G6
Liqui-Treat ASM-942 .... G6
Liqui-Treat CL-1038 .... G6
Liqui-Treat CL-1946 .... G6
Liqui-Treat CL-8202B G6
Liqui-Treat CL-8377 .... G6
Liqui-Treat CL-8475 .... G6
Liqui-Treat CL-995 .... G6
Liqui-Treat FAL-1051 .... G6
Liqui-Treat FAL-1343 .... G6
Liqui-Treat FAL-2019 .... G6
Liqui-Treat FAL-2133 .... G6
Liqui-Treat FAL-2166 .... G6
Liqui-Treat FAL-366D G6
Liqui-Treat FAL-531 .... G6
Liqui-Treat FAL-800 .... G6
Liqui-Treat FASM-1051 .... G6
Liqui-Treat FASM-1099 .... G6
Liqui-Treat FCL-1039 .... G6
Liqui-Treat FCL-1159 .... G6
Liqui-Treat FCL-1430 .... G6
Liqui-Treat FCL-1684A G6
Liqui-Treat FCL-1954 .... G6
Liqui-Treat FCL-2105 .... G6
Liqui-Treat FCL-2122A G6
Liqui-Treat FCL-2173 .... G6
Liqui-Treat FCL-2367 .... G6
Liqui-Treat FCL-248A G6
Liqui-Treat FSM-1954 .... G6
Liqui-Treat SM-1577 A G6
Neutrafilm 463 .... G7
Neutrameen NA-3 .... G6
Neutrameen NA-4 .... G6
Neutrameen NA-9 .... G6
Octameen G6
Orocol 16 .... G7
Orocol 206-F G7,G5
Orocol 384 .... G6
Orocol 68 .... G6
Orocol 68-B G6
Orocol 807 .... G2,G5
Permacol G6
Permacol 100 .... G6
Permacol 110 .... G6
Poly-Floc 3 .... L1,G1
Poly-Sperse 100 .... G6
Poly-Sperse 200 .... G6
Poly-Sperse 300 .... G6
Poly-Sperse 400 .... G6
Polymer 1100 .... G6,L1
Polymer 1100P G1
Polymer 1110 .... L1
Polymer 1110P G1
Polymer 1115L L1
Polymer 1115LP G1
Polymer 1120 .... L1
Polymer 1120P G1
Polymer 1125L L1
Polymer 1130P G1
Polymer 1150P G1
Polymer 1160 .... L1
Polymer 1160P G1

Polymer 1170 .... L1
Polymer 1175 .... L1
Polymer 1180 .... L1
Polymer 1185 .... L1
Polymer 1190 .... L1,G1
Polymer 1195 .... L1
Polynodic 601 .... G7
Polynodic 602 .... G7
Polynodic 603 .... G7
Polynodic 604 .... G7
Polynodic 605 .... G7
Polynodic 606 .... G7
Polynodic 607 .... G7
Polynodic 608 .... G7
Polynodic 613 .... G7
Polynodic 614 .... G7
Polynodic 615 .... G7
Polynodic 616 .... G7
Polynodic 617 .... G7
Polynodic 618 .... G7
Polynodic 619 .... G7
Polynodic 620 .... G7
Polynodic 621 .... G7
Polynodic 622 .... G7
Polynodic 631 .... G7
Polynodic 632 .... G7
Polynodic 633 .... G7
Polynodic 634 .... G7
Polynodic 636 .... G7
Polynodic 637 .... G7
Polynodic 638 .... G7
Polynodic 639 .... G7
Pre-Film 106 .... G7
Pre-Film 108L G7
Pre-Film 807 .... G7
Prekleen 346 .... A1
Slimicide A-9 .... G7
Slimicide B-7 .... G7
Slimicide C-30 .... G7
Slimicide C-38 .... G7
Slimicide C-7 .... G7
Slimicide DE-364 .... G7
Slimicide J-12 .... G7
Slimicide J-9 .... G7
Slimicide 242 .... G7
Slimicide 508 .... G7

Bi-Chem Industries, Inc.

All Purpose A4
Bac-T Liquid Live Bacteria L2
Bi-O-Solv Concentrate C1
Clean-Out L1
Contact Concentrate F2
D-Grease Concentrate K1
D-Solv K1
D-Tar Gel K1
Duz-It-All C2
Green Genie C1
Grime Buster A1
Hand Dandy E4
Hot Stuff L1
Hot-Shot L1
Mixer A1
Odor-Go C1
Ox-Clean A3
Rapid Foam C1
Safe Solv K1
Strip It C1
Trizyme L2
Un-Flam K1

Bi-Chem, Inc.

Ammo All Purpose Cleaner
   Ammoniated C1
APC-101 .... A4,A8
Bi-Cide F1
Bi-Dry Dry Lubricant H2
Bi-Lube H2
Bi-Pine C1
Boiler Treat-1 .... G6

BPC-17 Brake & Parts Cleaner .. K1
Control-B F2
Down & Out L1
Ez-Klean A3
EPD-19 Engine Degreaser K1
Formula 15 .... F1
FW-24 .... A4
King-Koil P1
LWG White Grease H2
Melty Ice Melt C1
Mighty Brite Stainless Steel
   Cleaner A7
Mighty Klean A3
Mr. Slick Silicone Lubricant
   Spray H1
MG 69 .... H2
Nu-View Glass Cleaner C1
Oven-Mite A8
Piney C1
Protector Guard Barrier Cream . E4
QDC-27 .... D1
R.I.S. Residual Insecticide F2
Scat A1
Scrub-A-Dub Bath & Tile
   Cleaner C2
Steam Treat-2 .... G6
SS-50 Safety Solvent I0
Tower All NC G7
Treatall-3 .... G7
Triple-Action C1
VMR Vandalism Mark
   Remover K1

Bickco Enterprises, Inc.

Astrodet A4
Bio-Syn C1
Blue Whiz C2
BWT-101 .... G6
Coco-40 Hand Soap E1
Command A1
CWT-300 .... G7
CWT-500 .... G7
Descaler One A3
Dish-A-Clean A1
Flash Klenze LC-12 .... A3
High Sudsing-Phosphoric Acid
   Detergent A3
Husky A4
Klink A1
LDS-20 .... A1
Porcelain Bowl & Tile Cleaner . A3
Quat-22 .... D1
Repro Solv A1
Repro Solv Concentrate
   Cleaner-Degreaser A1
Repro-Chlor C1
Steam-O-Clean A1
Tridet A4

Bidall Corp.

Bacteria Cultures L2
Clean And Thaw A5
Commander A4
Concentrated Odor Control C1
Contact Cleaner K2
D-S-D Disinfectant D2
Degreaser K1
Degreaser And Motor Cleaner .. K1
Deodorant Blocks C1
Dual Action Odor Control C1
Electro-Spray Concentrate C1P
Enzymes 300 .... L2
Foaming Cleaner C1
General Purpose Liquid
   Cleaner A4
Germidial Disinfectant
   Sanitizer & Deo. C1
Granular Deodorant C1
Grease Eliminator L1
Heavy Duty Industrial Cleaner .. A4

46

47

Desol...................................A1
Ener-Guard.......................... B1
General Purpose Cleaner LBG..A1
Guardall Iodine Disinfectant......D2
Heavy Duty Cleaner HAC-20.... A2
Heavy Duty Cleaner HAC-67.... A2
Heavy Duty Cleaner HAC-67-
M... A2
Heavy Duty Cleaner HAC-89.... A2
Heavy Duty Liquid Cleaner
HAC-25 .... A2
Heavy Duty Liquid Cleaner
HAC-48 .... A2
Laundri-Stat T-511F B1
Laundri-Stat T-522 .... B1
Liqui-Foam 100 .... A1
Liquid General Purpose
Cleaner STL-5 ....A1,H2
Meck...................................... A1
Shackle Cleaner No. 55 .... A1
Silicone Y-2505....................... H1
Tex-Care................................. B2

Bio-Serv Corporation

Bio-Serv Diazinon - 4lb
Emulsifbl............................ F2
Bio-Serv Diazinon - 4lb Oil
Solution.............................. F2
Bio-Serv Emulsifiable Spray
Concentrate F1
Bio-Serv Oil Concentrate 1 ...... F1
Bio-Serv Pyrethrin Spray
Concentrate F1

Bio-Tek Industries, Incorporated

"We-Care" Hospital
Disinfectant C1
Anti-Sep E1
Bactrol Iodophor Germicide...... D2,E2,Q6,E3
Bio-Dyne D2
Black Disinfectant (Phenol
Coeficient 2)........................ C1
Black Disinfectant (Phenol
Coeficient 5).......................... C1
Compound AF-H A1
Eggs-It Egg Shell Dissolver....... L1
Feather-Pik............................ N1
Formula-7............................... F1
Glo-Brite C2
Grip Dressing P1
Handi-Kleen Powder A1
Handi-XL................................ A1
Klear Out Drain Opener........... L1
Kleen Aid Disinfectant Foam
Cleaner................................ C1
Kleen-N-Brite Emulsion Bowl
Clnr.................................... C2
Lemonee Concentrated
Disinfectant C2
Lim-Germ C2
Lube-It................................... H2
Luster A3
Magnet Malathion Fly Bait F2
No-Tox F1
Odor Neutzr Lime For Time
Dispn. Units C1
Odorzene C1
Purge-It L1
Quatsan D1,Q3
Quick Kleen L1
Rid-It F2
Shelf-Life............................... D2
Shelf-Life Concntrtd Liq.
Santzr Germ......................... Q3
Tek Eze H1
Tek EB-115 Egg Destainer Q2
Tek EC-111 Chlorinated Egg
Wash Q1
Tek EC-113 Non-Foaming Egg
Wash Q1

Tek ED-117 ....Q5,N1,N4,N2
Tek ET-C121 Chlorinated Tray
Wash A1
Tek ET-119 Tray Wash.............. A1
Tek Lemon C2
Tek OS-30 Super Conc. Oil Sol.
Insect.................................. F2
Tek Phene A4
Tek SF-2 Super Fog Insect
Spray................................... F1
Tek SH-8 Liquid Insect Spray ... F1
Tek SR 6 Residual Insect Spray F2
Tek W/E 5 Concntrtd Emlsfbl
Insecticide........................... F2
Tek Zyme L2
Tek 101 Hook and Trolley
Cleaner................................ A2
Tek 103 Tripe Cleaner N3
Tek 105 Smokehouse Cleaner A2
Tek 107 Shackle Cleaner A2
Tek 109 Hog Scald................... N2
Tek 123 General Purpose
Cleaner................................ A2
Tek 125 Gen. Cleaner Hvy Dty
Alkln A1
Tek 127 Acidic Cleaner A3
Tek 129 Retort Cleaner G5
Tek 131 Can Rinse G5
Tek 133 Gen. Purpose Chlor.
Cleaner................................ D1
Tek 135 Rust Stripper............... A3
Tek 137 Heavy Duty Acidic
Cleaner................................ A3
Tek 139 Scald Liq. Additive N1,N2
Tek 139 Scald Liq. Additive N1,N2
Tek 141 Odor Control C1
Tek 143 Malodor Controller C1
Tek 145 Odor Counteractant...... C1
Tek 147 Liq. Conc. H.D.
Alkaline Cl. A1
Tek 149 Foam Additive.............. A1
Tek 151 .... D2,Q4,E3
Tek 153 .... D1,Q4
Tek 155 .... D1
Tek 157 .... A1
Tek 159 .... A1
Tek 161 .... A1,Q1
Tek 163 .... A1
Tek 165 .... A2,Q1
Tek 167 .... Q5
Tek 169 .... A1
Tek 171 .... A1
Tek 173 .... A1
Tek 175 .... A1
Tek 177 .... A2
Tek 179 .... A1
Tek 181 .... A2
Tek 183 .... A2
Tek 185 .... A1
Tek 187 .... A1
Tek 189 .... A2
Tek 191 .... A3
Tek 193 .... A1
Tek 195 .... A1
Tek 197 .... A3
Tek 199 .... G5
Tek 201 .... A3
Tek 203 .... A3
Tek 205 .... A2
Tek 207 .... A2
Tek 211 .... A1
Tek 213 .... A2
Tek 215 .... A1
Tek 217 .... A1
Tek 219 .... A1
Tek 221 .... A1
Tek 223 .... D1
Tek 225 .... L1
Tek 227 .... A1
Tek 229 .... H2

Tek 233...................................N2,N3
Tek 235...................................A2
Tek 237...................................A2
Tek 239...................................E1
Tek 241...................................D1,Q3
Tek 251 Odorless Concrete
Cleaner................................ A4
Tek 253 Heavy Duty Liquid
Cleaner................................ A1
Tek 255.................................. G6
Tek 257 Low Foam Laundry
Dtrgnt................................. B1
Tek 261.................................. C1
Tek 263 Solvent Degreaser
Bulk.................................... K1
Tek 265 Liquid Steam Cleaner... A1
Tek 267 Liquid Steam Cleaner... A1
Tek 269 Steam Cleaner A1
Tek 271.................................. A1
Tek 273 Lime and Scale
Remover.............................. A3
Tek 275 Unscented Liquid
Hand Soap........................... E1
Tek 277 Concentrate A4
Tek 281.................................. A1
Tek 283.................................. A1
Tek 285.................................. A1
Tek 309.............................. D1,Q1,Q3
Tek 311.................................. C1
Tek 313.................................. G7
Tek 317.................................. C1
Tek 319.................................. A4
Tek 321.................................. H2
Tek 323.................................. A4
Tek 325.................................. A1
Tek 327.................................. C1
Tek 329.................................. E1
Tek 331.................................. A2
Tek 355.................................. G6
Tek 365 Bactericide.................. G7
Terg-Cide Cncntrtd Clnr.
Disinfnt............................... D1
Time Dispensing Insecticide...... F1
TEK-219 Oven Cleaner A8
TEK-263 Solvent Cleaner-
Degreaser (Aer.).................... K1

Biodet, Inc.

Cleaner Base JS 915 ....A1

Bionomical Chemicals & Services, Inc.

"Blast" Multi-Purp. Steam
Cleaning Conc. A1
"Max" Liquid Steam Cleaning
Concentrate A1
"Monster" Ultra Hot Sewer
Solvent............................... L1
Alpha 5000............................. A4,A8
AMC-9.................................... A3
APC-2..................................... A1
Big "C" Concentrate................. A4,A8
Bionomical RD-500.................. D2
BDX C1
C-34....................................... A4
Chain And Conveyor Belt
Lubricant H2
General Purpose Insecticide...... F1
GC-3 .... A1
GP-280.................................. A1
H.D. Alkalene......................... A2
Heavy Duty Form. Liq. Stm
Cleaning Conc. A1
Inhibited Chemical Cl. &
Oxidation Rmvr..................... A3
Lubricant TL Concentrate H1
LD-515 .... A1
MPD-550................................ A1
No. 1 General Smokehouse
Cleaner................................ A2

48

No. 23 Milkstone & Scale Rmvr
Conc...A3
No. 5 Heavy Duty Smokehouse
Cleaner.......................................A2
Organic Acid Detergent A3
Red Hot Sewer Solvent.............L1
S-311 ....A2
Superus A1
SGC-6..A1
SH-26 ....A2
T-4...A2
Task Master.................................E1
X-10 ....A1
150...A1

Biotechnics, Incorporated
Alkaline Deruster A2
BL-25 ....A1
C.I.P. Acid A3
Chlorinated C.I.P. Cleaner A2
Compound AD 8 ....A2
Compound AF-A A1
Compound AF-B A1
Compound AF-C A1
Compound IDC-30.....................D2
Compound K-260 ....A4,H2
Compound L 144.........................A1
Compound L 144NP....................A1
Compound MSP-10......................A1
Compound PS 1...........................A1
Compound PS-1 NP.....................A1
Compound PS2............................A3
Compound RLB 1.........................B1
Compound RLB-2 Special...........P1
Compound S 1..............................B1
Compound S 2..............................A1
Compound SD 9...........................C1
Compound SD 9B........................C1
Compound T 1..............................B1
Compound TL 40.........................H1
Compound TR 22.........................A2
Compound TRC 12.......................A2
Compound 175 D A1
Compound 175 DNP....................A1
Compound 175 F.........................A1
Compound 175 F NP....................A1
Compound 175 FJ........................A1
Compound 175 FJNP...................A2
D.C. 14..A1
Dairy Cleaner 14 NP..................A1
Dart 254 ....H2
Hi-Power B1
High A.C.....................................A1
High A.C. NP..............................A1
Liquid Bottle Wash A2
Liquid Hand Soap.......................E1
Liquid Hog Scald N2
LD 18 ....B1
Marvella M A1
Marvella P...................................A4
Power Scrub Concentrate A4
Smokehouse SS............................A2
Smokehouse SS NP.....................A2
Stainless Cleaner.........................A3
SS HD Blue NP A1
Tripe Cleaner CD.......................N3
3 D's A1

Bird-Archer Company, The
#1273..G6
Bacophos #321 ....G2
Concentrol T-50 ....G6
Concentrol T-70 ....G6
Formula 1105 ....G6
Formula 1495 ....G6
Formula 1828 ....G6
Formula 530................................G6
Formula 600-FD..........................G6
Formula 600-0.............................G6
Formula 600-0D..........................G6

Birko Chemical Corporation/ See Birko Corporation
Acid Brite No. 1..........................A3
Acid Brite No. 2..........................A3
Actec Inhibited A3
Actee No. 100..............................A3
Acto-140.................................N3,B1
Aluma-Tec No. 3.........................A1
Aluminate....................................A3
Anti-Foam S P1
Armour Detergent 103................A2
B.M.C. 24-X...............................A3
Bact-O-Stat.................................E1
Bi Tec 55H..................................A1
Bi-Chlor.......................................D1
Bi-Clor Rite.....................D2,B1,E3,Q4
Bi-Quat.....................D2,Q1,Q3,E3
Bi-Tec No. S.W. 80.....................A1
Bi-Tec No. 11..............................A1
Bi-Tec No. 13..............................A1
Bi-Tec No. 14..............................A1
Bi-Tec No. 15..............................A1
Bi-Tec No. 66..............................A1
Bi-Tec 1000...........................A2,N5
Bi-Tec 77 H...........................A1,B1
Bi-Tec 99-AC..............................A3
Birko Bleach B1
Birko Liquid Sour B1
Birko Sour.............................A3,B1
Birkolene C1
Birkream.....................................E1
Castor Lube C H1
Castor Lube No. 1.......................H1
Challange...............................A1,E1
Chelate #24-L.............................P1
Chelate 24L.................................A3
Chlorific D1
Chlorofoam D1
Chloromatic No. 2................A1,Q1
Chloromatic No. 3.......................D1
Chlorotec....................................D1
Chlorotec 99 ....D1
Cir Scald N5
Cir-Tec 18 ....P1
Cirtec CL.....................................D1
Cirtec CL-QR..............................A2
Cirtec QR....................................A2
Cirtec SP.....................................D1
Cirtec 18................................A2,N5
CL-33..
Denaturant G......................M1,M3
Diactolate.....................N3,N5,P1,B1
Egg Wash...............................Q1,Q4
Foam Kleen D1
Fomacid 50..................................A3
Galvanize Brightener A3
Genzolate B1
Hard Scald N2
Hi Chlor QD.........................D1,B1
Hi-Cap No. 13.............................A4
Hi-Cap 7-71................................A1
Hi-Chlor 3600-L....................D1,B1
Hydrated Lime N3
HW-44-L......................................B1
HWL-2M......................................B1
Liquid Smokehouse Cleaner.......D1
Liquik Bleach........................N3,N5
Liquik Char..........................M2,M4
Liquik CB.............................A4,A8
Liquik CC....................................A1
Liquik 1 ....A1
Liquik 10................................A4,A8
Liquik 100....................................A1
Liquik 125 ....A1
Liquik 2 (Nu Blu)........................A1
Liquik 2 C.S.................................A1
Liquik 250 ....A2
Liquik 3 ....A1
Liquik 4 ....A1

Liquik 5 ....A1
Liquik 50......................................A1
Meta Tec No. 1 ....A3
Meta Tec 2C................................D1
Meta Tec 2C Liquid....................A2
Metalist..A2
Metalist PW A2
Miri-Cal A3
Monlaun No. 2.............................B1
Monlaun 1 ....A1
Neutra-Foam A1
Neutra-Foam C............................A1
Neutra-Sol A4
Poly-Tec......................................B1
Purgit Foam A2
Purgit R.......................................A1
Rinsall..A1
Rustec...A3
Rustop A3
Scale Off......................................A3
Scalite SR....................................A3
Sewage Splitr Accelerator........L1
Sewage-Splitr...............................L1
Sewer Cleaner.............................L1
Shine-A-Line...............................A3
Sigman General Purpose
Cleaner.......................................A1
Sigman Heavy Duty Cleaner A2
Sigman Smokehouse Cleaner A2
Smok-Off......................................A2
Smoke-Zoff...................................A1
Super Alkali A1
Super Chlor...........................D1,G4,B1
Sutec FG......................................D1
Synchlorozene P1
Syntalazene 68 ....P1
SPDE 33 ....A1
Tec So #100 ....A1
Tecso C A1
Tintec..A1
Tripe Wash No. 1........................N3
Tripe Wash No. 2-P....................N3
Tripe Wash 3A N3
Trolleze H1
Trolleze H....................................H1
Turkleen Poultry Scald...............N1
Viking..A2
Waconite No. 91..........................A1
Waconite 44L...............................B1
XLNT-4U2 Liquid A1

Birko Corportion/ See Birko Chemical Corporation
Grez-Out B1

Birsch Chemicals, Ltd.
Accel..A1
Act-All A1
Auto Scrub Detergent A4
Blue Min C2
Bulk Drain Opener......................L1
C-Quest..A1
Carrol's Cleaner..........................A1
Concrete Cleaner........................C1
Demon..A1
Dish Glo.......................................A1
Gusto A1
Heavy Duty Powdered Cleaner.A1
Heavy Duty Prestige Cleaner A1
Hvy Dty Prestige Cncntrtd
Wax Strppr A4
Kem-Tronics A1
Lemon Odor Disinfectant-
Cleaner Coef. 7 ....A4,C1
New Kem-Tronics........................A4
Pynet..C1
S.R.S. Wax Stripper A4
S-Teem A1
Spray Silicone Release H1
Trophy..............................D1,Q1,Q3

Winners Choice A4
301 Steam Cleaner....................... A2
3412 Cleaner A1

Biscayne Chemical Laboratories, Inc.
Acti-Sol L2
Bam A4
Bisco 690 .... A1
BCL 55 .... K2
Oven Cleaner A8
Silicone Spray.............................. H1
Sterizone Disinfectant
  Deodorant C2
Tuff-Tape H2

Bishop-Wisecarver Corporation
Tigress Insecticide No. 4 .... F1

Bishops Sanitary Maintenance Supply
Degreaser/Emulsifier A1

Bison Cleaning Chemicals, Inc.
Bisoklor #108 .... A1
Bisoklor #1081............................ A1
Bison 203 .... A1
Bison 204 .... A1
Bison 262 .... H2
Bison 284-E A1
Bison 311 .... A1
Bison 530-2 .... A1
Bison 658 .... A2
Bison 731 .... A2

Bixon Chemical Company Division of A. S. L. Enterprises, Inc.
Antiseptic Liquid Hand Soap..... E1
Big Job G7
Bison 203 .... A1
Bixon Multi-Purpose Insecticide F1
Chloroflo C1
Compactocide Spray F1
Concentrate Wax Stripper #1.... A4
Convey H2
Cove Base Cleaner C1
Enzymes.. L2
Formula 120................................. A4
Fuel Oil Treatment...................... P1
Germicidal Cleaner C1
Hard Surface Cleaner.................. A4
Heavy Duty Degreaser................ A4
Heavy Duty Steam Cleaner A1
Kustomize..................................... C1
Liquid Hand Soap E1
Low Pressure Boiler
  Compound G6
Mint Disinfectant C1
Mint Disinfectant Coef. 5 .... C1
Minute Clean 450........................ A4
Minute Clean 460........................ C1
Minute Clean 470........................ C1
Minute Clean 480........................ A4
Neutral Cleaner I......................... A4
Neutral Cleaner II........................ A4
Neutral Cleaner 3 .... A4
Odor Control Deodorizer........... C1
Odor Counteractant Conc. C1
Pine Oil Disinfectant Coef. 3 .... C1
Pine Oil Disinfectant Coef. 5 .... C1
Pine Scent Disinfectant.............. D1
Pyrenone General Purpose
  Aqueous Insect. F1
Quat Disinfectant and Cleaner
  II.. D1
Residual Insect. w/Pyrenone-
  Diazinon Liq F2
Residual Insecticide with
  Baygon..................................... F2
Safety Solvent 1........................... K2
Safety Solvent 2........................... K2

Sanotize A4
Sludge Solvent P1
Smoke-House Cleaner A8
Super Safety Degreaser A4
Synval All Purpose Floor
  Cleaner.................................... A4
Water Soluble Cleaner
  Degreaser A1
Wax Stripper #2........................... A4

Black Enterprise, Inc.
Glide.. E1
Trolley Cleaner And
  Rustproofer A3
Trolley Stripper A3

Black Leaf Products Company
Black Leaf Cygon 2-E Systemic
  Insect...................................... F2
Septic Tank Cleaner L2

Black Magic Company
Black Magic Roach Bait............. F2

Black River Paper Co.
Bute... A4
Diamondquat D1
DEG-15 .... A1
HD-10 Heavy Duty Industrial
  Degreaser A1
LHC-2 .... A1

Blaine Chemicals/ & Industrial Supply Inc.
Foam-Kleen.................................. A1

Blair Incorporated
N-Clog Liquid Organic
  Digester L1

Blew Chemical Company
Active Cleaner No. 43 .... A4
Active M-15 Foaming Agent A1
Active No. 2 .... A1
Active No. 4 .... A1
Cleaner G.P. Liquid
  Concentrate A4
Liquid Soap No. 50..................... E1
Liquid Soap No. 52..................... E1
Liquid Soap No. 64..................... E1

Blockstanz Brothers Company
BB Antiseptic Liquid Hand
  Soap E1
Dish Glo.. A1
NBC Non-Butyl Deodorant
  Cleaner/Dgrsr........................ A1

Blu-Line Industries
Tru-Blu Concentrate A1

Blue Jay Chemical Corporation
AB-1199... K1
Digester Drain Opener L1
Emulsion Bowl Cleaner &
  Disinfectant C2
Non Combustible Emulsifier A4

Blue Line/ GAC Janitorial Supplies, Inc.
C-B-K F1
Nok-Out... L1

Blue Magic Div./ Roman Cleanser Co.
Easy Monday Bleach D1,B1

Blue White Chemical Co.
Silicone Lube H1

Bob Agate Chemicals
Pink Delite A1

Bobmarlin Supply Company
BML Action VIII A1
BML 20% Castile Liquid Hand
  Soap E1
H.D. #1 Packing House
  Cleaner.................................... A1
ME 66 Heavy Duty Cleaner A1

Boise Western Corporation
B & C Concentrate Toilet Bowl
  Cleaner.................................... C2
Clinging Bowl Cleaner C2
D-Zolve-It Sulfuric Acid Drain
  Opener L1
Delicate Castile Conc. Liq.
  Hand Soap E1
Dri-Mol Moly-Micro-Film H2
Electro-Kleen Safety Solvent..... K1
Extend New Power...................... P1
Fluorocarbon Lubricant All
  Purp. Dry Lub H2
Green Lightning Conc.
  Alkaline Drn. Opnr L1
Hi-P Contact Cleaner................. K2
HDC 115 Cleaner-Degreaser A1
Lotion Hand and Body Soap E1
Meat Sope Foaming Cleaner..... A1
Metal Brite Cleaner & Polish A7
Pink Creme Hand and Body
  Soap E1
Porcelain & Bowl P & B Clean . C2
Showcase Glass Cleaner C1
Silicone Spray.............................. H1
Super Clean Concentrated
  Cleaner.................................... A4

Bolotin Inc.
CSA Antiseptic Liquid Hand
  Soap 20% E1
Super Scope A1

Bonanza Chemical Co.
Leben's Saf-T-Step J1

Bonco Manufacturing Corporation
Algaecide-Slimicide No. 10........ G7
Aqua-Tex CT 44........................... G7
Aqua-Tex CT 55........................... G7
Aqua-Tex CT 88........................... G7
Bonco Algaecide Briquettes G7
Bonco BT-10................................. G7
Scale-Ox A3

Bond Chemical Company, Inc., The
230... G6
232... G3
234... G6
27P... G6
31P... G6
315... G6
320-L... G6
329... G2
332... G6
346... G6
363... G6
402BC.. G6
402JC... G6
42P... G6
47LX.. G6
96LC.. G6
98LX.. G7

50

Bonchem Sta-Chlor D2,Q4,B1,E3,P1
Bonchem Super Foam.................. A1
Bonchem Super Glo.................... A1
Bonchem T.O.T. A3
Bonchem T.R.C. A2
Bonchem Tandem......................... A1
Bonchem Tile Floor Cleaner...... A4
Bonchem Tripe Cleaner N3
Bonchem Tripe Cleaner
    Certified N3
Bonchem Triple C. A1
Bonchem Trolley Oil H2
Bonchem Truck Wash A1
Bonchem Twin A1
Bonchem Viscera Pan Cleaner... A2
Bonchem Viscous Cleaner No.
    I.. A1
Bonchem Viscous Low Foam
    Cleaner................................. A1
Bouchem 1022............................. A2
Bouchem 1024............................. A2
Bouchem 1064............................. A1
Bonchem 110............................... A1
Bonchem 220............................... A1
Bonchem 30% Liquid Caustic
    Chelated A2
Bonchem 372 General Cleaner... A1
Bonchem 440............................... A1
Bonchem 448............................... A1
Bonchem.453................................ A1
Bonchem 453 Improved A1
Bonchem 499............................... A1
Bonchem 499 with Color........... A1
Bonchem 50% Liquid Caustic
    Chelated A2
Bonchem 540............................... A2
Bonchem 550............................... A3
Bonchem 5505 L......................... A1
Bonchem 561............................... A2
Bonchem 582............................... B1
Bonchem 588 A A2
Bonchem 614............................... A2
Bouchem 64 .... A2
Bonchem 658............................... B1
Bonchem 672............................... B1
Bonchem 673............................... B1
Bonchem 674............................... B1
Bonchem 7005............................. A2
Bonchem 7010............................. A2
Bonchem 7025............................. A2
Bonchem 7030............................. A1
Bonchem 7060............................. B1
Bonchem 7065............................. B1
Bonchem 7070............................. B1
Bonchem 7075............................. B1
Bonchem 7205............................. A2
Bonchem 7215............................. A2
Bonchem 729............................... A2
Bouchem 7305............................. A1
Bonchem 7310............................. A1
Bonchem 7420............................. A1
Bonchem 7440............................. A1
Bouchem 7480............................. A1
Bonchem 7535............................. A2
Bonchem 7537............................. A2
Bonchem 779............................... A1
Bonchem 7915............................. A2
Bonchem 7920............................. A2
Bouchem 829............................... A2
Bonchem 831............................... A1
Bonchem 854............................... A1
Bonchem 880............................... A1
Bonchem 890............................... A2
Bonchem 90................................. A1
Bonchem 909 Foam Cleaner A2
Bonchem 915 Foam Cleaner A2
Bonchem 920 Foam Cleaner A2
Bonchem 925............................... A2
Bonchem 926............................... A1
Bonchem 940............................... A2

Bonchem 960............................... A2
Bonchem 999............................... B1
Bonshield ....:............... E1
Bonshield 1000............................ E1
Bonshield 2000............................ E1
Can Cleaner #1 .... A1
Chelaton No. 5............................ A2
CH-10-AP A1
CH-50-AP A1
Defoamer DCWH A1
DP-506....................................... A1
DP-509....................................... A1
DP-525....................................... A2
Feather Strip 100........................ N1
Floor Cleaner #1......................... A1
Floor Cleaner #5......................... A1
General Cleaner #1...................... A1
General Cleaner #4...................... A1
General Cleaner #6...................... A1
General Cleaner #8...................... A1
In Line Trolley Cleaner............... A2
Kontrol Special.....................:..... A2
L-143 Neutral Cleaner A1
Mikro-San................................... A3
Mikro-Stat A1
MR-42 Concentrate A3
Organic Sanitizer...................... D1,Q4
Packers Cleaner #1 .... A2
Packers Cleaner #15................... A1
Packers Cleaner #2..................... A1
Packers Cleaner #4..................... A2
Packers Cleaner #5..................... A1
Packers Cleaner #779 .... A1
Pan Clean 31............................... A2
PCC Alkali Cleaner A2
Right Heavy Duty Cleaner A4
Standard Detergent #101 .... A1
Standard Detergent #103 .... A2
Standard Detergent #104 .... A2
Standard Detergent #105 .... A3
Standard Detergent #106 .... A3
Standard Detergent #109 .... A3
SCC Special A1
Trolley Cleaner........................... A2
Trolley Cleaner Activator A1
Viscera Pan Washer Detergent.. A2

Bonner, W. C. Co., Inc., The
Super Bon-Chlor Briquets A1

Bonnie Chemspec
Autobrite A1
B.V.C. Plus A1
Bonnie Bon Sol A1
Bonnie Hghlnd Kleener
    Bacteriostatic......................... C2
Bonnie Joy A4
Bonnie Kleen 20% E1
Bonnie Kleen 40% E1
Bonnie Kleen-Well A1
Bonnie L-P-O.............................. L1
Bonnie Lass B1
Bonnie Mac-40............................ A3
Bonnie Mint Magic Disinfectant C2
Bonnie Mud-A-Way
    Concentrate L1
Bonnie Pine Odor Disinfectant .. C2
Bonnie Safe K1
Bonnie Sewer Sol........................ L1
Bonnie Steam-Off A.P. A1
Bonnie Steam-Off H. D. A1
Bonnie Sunny Day C2
Bonnie 3AK Cleaner................... A4
Brightaway A1
CDS Cleaner-Degreaser
    Solvent.................................. A4
De-Solv C1
Delight.. A1
E-Z Clean................................... A1

H.T. 20 .... A2
Industrial Space Spray F1
M.D.W. 45 .... A1
Phos-40 .... A3
Pot and Pan Cleaner A1
Sana Clean.................................. D1
Sanaclor...................................... D1
Sandy's Kleen Window & Glass
    Clnr C1
Sudsing Cleaner A6

Borco Chemicals, Incorporated
Real-Ease H1

Borden, Incorporated
Krylon All-Purpose Sil.Spy
    No.1325/1325A...................... H2
Krylon Crystal Clear No. 1301 .. P1
Krylon Crystal Clear No. 1302... P1
Krylon Crystal Clear No. 1304
    Protect. C P1
Krylon Crystal Clear No.
    1305A.................................... P1
Krylon No. 1327 Electric
    Motor Spy Clnr K2
Krylon No. 1332 Let-Go H2

Boss Labs
Boss Labs All Purpose Cleaner.. A1
Boss Labs Butyl Cleaner............. A4
Boss Labs Enzyme L2
Boss Labs Steam Cleaner........... A4

Bowie Janitorial Supplies
Heavy Duty Degreaser................ A1

Bowman Distribution/ Barnes Group Inc.
All Purpose Cleaner No. 19451.. C1
Antiseptic Hand Cleaner............. E4
Belt Dressing No. 21905-1 .... P1
Brake Cleaner No. 19440........... P1
Food Grade Silicone H1
Gasket & Decal Remover No.
    19475 .... P1
Glass Cleaner No. 21941............. C1
Graffiti Remover No. 19450....... C1
High Strength Thread Locker ... P1
Medium Strength Thread
    Locker P1
Pipe Sealant with Teflon P1
Rust Penetrant No. 21901........... H2
Silicone Lube No. 21900............. H1
Teflon Dry Lube No. 19462....... H2
Under-Coating No. 21903........... P1
White Lube Lithium Grease
    No. 21911.............................. H2

Boy-Ko Supply Co.
Insect Spray ..:............................ F1
Triple D A4,C1
Tuff-Stuff.................................... A4

Boyle-Midway Division
Easy Off (Aerosol) A1
Easy-Off...................................... A1
Sani-Flush................................... C2

Bozzuto's, Inc.
C-Clear C1
Clean-All C1
Floor Stripper C1
Microshield C1

Bradford-Park Corporation
B-P Mtl Cl Rust/oxide Rmvr
    Type R-1 Liq A1
B-P Mtl Cl Rust/oxide Rmvr
    Type R-1 Pst.......................... A1

499.. G6

Brandel Chemical Company, Inc.
Aluma-Kleen A1
AC Cleaner A3
Compound 100 .... A1
Compound 105 .... A1
Compound 106 .... A1
Compound 121 .... A4
Compound 122 .... A2
Compound 127 .... A2
Compound 137 .... A3
Compound 150 .... A1
Compound 175 .... A2
Compound 200 .... A1
Extra Kleen A2
Fog-A-Cide F1
Gelling Agent A1
Hi-Fome A1
HD General Cleaner A1
Kleen-All A1

Brandt Chemical Company
Brandt Formula 88 .... A1
K-510 Cleaner A1

Brant Bros. Cleaner & Equipment Co.
Molly B-3 .... A1
Molly B-5 .... A1
Molly B-6 .... A1
Molly Dal-Tex A1

Brawner Paper Company, Inc.
Val-U-Line Clear Lemon
    Dsnfctnt C2
Val-U-Line Glass Cleaner C1
Val-U-Line Lime Cleaner A4
Val-U-Line Pk Ltn Hnd
    Dshwsh Dtrgnt A1

Brayton Chemicals, Incorporated
F-M Fog Concentrate F1
Fly Bait... F2
P-B Insect Spray.......................... F1

Brenco Corporation, The See Brennan Chemical Company

Brennan Chemical Company
Brenco #490 .... G7
Brenco #491 .... G7
Brenco #573 Water Treatment
    Microbiocide G7
Brenco #588 .... G7
Brenco #594-M G7
Brenco #654 Boiler Water
    Treatment G6
Brenco Hi-Chlor-12 Liquid D2
Brenco No. 100 Economy
    Cleaner.................................... A1
Brenco No. 101 Suds-Clene....... A1
Brenco No. 102 General
    Cleaner.................................... A1
Brenco No. 104 General
    Cleaner.................................... A1
Brenco No. 110 General
    Cleaner.................................... A1
Brenco No. 120 General
    Cleaner.................................... A1
Brenco No. 130 Super Cleaner .. A1
Brenco No. 140 Laundry
    Detergent................................ B1
Brenco No. 150 Hi-Kleen............ A1
Brenco No. 200............................ A2
Brenco No. 208............................ A2
Brenco No. 210............................ A2
Brenco No. 220............................ A2
Brenco No. 222............................ A2
Brenco No. 225 Tripe Wash....... N3
Brenco No. 230............................ A2

Brenco No. 240 Sewer-Clene..... L1
Brenco No. 243 Poultry Scald ... N1
Brenco No. 245 Metal Brite A2
Brenco No. 246 Conveyer
    Roller Cleaner......................... A2
Brenco No. 247 Pasteurizer
    Cleaner.................................... A1
Brenco No. 248 Pasteurizer Boil
    Out Cl. P1
Brenco No. 250 Spray Clene...... A1
Brenco No. 300 Alumex A3
Brenco No. 310 Brite A3
Brenco No. 330............................ A3
Brenco No. 350 L.S.R. A3
Brenco No. 360 Lac A3
Brenco No. 375 Acid Descaler .. A3
Brenco No. 410............................ G7
Brenco No. 496 Cooling Tower
    Treatment G7
Brenco No. 496 Pasteurizer
    Water Treat............................. G7
Brenco No. 497 Cooling Water
    Treatment G7
Brenco No. 500 Sani-Chlor D1,Q4
Brenco No. 575 Concentrated
    Iodine...................................... D1
Brenco No. 585............................ G7
Brenco No. 600 Boiler Water
    Treat. G6
Brenco No. 600 SM..................... G7
Brenco No. 600W........................ G6
Brenco No. 602 Boiler Water
    Treatment G6
Brenco No. 606 Boiler Water
    Treat. G6
Brenco No. 608 Boiler Water
    Treat. G6
Brenco No. 651............................ G7
Brenco No. 651 Lo-Alk G7
Brenco No. 655............................ G6
Brenco No. 656 Boiler Water
    Treatment G6
Brenco No. 658............................ G6
Brenco No. 662............................ G7
Brenco No. 686............................ G6
Brenco No. 687 Return Line
    Treatment G6
Brenco No. 688 Steam Line
    Treatment G6
Brenco No. 689............................ G7
Brenco No. 704............................ G6
Brenco No. 900............................ G1
Brenco No. 908............................ G1
Brenco No. 960............................ G3
Brenco No. 999............................ G2
Brenco No. 999 FESI.................. G2

Brewer Chemical Corporation
Freezer Cleaner A5
Ultramar J. J. All Purp. Liq.
    Det. A1
Ultramar Liquid Dishmachine
    Det. A1
Ultrasan P1

Bricide Corporation
Bricide #532 .... A3
Bricide #624 .... A3
Bricide #696 .... A1
Bricide #931 .... A1
Bricide Formula #805 .... A1
Bricide Formula #807 .... A1
Bricide Formula #809................. A2
Bricide Formula #811................. A3
Bricide Formula #813................. A3

Bridgerland Vet Supply
Ultra-Aid A3

53

Brigade Industries Chemical Division
Fleet-Foam All Surface Cleaner A1

Briggs Products
Tripe Cleaner N3

Bright Chemical & Supply Co.
Aciterge No. 27 .... A3
Bright Sol Bowl Cleaner C2
Bright's-Cide F1
Chiller Cleaner............................ A4 -
Concrete Cleaner........................ A4
Hi-Kleen Heavy Duty Liq
  Steam Clnr A2
Liquid General Purpose
  Cleaner...................................... A4
Liquid Hand Soap E1
Luxury Hand Dish Wash
  Powder...................................... A1
Odorless Disinfectant D1
Shackle Cleaner A3
Sure Solve Blue Sewer Solvent . L1

Bright Ideas Chemical Corp.
Concentrated Inhibited Muriatic
  Acid Cln................................... A3

Brighton Chemical Corporation
Tuffy Power Foam Degreaser ... A1

Brilco Laboratories
Boiler Water Adjunct PLS......... G6
Colloid-Alk KVE G6
I-Sol Cooling Water System
  Additive.................................... G7

Brissman-Kennedy, Inc.
Breakthrough A1
BK 20% Antiseptic Liquid
  Hand Soap............................... E1
Foaming-Plus A1
Stainless Steel Cleaner & Polish. A7
Stat D1
Stat II D1

Britex Corporation
Brit-O-Lux.................................. A1
Brite Vac A2
Brite-Kal A1
Deep Fatfry Cleaner A1
Dependon A1
Dependon H................................ A2
Dri-O-Brite................................ A3
DOB A1
DOBN 1 .... A1
Easybright A1
Formula ABC A1
Formula BBC.............................. A2
Formula COB 1 .... A2
Formula COB 11 .... A1
Formula NBC A1
Formula OK-202......................... A1
Formula OK-242......................... A2
Formula OK-262......................... A2
Heavy Duty Klor-O-Suds.......... A1
HT-Alkali A2
I-O-Brite...................................... D1
Institutional Detergent Beads.... A4
Klor-O-Brite BDCD D1
Klor-O-Brite 16 Plus % D1
Klor-O-Brite 8 Plus % D2
Klor-O-Kleen CIP...................... A1,Q1
Klor-O-Kleen CME A1
Klor-O-Kleen TT A1
Klor-O-Suds............................... A1
Pack-O-Suds............................... A1
Phos-O-Brite A3
Pink Powder A1
Soapline A1

Britt-Tech Corporation
Heavy-Duty Detergent No. 710 A1
Kleen King No. 300 Concrete
  Floor Cleane A4
Kleen-King No. 799 Hvy Dty
  Indust. Det. A1
No. 120 Whitewall Tire Cleaner A1
No. 350 Liquid Detergent
  Additive.................................... A1
No. 720 General Purpose
  Detergent.................................. A1
No. 730 Hvy Duty Caustic-Free
  Detergent.................................. A1

Brothers Janitorial Supply
Brothers Coconut Oil Liquid
  Soap E1

Brown Chemical Company
Blue Pressure Spray Wash......... A1
Bro-Co Blast-Off Triple Cone
  Liq Ind. Cl A4
Bro-Co Command-O A4
Bro-Co Germicidal Cleaner....... D1
Bro-Co Hospital Disinf
  Deodorant C2
Bro-Co Pink Liquid Hand
  Dishwash Detrgnt A1
Chlorinated Mach. Dish Wash
  Comp A1
Deluxe Concrete Cleaner........... A4
Dermaterg 100A.......................... A1
Hand Dishwashing Compound
  No. 1 .... A1
HPN Pink Automatic Spray
  Wash A1
Low Suds Laundry Compound . B1
Machine Dish Compound No. 1 A1
Neutrovapor Steam Cleaning
  Comp. A1
Packing House Cleaner.............. A1
Quick Clean Concrete Cleaner... A1
Sani-Lube Food EQuipment
  Lubricant H1
Snow White Laundry
  Detergent.................................. A4,B1

Brown Pharmaceutical Co., Inc., The
Heliogen Sanitizer D1,Q6

Brown Supply Company, Inc.
Clean Up A1

Brownsey, L. M. Supply Co.
Automatic Floor Scrubber
  Cleaner...................................... C1
Bro-Brite Stainless Steel
  Cleaner...................................... A7
Brotol No. 120............................ A4
Hand Soap................................... E1
Mapp Concentrate Cleaner And
  Degreaser A1
Pi-No-Germ C1

Brulin & Company, Inc.
"Offer" Liquid Insect Spray....... F2
Bio-Zyme L1
Bowlaide..................................... C2
Bowlette..................................... C2
Brulin Cleanser A6
Brulinfoam.................................. E1
Brulinfoam A E1
Chyna-Bryte................................ A1
CDQ A4
CL600 .... A2
CL610 .... A2
De-Odor 3600 .... C1
Disinfectant Cleaner #6.............. A4
Epitaph Liquid Insecticide F1

54

k

55

CHB-101S................................G6
CHB-102................................G6
CHB-103................................G6
CHB-108................................G7
CHC-107................................G7

C & I Chemical Co.
CCC Coolex Coil Cleaner..........P1
Dynamite Drain Opener............L1
Mr Hero Hero Kleen..................A1
Razz-15.....................................F2
Super Wonder Neutral Cleaner . A4

C & K Manufacturing And Sales Co.
Zip-Saf-T-Kleen........................J1

C & M Chemicals, Inc.
Custom Culture.........................L2

C & N Brokerage Company
C & N Super...............................A1

C & S Chemical Supply Company
All Purpose Cleaner No. 48........A1
Special Boiler Formula No. 1 ... G6

C C C Chemical Corporation
Action Pack..............................L1
Du-All "C"................................A1

C D C Chemical Corp.
Odor-End..................................C1

C F S Continental, Inc.
605 On-Premise Laundry Break. B1
615 On-Premise Laundry
    Detergent.............................B1
625 On-Premise Laundry
    Chlorine Bleach.....................B1
640 On-Premise Laundry
    Liquid Break..........................B1
650 On-Premise Laundry
    Liquid Detergent....................B1
655 On-Premise Laundry
    Softener/Brtnr. B2
670 On-Premise Laundry
    Liquid Chlor Blch...................B1
690 On-Premise Laundry
    Liquid Softener......................B2
720 Manual Cleaner...................A1
733 Heavy-Duty Cleaner............A2
747 Super Active Manual
    Cleaner................................A1
750 Limestone Solvent Acid
    Cleaner................................A3
795 Concentrated All-Purpose
    Cleaner................................A4
828 Machine Dishwashing
    Compound.............................A1
888 Machine Dishwashing
    Compound.............................A2
900 Concentrated General
    Purpose Cleaner....................A4
944-Heavy Duty Stripper...........A4

C G S Industries, Inc.
CGS All Purpose Cleaner #1 .... A1
CGS Boiler Water Treatment
    #3..G6
CGS Boiler Water Treatment
    Compound.............................G6
CGS Butyl Cleaner #2...............A4
CGS Concrete Etching
    Compound.............................A4
CGS Hand Dishwashing
    Detergent.............................A1
CGS Reserve..............................D1
CGS Safety Solvent &
    Degreaser..............................K2

CGS Waterless Hand Cleaner.... E4
CGS-Power................................A4
Liquid Steam Cleaner...............A1
Truck Wash...............................A1

C H 2 O, Inc.
Product #315.............................G6
Product #3171...........................G6
Product #3212...........................G6
Product #3220...........................G6
Product #3262...........................G6
Product #3387...........................G6
Product #390.............................G6
Product No. 3153.......................G7
Product No. 3181.......................G6
Product No. 3213.......................G6
Product No. 3265.......................G7
Product No. 3267.......................G6
Product No. 3270.......................G6
Product No. 3281.......................G6
Product No. 3388.......................G6
Product No. 3396.......................G6

C P M Chemical Company
CPM..C1

C R C Chemicals USA
Antiseptic Hand Cleaner...........E4
Brakleen (Aerosol)....................K2
Brakleen (Bulk).........................K2
Co Contact Cleaner No. 02016 .. K2
Contact Cleaner No. 03070
    Purakleen.............................K2
CRC Marine Formula 6-66 No.
    06005..................................H2
CRC Marine Formula 6-66 No.
    06006..................................H2
CRC Marine Formula 6-66 No.
    06008..................................H2
CRC Marine Formula 6-66 No.
    06010..................................H2
CRC Marine Formula 6-66 No.
    06012..................................H2
CRC 2-26 Multi-Purpose Spray
    No. 02004............................H2
CRC 2-26 No. 02005/Multi-
    Purpose Spray......................H2
CRC 2-26 No. 02006/Multi-
    Purpose Formula..................H2
CRC 2-26 No. 02009..................H2
CRC 2-26 No. 02011..................H2
CRC 3-36 Multi-Purpose Spray
    No. 03004............................H2
CRC 3-36 Multi-Purpose Spray
    No. 03005............................H2
CRC 3-36 No. 03006/Multi-
    Purpose Formula..................H2
CRC 3-36 No. 03009..................H2
CRC 3-36 No. 03011..................H2
CRC 5-56 No. 05004/Multi-
    Purpose Spray......................H2
CRC 5-56 No. 05005/Multi-
    Purpose Spray......................H2
CRC 5-56 No. 05007..................H2
CRC 5-56 No. 05009..................H2
CRC 5-56 No. 05011..................H2
Electrical Quality Silicone No.
    02094..................................H1
Extreme Duty Silicone Lub.
    No. 03030............................H2
Heavy Duty Degreaser...............K1
Heavy Duty Silicone No. 05074 H1
Lectra Clean (Aerosol)..............K2
Lectra Clean (Bulk)...................K2
Lectra-Motive Cleaner
    (Aerosol).............................K2
Lectra-Motive Cleaner (Bulk)... K2
Marine Lectra Clean..................K2
Marine Lectra-Clean (Aerosol).. K2
Marine Silicone 06077................H1

Mechanical Brake&Clutch
    Cleaner (Bulk)......................K1
Mold Release & Lubricant
    #03069.................................H1
Rust Remover Concentrate No.
    3024....................................A3
3% E.D. Silicone 03091.............H2
3% E.D. Silicone 03092.............H2
3% H.D. Silicone 02096.............H1
3% H.D. Silicone 02097.............H1
3% H.D. Silicone 05057.............H1
3% H.D. Silicone 05058.............H1
3% H.D. Silicone 06059.............H1
3% H.D. Silicone 06060.............H1

C S P Industries, Incorporated
Aide..C1
CSP-202....................................L1
D Zolv.......................................L1
Liminate...................................L1
Tru-View...................................C1

C. M. C. Laboratories Company
CMC Liquid Detergent No. 100 A4
CMC Liquid Detergent No. 88.. A1

C.C.I.
Bowl Cleaner.............................C2
Brite Degreaser Super Conc
    Multi-Purp Cl.......................A1
Brite Lac-Stone Milk Stone
    Remover...............................A3
Brite Neutra-Clean....................A1
Brite Pink Lotion Hand Soap.....E1
Brite Stainless Steel Plsh & Surf
    Presrv..................................A7
Liquid Drain Opener.................L1

C.L.M., Inc.
Safety Solvent Motor Deg. &
    Equip. Clnr............................K1

C-E Minerals/ Combustion Engineering, Inc.
BAC-CAU-FP Alkalinity
    Control.................................G6
BAF-400FP Antifoam & Sludge
    Conditioner...........................G6
BCS-483PFP Closed System
    Corrosion Inhib.....................G7
BFA-200FP Filming Corrosion
    Inhibitor...............................G7
BNA-500FP Neutralizing
    Corrosion Inhbtr...................G6
BNA-600FP Neutralizing
    Corrosion Inhbtr...................G6
BOS-470FP Oxygen Scavenger . G6
BSI-410FP Corrosion And
    Scale Inhibitor......................G6
BSI-914FP Corrosion And
    Scale Inhibitor......................G6
CEC-KLN..................................A1
CI-100AFP Corrosion Inhibitor. G2
SI-100FP Scale Inhibitor..........G2,G5

C-K Distributing, Inc.
Marvel Cleaner Degreaser..........A1
Marvel Dishwash Compound
    M-DC #10..............................A1
Marvel Extra Hvy Dty Clnr.
    Degreaser..............................A1
Marvel Fat Foam Heavy Duty
    Foaming Clnr.........................A1
Marvel Frock Cleaner M-FC 32 A1
Marvel Heavy Duty Acid...........A3
Marvel Liquid K #21..................A1
Marvel Shroud Cleaner M-SC
    #105-1..................................B1

56

Cadco, Incorporated
Cadeo Egg Washing Compound Q1

Caesar Chemical Manufacturing
Brite-N-Kleen.............................. A4
Penny Trate A4

Cagle Associates, Inc.
Aerosol Insecticide 5000............. F1
Formula 410 Potable Water
   Treatment G7
200 Boiler Water Treatment...... G6
230 Boiler Water Treatment...... G6
260 Boiler Water Treatment...... G6
270 Boiler Water Treatment...... G6

Cajun Janitor Service & Supply
Green Velvet A1

Cal Chem Company, Incorporated
Formula #122-76........................ Q1

Cal Clean
All Purpose Cleaner.................... A1

Cal-Maine Foods, Inc.
CMF #22 Hand Soap Lotion..... E1
CMF #33 Defoaming Agent...... A1
CMF #44 Concentrated
   General Purpose Cl................. A1
CMF #55 Egg Washing
   Compound................................. Q1

Cal-Pac Chemical Company, Inc.
Cal 1142.................................... A2
Cal 1178.................................... A3
Cal 1618.................................... A2
Cal 1620.................................... A1
Cal 186...................................... A1
Cal 240...................................... A3
Cal 262...................................... A2
Cal 32.. A1
Cal 675...................................... A2
Cal 86.. A1
Cal 92.. A1
Glows A1
Kleen-All.................................... A1

Cal-Tek Industries
Advert G6
Aluminite................................... A2
Aqua-Cal A1
Cal A-76 .... A3
Cal AC-1 .... A1
Cal AC-2 .... A1
Cal B-V-3 .... Q1
Cal Bars A1
Cal Bee A3
Cal CT-55.................................. A2
Cal CT-60.................................. A2
Cal Defoam D-F-S..................... Q5
Cal Defoam D-F-6 .... Q5
Cal Det...................................... B1
Cal Fish & Poultry Plant Cl....... A2
Cal Foam................................... A1
Cal Food Plant Cleaner A2
Cal Go Special #3...................... A2
Cal Go Special #4...................... A2
Cal Klor D1
Cal Klor Colored D1
Cal Kwik A1
Cal L-P-6................................... A1
Cal L-P-7................................... Q1
Cal Lubit A1,H2
Cal M-3-2 .... A1
Cal MSR 2 .... A3
Cal Packers Cleaner A1
Cal Pine-Clean........................... C2
Cal Quat B D1

Cal R-E-2 .... Q1
Cal San C A1
Cal 140...................................... A2
Cal 2751.................................... A1
Cal 33.. G6
Cal 40.. A2
Cal 40 & B................................. A2
Cal 415...................................... A1
Calex N-C A1
Calex 5-A-2 .... A2
Calex 5-B................................... A1
Cleanite D A4
D-Scale 2................................... A3
D-Tar... C1
Dish Shine A1
Drain Opener L1
Du-Cal....................................... A1
Du-Cal #2.................................. A1
Electronic A2
Electronic C A1
Electronic X............................... A2
Galvanizing Brightener.............. A3
Han Cal C C2
Impact.. A2
Lift-Off A1
Liquid Hand Soap E1
Liquid Speedie A1
Loob Supreme A1,H2
Lubit BJ..................................... A1,H2
Maid Kleen A6
Packers Cleaner HD A2
Packers Cleaner HD2 .... A2
Packers Cleaner HD3 .... A2
Pan-Glow A1
Powdered Acid Concentrate....... A3
Red Devil................................... L1
Rinse Kleer A1
Rust-Off P A2
Rust-Off 2.................................. A3
Rustgoe...................................... A2
Sof Suds.................................... A1
Speedie A1
Steamite A4
Steamite "D".............................. A1
Steamite Aviation A1
Steamite NL............................... A1
Steamite Z.................................. A1
Super Cal C................................ A2
Super Cal 4 .... A2
Teklene...................................... A2
Universal A1
Universal 4 .... A1
Universal 5 .... A1
XL-30... A1,H2

Calar Chemical Co.
APC Concentrated All Purpose
   Cleaner..................................... A4
Busol Heavy Duty Degreaser A4

Calfonex, Incorporated
AG-51.. K2
Calfonex CC............................... K2
Calfonex Silicone....................... H1
Pre-Clean #10 Aerosol K1
Pre-Clean #10 Bulk.................... K1

Calgon Corporation
Acid Cleaner 408....................... A3
Acid 817.................................... A3
Alkaline Cleaner 428.................. A2
Antifoam C-1 .... G6
Approve..................................... E1
Aura.. A1,Q1
Bactergent D2,E3
Banox No. 1-P G7,G5,P1
Banox No. 1-PF......................... G7,G5
Banox No. 10 .... G7,G5
Big Cat A1,H2
Boiler Treatment HW-CD......... G6

Boiler Treatment SW-CD.......... G6
Boiler Water Treatment HW-
  200.. G6
Boiler Water Treatment ZS-200 G6
Burosil G6
BA-10 Antifoam G6
BA-11FP Antifoam G6
BA-119FP Antifoam & Sludge
  Contl Treat............................... G6
BL-409 Alkalinity Control........ G6
C-39 Corrosion Inhibitor G1,G5
C-7 Corrosion Inhibitor G1
C-8 Corrosion Inhibitor G1
C-979 Powdered Acid Cleaner .. A3
Cal Clean.................................... A1
Calade... A1
Calgolac..................................... A1,N4
Calgon (glass)............................. A1,G2
Calgon #3 Conveyor Lube H2
Calgon Pump Protector H2
Calgon Unadjusted Crushed A1,G6,G2
Calgon Unadjusted Glass............ A1,G6,G2
Calgon Unadjusted Powder........ A1,G6,G2
Calgon Water Conditioner G1
Calgon Waterless Hand Cleaner E4
Calgon 109 Q E1
Calgonite A1
Calgosil G7
Clor-Ade A1
Composition TG......................... G7,G5
Composition TG-10..................... G7,G5
Conditioner 206 .... A1,N4,G2
Conditioner 412 .... A1,G2,N4
Cryosan D1
CB-125 Boiler Treatment........... G6
CB-150 Boiler Treatment........... G6
CB-260 Boiler Treatment........... G6
CB-265 Oxygen Scavenger......... G6
CB-315 Condensate Treatment .. G6
CC-280 Corrosion & Scale
  Inhibitor................................... G7
CC-400 Silt Control Treatment.. G7
CC-800 Plates............................. G7,G2
CC-850 Crystals.......................... G2
CL-246 Cooling Water
  Treatment G7
CL-427 .... D1
CL-447 Cooling Water
  Treatment G5
CL-49 Can Cooker Treatment ... G5
CL-50 Corrosion and Deposit
  Inhibitor................................... G6,N4,G2,G5
CL-56 Corrosion Inhibitor......... G7
CL-70 Corrosion and Scale
  Inhibitor................................... G7
CL-74 Corrosion Inhibitor......... G5
CS-420 Wash Sanitizer............... D2,Q1,Q2,Q4
Degreasing Solvent K2
Detergent 6152 .... A1
Dishmate..................................... A1
Economy Powdered Acid A3
Electric Motor Degreaser........... K1
Electrical Contact Cleaner K2
F-95 Filming Corrosion
  Inhibitor................................... G6
Fantastic A1
Food Grade Silicone.................. H1
Formula CK................................ A1
FA-114 Foam Additive A1
FD-1.. A1,Q5
FF-108 .... E2
FP Burolock G6
FP Dispersive G6
FP-89 Oxygen Scavenger G6
Gas Leak Detector...................... P1
General Cleaner Concentrate A1
H-106 Microbiocide.................... G7
H-204 Microbiocide.................... G7
Hand Cleansing Lotion E4
Hel-Cat A1

Hi-Lo A1
Hi-Suds....................................... A4
Hi-Time A1
Hy-Soil Concentrate................... A1,Q1
Industrial Waterless Hand
  Cleaner..................................... E4
Instant Calgon............................. B1
Instant Dry Bleach...................... B1
Instantreat................................... G2
Iron And Rust Remover.............. P1
JA-224 Gel Additive................... A1
Kleer N'Kleen............................. A1
L C 426 Concentrate................... A1
Liquid Economy Cleaner A1
Liquid Ice Machine Cleaner....... A3
Liquid Scale Dissolver A3
Liquid Soak-Eze........................ A1
Liquid Steam Cleaner A1
Lube 423..................................... H2
Lube 440..................................... H2
Lubricant 411............................. H2
LP-15S.. G6
LP-159S Boiler Treatment.......... G6
LP-209 Boiler Treatment............ G6
Maxade A2
Microbiocide 22......................... G7
Micromet.................................... A1,G1
Micromet Crystals...................... G2
Micromet Plates.......................... G7,G2
Micromet Plates-Formula 18G... G7
Micromet 6R............................... G2
No. 11 Algaecide........................ G7
No. 33 Algaecide........................ G7
No. 340 Liquid Scale Inhibitor .. G7
NL-100 .... G6
NL-80 Condensate Corrosion
  Inhib... G6
NL-90 Condensate Corrosion
  Inhib... G6
NLD-909 Condensate
  Corrosion Inhibitor.................. G6
P-2000 .... A2
P-2003 .... A2
P-3576... K1
Pax Solv E4
Paxit E4
Payload....................................... A1
Penetrating Lubricant H2
Pinnacle A1
Plumclean Acid A3
Powdered Acid Cleaner A3
Power Wash Compound A1
Powr-Clor A1
Powr-Jet A1
Resolve.. A1,Q1
S-125 .... A4
S-33 .... A3,Q2
S-44 .... A1
S-49 .... K1
S-88 .... A1
S-88-L... A1
Safe'N Sure E1
Scale Remover A3
Scale-Gon A3
Seasontreat G2
Skin Care Lotion E4
Soak Cleaner 737....................... A2
Soak-Eze..................................... A1
Sparkleen A1
Special Calclean HD................... P1
Spot Guard A1
Spray Off..................................... A1
Strip-It-Special........................... C3
Strip-Tee..................................... A8
Sum Totil.................................... D2
Super-Kleen A1,B1
Surf-Kleen A1,B1
Syn-Cide N. R. D2,E3,Q3
Syn-Dek...................................... A1
Syn-Sol D1,Q1,Q2

SL-34 .... G6
SL-500FP G6
SMS-422 .... D2
SP-763... A1
SW-3 .... A2
SW-3 Gel..................................... A2
Thermo-Trap P1
Thick 'N Quick D1
Totil D1
V-Belt Dressing P1
V-419 :.. Q1
Vantage A1,Q1
Weltone P1
Zeotone P1
Zephyr A1

California Cleaning Products Inc.

Caloid Super Degreaser.............. A1
Heavy Duty Descaler A3
New Life Vinyl Cleaner A4
Soak Tank Compound A1
Steam Cleaning Powder A2

California Soap Company, Inc.

Hilite Suds................................. A1

Caljen Sales Company

Fast Clean A4,A8
Gleem A7

Calla Chemical Company/ A Div. of Diversified Chemicals Corp.

Calla 301A Lemon A4
Calla 305..................................... A1

Calusa Chemical Company

Calchlor Concentrate.................. B1
Fleece Powder............................. B2
Indus. H D.................................. B1
Soapthetic B1

Calvert Chemical Company, The

Cal-Phos TCL............................. D1

Camco Products, Incorporated

No Foam A1
No. 59 Heavy Duty Cleaner A4,E1,A8
Spray Clean 81............................ A1

Cameo Paper & Supply Co.

Cleaner/Degreaser A1

Camie-Campbell, Inc.

Camicide 2 In 1 .... F1
Camie Formula #666 .... H1
Formula #110 Camie Sil. Lub.
  & Mold-Rel............................... H1
Formula #410 Silicone Spray H1
30-30 Heavy Duty Anti-Rust H2

Camp Laboratories, Incorporated

Concentrated Pyrethrum
  Compound F1
DeGerm....................................... D1
Digest C1
Glo-tone Liquid Hand Soap E1
Grease-No-More A4
Grease-No-More Spcl Formula . A1
GTC Enzymes L2
Odor-Solv................................... C1
Organic Digester L1
S.D. #40...................................... A1
Sanicamp Deodorant................... C1
Sanitizer Detergent..................... Q1
SP #164....................................... A1

Campbell Chemicals, Inc.

New Spr Camicide Inset Sp.
  Form #100 .... F1

Cardinal Mill Spray.....................F1
Cardinal Pimalarin #1
  Emulsifiable.............................F2
Cardinal Pimalarin #2 Non-
  Emlsfbl....................................F2
Cardinal SF-1 Insecticide F1
Cardinal SF-2 Insecticide F1
Concentrated Insecticide X-10-1 F1
Concentrated NE 10-1
  Insecticide...............................F1
General Purpose Fogging
  Insecticide...............................F1
Mill Fogging Spray No. 3 ....F1
Mill Kill Insecticide F1
Premium Grade Card-O-Sect F1
Premium Grade Card-O-Sect
  #25...F1
Premium Grade Cardinal SF-2
  Insctcd. F1
Premium Grade Cncntrt. NE-
  10-1 Insctcd............................F1
Premium Grade Cncntrtd.
  Insctcd. NE 5-1 ....F1
Premium Grade Mill Spray #3..F1
Premium Grade Mill Spray #35 F1.
Residual Insecticide....................F2
Stored Grain Protectant F2
Swifts Ready to Use Fly Spray
  #2...F1
Thermal Fogging Insecticide
  Type A F1
Thermal Fogging Insecticide
  Type M....................................F1
3% Malathion Insecticide F2

Cardinal Chemical Company
Cardite #422................................A8
Lustreen #HWF...........................A1

Cardinal Chemical Company
Cardinal DDVP-Ronnel
  Emlsfbl. Concntrt. F2
Cardinal Spray #3610.................F1
Diazinon 4E Insecticide.............F2
Emulsifiable Spray Concentrate
  #101 ....F1
Insect Spray "B" F1
Insect Spray "D".......................F1

Cardinal Paper Company
All-Purpose Lemon Scented
  Cleaner....................................A4

Care Chemical Corp.
Care Concentrate.........................A1
Care Three D. Fungicide NP.....D2
CD Cleaner Disinfectant
  Deodorizer Fngcd D2
Freezer Cleaner A5
Heavy Duty Industrial Cleaner..A4
Lemon Hand Soap E1
Non Butyl Cleaner A1
Stericide Disinfectant Defilmr
  Deo Fngcd D1

Care Industrial Chemical Co.
Bowl & Bathroom Kleaner.........A3
Grease Clean................................A4
Grease Off...................................C1
Remove All..................................C3
Safety Solv..................................K1

Carey Salt/Div. of Interpace Corp.
Brine Block P1
Coarse Rock Salt.........................P1
Fine Rock Salt.............................P1
Fluffo Salt P1
Granulated Salt Coarse...............P1
Granulated Salt Food Grade......P1

Medium Rock Salt.......................P1
Rancher's Mixing and Stock
  Salt...P1
Soft'ner Pellets............................P1
Soft'ner Salt Coarse Solar P1
Soft'ner Salt Extra Coarse Solar P1
Soft'ner Southern Louisana
  Coarse P1
Soft'ner Southern Louisana Ex.
  Coarse.....................................P1

Cargill, Incorporated
All Purpose Evaporated Salt......B1,P1
Brine Bits Water Softener Salt...P1
Hgh Prty Lousi. Soft. Wtr
  Rock Salt.................................B1,P1
Hi-Grade Evaporated Salt...........P1
High Purity Northern Wtr Soft. P1
Iodine Disinfectant D2
Mix-N-Flo Evaporated Salt........P1
Top-Flo Evaporated Salt.............P1
Water Softener Salt Pellets.........B1,P1

Carig Company
Industrial Clnr. 101 Non Toxic
  All Purp.................................A1
Industrial Clnr. 102 Non Toxic
  All Purp..................................A1

Carl Kaster Company
Anti-Foam K...............................Q5
Comet A1
Hog Scald Compound.................N2
Hy-Clor D2,Q4
Jupiter..A2
Lunar A1
Mars...Q1
Mercury......................................A2
Meteor A2
Orbit...A3
Pluto A2
Smoky...A2
Stellar...A2

Carlan Corporation
1-M Plus Grs Trap Sol. & Drn
  Fld Renvt L1

Carmel Chemical Corporation
Banish C1
Formula F-2................................F1
Formula F-3................................F1
Formula F-5................................F1
Formula F-500............................F1

Carnation Chemicals Corporation
Dyna-Brite..................................A3
Dyno Brite Special.....................A3
Dyno-Boil Out............................A2
Foam-Mate..................................A1
Hi-Chlor-GP A2

Carnegie International Chemicals, Inc.
CIC F 101 ....A3
CIC F 103 ....A1
CIC F 104 ....A2
CIC F 105 ....A1
CIC F 106 ....A6
CIC F 107 ....A1
CIC F 108 ....A1
CIC F 109 ....A2
CIC F 110 ....A2
CIC F 111 ....A3
CIC F 112 ....A1
CIC F 113 ....A2
CIC F 114 ....A4
CIC F 115 ....A1
CIC F 116 ....A1
CIC F 117 ....A2

Pure Pine Oil Emulsifiable C1
PC 963 Hand Dishwashing
 Detergent................................. A1
Quat Sanitizer D2,E3
Ready-for-use Liquid Hand
 Soap E1
Round Deodorant Block........... C1
S-1000 .... A1
Solvs-it....................................... A1
Spray Solvs-It C1,A8
Stainless Steel Cleaner A7
Strongarm Concrete Floor
 Cleaner..................................... C1
Sul-Fury L1
Sunshine Laundry Detergent B1
Super Ocide Hospital Clnr.-
 Disinfectant C1
Super Six A4
T Mint Disinfectant................... C2
T Pine Pine Odor Disinfectant
 Five.. C1
Ten Pine Odor Cncntrtd
 Disinfectant C1
Thick Pink Hand And Body
 Soap E1
Thick Thick Thick Porcelain &
 Bowl Clnr.................................. C2
Thrifty Mint Disinfectant C2
Thrifty Pine................................ C1
Triple Concentrated Hand Soap E1
Ultra Solve................................. K1
Verde A1
Winda-Shine............................... C1
Wired Toilet Bowl Deo. Block.. C1
Zzzott! Dry Chlorine Bleach..... B1
389 Cleaner A3
40% Coco-Castile Soap E1
40% Pure Coconut Oil Hand
 Soap E1
922 Institutional Machine
 Dishwash A2
944 Laundry Detergent.............. B2
946 Non-Phosphate Laundry
 Detergent................................. B1

Casco Chemical Company
Boiler Compound Solution #1...G6

Cato Oil and Grease Company
Mystik Anti-Leak Industrial Oil. H2
Mystik FG-0 Food Machinery
 Grease.. H1
Mystik FG-2 Food Machinery
 Grease.. H1
Mystik FG-70 Fd Ma. Oil Code
 1982... H1
Mystik Hi-Temp JT-6 .... H2
Mystik JT-7 Multipurp Gear
 Lub.SAE80-90........................... H2
Mystik JT-7 Multipurp Gear
 LubSAE90-140........................... H2
Mystik SSG/825 Gear Lub.
 SAE 80W-140............................ H2
Wanda Lubrizide 22A................. H2

Cavalier Chemical Company, Inc.
AD-5100..................................... A4
Bake Pan Cleaner HH................. A1
Can Washing Compound A1
Car Wash Liquid No. 22............ A4
Cavalier All Purpose Cleaner &
 Degreaser A4
Cavalier FP-109 .... A1
Cavalier LC 1152 .... B1
Cavalier Silicone Spray H1
Chlorinated Dish-O-Matic A1
Concrete Cleaner........................ A4
CIP 17.. A2
Dual Chlor D1,Q4
Flash A4

Formula FP-109.......................... A1
Formula FP-1408........................ Q1
Formula H-334........................... A1
Formula M-119 .... A1
Formula No. 605 .... D1
Formula 1145 .... B1
Formula 1427 .... A3
Formula 1450 .... C1
Formula 301 .... A1
Formula 341 .... A1
Formula 342 .... A1
Formula 900 .... A4
Formula 901 .... A1
Formula 902 .... A1
Formula 911 .... A1
Freeze-Kleen.............................. A5
Glasswashing Compound............ A1
Greasekicker A8
H-339 .... A1
Hand Dishwashing Detergent
 340.. A4
Heavy Duty Floor Cleaner A4
Hy Pressure Cleaner A2
Hydro-Chlor D1
Iodex D2
IND-1510.................................... A1
Jester NF.................................... A1
Jester OF.................................... A2
Liquid Concentrate 125 .... A1
Liquid Concentrate 126 .... A1
Liquid Drain Solvent L1
LC-1151..................................... B1
Magic... A1
Meat Plant Cleaner No. 50......... A1
Meat Plant Cleaner No. 7.......... A2
Meat Plant Cleaner No. 70..:..... A2
Mint-O-Green C1
New G-R Germicidal Rinse D2
N1H-15....................................... D1,Q4
Oven Cleaner And Degreaser A8
Pantastic A1
Pantastic with Germicide D1
Pantex A1
Peak A3
Peak LF...................................... A3
Prompt A1
Purge A1
Rinse-It A1
Sani Sol A1
Sanolac #5 .... A1
Sanolac #7-0 .... A2
Scale & Film Remover A3
Sewer Solvent L1
Speedy A4
Spraize....................................... A4
Steam Cleaner AD 1235 .... A2
Steam King A1

Cavalier Industries, Incorporated
Cavalier Concentrate A1
Cavalier Fogging Spray.............. F1
Cavalier H.D. Drain Opener L1
Cavalier Lanolin Waterless
 Hand Cleaner E4
Cavalier Orange Concentrate A4
Cavalier Power Clean................. A1
Chain & Cable Lubricant........... H2
KT-21 .... K2
Metal Brite A7
New Power Battery Terminal
 Cl & Prtctr P1
Oven Kare A8
Possit-Grip P1
Power Foam C1
Safety Solvent K1
Sani-Lube H1
Silicone Spray H1
Wyte-Ease H2

61

Cayuga Rock Salt Company, Inc.
High Purity Northern Wtr
Softnr Salt B1,P1

Cecil H. Jarrett Company, Inc.
B-24 Safe Fireproof Cleaner A1
D.C.D. Disinfectant Cleaner A4
Jarrett's It All Purpose Cleaner . A1

Cello Chemical Company
Cello Cleaner-Disinfectant-
Deod. D1
Cello Pink Deodorant Liq.
Hand Soap #506 .... E1
GRL All-Purpose Clnr
Concentrate A4
Habit Deodorant Lotion Hand
Soap #27 .... E1
Industrial Cleaner 55 .... A4
Multi-Purpose Insect Killer F1
Veltone Antibacterial/Deo.
Hand Soap E4

Cem-Tex, Incorporated
Super 71 .... A1

Cenol Company
Dry Fly Killer (Green) F2
Fogging Spray SBP-1382 Kill
Quick F1
Food Plant Pressurized Insect
Spray F1
Hi-Pressure Industrial Aerosol
Bomb F1
Industrial Insect Spray F1
Kill Quick Concentrate F1
Mill Spray F1
One-Shot F2
Professional Insect Killer Bomb . F1
Prolin Bait Station for Rats and
Mice F3
Prolin Ready To Use Mice
Station F3
Pyrethrum Pro Concentrate F1
Roach & Household Insect
Spray F2
Tensite Residual Spray with
Baygon F2
Water Soluble Warfarin F3

Center State Petroleum
FM White Grease No. 2 .... H1

Central Arkansas Industrial Service Inc.
Cais 701 .... G6
Cais 703 .... G6
Cais 704 .... G6
Cais 710 .... G6
Cais 731 .... G6

Central Chemical Company, Inc.
#25 Concentrate A4
"44" Deluxe A2
"44" M A1
Bactro-Cen D1
C.C. Concentrate A4
Cen-Kill F1
Cen-Quat D1
Cen-Solve A4
Conc. Complete BWT
Chemical Solution G6
D-Klog L1
Derma-Cen E1
Fligone F1
Fligone II F1
Fligone Spray Insect Killer F1
Nufome Concentrate A1

Pheno-Cen C1
Porcenal C2
Power-Cen A1
Premier 128 .... D1
Ready to Use Surgical Liq.
Soap E1
Sani-Cen C2
Surgical Liquid Soap 40 .... E1
WT-96 Special A1

Central Chemicals
CB-20 .... G7
CB-200 .... G7
CB-222 .... G6
CB-81 .... G6
CBP-34 .... G6
CW-13 .... G7

Central Cleaning Service & Supply
Fogging Spray Concentrate F1
R.I.S.-15. .... F2

Central Janitors' Supply Co.
A-101 Clean All A1
Concrete Cleaner C1
Liquid Soap E1
Low Suds Laundry Detergent ... B1
Spray All A4
Vista A4

Central O-B Products Company, Inc.
Kil-O-Mist F1
Kil-O-Mist No. 315 Concentrate F1
O-B Cleaner A1
O-B Heavy Duty Cleaner A1
O-B Liquid Detergent A1
O-B Meat Hook Cleaner A2
O-B No. 101 Smoke House
Cleaner A2
O-B No. 350 Smoke Stick
Cleaner A1
O-B No. 356 Smoke Stick
Cleaner A1
O-B 15% Liquid Soap E1
O-B 990 Power Wash A1

Central Petroleum Company
Blaster H.D. Concentrated
Cleaner A1
George A1

Central Products, Inc.
Advance Lemon Germcdl Clnr
& Disnft C2
Blast! Bowl Cleaner C2
Bright Dishwash A1
Central #20 Steam Cleaner
Liquid A2
Central #25 All Purpose Heavy
Duty A2
Central Centralcide No. 520 ...... F2
Central Chemical Cleaner No.
105 .... A3
Central Coil Cooler A4,P1
Central Disinfectant C1
Central Fog-O-Spray No. 522 .... F1
Central Super Heavy Duty
Cleaner A1
Central Yellow Tiger Concrete
Cleaner A4
Central Yellow Tiger No. 705 ... C1
Centralcide No. 520 .... F2
Expedite A1
Fog-O-Spray F1
Fog-O-Spray No. 522 .... F1
Gentle Hand Soap E1
Germalene Bacteriasol Grmcdl
Clnr No.200 .... C1
Germalene Bacteriasol Grmcdl
Clnr No.202 .... D1

Gérmalene II Bacteriasol
Germicidal Clnr D1
Hand Soap E1
Laundry White B1
Pure Pine C1
Spin Safety Solvent K1
Super Nu Kleen A4,E1
Super Quick Kleen A1
Un-Clog Drain Opener L1

Central Soya
Gilt Edge Rat Bait F3
Gilt Edge Rat Bait Thro Pac F3
Gilt Edge Sugar Fly Bait F2
Master I-0 .... D2
Master Mix Blue Death Rat
Bait F3
Master Mix Blue Death Throw
Pak F3
Master Mix Fly & Mosquito
Insect Klr F1
Master Mix Mouse&Rat Bait
Throw Pak F3
Master Mix Pro-White Egg
Detergent. Q1
Master Mix Rat Bait F3
Master Mix Sugar Fly Bait F2

Century Chemical Company
Century Jet #403 .... A1
Century Jet #404 .... A1
Jet Quick #400 .... A4

Century Chemical Corporation D Of Century Papers, Inc.
Act. C1
Centurion Cleaner C1
Cleaner-Degreaser A4
CC Concentrate C2
CC Deodorant Blocks C1
CC Glass Cleaner C1
Dina-Mite Sulfuric Acid Drain
Opener L1
Gran A1
Liquid Lite-Ning A1
Mint Disinfectant C2
Mint Disinfectant C2
Mr. Jon C2
Mr. Jon's Helper Deodorant
Blocks C1
Multi-Quat 200 .... D2
MPC Multi-Purpose Cleaner A4
Pine Scent Disinfectant C1
Pine Scent II Disinfectant C1
Pink Lotion A4
Pink Petal A1
Pure Pine Oil Disinfectant C1
Super Mr. Jon Instant Bowl
Cleaner C2
Thrifty Hand Soap Ready to
Use E1
Thrifty Hand Soap Ready To
Use E1
Wil MPC Concentrate A4
Wow L1
40% Liquid Toilet Soap E1
40% Liquid Toilet Soap E1

Century Laboratories, Incorpo
Action-Flo C2
Aqua Tower Slimicide
Algaecide. G7
Artic Clean A5
Award Skin Cleanser E1
Big Red L1
Born Bard Super Flying Insect
Bomb F2
C.L. 600 Hospital Disinfectant . C1
Cen-O-Bate. C1

C-Deep A4
C-Treat A4
C-3271 .... A3
C-700 .... G6
Cerfact 100 .... A1
Cerfact 200 .... A7
Cerfact 800 .... L1
Cerfact 900 .... A1
Cerfact 920 .... K1
Cerfact 930 .... H2
Cerfact-CX110 .... P1
Cerfactant Power Action
 Detergent A4
Cerfactol A4
Cerfcide F1
Cerfcide Aerosol F1
Cerfcide Concentrate F1
Cerfkill Formula B F2
Cerflan E1
Cerflash A4
Cerfmaid A1
Cerfox A3
Cerfsan C2
Cerfsect F1
Cerfsil H2
Cerfsol C2
Cerfsteam A1
Cerftize D1
Cerftrol C1
Cerfzyme L2
Cergent E1
Cerphene A4
Cir A1
Cicen & Shine C1
Cleenz E4
Concrete Renovator A4
CX-220 .... K1
D-Scale 211 .... A3
Dishcleen A1
Dizzolv L1
E-Z Strip (Aerosol) C3
Easy Kill F3
EZ Strip C3
Fact A4
Foam Out A1
FML H1
Grab P1
Hell on Ice C1
HP 10 .... A4,A8
Launder Cerf B1
Liqui-Broom P1
Lubefact EP-AA H1
Lubeloc H2
Moto-Cerf K1
Poly-Clean C2
Pro-Cream E4
Pronto-Cleen A3
Proof K1
Protecto-Shield P1
Radiant C1
Rapid A4,A8
Re-Pel C1
Rout C2
Rust Off A3
S.S. 250 .... L1
S.T. 101 .... C1
Safix C2
Sanicerf C1
Sila-Cerf H1
Silt-X CT G7
Solvent 99 .... K1
Spotless A1
Strike C1
Strip-It K1
Suavesito E1
Termafog F1
Viro-Cerf II C1
X-Scale A3

Certified Chemical Company
Certified D.C.D. A4

Certified Chemical Products
Generation II A1

Certified Laboratories/ Division of NCH Corporation
'76 Heavy Duty Cleaner A2
Accel H1
Aero-Strip C3
Algon-X G7
Aqua-Sol A1
Bandolero K1
Bayside C2
Bexane Base A and Hardener P1
Bio-Last L1
Boilex G7
Bolt-Off H2
Bug-X F2
C.A.-N.F. L1
C-W Concentrate A1
Certamate F2
Certi-Chlor C1
Certi-Chlor Concentrate C1
Certi-Etch A3
Certi-Fog Concentrate F1
Certi-Magic C2
Certi-Mist F1
Certi-Sect F2
Certicide F1
Certistaph A4
Certistrip C3
Certisuds A4
Certop SAE 80W-90 Gear Oil ... H2
Certop SAE 85W-140 Gear Oil . H2
Commend H2
Cond-X P1
Courier H2
CA-47 .... L1
CCL-24 .... H2
CCL-500 .... H2
CCL-600 .... H1
CE-27 .... A2
Darathene K2
Darathene Aerosol K2
Decona G7
Deolite C1
Deputy K1
Double Action Odor Digester ... C1
Dri-Lube H2
Duo-Commander Hot Water
 Cleaner A2
Duotrex Liquid Cleaner and
 Descaler A3
Dylek Cleaner & Degreaser K2
Dysh A1
DP-70 .... P1
Emulside F2
Eze-Sheen C1
Five-Flo H2
Free H2
Germa-Cert A4
Glad A1
Gone C1
Grrr A1
Hi-T 3500 .... A1
Hi-Top H2
Hold-Fast P1
HC-200 .... F1
I-OX Ionized Penetrating Fluid . H2
Ice Breaker Liquid Ice Melter .. C1
Ice Pellets C1
Jix Foamy Oven Cleaner A8
Kattle-Sect F1
Kustom K2
L.P.O. Liquid Pipe Opener L1
Liqua-Dysh A1
Liqua-Sect F1
Lube-Trac H2
Luxury Concentrated Hand
 Soap E1

Luxury-CPC E1
Micro-Pel (aerosol) C1
Mint-A-Lene Disinfectant C1
Mitt Grit E4
Multoil 10W/40 .... H2
Multoil 5W/20 .... H2
New Gly San C1
New Gly-San (aerosol) C1
New Lok-Cease (aerosol) H2
New Lok-Cease (bulk) H2
No. 77 Steam Cleaning
    Compound A1
Now C2
Nu-Bowl C2
Nuzair Filter Spray Coating P1
NM-40 Cleaner A4
NM-50 Cleaner and Descaler A3
Odorless Vysol A1
Orbit A4
Packing House Cleaner A2
Perform-All H2
Pinalene C1
Punch Water-Soluble Industrial
    Cleaner A4
Ream L1
Reason A1
Resolve H1
Romaire B C1
Romaire M C1
S.S.C.-14 .... A7
Sabre F1
Sabrecide F1
Saf-Sol (Bulk) K1
Sav-It P1
Scat K1
Slip H2
Sparkle A1
Spring E1
Steam-Off A2
Sting X F2
Stisect F1
Stop Foam P1
Sunshine C1
Sure Flow L1
Sure-Fom Liquid Foam
    Additive A1
Syn-Fog F1
SPR-50 .... A2
T-A-T P1
T-Lube H1
Task E1
Therma-Lube H1
Triple-XXX F3
Twin-Lube H2
TLF-44 .... H1
Ultra-K Aerosol Bathroom
    Cleaner Dsnf. C2
Ultra-K Concentrate C2
Vast C1
Viabac Part A & Part B L2
Vos-Ban Insecticide F2
Watermatic P1
Well-Done Concentrated Oven
    Cleaner A8
Wildcat 3-C Heavy Duty
    Cleaner A1
Wildcat 3C Super-X Cleaner A1
WT-21 .... G6
WT-22 .... G7
WT-31 .... G6
WT-33 .... G7
WT-44 .... G7
WT-51 .... G7
WT-61 .... G6
WT-88 .... G7
X-433 .... H2
Zymex A4
106 C-Cleaner C1
3-C Wildcat Germ. Food Plant
    Clnr D1

3-D Odor Control C1
842 Industrial Cleaner A1

Cesco, Incorporated

Al Clean #4 .... A1
AD A2
Chloro-Clean A1
Dry Det A1
E-2000 G A1
Neutralizer No. 1 .... A1
ND-251 .... A1
ND-255 .... A1
ND-260A A1
ND-278 .... A1
RS-100 .... A2
Saf-Steam A1
Steam Clean A1
Steam Power A2
Steam Terg A1
20-A A3
30-P A2
707 .... A2
707-S A2
740-R A2
746 .... A2
778 .... A1

Cessco, Incorporated

Bait-Tox Ready Mixed Rat Bait
    (Block) F3
Bait-Tox Ready Mixed Rat Bait
    (Meal) F3
Cessco Brand Aerosol
    Insecticide F1
Cessco Econoline Aerosol F1
Cessco Pro. Type Aerosol F2
Cessco 5 Aerosol Insecticide F1
Cessco 7 Aerosol Insecticide F1

CeCO2 Chemical Company

Astronaut D-1 (Aerosol) K1
Astronaut DI (Bulk) K1
Attacked A8
Blitz Disinfectant Foam Cleaner C1
Ce-Ox C1
Ce-Power A4
Clean N'Cold A1
Digestant 250 .... E1
Dizzolv A1
Formula 440 .... A1
Guardian H2
Knock Out F2
Marvel Industrial Degreaser A4
Nice & Dry P1
Odor-Ban C1
Quick Drane L1
Release H2
Residual Insecticide F2
Shimmer C1
Sila-Spray Silicone Lubricant H1
Stericide D1
Treatment G6
Triple Threat A4
White & Tite Lubricant H2

Challenger Products Co., Inc.

Bio-Sure L2

Champion Chemical Company

Blast-Off A4,A8
Dynamite Neutral Detergent
    Clnr C1
Pine-Odor Disinfectant C1
Product No. 115 .... A3
Product No. 135 .... A1
Product No. 145 .... A1
Product No. 155 .... A1
Product No. 160 .... C1
Product No. 75 .... C1

Product No. 80 .... K1
Solv-It A4,A8
Twister A4

Champion Maintenance & Supply,

Admire A1
Ammoniated Wax Stripper A4
Aquacide F1
Bole Aire C2
Champine C1
Citrosene 10 .... C1
Cocoanut Oil Hand Soap
    Concentrate E1
Devil L1
Drain Opener L1
Dual-27 .... D1
Efficient D1
Emulsion Bowl Cleaner C2
Fix F2
Flush Drain Opener L1
Fogging Spray Concentrate F1
Heavy Duty Clnr. (Pine Odor) .. C1
Hydra F1
Kleer Vue Window Cleaner C1
Macbrite A1
Metbrite C1
Mountain A1
Mr. Brite C2
N-Odor C1
Palm E1
Perfumed Deodorant Blox
    No.16In Hang Ups C2
Perfumed Deodorant Blox
    No.24In Hang-Ups C2
Perfumed Deodorant Cakes C2
Perfumed-Wired Bowl Blox C2
R.I.S.-15 Residual Insecticide F2
Red Hot L1
Saphene II E1
Sharp C2
Sol-Clean A4
Super Solv K1
Synthetic Cleaner Concentrate .. A4
Tops A1
Vita Pine (Phenol Coef. 5) C1
Vita-Pine C1
6/49 Concentrate A4

Champlin Petroleum Company

Beta 2W H1
Deluxe Motor Oil SAE 10W H2
Deluxe Motor Oil SAE 20W &
    20 .... H2
Deluxe Motor Oil SAE 30 .... H2
Deluxe Motor Oil SAE 40 .... H2
Deluxe Motor Oil SAE 50 .... H2
General Industrial Oil A H2
General Industrial Oil B H2
General Industrial Oil C H2
General Industrial Oil D H2
General Industrial Oil E H2
Hydrol Oil Heavy HD H2
Hydrol Oil Light HD H2
Hydrol Oil Medium Heavy HD H2
Hydrol Oil Medium HD H2
Inhibited Turb. & Hydra Oil
    Hvy H2
Inhibited Turb. & Hydra. Oil
    Light H2
Inhibited Turb. & Hydra. Oil
    Med. H2
Inhibited Turb. & Hydra. Oil
    Med. Hvy H2
Kappa M H1
MGX Gear Lubricant #3 .... H2
MGX Gear Lubricant #5 .... H2
MGX Gear Lubricant #6 .... H2
MGX Gear Lubricant #8 .... H2

Chapman Chemical Company

Millerite Mark IV Clear P1
Pen-A-Seal P1
PQ 8 .... P1
Q-San Ready-To-Use P1
Sealtite 60 Clear....................... P1

Chardon Laboratories, Inc.

AC-74 .... A3
AF-16...................................... G6
Big C.. A4
BF-10....................................... G6
BF-5... G6
BG-10 .... G6
BH-10....................................... G6
BJ-10.. G6
BX-2... G6
BX-3... G6
BZ-10....................................... G6
BZ-11 .... G6
BZ-5... G6
Chardon 150............................ A4
Chardon 152............................ A1
Chardon 58.............................. A1
Chardox 202 Insecticide........... F2
Chardox-808............................. F1
Condex..................................... P1
Conquest.................................. A4
CF-6... G6
CK-8 .... G6
CTA-711 .... P1
Deo Sept C2
Deocide D2
DA 10....................................... G6
DA-11....................................... G6
Ecolo-Power............................. A4
Exotherm.................................. L1
ERA III..................................... A1
Foam Process........................... A1
FPC No. 7................................. D1
Germactin................................. E1
Glassy C1
Granular-Therm........................ L1
JN-3 .... G6
JN-5 .... G6
JN-7 .... G6
JX 3 .... G6
JX-5.. G6
JX-7.. G6
Kitchen King A4
LPWD....................................... A1
M-6 Cleaner A1
One-Two-Plus C1
Ox-Sol...................................... A3
PS220....................................... A2
PS440....................................... A2
PS880....................................... A2
RL-21.. G6
Sani-Safe.................................. C2
SW-06 .... G7
X-Tox 404 Insecticide F1
X-Tox 606 Insecticide F1
X-Tox 808 Insecticide F1
Zero A5

Charles A. Crosbie Laboratories, Inc.

CC-1000 Hydrocarbon
　　Dispersant............................ K1
CC-2000 Paint And Coatings
　　Stripper................................ C3
CC-3000 Deoxidized A3

Charles E. Hofmann Water Treating Company, The

Boiler Water Treat. Form. No.
　　1002................................... G6
Boiler Water Treat. Form. No.
　　1003................................... G6
Boiler Water Treat. Form. No.
　　1010................................... G6

Boiler Water Treat. Form. No.
　　1052................................... G6
Boiler Water Treat. Form. No.
　　1065................................... G6
Boiler Water Treat. Form. No.
　　1808................................... G6
Boiler Water Treat. Form. No.
　　1809................................... G6
Boiler Water Treat. Form. No.
　　1811................................... G6
Boiler Water Treat. Form. No.
　　1819................................... G6
Boiler Water Treat. Form. No.
　　1871................................... G6
Cooling Water Treat. Form.
　　No. 1005............................. G7
Return Line Treat. Form. No.
　　850..................................... G6
Water Treatment Formula No.
　　1001A................................. G6
Water Treatment Formula No.
　　1020................................... G6
Water Treatment Formula No.
　　1605................................... G6
Water Treatment Formula No.
　　900..................................... G1

Charles G. Stott Co.

Depart.................................... A1

Chase Products Company

BAC-Shield Spray Disinfectant
　　& Deodz. C1
Kild Germicidal Odor-
　　Suppressant Disnf. C1
Lemoh Glycolized Air Sanitizer
　　Deodzr................................. C1
Spraypak Battery Cleaner And
　　Protector.............................. P1
Spraypak Belt Dressing Aerosol P1
Spraypak Deo-Germ
　　Disinfectant C1
Spraypak Dri-Film Graphite
　　Lubrcnt (Aer) H2
Spraypak Dry Lubricant &
　　Release Agent H2
Spraypak Dubl Duty Germ
　　Odor Supp............................ C1
Spraypak Flush Off
　　Degreaser(Aerosol) C1
Spraypak Foaming Cleaner C1
Spraypak Four Way Action
　　(Aerosol)............................... H2
Spraypak Glass Cleaner Spray... C1
Spraypak Indian
　　Spc.Glycolzd.AirSan.Deo...... C1
Spraypak Instant Foamer
　　(Aerosol)............................... C1
Spraypak Kleen Hand............... E4
Spraypak Lemon Fresh Glycl
　　Air San & Deo....................... C1
Spraypak Penetrating Oil
　　(Aerosol)............................... H2
Spraypak Quick-Kil Insecticide . F1
Spraypak Silicone Lubricant H1
Spraypak Spray Disinfectant..... C1
Spraypak Spray Disinfectant &
　　Deodorizer C1
Spraypak Stainless Steel
　　Cleaner................................. A7
Spraypak Stnls. Steel Clnr.
　　Aerosol................................. A7
Spraypak Surface Disinfectant
　　& Deo................................... C1
Spraypak Wasp Long Range Jet
　　Spray.................................... F2
SprayPak Cool Mint Glyclzd
　　Spy AirSantzr C1

Chaska Chemical Company, Inc.

Ambush A4
Aqua Lotion.............................. E1
Aqua Power A1
Aqua Treat................................ G2
Bleach Rite............................... B1
Bottle Brite............................... A2
BMC-100 .... A1
C-S-C.. A2
Can-D....................................... A3
Can-D Dairy Hi Foam............... A3
Char-Go A2
Chaska C.I.P. A2
Chaska Dine............................. D2,E3,Q6
Chaska-Chlor............................ A1
Chaska-Chlor HD...................... A1
Chaska-Q.................................. D1,Q3
Chaska-San............................... D1
Chaska-Sol (Regular) A3,Q1
Chaska-77 .... A1
Chief Chaska............................ A1
Chlor Chief............................... A1
Clor 12..................................... D2,Q4,G1,E3
Dairy Can-D Formula A3
Defoamer.................................. Q5
Del-Chlor A1
Destainer B1
Dia Residual.............................. F2
Dine-O-Mite D2,Q6
Drain Rite L1
Du-EE A1
Eco-Sol A3
Foam Rite A1
Foam-Chlor............................... A1
G-G All Purpose A1
G-P Cleaner A2
Grease Go A1
GG Chelated Heavy Duty A1
GG Chlorntd Liq. Hvy Dty
　　Detergent.............................. A1
GG Heavy Duty Caustic............ A2
GG Hvy Dty Liq. Alkaline
　　CIP Detrgnt A1
GG Hvy Dty Liquid Alkaline
　　Detergent.............................. A1
GG Manual Dtrgnt (Phosphate
　　Free).................................... A1
GG New H.D. Non-phos Liq
　　Alk. Detrgnt.......................... A1
GG Special Acid Detergent....... A3
GG Special Manual Detergent .. A1
GG Special 8 .... A1
GG 8 (Phosphate Free) A1
H. D.-6 Regular......................... A3
H-D-6 Controlled Suds.............. A1
Heavy Duty Bowl Cleaner......... C2
Hi-Lite B1
Hi-Speed H2
K-P... A1
Kelly Foam A1
Kelly Green A1
Kelly Klear A1
Kelly Klor A1
Kelly-Sol A1
Kik-1... F1
Kik-2... F1
Kik-3... F1
Kontrol Q1
KIK-25 .... F1
Licq Quid A1
Liqui-Date A1
Liquid Can-Clean A1
Liquid E-G................................ Q1
Liquid Nu-Kee.......................... P1
Lube-Rite H2
Lube-100 Deluxe Lubricant...... H2
LMC-50..................................... A1
M-G A1,Q1
Mandate.................................... A1
Meat Packers Special A1

65

New Chaska Sol A3
New Controlled Suds Liq. Can
  Clnr. A1
Nu-Chlor A1
Nu-Lube H2
Nu-Wet B1
P.A.D. A1
Phen-Nal A4
Pink Chlor-Chief A1
Pow D1
Q-K D2
Release A1,B1
S.M.C. A1,Q1
Saf-Gard 40% Concentrate E1
Scal-Dee N1
Scald-Rite N1
Ska-Lof A3
Smokehouse Cleaner A2
Smother C1
Solv-It A1
Special D A1,Q1
Super Alkali N4
Super Alkali 30 .... A2
Super Alkali 30-C A2
Super Alkali 55 .... A2
Super Chief A1
Super Chief Egg Washing Q1
Super G.P.C. A1
Super Wet A1
Super-Kee A1
Super-Lime-Solv A3
Super-Solv A3
Sure-Cote Remover C1
Sure-Cote-I P1
Sure-Cote-O P1
T"n"T A1
Tac-San A4
Tac-San Plus D2
Trust B1
Vari-Chlor A1,Q1
Wall'n Floor Cleaner A4
Wet-O-Wash N4
Wipe Off A1
WA-10 .... A1
X-Scale A3
X-Tra Soft B2
X-Tra Wet B1

Chattanooga Products Company
AAA All Purpose Powdered
  Cleaner A1
C-P Neutral Cleaner A1

Chem America
Dri-Mol Moly-Micro-Film H2
Hi-P Contact Cleaner K2
Insect Spray F1
Showcase Glass Cleaner C1
Wyte-Ease Lithium Base White
  Grease H2

Chem Cal Supply Company
101 Cleaner Degreaser A1

Chem Care
CC-60 Safety Solvent And
  Degreaser K1

Chem Clean Janitorial
Greasolv A4
Non Butyl Degreaser A1
White Magic B1

Chem Mark
Laundry Liquid Fabric Softener B2

Chem Pride, Inc.
All Purpose Cleaner A1
Chem Pride No. 580 .... L2

Hard Surface Cleaner No. 400 ... A4
Hy-Gro L2
Liquid Hand Soap No. 470 .... E1
Odor Control Granules #600 ..... C1

Chem Spec Industries
Hi-P Contact Cleaner K2
Sani-Lube H1

Chem Tech Ltd.
Chem Tech 1011 .... D1

Chem-Brite Industries, Inc.
Attack Liquid Scouring Creme .. C1
Bowlex Bowl Sanitizer C2
Break-Thru Powder L1
Deo-Gran C1
Electro-Solv Safety Solvent Cl
  & Dgrsr. K1
Emulsifiable Spray Concentrate . F1
Formula 226 Heavy Duty
  Cleaner A4
Formula 227 Neutral pH
  Cleaner A1
Formula 408 Descaler And
  Delimer A3
Formula 902 Mint Odor
  Disinfectant Clnr C1
Gentle Liquid Hand Soap E1
Insecticide F2
Lift Station Cleaner L1
Liquid Hand Soap E1
Melt-Away Ice Melt C1
Neutri-Clean A1
Nibble L2
Pow Powdered Cleaning
  Compound A1
Rallye Oven Cleaner A8
Release Liquid Drain Opener L1
Removes All Heavy-Duty
  Floor Cleaner C1
Rust Off A3
Score A4
Sil Silicone Spray H1
Softness Powdered Hand Soap .. C2
Strip-Ease A4
Swat AA Grade Fly Spray F1
Tempered Organic Digester L1
Three-Gly C1
Virucide D1
Whiz A1

Chem-Care Inc.
Chem-Foam A1

Chem-Clean, Incorporated
All Purpose Detergent A1
Chem-Clean Spray Clean 568 ..... A4
Chlorinated Non-Foaming Egg
  Wash Q1
Germ-X C1
Lemo-Clean-Dis A4
S & A Metal Clean A1
Spray Clean A4
Spray Clean 568-Low Suds A4

Chem-Co Diversified Corporation
054 Food Freezer Cleaner A5
407 Liquid All Purpose Cleaner A4
409 Alkaline Foam Additive A2

Chem-Fab-Co
Kleen Brite C1
Metal Brite Stainless Steel
  Cleaner A7
Speedi C1
Zip C2

Chem-I-Matic, Inc.
Pyranha 1-10 Livestk. Insect.
  Spc. Spy. F1

Chem-Lab International
Boil-Away A1
Brite All A3
GTK C1
White N' Bright B1

Chem-Lube Corporation
K-108 .... A1
K-113 .... A2
K-12 Cleaner A1
K-13 .... A1
K-21 General Purpose Cleaner .. A2
K-22 .... A1
K-27 .... A3
K-28 .... A3
K-3 Cleaner A2
K-34 Liquid Cleaner A4
K-37 .... C3
K-4 Soak Cleaner A2
K-45 Soak Cleaner A2
K-48 Chlorinated All-Purpose
  Cleaner A1
K-5 .... A2
K-51 Cleaner A1
K-510 Cleaner A1
K-56 Cleaner A4
K-57 .... A3
K-6 Scald N1,N2
K-70 .... A1
K-75 .... A2
K-77 .... A1
K-78 .... L1
K-8 .... A1
K-80 .... B1
K-82 .... G2
K-87 Floor Cleaner C1
K-89 .... A1
K-94 .... A1
K-95 .... A1

Chem-Masters Corporation
AC-407 Series-Powder G6
AC-700 Series Liquid G6
DAK G7
FOS-8 Liquid G6
FOS-8 Powder G6
Internal Coagulant Liquid G6
Scivol Internal Coagulant-
  Liquid G6
Scivol-Liquid G6
Septa Phos Solution G6
Special Phosphate Powder G6
Special Powder G6

Chem-Power
Auto Foam A1
Boiler Wtr Trtmnt Conc. Form.
  #18-01 .... G6
Boiler Wtr Trtmnt Conc. Form.
  #18-02 .... G6
Boiler Wtr Trtmnt Conc. Form.
  #18-03 .... G6
Boiler Wtr Trtmnt Conc. Form.
  #18-04 .... G6
Boiler Wtr Trtmnt Conc. Form.
  #18-05 .... G6
Boiler Wtr Trtmnt Conc. Form.
  #18-06 .... G6
Boiler Wtr Trtmnt Conc. Form.
  #18-07 .... G6
Boiler Wtr Trtmnt Conc. Form.
  #18-08 .... G6
Boiler Wtr Trtmnt Conc. Form.
  #18-09 .... G6
Boiler Wtr Trtmnt Conc. Form.
  #18-10 .... G6

Duco-72 .... D1
LC-55 Steam Cleaner A2
Remov-Ox A3
Sewer Con C1
WS-309 .... A1
40% Cocoanut Oil Hand Soap ... E1

Chem-Sultants, Inc.
Heavy Duty Degreaser A1

Chem-Tab Chemical Corp.
Ammonia Cleanse C1
Pro-Clean A4
Workhorse A4

Chem-Tech, Incorporated
CD/400 .... A4

Chem-Tek
Calibre Degreaser Cleaner Wax
 Stripper A1

Chem-Tex, Incorporated
Am-O-Suds C1
AS-501-B C1
Bac-San Bowl Cleaner C2
Black Disinfectant C1
Clear Tone Mist Cleaner A4
Clear Tread C1
Crestline Neutral Cleaner A4
Deo-Drane C2
Derma-San E1
Pine Oil Disinfectant C1
Purge L1
PLS A1
R-502 .... A4,A8
Saf-O-Solv K2
Syn-O-Pine C1
Vita-Pine C1
X-IT A1

Chemagro Agricultural Division
Mobay Chemical Corporation
Baygon 1.5 .... F2
Baygon 2% Bait Insecticide F2
Baygon 70% Wettable Powder .. F2
Baytex 4 .... F2
Entex 4 .... F2

Chemaide Products
Clean-All A3

Chemation, Incorporated
Chema-D-Degreaser FP A1

Chemax Corporation
A.I.O. 1500 .... G6
Bravo Concentrate Liquid Skin
 Cleaner E1
Shackle Cleaner A1

Chemax Inc.
Ace 572 .... A1
Bolt 569 .... A4
Cut 414 .... A1
Deck 566 .... C1
Deft 413 .... A1
Draw 476 .... D2
Foam Jack-559 .... A1
Foam 849 .... A1
Fold 522 .... A2
Halt 510 .... A4
Jack 570 .... A1
Kick 551 .... A2
Neat-610 .... B1
Part 560 .... A3
Phostex 561 .... A3
Run 888 .... H2
151 .... G6

158 .... G6
166 .... G7
169 .... G6
172 .... G6
174 .... G6
177 .... G6
183 .... G6
193 .... G6
234 .... G6
263 .... G7
270 .... G6
273 .... G6
279 .... G6
447 .... G2
511 .... A1

Chemclean Corporation
Chem-Zime L2
Chemclean #250B A2
Chemclean #301A P1
Chemclean #35 .... A4
Chemclean #36 .... A1
Chemclean #500 .... A3
Chemclean #502 .... A3
Chemclean #621 .... P1
Chemclean #77C A1
Chemclean #81C A2
Chemclean Deodorizer M C1
Chemclean 310-S A4
Chloro-Clean ALN A1
Drain Cleaner #80 .... L1
Handi-Clean EX A1
Supaclean A1
SC Cleaner A1
T-650 .... A1
T-655 .... A4
T-660 .... A1
TW Liquid Cleaner A1

Chemco Chemical Company
All Purpose Dairy Cleaner A1
Alu-Lube E.P.G.-AA H1
Be-Gone F2
Be-Gone Residual Insecticide F2
Big Orange A4,A8
Biothrin F1
Borax Hand Soap C2
Car Wash A1
Chem-Slove K1
Chemco #1 .... A4
Chlorinated Machine Dishwash . A1
Clean Up A1
Con-Clean A3
Concrete Cleaner A4
CD-80 .... K1
CD-80 Rust Solv H2
Derusting Cold Vat Compound . A2
Fast-Flow L1
Foaming Glass Cleaner C1
Food Plant Cleaner A1
Hand Dishwash A1
Heavy Baby E4
Hi Foam C1
Hospital Spray Disinfectant &
 Deodorant C1
Hot Vat Compound A2
Insect-O-Fog F1
Laundry Powder B1
Longone F1
Lovit E1
Moisture Guard H2
OD-40 .... C1
Packing House Cleaner A1
Pinklor D1
Pro-Cleaner E1
Rubber Belt Dressing P1
Sanitary Lube H.D. H1
Sewer Pipe Cleaner L1
Silicone Spray H2
Silicone Spray (Aerosol) H1

Steam Cleaner-Heavy Duty A2
Steel Brite.................................. A7
STSC-1050................................. A1
Zor-O-Cide................................. F1
500 Diazinon Residual Spray F2

Chemco Industries, Incorporated
Chemco #611............................ A1
Chemco #856............................ A1
Chemco #857 Super
Concentrate............................. A1
Chemco BB & G C2
Chemco Chain Lubricant H2
Chemco Coconut Hand Soap E1
Chemco Liquid Drain Solvent... L1
Chemco S-25.............................. A1
Chemco Super Acid Cleaner...... A3
Chemco Unicide......................... D1
Chemco Vat Dip A2
Chemco 800.............................. A4
Emulsion Bowl-Kleen................. C2
Heavy Duty Odorless Concrete
Clnr.. A4
NTI-25...................................... F1
Rid All Disinf. Deodorant
Sanitizer................................... D2
Uniclean.................................... A1

Chemco Products
E-Z-Steam................................. A1

Chemco Products Company
Chemco #2910.......................... A2
Chemco #2913.......................... A3
Chemco #2929.......................... L1
Chemco #387............................ A1
Chemco #429............................ A1
Chemco #56.............................. A2
Chemco #64.............................. A1
Chemco Alk-Rus A2
Chemco Bonanza....................... A2
Chemco Chem-Chlor A1
Chemco Chloro-Brite................. A1
Chemco Chloro-San................... A1
Chemco Con-Syn A1
Chemco Drain Opener............... L1
Chemco Foamzall...................... A1
Chemco Heavy Duty Degreaser A4
Chemco Liquid Hand Soap....... E1
Chemco Lubrite.......................... H2
Chemco Lubrite 75.................... H2
Chemco LC-U B1
Chemco MR-1 .... A3
Chemco MR-2 .... A3
Chemco MSR A3
Chemco Packers General A1
Chemco Power Kleen................. C1
Chemco Power-Chem................. A1
Chemco Quik-Wipe................... A4
Chemco Rinse Add G1
Chemco Steam Elite................... A1
Chemco Synterg A4
Chemco Target.......................... A1
Chemco Water Based
Degreaser A1
Chemco 101 .... A2
Chemco 125 .... A1
Chemco 16 .... A1
Chemco 19 .... A8
Chemco 20 .... A2
Chemco 21 .... A1
Chemco 25 .... A3
Chemco 301 .... A1
Chemco 35 .... A1
Chemco 40 .... A2
Chemco 44 .... N4
Chemco 51 .... A1
Chemco 62 .... A1
Chemco 99 .... A2
Chemeen 100.............................. G6

Chemquest 101........................... G7
Dynamist A1
Enerex 201 .... G6
Enerex 202 .... G6
H T S A2

Chemco Products, Incorporated
Bo-Chem G6
Brute.. A1
C-13.. Q4,N4,G1,E3
Chembrite 1000......................... A1
Chemcide................................... D2,Q4,N4
Chemclean X-40 .... A1
Chemclean X-40-C A1
Chemclor................................... D1,Q4
Chemco Bar Lube H2
Chemco C D-D-B D1
Chemco Egg Wash..................... Q1
Chemco Swish A4
Chemco-NS A4
Chemdex.................................... A1
Chemglo Liquid......................... A1
Chemlac.................................... A3
Chemlac Concentrate A3
Chemlube #1 .... H2
Chemlube #2 .... H2
Chemrak #24 .... A1
Chemrak #25 .... A1
Chemrak #26 .... A2
Chemtek A1
Chemtile A3
Chemtron A1
Dyna-Chem A1,B1
Dyna-Chem No. 1 .... A1
Express Cleaner A1
Flo-Chem A3
Foaming Agent A1
Goliath...................................... A2
Lift-Off..................................... A1
Lift-Off Phosphate Free............. A1
Pro-Chem No. 1 .... A1
Pro-Chem No. 3 .... A1
Prochem A4
Ready to Use Liquid Soap E1
Regular Liquid Soap
Concentrate.............................. E1
Remove A2
Spray Chem No. S A1
Spraychem.................................. A2
Sterichem A1
Sterichem No. 1......................... A2
Sterichem No. 2......................... A2
Super Glo-Chem........................ A1
Super Liquid Soap Concentrate. E1
Tru-Brite.................................... B1

Chemcor, Inc.
B-405... G6
B-520... G6
B-610... G7

Chemcraft Industries, Inc., The
Bio Phene A4
Bio Quest D1
Bowl-Tek................................... C2
Chemcraft Formula 311.............. A4
Chemcraft Saniteen Cleaner A1
Chemcraft Stainless Steel
Cleaner..................................... A3
Chemi-Jel Floor Cleaner A4
Formula 312 Floor & Wall
Cleaner..................................... C2
Formula 538 .... A2
Formula 547 .... A2
Kleenaide 215 OD E1
Kleenaide 220 OD E1
Kleenaide 240 OD E1
Neo Concentrate......................... A1
Neo Solve.................................. A4
Synthelene Cncntrtd Detergent
XX... A4

Chemec Co.
Grad A A1

Chemesco of Virginia
Activ VIII.................................. A1

Chemetron Corporation
Cee-Bee C-100 .... A1
Cee-Bee C-311 .... Q1
Cee-Bee C-312 .... Q1
Cee-Bee Emulso Clean 20 .... A1
Cee-Bee Powersan..................... D1

Chemex Chemical Corporation
100.. A2
105.. A2
106.. A2
125.. A2
180.. A1
630.. A3
631.. A3
901.. A3
910.. A1
920.. A1
933.. A1
967.. A1

Chemex Chemicals & Coatings Co.,Inc.
Chemex Stainless Steel Cleaner
& Polish................................... A3
Chex-Ston 780............................ A3
Go-Go....................................... E1
Omni-Cocoanut 40 Concentrate. E1
Omni-G10 Bowl Cleaner C2
Omni-Solv 99 .... K2
Omni-Super Concentrate A1
Omnimene Concentrate C1
Pan-Aid A1
Pastel Hand Dish Wash A1
Pine Odor Cleaner..................... C1
Pink Pearl Hand Soap
Concentrate.............................. E1
Porse-Brite................................ C2
PLH-74 Pink Lotion Hand
Soap E1
Steam Power.............................. A1
Ston-Ease 781 .... A3

Chemex Industries, Incorporated
All-Kleen Stripper Degreaser ... A4,K1
Blue Streak A4
Bowl Master C2
Chemo Cide D1
Iodorite Concentrate D1
Scent-O-Lemon.......................... A4
Skip Concrete Cleaner A4
2300-FPP A3
2386-FPP A1
2399-FPP Egg Cleaning
Compound................................ Q1
2414-FPP A1
2426-FPP A2
2434-FPP A1
2458-FPP A2
2675-FPP A1
2995-FPP A3
2997-FPP A1

Chemical & Engineering Associates/ A Division of Pro-Chem, Inc.
Chlorinated Alkaline Cleaner
PC-98 .... A2
Chlorinated General Purpose
Cl. PC-48 .... A1
Feather Softener PC-8 .... N1
Foam-AD PC-10 .... A1
General Purpose Cleaner PC-24 A1

68

HD Caustic Cleaner PC-521 ...... A2
Liquid Acid Cleaner PC-75........ A3
PC-1 .... A1
PC-2 .... A1
PC-221 .... A2
PC-47 .... A2
PC-52 .... L1
PC-56 .... A1
PC-6 .... A1
PC-61 .... A1
PC-78 .... A1
PC-93 .... A1
PC-99 .... A2
PCL-12 .... A2
PCL-14 .... A1
PCL-92 .... A1
PCL-95 .... A3

Chemical Associates, Incorporated
Econowash A1
Fleet Swipe A1
Foam-A-Way A8
Formula I.L.D. B1
Formula XDL A1
Formula 100K A1
Formula 109A A1
Formula 110 .... A1
Formula 110NF Q1
Formula 130 .... A1
Formula 132 .... A2
Formula 138 .... A1
Formula 140 NP A4
Formula 166 .... A1
Formula 182 .... A2
Formula 203 .... A1
Formula 209 .... A1
Formula 360 .... A1
Formula 405 .... A3
Formula 529 .... A4
Formula 555 .... A1
Formula 60 NF A2
Jet Swipe A1
No Phos Swipe A1
Super Egg Wash Q1
Super No Phos A1
Super-Tergent A1
Swipe A1
Swipe Brite C2

Chemical Control, Incorporated
BL-440 Boiler Water Treatment G6
BP-440 Boiler Water Treatment G6
CP-440 Boiler Water Treatment G6
FL-440 Boiler Water Treatment G6
HL-440 Boiler Water Treatment G6
HL-441 Boiler Water Treatment G6
NL-442 Cooling Water
   Treatment G7
NP-442 Cooling Water
   Treatment G7
VL-440 Potable Water
   Treatment G1
VP-440 Potable Water
   Treatment G1

Chemical Discoveries, Inc.
Sea-Wash Cleaner and
   Dispersant A4

Chemical Dynamics Corporation
K-50 Heavy Duty-Multi-
   Purpose Cleaner C1
K-50 Hot or Cold Water Dry
   Bleach B1
K-50 Hot or Cold Water
   Laundry Detergent B1
K-50 Light Duty-Multi-Purpose
   Cleaner A1

Chemical Engineering Co., Inc.
Chenco Type LPB-M BWT G6
Chenco Type PSB-M Pwdrd
   BWT G6

Chemical Engineers
CE-R-5945 .... H1
CE-0633 Inorg. Cmplx Food
   Mach. Grs. H1
CE-066 .... A1
CE-075 .... A2
CE-1182 .... H2
CE-1255 Special Lube Grease... H2
CE-129 .... A3
CE-137 .... A1
CE-1469 .... H2
CE-157 .... H1
CE-185 .... B1
CE-189 .... N1
CE-190 .... A4
CE-216 .... A4
CE-2304 All Purpose Lube H1
CE-256 .... A1
CE-265 .... A2
CE-316 .... A1
CE-336 .... Q1
CE-437 .... A1
CE-465 .... A1
CE-5104 .... H2
CE-562 .... A1
CE-620 .... A3
CE-625 .... C2
CE-652 .... A1
CE-7135 .... H1
CE-727 .... A1
CE-765 .... A1
CE-816 .... A1
CE-820 .... A3
CE-855 .... L1
CE-904 .... A2
CE-9255 .... H2
CE-9533 .... H1
Razorback Red 452 .... A1,E1
8000 Spraymatic Gear Film........ H2

Chemical Finisher, Inc.
CFS-10d Q1,Q2

Chemical Industries, Incorporated
A.A.I. All Purpose Cleaner C1
Blue Diamond Stripper A4
Chem Strength "O" A1
CII Saf-T-Solvent K2
CII Hard Surface A1
Dual Insect Spray F1
End-Con P1
Germicidal Cleaner D1
Glass & Frame Cleaner.............. C1
Hy-Gro Liquid Live Micro-
   Organisms L2
Knock-Out C1
Liquid Hand Soap E1
Liquid Piranha Liq. Live
   Micro-Organisms L2
Mint Disinfectant Coef. 5 .... C1
No. 25 Insect Spray.................... F1
Non-Flammable Safety Solvent . K2
Oven-Grill & Fat Fry Cleaner ... A1
Pine Cleaner C1
Pine Deodorant C1
Pink Lotion Hand Soap E1
Safety Solvent K1
Sani-Lube Food Equipment
   Lubricant H1
Silicone Spray H1
Sparkle C1
Super Concentrate Cleaner And
   Degreaser A1

Tef-Tape.. H2
Teflon Spray H2
Truck Wash A1
Window Cleaner C1

Chemical Laboratories
Grease-Strip Degreaser Cleaner. A4,A8

Chemical Lubricants Company
Chem Cool WW Food Machine
   Grease... .... H1

Chemical Maintenance, Inc.
Conquer A4
CM-11 .... E1
CMI Big M Mark Remover K1
CMI Blast Out Drain Opener L1
CMI Dr. One Oil Soluble Spot
   Remover K1
CMI Lo-Foam A4
CMI Mountainaire Lemon
   Fragrance C1
CMI Quick Clean A2
CMI Surgical Hand Soap E1
CMI Treat Concentrated
   Porcelain Cleaner C2
CMI 100 Plus.............................. A1
Dia-Tox Roach & Ant Killer F2
Dia-Tox Roach Spray................. F2
Ezy-Foam A1
Formula 620 .... D2
Germ-O-Kill................................ D1
Ice-Off C1
Limes Off A3
Liqui-Zyme L2
Lo-Foam A4
Mountainaire Bouquet
   Fragrance C1
Mountainaire Spice.................... C1
No Bind Silicone Lubricant....... H1
Phen-O-Kill................................ A4
Phen-O-Kill 256 .... A4
Pyro-tox Super Strength
   Insecticide F1
Pyro-Tox Big Boy....................... F1
Pyro-Tox Contact Insecticide ... F1
Pyro-Tox Insect Killer F1
Pyro-Tox 710 .... F1
Pyro-Tox 710 Contact Insect.
   (Fogger).................................. F1
Pyro-Tox 710 Contact
   Insecticide F1
Quick Clean A4
S.O.B. Meat Department
   Degreaser A4
Stainless Steel Cleaner A7
Super 100-Plus A4
Surgical....................................... E1
Swift Clinging Bowl Cleaner C2
SD-632 .... K1
The Grabber A3
Triple Threat Bowl Cleaner....... C2
Twin Disinf. Deodorant &
   Sanitizer................................. C2
X-Cell A1
X-Cell MP A1

Chemical Packaging Corporation
Ammoniated Oven Cleaner........ A8
Block Out (Aerosol).................... C1
Chem-Cap Safety Solvent.......... K1
Chemi-Cap De Aqua Lube
   (Aerosol)................................. H2
Chemi-Cap Chain & Cable....... H2
Chemi-Cap Dual Synergist
   Insctcd F1
Chemi-Cap Glass Cleaner......... C1
Chemi-Cap Insect Spray F2
Chemi-Cap New Power.............. P1
Chemi-Cap No. 25 Insect Spray F1

69

Chemi-Cap Oven Protector........ A8
Chemi-Cap Sani-Lube (Aerosol) H1
Chemi-Cap Sani-Lube (Bulk)..... H1
Chemi-Cap Silicone Spray......... H1
Chemi-Cap Skin Shield Cream:.. E4
Chemi-Cap Spray Silicone
  Release... H1
Chemi-Cap Sur-Grip.................. P1
Chemi-Cap Touch-Up All Purp.
  Clnr... C1
Chemi-Cap Wyte-Ease
  (Aerosol)..................................... H2
End-Con P1
Hi-P... K2
Lanolin Waterless Hand
  Cleaner... E4
Silicone Release (Bulk) H1
Stainless Steel & Metal Polish.... A7
TEF-Tape..................................... H2
TFE Fluorocarbon Lubricant H2

Chemical Products Company
Ems 606.. A4
Ems 650 Neutral Fragrance D1
Ems 680.. A4

Chemical Products Company
Aqua Solve.................................... A1
Conquest....................................... A1

Chemical Products Company, Inc.
Chem 655 .... P1

Chemical Products Corporation
Sodium Metasilicate,
  Pentahydrate.............................. A1

Chemical Products Inc.
Super Dix-O-Solv........................ A1

Chemical Products Mfg. Corp.
Chempro Clean............................ A2

Chemical Research & Development Co.
Gold-End Fly Bait F2

Chemical Research Corp.
All Purpose Cleaner.................... A1
Jewel.. A1
Lemonee.. A4
Liquid Super "C" A4
Neutral Cleaner A4
Pine-O.. C1
Pink Neutral Cleaner.................. A4
Super Z.. A4

Chemical Research Products
Blaze-Off....................................... A4
Cope FG C1
CRP Drione.................................. F2
Golden Concentrate X................. A1
Hi-Purity Contact Cleaner K2
Sani Suds Detergent.................... A1
Star Liquid Hand Soap E1
Super Dri P1

Chemical Resources Inc.
CR-44 Concentrate...................... A1

Chemical Sales & Service Co.
Detergent No. 23......................... B1
Detergent No. 980....................... A3
Smokehouse Tree Cleaner No.
  751...A2
Tripe Cleaner N3

Chemical Service Division of American Chemmate Corp.
Aerocide Odor Control
  Granules...................................... C1
Amplifex C1
ATA 105 Space & Contact
  Insecticide.................................. F1
ATA 107 Penetrating Release
  Agent.. H2
ATA 111 Air Refreshant C1
ATA 112 Air Conditioning
  Spray.. C1
ATA 113 Air Restorer................. C1
ATA 114 Air Reconditioner C1
ATA 115 Utility Foam Cleaner. C1
ATA 117 Stainless Steel Polish.. A7
ATA 118 Silicone Spray............. H1
ATA 119 Foam Oven Cleaner... A8
ATA 121 Moisture Barrier......... H2
ATA 122 Tar Solvent................. K1
ATA 127 Safety Solvent K2
ATA 129 Heavy Duty Foaming
  Degreaser C1
ATA 130 Mint Air Refresher ... C1
Big Power A1
Bionite D2,Q3
Bol-Trol C2
Break-Thru A1
Butylan A4
Chlortrol B-74.............................. C1
Chlortrol Plus C1
Cocinate....................................... A4
Creamylin E1
CS Concentrate............................ A4
CS Hand Soap E1
CS Re-Move N1
CS-100 .... F1
CS-102 .... F1
CS-42B Chlortrol B..................... C1
CS-44 .... L1
C3-58A A7
D-Lease H1
D-Lime A3
Deo-Drane C2
DeoVent C1
Dermene E1
Di-Mercid.................................... L1
Di-Penval C1
Doil K1
Enzymat L2
Fast Oven Cleaner A8
Fast Oven Cleaner A8
Filter-Con P1
Flo-21 .... L1
Formula 101 Cleaner
  Disinfectant A4
Formula 93 .... C1
Glass Glint C1
Hand Soap Unscented................. E1
Hi-Buty....................................... C1
Industrial Freezer Cleaner #1.... A4
Liquid Aerocide C1
Liquid Swab................................. C2
Malatrol C1
Metronic K2
Need.. C2
Nu-Power A4
On The Spot Cleaner.................. C1
Open L1
Pine Odor Disinfectant C1
Power Clean A1
Prode A2
Prodrane L1
Prudent H2
Return Line Treatment No. 154 G6
Rus-T-Race A3
Safe-T-Signal............................... K1
Safety Solvent & Degreaser K2
Silicone Lubricant H1

Sno-Steam C1
Stainless Steel Clnr. & Polish..... A7
Steel Safe.................................... A7
Suede E1
Syno-Clean A4
Syntrox C2
Telamine C1
Toxene F1
Tri-Dex-An A1
Tri/Ban....................................... D1
Trimonide.................................... A8
Uni-Strip..................................... A4
Zing A1
15% Liquid Hand Soap Scented E1

Chemical Service Company
Boiler Treatment No. 192.......... G6
Cooling Water Treatment No.
  121... G7
Dry Acid No. 41 .... A3
Organic A G6
Oxygen Control Treatment M ... G6
Phosphate Treatment No. 21...... G6
Phosphate Treatment No. 22...... G6
Return Line Treatment MC50 ... G6
Zeolite Treatment No. 181 .... P1

Chemical Services
Antiseptic Liquid Hand Soap.... E1

Chemical Solvent Company
Coil-X-Cleaner............................. A1
CSCO B2000............................... G6
Distilex P1

Chemical Specialties
Step One...................................... A1
Step Two..................................... A1

Chemical Specialties Company
Break-Away A4
Pine-O C1
Pink Neutral Cleaner A4

Chemical Specialties Corporation
Yellow Label 444 Insecticide F1

Chemical Specialties Inc.
Kemsuds Klene C2
Kemsuds Solu-Base Industrial ... C2
Phlo pH-Balanced Liquid Skin
  Cleanser E1
Phlo Antimicrobial Lotion Skin
  Cleanser E1

Chemical Specialties, Inc.
Blast Liquid Drain Opener........ L1
Brite-White Bowl Cleaner.......... C2
Chain Lube Jell (Soft)............... H2
Chem Spec Food Plant Insect
  Bomb... F1
Chem Spec Hgh Grd Space
  Spry Insctd F1
Chem-I-Zyme Enzyme Sewer
  Digester L2
Coal Tar Disinfectant................. C1
Coconut Oil Hand Soap............. E1
F-22 General Cleaning Powder . A1
General Purpose Detergent A1
Germicidal Detergent D1
Hvy Dty Wax Stripper &
  Grease Rmvr............................. A4
Kwik Heavy Duty Detergent A1
O-Do-Nox Sewer Slvnt &
  Neutralizer................................ C1
Phlo Liquid Lotion Soap........... E1
Phlo Waterless Hand Cleaner ... E4
Pine Odor Disinfectant C1
Scalder & Blancher Compound . A2

70

IS-33............................F2
IS-66............................F1
IS-77 Insecticide F2
K&L Kleaner A3
Kleener/Sanitizer 60 ....D1
KS-30 ....A4
KS-32 ....A1
L.D.C. B1
LC-2085 ....H2
LC-2087 ....H2
MB-213 ....A3
MB-334 ....A3
MB-337 ....A3
P & P Detergent A2
Pot & Pan Detergent...............A1
Quatri-Kleen....................D1
QS-250 ....D1
QS-260 ....D2,E3
RS 1115 ....A2
RS 1115L A2
S-1...............................A4
San-Syn A4
Skeen A4
Slip-It...........................H2
Sta-Put Bowl Cleaner C2
Swav-It Detergent Bowl
  Cleaner.........................C2
Synpy-Diazinon Emulsifiable
  Concentrate....................F2
SC-425 ....A1
SC-425T A1
SW-289 ....A4
SW-294 ....A2
SW-589 ....A1
SW-83 ....A2
SW-90 ....A2
U-Neek A1
UN-181 ....A1
UN-188 ....A1
XEC-18............................C1
XEC-27............................A4
XEG-25 ....C1
1165 Low Suds A1

Chemical Technologies, Inc.

AG-Kleen.........................A1
BC-R.............................A1
Caress...........................A1
Cut..............................A1
CT-10 Super Base Concentrate..A4
CT-2 Steam Cleaner...............A1
CT-20............................A4
CT-4.............................A1
CT-40 Pot & Pan Cleaner..........A1
CT-5 Hand Soap...................E1
De-Carb..........................A1
Drudge A1
Formula 9 Hand Cleaner A1
Gre Solv.........................A1
Grease Off.......................A1
H + Liquid Drain Opener L1
Hy-Foam A1
Lime Gone C1
MSR..............................A3
Ocean Power A4
S.O.C. A1
Smoke-Off........................A8
Super-L Dish Cleaner A1
Task Master......................A4
Tempest..........................A1
Velvet Touch.....................B1
X-Degreaser A1

Chemical Testing Corporation

Compound B BS 173................G6
Compound B BS 333................G6
Compound B BS 450................G6
Kemcatalyst VS-25................G6
Kemcolloid.......................G6
Kemcolloid-S G6

Kemcor-A G6
Kemcor-S G6
Kemcor-SX G6
Kemfos G6
Kemfos-SD........................G6
Kemosect.........................F1
Kemtest L4K G6
Kemtest L6K G6
Kemtest PR 360...................G6
Kemtest PS 900 ....G6
Kleen-Flo L1
Oxite-L G6
Oxite-M G6

Chemical Treatment Company

BL-110 Boiler Water Treatment G6
BL-1353 Boiler Water
  Treatment G6
BL-1545 Steam Line Treatment. G6
BL-1546 Steam Line Treatment. G6
BL-1547 Steam Line Treatment. G6
BL-156A..........................G6
BL-160 Boiler Water Treatment G6
BL-171 Boiler Water Treatment G6
BL-174 Boiler Water Treatment G6
BL-176 Boiler Water Treatment G6
BL-8631 Boiler Water
  Treatment G6
BL-8641 Boiler Water
  Treatment G6
BL-8671 Boiler Water
  Treatment G6
BL-8681 Boiler Water
  Treatment G6
BL-8691 Boiler Water
  Treatment G6
Chemical Treat. BL-128 Clsd
  Sys Treat G7
Chemical Treat. BL-8621 Boiler
  Wtr Treat G6
Chemical Treat. CL-435 Cool.
  Wtr. Treat. G7
Chemical Treatment B-110........G6
Chemical Treatment B-121........G6
Chemical Treatment B-141........G6
Chemical Treatment B-142........G6
Chemical Treatment B-144........G6
Chemical Treatment BL-130 ......G6
Chemical Treatment BL-131 ......G6
Chemical Treatment BL-133 ......G6
Chemical Treatment BL-134 ......G6
Chemical Treatment BL-135 ......G6
Chemical Treatment BL-136 ......G6
Chemical Treatment BL-137 ......G6
Chemical Treatment BL-139 ......G6
Chemical Treatment BL-151 ......G7
Chemical Treatment BL-152 ......G7
Chemical Treatment BL-153 ......G6
Chemical Treatment BL-1540 .....G6
Chemical Treatment BL-1541 ....G7
Chemical Treatment BL-1542 .....G6
Chemical Treatment BL-1543 ....G6
Chemical Treatment BL-1544 .....G7
Chemical Treatment BL-156 ......G6
Chemical Treatment BL-161 ......G6
Chemical Treatment BL-162 ......G6
Chemical Treatment BL-163 ......G6
Chemical Treatment BL-166 ......G6
Chemical Treatment BL-170 ......G6
Chemical Treatment BL-170C .. G6
Chemical Treatment BL-172 ......G6
Chemical Treatment BL-175 ......G7
Chemical Treatment BL-184 ......G7
Chemical Treatment BL-197 ......G6
Chemical Treatment C-218........G7
Chemical Treatment CL-135 ......G7
Chemical Treatment CL-153 ......G6
Chemical Treatment CL-1532 .. G6
Chemical Treatment CL-2152 .....G7
Chemical Treatment CL-216......G7

Chemical Treatment CL-402 ...... G7
Chemical Treatment CL-406 ...... G7
Chemical Treatment CL-407 ...... G7
Chemical Treatment CL-408 ...... G7
Chemical Treatment CL-410 ...... G7
Chemical Treatment CT-50 .... G2
Chemical Treatment CT-501 ...... N1
Chemical Treatment CT-55 .... G2
Chemical Treatment CT-57 .... G2
Chemical Treatment CT-708 ...... G2
Chemtreat BL-1356 Boiler
   Water Treatment G6
Chemtreat BL-8622 Boiler
   Water Treatment G6
CL-1406 Cooling Water
   Treatment G7
CL-417 Cooling Water
   Treatment G7

Chemical Waste Removal, Inc.
Liquid Fire Drain Opener L1

Chemical Water Treating Engineers, Incorporated
Kem Boiler Water Treatment
   CP G6
Kem Boiler Water Treatment
   MP G6
Kem Boiler Water Treatment
   PCP G6

Chemical-Ways Corporation
Acid Clean 6A A3
All-Brite M A3
Bottle Bright #70 .... A2
Can Wash #10 .... A1
Champ 1 .... A4
Clean #110 .... A2
Clean #144 .... A2
Clean #345 .... A3
Clean #345-M A3
Clean #440 .... A1
Clean #445 .... A1
Clean #60-L-2 .... A1
Clean #74 .... A2
Clean #747 .... A2
Clean #85 .... A1
Clean #850 .... A1
Clean #898 .... A1
Clean #90 .... A2
Clean Bright #40 .... A3
Clean Bright #50 .... A3
Clean Bright #60 .... A3
Clean C-500 .... A2
Clean NC-Plus A1
Clean PCF A2
Clean PCF-2 .... A2
Clean Solv 10 .... A4
Conc WW C1
Copper Bright A3,A6
CWC-1000 Q A4
CWC-119 .... A2
CWC-1620 .... A2
CWC-2000 P A4
CWC-246 .... A3
CWC-3000 P A4
CWC-346 .... A3
CWC-421 .... A2
CWC-46A A1
CWC-5000 Q A2
CWC-680 .... A1
CWC-709 .... A1
CWC-765 P. F. A2
CWC-826 CFQ A2
CWC-891-C A2
CWC-946A A1
CWC-950 .... A1
Drainex L1
Egg-Wash D Q1

Egg-Wash L.F. Q1
Gen Clean #404 .... A4
Gen Clean G A1
GBP 1000 .... A1
Hand Soap #46 .... E1
Kettle Clean D-1 .... A2
Kitchenaid Oprtng Dsbw. Disp.
   Fld #104 .... A1
Liqui Clean 20 .... A1
Lube 65 .... H2
Pik-Slik N1
PC-600 .... A1
PC-601 .... A1
PCF-3 .... A2
Quaternary #1 .... D1
Release Agent MH A1
Safeclean A1
Safety Solvent 111 .... K1
San Clean #10 .... A1
San Clean #20 .... A1
San Clean #2000 .... A1
San Clean A A1
San Clean N A1
Scald-47 .... N2
Spray Clean #2 .... A1
Spray Foam A1
Steam L-276 .... A1
Strip R-14 .... A2
Super Derust A2
SP-400 .... A2
SS #1000 .... A1
Tempo A4
Tripe Clean N3
10-Terg A3

Chemicals Unlimited, Inc.
Speed Sol Extra A4

Chemidyne Corporation
CD-10 .... A2
CD-100 .... A2
CD-115 .... A2
CD-115 SH A2
CD-130 .... A2
CD-150 .... A2
CD-150 SH A2
CD-20 .... A2
CD-200 General Cleaner A1,Q1
CD-200NP A1,Q1
CD-210 Gen. Clnr. Low Foam .. A1
CD-210 General Cleaner A1,Q1
CD-210NP A1,Q1
CD-212 .... A1
CD-222 .... A1
CD-230 .... A1
CD-250 Liquid General Cleaner A1,B1
CD-260 .... A1
CD-262 .... A1
CD-266 .... A1
CD-280 Liquid Hand Cleaner E1
CD-281 Liquid Hand Clnr.
   Concntrt E1
CD-283 Liquid Hand Cleaner E1
CD-284 .... E1
CD-285 .... E1
CD-286 .... E2
CD-287 .... E2
CD-288 .... E2
CD-300 .... A1,Q1
CD-300NP A2,Q1
CD-310 .... A1,Q1
CD-320 .... A1,Q1
CD-322 Egg Wash Q1
CD-330 .... A1
CD-332 .... A1
CD-333 .... A1
CD-340 .... A1
CD-344 .... A1
CD-345 .... A1

CD-345NP A1
CD-349 Gambrel & Trolley
   Cleaner A1
CD-350 .... A1
CD-352 .... A1
CD-380 Shroud Wash B1
CD-400 .... A3,B1
CD-400NP A3
CD-401 Liquid Acid Cleaner A3
CD-401NP A3
CD-402 .... A3
CD-403 Liquid Acid Cleaner A3
CD-404 .... A3
CD-410 Cncntrtd Acid Cleaner . A3
CD-410NP A3
CD-420 Pwdrd Acid Cleaner A3
CD-430 .... A3
CD-45 Floor Cleaner A4
CD-450 Liquid Acid Cleaner A3
CD-455 .... A3
CD-46 Floor Cleaner A4
CD-484 .... A3
CD-485 .... A3
CD-500 .... A2
CD-500NP A2
CD-505 .... A2
CD-506 .... A2
CD-510 .... A2
CD-510NP A2
CD-520 .... A2
CD-522 .... A2
CD-530 .... A2
CD-530NP A2
CD-532 Heavy Duty Cleaner..... A2
CD-533 .... A2
CD-540 .... A2
CD-543 .... A2
CD-543C A2
CD-543NP A2
CD-544 .... A2
CD-545 .... A2
CD-550 .... A2
CD-550NP A2
CD-551 Heavy Duty Cleaner..... A2
CD-552 Heavy Duty Cleaner..... A2
CD-553 .... A2
CD-554 .... A2
CD-555 .... A2
CD-559 Smoketree Cleaner A1
CD-560 .... A1,Q1
CD-561 .... A1
CD-562 .... A1,Q1
CD-563 .... A1
CD-564 Liquid Heavy Duty
   Cleaner A2
CD-565 .... L1
CD-566 .... L1
CD-567 .... A1
CD-568 .... A2
CD-569 .... A1
CD-570 .... A2
CD-580 .... A2
CD-590 .... H1
CD-592 .... H1
CD-593 .... H1
CD-594 Trolley Oil H1
CD-600 .... D2,G4,E3
CD-610 .... D1
CD-620 .... Q6,E3,D2
CD-630 .... D2,E3
CD-640 .... D2
CD-650 .... D2
CD-701 Powder Booster............ A1
CD-705 Detergent Additive A1
CD-706 Detergent Additive A1
CD-707 .... A1
CD-710 .... A1,N1,Q5
CD-711 Defoamer P1
CD-712 .... P1
CD-720 .... A1,Q1

Statix Anti Static Detergent A1
Sulfam Sulfamic Acid Cleaner
& Descaler.................................. A3
Suresect Extra.............................. F2
Suresect MST Concentrate F1
Suresect MX Concentrate........... F1
Suresect P-3 .... F1
Suresect P-5 .... F2
Suresect Prime F1
Suresect XV F1
Suresect 25 .... F1
Syntro-200................................... A2
Trax-XD Acid Cleaner and De-
Ruster.. A3
Triumph-NT Chlorinated Alkali
Cleaner.. A2
Truterge Conc. Geheral Purp.
Clnr. A1
V. F. D-Soiler Mild Detergent .. N4
Virex Q6
Yello-X Dry Deoxidizing
Bleach B1
Zero-Bac...................................... E2

Chemola Corporation
Desco 50 Pure.............................. H1
NPQD Concentrate...................... A1
Tap.. A1
745 Hand Cleaner E4
745 Washing Creme E1

Chemopharm
Blue Laundry Concentrate B1
BC-42 .... A1
Ceramo C2
Charm... A1
Charm Extra Mild A1
Charm Super A1
Concrete Cleaner #103 .... A1
Extra LC..................................... B1
EZ Steam A1
Floor Scrub................................. A4
Foaming Cleaner A4
GP Chlorinated Cleaner A1
Hand Cleaner & Conditioner..... E1
Heavy Duty Chlorinated
Cleaner.. A1
MR-30.. A3
MR-40.. A3
Proud III With Amonia.............. C1
Proud Super A4
Proud Triple A4
Ruf Nek's Pal.............................. A1
Scale Remover............................. A3
Scale Remover XX A3
Steam Cleaner HD A1

Chempace Corporation
All-Brite....................................... B1
Blue Giant A1
Deo-Gran C1
Foamy.. C2
Force L2
Gleem C1
Gleem C1
H-D 321.. A4
Heet... L1
Hercules...................................... C1
Lic-It... C1
Metacon....................................... A3
OAD... A3
Pen-Lube..................................... H2
Pen-Lube..................................... H2
Pink Panther A1
Powdered Bleach B1
Power Foam C1
Power Kleen A1
Red Fire....................................... L1
Sewer-Cide................................... L1
Solo San 510 .... D1

Super Uni-Fog F2
Super Uni-Terg............................ A4
Triple Threat A1
Uni Wash (Powder) A1
Uni-Chlor.................................... C1
Uni-Quat 14 .... D1
Uni-Terg...................................... A4
Uni-Zyme L2
X-O Therm C1

Chempar Chemical Company, Inc.
Rozol Blockette F3
Rozol Blue Tracking Powder .... F3.
Rozol Canary Seed Mouse Bait . F3
Rozol Paraffinized Pellets.......... F3
Rozol Rat & Mouse Killer F3
Rozol Rat & Mouse Killer(Plstc
Thrw Bgs) F3
Rozol Ready-To-Use Rat &
Mouse Bait F3

Chemplast, Incorporated
Fluoro Glide Chemically Pure
Grd. H2
Fluoro Glide Film Bonding
Grade... H2

Chemrite Corporation
CR 355 Sewer Guard Sewer
Solvent.. L1
CR 727 Steel Rite A7
CR 800 Oven Cleaner A8
CR-103 Glucorite A1
CR-105 Powerite 90 .... A4,A8
CR-106 Powerite A2
CR-107 Cleanrite A1
CR-110 Megasol Cleaner A4
CR-111 Creterite C1
CR-117 Allrite A4
CR-118 Solvrite A1
CR-119 Steamrite H.D................. A1
CR-120 Blastrite A2
CR-121 Super Solvrite A1
CR-128 Treatrite........................... A3
CR-133 Preprite............................ A3
CR-161 Aciderite Powder A3
CR-162 Sulfamorite A3
CR-164 D-Rustrite A2
CR-168 Drain Rite L2
CR-171 Dedrite.............................. F2
CR-172 Vaporite with Cherry ... F2
CR-173 Dedrite.............................. F2
CR-174 Vaporite............................ F1
CR-180 Econorite.......................... A1
CR-182 Maskrite C1
CR-300 Chlorite............................ A1
CR-301 Chlor-Brite A1
CR-303 Panrite Spraywash.......... A1
CR-303-E Panrite E A1
CR-304 Econo Foam A1
CR-305 Foamrite A1
CR-307 H.D. Panrite Spray......... A1
CR-308 Power Foam A1
CR-309 Breakrite A4
CR-313 Iodorite............................. D2
CR-316 Sanirite D1
CR-317 Handrite............................ E1
CR-320 Jelrite A1
CR-321... A1
CR-322... A6
CR-323... A3
CR-324... A2
CR-325... A2
CR-326... A2
CR-327... A2
CR-328... A1
CR-329... A4
CR-330... A3
CR-335 Chlorinated Equipment
Cleaner.. A1

CR-336 Iodobac............................ D2
CR-337 Iodosan LF Q6
CR-341 Luberite H2
CR-342 Megarite A1
CR-344 Liquid Luberite.............. H2
CR-346 Econopower..................... A2
CR-350 Bio-Thanarite D2,Q3,E3
CR-352 Ciprite.............................. A2
CR-353 Alkarite (Liquid)............ A2
CR-371 Egg Wash......................... Q1
CR-372 Liquid Egg Wash Q1
CR-374 Egg Destainer Q2
CR-376 Antifoam-S Q5
CR-377 Antifoam.......................... Q5
CR-390 Vapo-Syn F1
CR-400 Neutra Cleanrite............. A1
CR-403 Vacrite............................... A4
CR-409 Unirite.............................. A1
CR-410 Extrarite K1
CR-411 Alumasafe......................... A1
CR-500 Boilerite-SF G6
CR-501 Boilerite H F G6
CR-503-F Streamlinerite-F G6
CR-504 Towerite............................ G7
CR-505 Algaerite........................... G7
CR-505 Plus Algaerite................. G7
CR-507 D-Scalerite A3
CR-509 Algaerite B....................... G7
CR-600................................. H2
CR-600-EP H2
CR-602.................................. H2
CR-603-EP H2
CR-604................................. H2
CR-605.................................. H2
CR-606................................. H1
CR-607-EP H1
CR-711 Neutrite A4
CR-718 Super Bowlrite................ C2
CR-723 Electrorite K2
CR-807 Silicone Spray................. H1
CR-808 Fluorocarbon
Lubricant H2
CR-811.. E4
CR-824 Food Equipment
Lubricant H1
CR-901 Chemrite XL.................... A1
CR-931 Liporite............................ A8
CR-933 Clingrite........................... A8
CR-973 Pot-Rite Unscented A1

Chemrock Chemical Corporation
#201 Action Concentrated
Liquid Detrgnt A1,E1
"Heavy Duty" A1
Blue Clean A1
Chemrock Smoke House
Cleaner.. A2

Chemscope Corporation
Aerosol Insecticide 150.............. F1
Aerosol Insecticide 300.............. F1
Aerosol Insecticide 600.............. F1
Anti Seize.................................... H2
Apple Blossom Room
Deodorant (Liquid) C1
Aqua Not Waterless Hand
Cleaner.. E4
Blast Off...................................... A1
Bouquet Room Deodorant
(Liquid)....................................... C1
Chain & Gear Lubricant............ H2
Chem Solv.................................... A4
Clean-N Lube H2
Diazinon 500 Insecticide............ F2
Diazinon 500 Residual Spray ... F2
Dri Lube...................................... H2
Electro Spray............................... K1
Emulsion Degreaser K1
Evergreen Room Deodorant
(Liquid)....................................... C1

75

Fanny Foster's Emuls. Bowl
Cln & Disinf C2

Cheney Lime & Cement Company
Cheney's Hydrated Lime N3

Cherokee Chemical Co., Inc.
Alk-Rus A2
AAA 100-L Q1
AAA-Shell Egg Detergent Q1
AAA-203 .... Q1
AAA-302 Destiner Q2
AAA-700 E.W. Defoamer Q5
Bonanza HW A2
Chain and Conveyor Lube H2
Cherokee 1043 .... A1
Chlorb-San A1
Con-Syn A1
Dock Cleaner A1
Egg-Kleen Q1
Egg-Kleen Plus Q1
Hasa Anti-Foam Q5
Hasa Chlor A1
Hasa Chloro-Brite A1
Hasa Smokehouse Cleaner A3
Hasa 101 .... A2
Hasa 2418 .... L1
Hasa 2910 .... A2
Hasa 2913 .... A3
Hasa 2929 .... L1
Hasa 387 .... A1
Hasa 387 LD N1
Hasa 391 .... Q1
Hasa 429 .... A1
Hasa 485 .... A2
Hasa 64 .... A1
Hasaquat 200 .... D1
Hydro Pan A1
Hydro-Pan Chlorinated A1
Keechem MR-2 .... A3
Keechem 16 .... A1
Keechem 19 .... A8
Keechem 20 .... A2
Keechem 21 .... A1
Keechem 25 .... A3
Keechem 301 .... A1
Keechem 301-A A4
Keechem 40 .... A2
Keechem 45 .... A3
Keechem 51 .... A1
Keechem 56 .... A2
Keechem 62 .... A1
Keechem 80 .... N3
Keechem 80 R N3
Keechem 80 W N3
Keechem 99 .... A2
Liqui-Shine With Chlorine A1
Liquid Supreme A1
LC-U B1
MR-1 .... A3
Packers General A1
Peel Add N4
Power Chem A1
Power Chem-S A4
Power Kleen C1
Quik Wipe A4
Steam Con A2
Steam Con Liquid A1
Steam Elite A1
Synterg A4
Target A1
Whiz Bang A4
1021 General A1

Chesterton, A. W. Company
Anti-Seize Compound (Aerosol) H2
Anti-Seize Compound (Bulk) H2
Belt-Flo P1
Chain Drive Pin & Bushing
Lub. Aerosol H2

Chain Drive Pin & Bushing
Lub. Bulk H2
Detergent Lubricating Oil
(Aerosol) H2
Electric Motor Cleaner K2
Electro Contact Cleaner K2
Gear Reducer Oil H2
Goldend Paste P1
Industrial & Marine Solvent A1
Nickel Anti-Seize Compound,
Bulk&Aer. H2
Penetrant and Lubricant H1
Phosphate-Free Cleaner #360 .... A4
Premium Detergent Lubricating
Oil (Bulk) H2
Pump, Valve and Machinery
Cleaner K2
Rustsolvo H2
Silicone Lubricant H1
Spraflex Aerosol H2
Spraflex Bulk H2
Spragrip Aerosol P1
Sprasolvo H2
Style 1722 Multi-Service
Packing P1
Style 322 White-Lon Multi-
Ser.Packing P1
White Grease for Food Mach.
w/Teflon H1

Chevron Chemical Company
Ortho Dibrom Fly & Mosquito
Spray F2
Ortho Fly Killer D F2

Chevron Oil Co. See Chevron U.S.A. Inc.

Chevron U.S.A. Inc.
Chevron ATF Special H2
Chevron Cylinder Oil 460X H2
Chevron Cylinder Oil 680X H2
Chevron Delo 100 Motor Oil
SAE 10W H2
Chevron Delo 100 Motor Oil
SAE 20W-20 .... H2
Chevron Delo 100 Motor Oil
SAE 30 .... H2
Chevron Delo 100 Motor Oil
SAE 40 .... H2
Chevron Delo 100 Motor Oil
SAE 50 .... H2
Chevron Delo 200 Motor Oil
SAE 10W H2
Chevron Delo 200 Motor Oil
SAE 20W-20 .... H2
Chevron Delo 200 Motor Oil
SAE 30 .... H2
Chevron Delo 200 Motor Oil
SAE 40 .... H2
Chevron Delo 200 Motor Oil
SAE 50 .... H2
Chevron Delo 400 Motor Oil
SAE 10W H2
Chevron Delo 400 Motor Oil
SAE 20W-20 .... H2
Chevron Delo 400 Motor Oil
SAE 30 .... H2
Chevron Delo 400 Motor Oil
SAE 40 .... H2
Chevron EP Hydraulic Oil MV H2
Chevron EP Hydraulic Oil 32 .. H2
Chevron EP Hydraulic Oil 46 .. H2
Chevron EP Hydraulic Oil 68 .. H2
Chevron EP Industrial Oil
100X H2
Chevron EP Industrial Oil
150X H2
Chevron EP Industrial Oil
220X H2

Chevron EP Industrial Oil 46X . H2
Chevron EP Machine Oil 10 ..... H2
Chevron EP Machine Oil 100 ... H2
Chevron EP Machine Oil 150 ... H2
Chevron EP Machine Oil 22 ..... H2
Chevron EP Machine Oil 220 ... H2
Chevron EP Machine Oil 32 ..... H2
Chevron EP Machine Oil 320 ... H2
Chevron EP Machine Oil 46 ..... H2
Chevron EP Machine Oil 68 ..... H2
Chevron FM Lubricating Oil
100X H1
Chevron FM Lubricating Oil
105X H1
Chevron FM Lubricating Oil
32X H1
Chevron FM Lubricating Oil
460X H1
Chevron Heavy Duty Grease-1 . H2
Chevron Heavy Duty Grease-2 . H2
Chevron Industrial Grease
Light H2
Chevron Industrial Grease
Medium H2
Chevron Lily White Petrolatum H1
Chevron Marine Oil 150X H2
Chevron Marine Oil 220X H2
Chevron Motor Oil SAE 10W .. H2
Chevron Multi-Motive Grease 1 H2
Chevron Multi-Motive Grease 2 H2
Chevron NL Gear Compound
100 .... H2
Chevron NL Gear Compound
1000 .... H2
Chevron NL Gear Compound
150 .... H2
Chevron NL Gear Compound
1500 .... H2
Chevron NL Gear Compound
220 .... H2
Chevron NL Gear Compound
2200 .... H2
Chevron NL Gear Compound
320 .... H2
Chevron NL Gear Compound
460 .... H2
Chevron NL Gear Compound
68 .... H2
Chevron NL Gear Compound
680 .... H2
Chevron OC Turbine Oil 100 ... H2
Chevron OC Turbine Oil 150 ... H2
Chevron OC Turbine Oil 220 ... H2
Chevron OC Turbine Oil 32 ..... H2
Chevron OC Turbine Oil 46 ..... H2
Chevron OC Turbine Oil 68 ..... H2
Chevron Pinion Grease MS H2
Chevron Poly FM Grease 0 .... H1
Chevron Poly FM Grease 1 .... H1
Chevron Poly FM Grease 2 .... H1
Chevron Polyurea EP Grease 0. H2
Chevron Polyurea EP Grease 1. H2
Chevron Polyurea EP Grease 2. H2
Chevron Refrigeration Oil 32 .... H2
Chevron Refrigeration Oil 68 .... H2
Chevron Snow White
Petrolatum H1
Chevron Soluble Oil H3
Chevron Special Motor Oil
SAE 20W-20 .... H2
Chevron Special Motor Oil 30 .. H2
Chevron Special Motor Oil 40 .. H2
Chevron SRI Grease 2 .... H2
Chevron Universal Gear Lub.
SAE 80W-90 .... H2
Chevron Universal Gear Lub.
SAE 85W-140 .... H2
Chevron White Mineral Oil No.
7 .... H1
Standard Soluble Oil H3

Chicago Research Laboratories, Inc.
C-94..A1
Centennial II..................................D1

Chicago Sanitary Products Co.
Krag Cleaner #100......................A1
Krystokleen...................................A1
Thorokleen....................................L1
Velvene Castile.............................E1

Chloral Chemical Corporation
Formula BF Super B1
Soft-D-Dry.....................................B2
Super Solite Dry Bleach............B1
Super Spec. Sour W....................B1
TX 100..B1

Chocola Cleaning Materials, Inc.
Sprayaway......................................A4

Christner Industries
C-1000 Dishwashing Detergent.A1
C-1040 Toilet Bowl Cleaner.......C2
C-1060 Hand Soap.......................E1
C-1120 Concentrated Cleaner A4
C-503 De-Limer............................A3
Dasco C-511 C-A Concentrate ..A1
Dasco DG-82 Degreaser K1
Dasco Super-Solv.........................K1
Dasco Window Shine Glass
    Cleaner.....................................C1
Super Bowl Toilet Bowl
    Cleaner.....................................C2

Christy Company, Incorporated
Centraz Egg Wash Q1
Centraz Heavy-Duty Floor
    Cleaner.....................................A4
Chlor-Concentrate.......................D2,Q4
Chlorinated Tank & Pipe Line
    Clnr...A1
Heavy Duty Acid Detergent
    Cleaner.....................................A3
Super Cleaner A1

Chromate Industrial Corporation
Chromate C200 Sheer Slik H1

Chronley, J. F. Manufacturing Co.
Heavy Duty Cleaner....................A1
Quat Germicidal Cleaner............D1 .

Church & Dwight Co., Inc.
Arm & Hammer Baking
    Soda(Sodium Bicarb.).........A1,N1,L1,G5
Church & Dwight Sodium
    Bicarbonate USP A1,N1,L1,G1
Con-Sal A1,N1,G6,N2,
                                                                N3,G5
Sodium Carbonate
    Monohydrate U.S.P................A1,N1,G6,N2,
                                                                N3

Chute Chemical Company
Chlor-Ten.......................................A1
Chute Sanitizer-Cleaner D2
Clean-Shell....................................Q1
Clean-Shell Type I........................Q1
Clean-Shell Type II.......................Q1
CCC-20 Caustic Cleaner A4
CCC-8-Liquid Caustic Cleaner ..A4
E-Z Chlor.......................................D1
Gly-Phos Acid Dairy Cleaner A3
Institutional All-Purpose
    Cleaner.....................................A4
Institutional Oven Grill & Fat
    Fry. Clnr..................................A8
Iodosan Medicated Hand Wash .E1

Iron Stain Remover.....................A3
Light Duty Caustic Cleaner A....A1
Liquid Food Plant Cleaner
    Type A A1
Liquid Food Plant Cleaner
    Type B A1
Liquid Metal Brightener A.........A3
SPF Chlorinated Dairy Cleaner.D1
SPF Dairy Pipeline And Bulk
    Tank Clnr.................................A1
SPF Heavy Duty Caustic
    Cleaner.....................................A1
SPF Heavy Duty Caustic
    Cleaner-80................................A2
SPF Heavy Duty Cleaner-Type
    I..A1
SPF Heavy Duty Cleaner-Type
    II...A1
SPF Liquid Hand Soap...............E1
SPF-HD Caustic Cleaner-40 ......A2

**Ciba-Geigy Corporation/ Agricultural
Division**
Diazinon 4E ....F2 -
Diazinon 4S...................................F2
Diazinon 50W F2

Cincare Chemicals Incorporated
CC-10...G6
CC-12...G6
CC-14...G6
CC-18...G6
CC-20...G6
CC-22...G6
CC-28...G6
CC-32...G6
CC-34...G6
CC-36...G6
CC-40...G6
CC-42...G6
CC-44...G6
CC-80...G6

Cincinnati Brush Manufacturing Co.
Hi-Performance Cleaner &
    Degreaser..................................A4
Mohawk Floor Cleaner A4
Mohawk Liquid Hand Soap
    15%...E1
Mohawk Liquid Hand Soap
    20%...E1
Mohawk Liquid Hand Soap
    40%...E1

Circle Research Laboratories
Dráin Safe......................................L2
Grees-Out.......................................L2

Citation Equipment & Chemicals
Blast Off...A2
Steam Off A1

Citation Manufacturing Co., Inc.
Brush Magic...................................A1
Coil Doctor....................................A3
Dyna-Kleen....................................A1
Dyna-Steam Hvy Dty Powder ..A1
Dyna-Steam Medium Duty
    Powder.....................................A1
Dyna-Wash....................................A1
Dyna-White....................................A1
Master Wash..................................A1
Steam Heavy Duty Liquid A1
Super Power Plus.........................A1

City Janitor Supply Company
Industrial Degreaser/Cleaner A4

City Supply and Paper Company, Inc.
Cipco Liquid Hand Soap............E1

City Wide Chemical Co., Inc.
All Pine Surface Cleaner C1
All Purpose General Degreaser.A4
Automatic Scrubber
    Concentrate..............................A4
Deodorizer And Sanitizer...........C1
Establishment................................A4
Gelled Degreaser..........................K1
Granulated Concrete Cleaner....C1
Granulated Powdered Hand
    Cleaner.....................................C2
Hand-Aid.......................................E1
Klean All..A1
Lemon Scented Germicidal
    Detergent..................................C1
Lotionized Hand Dishwashing
    Liquid Det.................................A8
No. 641 Germicidal Cleaner.......D1
Penefoam.......................................A1
Power Degreaser...........................A1
Quick Klean A4,A8
Sanirinse.......................................D2
Solv-It..A4
Steam Cleaner...............................A1
Super Quick Klean A4,A8
15% Coconut Oil Soap...............E1
40% Coconut Oil Soap 40%.......E1

Claiborne Products
Fran-O-Lite....................................J1

Claire Manufacturing Co.
Automatic Insect Bomb..............F1
Claire Disinfectant Bathroom
    Cleaner.....................................C2
Disinfectant Spray........................C1
Fast Kill Roach & Ant Killer ...F2
Food Contact Silicone Spray.....H1
Grill and Oven Cleaner A8
Hospital Grade Disinfectant.......C1
Mister Jinx....................................C1
Peni-Lube Penetrating Oil H2
Silicone Spray...............................H1
Solvent Cleaner & De-Greaser ..K1
Spray Lubricant............................H1

Clara's
Institutional Liq Drain Pipe
    Opnr & Mntr.............................L1
Liquid Hand Dish Wash.............A1

Clark Products, Inc.
165 Steel Bright Lemon
    Scented A7

Clarkson Laboratories, Inc., The
Can Cleaner C A1
Circulation Cleaner A2
Clarco Freezer Cleaner................A5
Clarco RCR A3
Clarco Sept D1
Clarco Steamrite HDB A1
Clarco Steamrite MD...................A1
Clarco 1044 ....A1
Clarco 1171 ....A4
Clarco 1171 A................................A1
Clarco 1277 ....A1
Clarco 14..A2
Clarco 260 ....A4
Clarco 310 ....A1
Clarco 310-B A1
Clarco 440 ....A1
Clarco 570 ....A1
Clarco 725-c...................................A1
Clarco 9124 ....A4
Claro 727 ....A1

78

Cline-Buckner, Incorporated
New Purge II F1
Purge F2
Purge Air Sanitizer C1
Purge CB-40 Aerosol Insect
  Killer F1
Purge Instant Fogger F1
Purge III F2
Purge Super Odor Neutralizer ... C1

Cling-Surface Company
Belt Preservative P1
Chain & Cable Lubricant H2
Moly Dry Film Lubricant H2
Moovit Super Penetrating
  Lubricant H2
Rust Remover Liquid A1
Rust Remover Paste A1
Spray Belt Dressing P1
TFE Dry Film Lubricant H2

Cloroben Chemical Corporation
Cloroben PT C1
Greastrol C1

Clorox Company, The
Clorox D2,Q4,C2,N4,E3
Formula 409 All Purpose
  Cleaner C1
Institutional Cleaning &
  Panwash Conc. C1

Clorox Food Service Products Div., The
Clorox Institutional Degreasing
  Cleaner A4,A8
Institutional All Purpose
  Cleaner A4
Institutional Stainless Steel
  Cleaner A7

Clover Chemical Company
Cleaning Compound #18 .... A1
Clovasuds A1

Clover Chemical Corporation
Clover Clean A1
Clover Drain L1
Clover Hand Soap E1

Clyde Chemical Company
Sanitizing Agent D2

Coastal Chemical Company/ Div. Coastal Industries, Inc.
Bi Quat Disinfectant Cleaner D1
Gypsy C1
Hi Brite C1
Instant Degreaser C1
Jab C1
Jab Odorless Floor & Wall
  Cleaner A1
Jab 500 HD C1
Mello Hand Cleaner E1
Sani Jet Germicidal Cleaner D1
Steamex 260 .... A1
Sure Plant Pellets L1
Sure Renovator A4,A8
Sure Steam Cleaner A1
Sure Wax Stripper C1

Coastal Engineering Corporation
Ceco Cide PQ G7
Ceco Sperse G6
Ceco 120 L G6
Ceco 120 S G6
Ceco 131 L G6
Ceco 131 S G7

Ceco 150 SLS G6
Ceco 170 .... G6
Cecotrol-1283 .... G7

Coke Chemical Company
CRW Commercial Cleaner A1

Colberg Chemical Company
Aldran A1
El-Cid A4

Cole Chemical & Supply, Inc.
Flow Prilled Drain Opener L1
JP-30 Insect Spray F1
Liquid Conveyor Lubricant H2
Sani-Power The Reserve Power
  Detergent A4
Silky II Liquid Hand Soap E1
Silky Liquid Hand Soap E1
Stay-Power Residual Insect
  Spray F2
Sunshine Porcelain Cleaner C2
Swat Insect Spray F1
Wipe Out Spray Concentrate F1

Cole Chemical Co.
Heavy Duty Wtr Soluble Saf
  Solvt RX-100 .... A1

Colgate-Palmolive Company
Ajax with Chlorine Bleach A6
Ajax with Oxygen Bleach A6
Ajax Professional Window
  Cleaner C1
All Purpose Liquid Cleaner A4
Arctic Syntex M A1
Arctic Syntex M Beads A1
Arctic Syntex M-Beads (No
  Phosphate) A1
Quik-Solv Spray Cleaner C1
Super Ben Hur A1

Colonial Chemical Corporation
Formula No. 101 G.P.C. Gen
  Purp. Liq Cln A1
Formula No. 102 Cleaner-
  Degreaser A1
Formula No. 104 Gen. Purpose
  Liq. Clnr. A1
Formula No. 112 15% Liquid
  Hand Soap E1
Formula No. 113 20% Liquid
  Hand Soap E1
Formula No. 114 Liq. Hand
  Soap 40% Conc. E1
Formula No. 151 Laundry
  Detergent B1
Formula No. 153 Powdered
  Bleach B1
Formula No. 180 Inhibited
  Steam Clnr. A1
Formula No. 181 Hvy Duty
  Steam Cleaner A1
Formula No. 212 Liquid Acid
  Cleaner A3
Formula No. 226 Drain Cleaner L1
Formula No. 307 Gen. Purpose
  Cleaner A2
Heavy Duty Liquid Steam
  Cleaner A1
Liquid Pressure Wash
  Detergent A1

Colonial Products, Inc.
CP Pes-Ta-Rest Industrial
  Insect Spray F2
CP Pes-Ta-Rest Insecticide
  Concentrate F2

Colonial Refining and Chemical Co.
Kleen Sweep C1

Colonial Research Chemical Corp.
CRC-100 .... A3
CRC-12 Safety Solvent K1
CRC-13 Insecticide Concentrate F1
CRC-18 .... E1
CRC-19 Ice Melting Pellets C1
CRC-300 .... C1
CRC-302 .... G6
CRC-36 .... A1
CRC-37 .... D1
CRC-39 .... A4
CRC-45 .... C1
CRC-49 .... C2
CRC-51 Bowl Cleaner C2
CRC-60 .... L1
CRC-63 Deodorant Granules C1
CRC-64 .... C1
CRC-71 Mud Remover L1
CRC-77 .... A1
CRC-91 Enzymes L2
CRC-96 .... C1
CRC-98 .... C1

Colony Chemicals, Inc.
CD-130 Disinfectant Cleaner D1
CD-133 Pine Odor Disinfectant. C1
CD-134 Lemon Deodorant
  Cleaner C1
CD-136 Sanni-Rinse D2
CH-120 Coconut Oil Hand
  Soap E1
CL-100 Cleaner Degreaser A1
CL-101 All Purpose Cleaner A4
Fogging Insecticide F1
Super Diazinon Spray F2

Colorado Chemical Company
AA A1
AC 300 Acid Cleaner A3
AC-P-350 Acid Powder A3
Chain Lube #1 .... H2
CCC Cleaner A6
Grease-Away A1
GC-104 Cleaner A1
GC-105 Chlorinated Cleaner A1
HD-202 Alkali A2
HD-203 Drain Pipe Opener L1
HD-206 Alkaline Cleaner A2
HD-208 Chelated Alkali A2
HD-209A Chelated Alkaline A2
HDL-271 Chelated Alkali A2
L-100 Laundry Compound B1
Right Hands Concentrated
  Hand Soap E1
Slash Procelain Cleaner C2
Triple C Laundry Carbonated
  Alkali B1
Triple C Laundry Complete
  Syn. Low-Suds B1
Triple C Laundry Liquid
  Builder B1
Triple C Laundry Liquid
  Detergent B1
Triple C Laundry Liquid
  Fabric Softener B2
Triple C Laundry Powder
  Fabric Softener B2
Triple C Professional Drain
  Opener L1

Colt Chemical Company
APC 25 .... A4
H-D Butyl A4
H-D Granules A2
Liqui-Steam LD A1
Perki C1

Rusticide A4
Ultra-Sol A1

Colt Chemical Manufacturing Co. Inc.
Supreme A4,A8
Track & Chain Liquid
Conveyer Lubricant H2

Columbus Janitor Supply Co.
C J Super 10 Degreaser/
Cleaner A1
C J Super 10 Non Combustible
Emulsifier A1
C J Super 10 Steam Cleaner A4
Lemon Hand Soap E1
575 Hi-P Contact Cleaner K2
576 Silicone Spray H1

Com-Pak Chemical Corporation
Chem-I-San A1

Comark, Inc.
Comark Mark VIII Plus Dgrsr
& Stripper A4
Comark Power A4
Comark Power, Neutral
Fragrance D1
Comark-Foam All Cleaner A1
Comark-Freezer & Locker
Cleaner A5
Comark-General Purpose
Cleaner A4
Comark-Low Foam Detergent .. A4
Comark-Phenolic Clnr-Disinf.
Form. 4 .... A4
Comark-Quaternary Cleaner-
Dis-Citrus A4
Comark-Quaternary Cleaner-
Dis-Regular D1
Comark-Quaternary 256 Lemon
Cl/Disinf. A4
Comark-Quaternary 256 Neutral
Cl/Disinf. D1
Comark-Safety Solvent K1
Comark-Steam Cleaner A1
Comark's Utidyne D2

Combined Sales Co.
All Purpose Cleaner A4
Concentrated Floor Cleaner A1
CS Co. Chlorinated Bleach
Cleanser A6
Glycol Air Sanitizer C1
Heavy Duty Glass Cleaner C1
Hospital Disinfectant
Deodorant C1
Industrial Drain Opener L1
Lemon Hand Soap E1
Lotion Hand Soap E1
No Fume Liquid Drain Opener . L1
Pot and Pan Cleaner A1
Residual Insecticide F2
Spray and Wipe Foaming
Cleaner C1

Comet Manufacturing Corporation
C-11 Concentrate A1
Chick-Klean No. 631 .... A4
Com-O-Phene D1
Com-O-Terg D1,Q3
Comet Emulsion Bowl Cleaner . C2
Comet Kreme E1
Comet No. 541 .... A4
Comet No. 88 .... D1,Q3
Comet R-55 .... A4
Comonite A1
Concrete Cleaner A4
CMC All Purpose Cleaner C1
CMC Com-O-Fect Surface
Disinf. Deodrnt C1

CMC Glass Cleaner C1
CMC Metal Polish & Tarnish
Remover H3
CMC Roach and Ant Killer F2
CMC Safety Solvent K1
CMC Silicone Spray H1
CMC Spray De-Icer C1
CMC Super Mechanic H2
CMC Wasp & Hornet Spray F2
Fog-Out F1
Formula 666 .... A2
Kling-On Thixotropic
Concentrate C2
Liquid Hand Soap No. 10 .... E1
Metal Polish & Tarnish
Remover A7
Oven Bright A8
P-55 General Purpose Cleaner ... A1
Pine Oil Disinfectant C1
Pipe Shield P1
Porc-O-Nam T/C Bowl/Tile/
Porcelain Clnr. C2
Safety Solvent K1
SS-44 .... L1
TFE-Lube H2
XX Liquid Steam Cleaning
Compound A2
333 Cleaner A2

Commercial and Industrial Products Co.
Cipco Liquid Hand Soap E1

Commercial Chemical Company
BT 75 .... A1
Grill & Oven Cleaner A4

Commercial Chemical Company
#201 Triple Action Cleaner A1

Commercial Chemical Corporation
Orange Blossom Germicide C2

Commercial Chemical Products, Inc.
CCP Econoclean A4
CCP Heavy Duty Degreaser A4,A8
CCP Heavy Duty Detergent A4
CCP Tripe Cleaner N3

Commercial Cleanser Company
Formula MP-10 Powdered
Hand Soap C2
Formula 101 Cleaner A1
Sani-Guard D2
Sof-Sope Lotionized Hand
Cleaner E1

Commercial Industrial Programs, Inc.
LKO 140 Boiler Compound &
Cleaner G6
LKO 149 Water Tower
Treatment G7
LKO 176 Boiler Water Adjunct G6

Commercial Maintenance Supply
FPA Foamy A1

Commercial Mechanisms, Inc.
Formula 1000 .... A1
Formula 2000 .... A1
Formula 4000 .... A1
Formula 6000 Detergent-
Sanitizer D1

Commercial Supply Company
Terrific A1

Commercial Supply Corporation
Applause E4

Grease Gun H2
Handy E1
Insect Kill F2
Multi-Purpose Space&Contact
Spray F1
Phase I A1
Prudent H2
Syntrox C2
Taskmaster C2

Community World, Inc.
Ease A4
Sparkleen Automatic
Dishwashing Compound A1
Sun-Fresh Laundry Compound . A1,B1

Compact Chemical Corporation
Formula #301 Concentrated
Stone Cl. A1
Formula 1700 Production
Equipment Cln. A1

Con-Chem Specialities/ A Div.of Continental Warehouse of MO,Inc
Big Blast A1
Chemtronic A1
Fast Klean A3
Fogging Spray Concentrate F1
KP A1
L-Tox Spray F1
Liquid Steam Cleaner A2
R.I.S.-15 Residual Insecticide
w/Baygon F2
Remuv-Ox A3

Con/Chem, Incorporated
Cono/Clean 7 .... A3
Stop Rush Dead P1

Conco Chemical International Inc.
Conco All Purpose Cleaner A4

Condor Chemical Company
B-198 .... G6
CA-104 .... G6
CD-119 .... G6
CD-500 .... G6
CD-501 .... G6
CD-503-10 .... G6
CD-503-20 .... G6
CD-503-30 .... G6
CH-101 .... G6
CL-102 .... G6
CP-103 .... G6
CP-106A G6
CP-249 .... G6
CP-260 .... G6
CP-261 .... G6
CP-265 .... G6
CPO-268 .... G6
CS-100 .... G6
CS-267 .... G6
CT-339 .... G7
CT-3401 .... G7

Conklin Company, Incorporated
Bontz A4,A8
Care E1
Crust Buster C2
Dike G7
Gear Glide H2
Luboil H2
Mox A4
Rust Not H2
Rust Not (Aerosol) H2
Rust Off A3
S.D. II. A1,E1
Safe G7
Sanox Cleaner Disinf-Deod-
Fungicide D1

Spring Dew.................................. A1,E1
Spring Dew Concentrate........... A4
Waush B1
Wex.. A4

Conley Chemical & Supply Co.

Concentrate 70 Kitchen
    Detergent.................................. A1
Coronet...................................... A1
Crown All-Purpose Cleaner...... A4
Crystalite Heavy Duty Floor
    Cleaner.................................... A4
Spra-Kleen A4

Conner Chemical Company

Am-O-Cide.................................. D1
Big Brute A1
Conn-O-Fog F1
Hot Stuff Sewer Solvent........... L1
Hydrocide F1
Kleen Koil................................... P1
Ox-Off.. A3
Shazam A1
Tank Guard................................ G7
Whiz Kid A4

Connor Corporation

X-Stream A1

Consan Pacific Inc.

Physan 20 .... D2,Q3

Consolidated Chemex Corporation

C-102-X Consolidated Chemex
    Clnr... A1
Chemex 44X............................... A4
Chemex 47.................................. A3
Chemex 48.................................. A1
Consolidated Chemex Cleaner
    611 .... A1

Consolidated Chemical Co., Inc.

Crystal Concrete Cleaner A4
Emülsion Spray Formula 79....... A1
Liq. Steam Clung Compnd No.
    3457.. A2
Liq. Steam Clnng Compnd No.
    522.. A1

Consolidated Chemical Corp.

Consolidated Kleans It............... A1

Consolidated Chemical Corporation of Florida

Consol Acid C-2000 .... A3
Consol AP-10.............................. A1
Consol Concrete Cleaner........... A4
Consol CC-107............................ D1
Consol Germicidal Cleaner D1
Consol LS-227 .... A1
Consol SS-333 .... L1
Consol T-83................................ A2
Consol Tripe Cleaner N3
Super Kill Concentrate F1

Consolidated Chemical, Inc.

All Purpose Steam Cleaner A2
C.D.S.. D1
Creo-Carbol C2
Di-Clorothan F2
Econ Cleaner A4
FS-II E1
Germio-Cide D1
Germo-Dine D1
Germo-Phene D1
Germo-Sol.................................. C1
Insectrol Liquid Insecticide....... F1
Mint-O-Sol C2
Paramount Soapless Detergent .. A1

Pearl Liquid Soap...................... E1
Pine.. C1
Pink Satin Lotion Hand Soap E1
Quick Clean A1
Sanitize C1
Sno-Clean A3
Sno-White Egg Washing
    Compound Q1
Sparkle Concentrate................... A1
Staphosan E1
Top-Quat A4
Vermingo F1
Vol-O-Cide................................. A4

Consolidated Laboratories, Inc.

Bye Bye Rat................................ F3
C L Concentrate.......................... A1
C L Floor Cleaner....................... C1
Con Chlor Concentrate............... C1
Conzyme L2
CL Safety Solvent...................... K1
Germ Brite D1
Glisten A1
Kleen Out................................... L1
L.D.O. L1
Lightning.................................... C2
Odor Gran.................................. C1
Pinky.. C2
Rid Ice....................................... C1
Sewer Sweet C1
Super CL 666.............................. L1
Vanish.. C1

Consolidated Marketing, Inc.

Air-O-Mint.................................. C2
Ambush Residual Insecticide F2
Biocide....................................... D1
Blen Liquid Detergent A4
Chlorinal.................................... A1
Foam-O-Cide Disinfectant
    Foam....................................... C1
Fury Clean 101 Industrial
    Surfactant A1
Fury Clean 181 Press Roller
    Wash....................................... A4
Fury Clean 401 Steam Cleaner
    Conc.. A4
Fury Clean 403 Fireproof
    Degreaser................................ A4
Fury Clean 405 Meat&Poultry
    Eq. Cl...................................... A1
Fury Clean 499 Steam Clnr
    Conc.. A4
Fury Clean 703 Krete Clean A4
Fury Clean 705 Supreme
    Concrete Cl. A4
Fury Clean 727 Sewer Pipe
    Cleaner.................................... L1
Fury Clean 728 Food Plant
    Cleaner.................................... A1
Fury Clean 729 Dairy Cleaner... A1
Fury Clean 732 Power Steam ... A2
Fury Clean 737 .... A1
Fury Clean 737-A....................... A1
Fury Clean 747 Bake Pan
    Cleaner.................................... A1
Fury Clean 748 Clean House A1
Fury Clean 750 Clini Clean....... A1
Fury Clean 756 Spray Sanitizer . D1
Fury Clean 757 Chlor Dairy
    Pipe Cl.................................... A1
Fury Clean 761 Spray Wash
    Deluxe..................................... A1
Fury Clean 763 Kleen'N Shine .. A1
Fury Clean 764 Hard Wtr Veh.
    Wash....................................... A1
Fury Clean 765 Super Spray...... A1
Fury Clean 766 Waxy Blue
    Spray Wash............................. A1

Fury Clean 768 Powdered Car
    Wash....................................... A1
Fury Clean 925 Indust. Silcn.
    Spray....................................... H1
Fury Clean 975 Skin Guard E4
Fury-Clean #740 Chlor
    Machine Wash Cmpd. A1
Fury-Clean #801 Industrial
    Odor Control C1
Fury-Clean No. 940 Motor
    Degreaser................................ K1
Glad Hand Waterless Cream
    Hand Cleaner E4
Glisten Stainless Steel Cleaner... A7
Harmony #150 Detergent A1
Harmony #175 Detergent A1
Harmony Air-O-Mint................. C2
Harmony All Purpose Soap
    Powder A1
Harmony Ammoniated Oven
    Cleaner.................................... A8
Harmony DeSoil.......................... B1
Harmony Dish Kleen A4
Harmony Fluffy Detergent A1
Harmony Frosty Rose E4
Harmony Germicidal Cleaner D1
Harmony Glass Cleaner.............. C1
Harmony Insecticide F1
Harmony Kar Kleen A1
Harmony Mini Suds B1
Harmony Pine Oil C1
Harmony Pink Dish Kleen A4
Harmony Remove-It A4
Harmony Sani-Lube Food
    Equip Lub. H1
Harmony Scent-O-Cide D1
Harmony Shine-It........................ A4
Harmony Smooth 40% E1
Harmony Spray Sanitizer D1
Harmony Stainless Steel &
    Metal Polish A7
Harmony Super Concentrate...... A4
Harmony Super Surfactant........ A4
Harmony Surgical Liquid Hand
    Soap.. E1
Hygrade Liquid Detergent......... A4
Hygrade Wax Remover A4
Kleen All..................................... A1
Lemon Disinfectant No. 10 .... C1
Lemon Disinfectant No. 39 .... C1
Liquid Toilet Soap 40% E1
Lubon H2
Mister Green Kleen.................... A4
Ms. Red-Y Cleaner.................... A4
Oven-Glo Oven & Grill
    Cleaner.................................... A8
Pink Lotioned Liquid
    Detergent................................ A4
Sani Shine.................................. A4
Shield Moisture Guard............... H2
Shine Metal Polish..................... A7
Skin Guard Barrier Cream E4
Slick Silicone Lubricant............. H1
Smooth Forty E1
Smooth Twenty.......................... E1
Sparkle Glass Cleaner C1
Spicy Air Sanitizer.................... C1
Sprasaf Engine Degreaser.......... K1
Steri Pine C1
SR-158-SZ.................................. A1
Technical APC............................ A4
Tile Brite Bath & Tile Cleaner... C2
Twist Lemon-Lime Air
    Sanitizer.................................. C1
Victory Cleaner A4
White Lightning C1
Zap... A4

81

Consolidated Protective Coatings Corp.

Consolite Cledide D1
Consolite Hand-Kleen E1
Consolite Hvy Dty Clnr No. 33 A1
Goodyear All Purpose Cleaner .. A1
Greeskut A4

Consolidated Research Corporation

CRC Acid Cleaner .... A3 ·
CRC Concrete Cleaner
  Odorless A4
CRC Slick Streak A1
CRC Spark-It A1
CRC Super Insecticide
  Concentrate F1
CRC 212 Sewer Solvent L1

Consolidated Water Treatment Corp.

GB-519 .... G1
GB-521 .... G1

Contact Industries, Inc.

A.P.C. C1
Anti-Seize H2
Bowl Cleaner C2
Chain & Cable H2
Ease Off Fluorocarbon H2
Electra-Solve .:........................... K2
Filtrol Treatment P1
Foam Metal Cleaner & Polish A7
Fresh Mint Glycolized Air
  Freshener C1
Germicidal C1
Ice Away C1
Non Conduct \............................ H2
Oven Cleaner A8
Oven Guard P1
P-30 .... F1
P-30 Cherry F2
Pipe Coat P1
Pure Lube H1
Quick Slip Silicone H1
Roach & Ant Killer F2
Stainless Steel Cleaner/Polish A7
Tile and Grout Cleaner C1
Trouble Spots A4
Water Soluble Degreaser C1
X-Term Indoor Fogger
  Insecticide Spray F1

Continental Chemical & Supply Co.

Boilerite-100 .... G6

Continental Chemical Company

A-K No. 46 Liquid Cleaner A3
Algex No. 829 .... G7
Aqua-X No. 800 .... G6
Aqua-X No. 802 .... G6
Aqua-X No. 804 .... G6
Aqua-X No. 808 .... G6
Automatic Floor Scrub No. 170 A4
AK No. 41 .... A3
Chain Lube No. 952 .... H2
Chain Lube No. 953 .... H2
Chain Lube No. 954 .... H2
Chem-X No. 820 .... G6
Chem-X No. 821 .... G6
Chem-X No. 822 .... G6
Chem-X No. 823 .... G6
Chem-X No. 824 .... G6
Chem-X No. 825 .... G6
Chem-X No. 832 .... G6
Chem-X No. 836 .... G6
Chem-X No. 837 .... G6
Chem-X No. 838 .... G6
Chem-X No. 839 .... G6
Cold Room Cleaner No. 174 ...... A5
Compil-X No. 865 .... G7

Compil-X No. 866 .... G7
Compil-X No. 870 .... G7
Compil-X No. 875 .... G2
Compil-X No. 880 .... G7,G3
Compil-X No. 883 .... G7
Compil-X SS-1 .... D2
De-Ox No. 855 .... G6
Descaler No. 905 .... A3
Descaler No. 906 .... A3
Det-San No. 175 .... A4
Det-San No. 181 .... D1
Foam Additive No. 941 .... A4
Ice-Away C1
K-Loid No. 850 .... G6
K-Loid No. 851 .... G6
Klenosan No. 144 .... A1
Klenosept A4
Klenosolv No. 148 .... A4
Klenosolv No. 148-B A4
Lube No. 6 .... H2
Meta Klen #50 .... A3
Meta Klen #55 .... A3
Neutra-San No. Z-5013-1 WS C1
Neutro-San No. 415 W.S. C1
Neutro-San No. 425 W.S. C1
Neutro-San No. 426 W.S. C1
Neutro-San No. 429 W.S. C1
Neutro-San No. 434 O.S. C1
Neutro-San No. 434 W.S. C1
Neutro-San No. 435-B C1
Neutro-San No. 436 O.S. C1
Neutro-San No. 437 W.S. C1
Neutro-San No. 438 .... C1
Neutro-San No. 440 W.S. C1
Neutro-San No. 442 O.S. C1
Neutro-San No. 442 W.S. C1
Pad No. 901 .... L1
Pad No. 902 .... L1
Pan Washing Detergent No.
  291 .... A1
Pan Washing Detergent No.
  291-NP A1
Powdered Detergent No. 247 ... A1
Powdered Detergent No. 281 ... A1
Powdered Detergent No. 284 ... A2
Powermist No. 164 .... A2
Rack Wash A2
Samoclean K1
Samoclene Formula 122 .... K1
Samoclene Formula 128 .... K1
Samoclene Formulla No. 125 .... K1
Samoclene No. 126 .... K1
Shell Brite Q1
Tile-O-San A6
Vol-Am No. 860 .... G6
Zeo Klen No. 936 .... P1

Continental Chemical Company

Con-O Odor Control Agent G1
Con-O OCA Acid A3
Con-O Sodium Hypochlorite
  12.5% D2
Con-O Super OCA G4
Con-O-Clean N-San A4
Con-O-Descaler A3
Con-O-Digester A3
Con-O-Insect Spray F1
Con-O-Liquid Hand Soap E1
Con-O-Mill Spray F1
Con-O-San-O-Clear D1
Con-O-1346 .... A3
Con-O-1360 .... A4
Con-O-1389 .... A2
Con-O-1405-P A2
Con-O-1555 .... L1
Con-O-250 .... A3
Con-O-255 .... A1
Con-O-400-P A1
Con-O-805 .... A1

Continental Chemical Company

Bleach Bright N3
Chloro-Clean A1
Dock Clean A4
Floor Cleaner HD A2
Foam A Way A1
Packers All Purpose A1
Pan Bright A3
Roller Hook Cleaner A2
Smokehouse Cleaner A2
Smokehouse Cleaner 53 .... A2
Super Steam A2
Tripe Clean N3

Continental Chemical-Westbury, Inc.

Con-Clean A4
Con-Solv A1
Con-Solv "W" A4
De-Con-Gest "TD" L2
Safety Solvent K1
Super Heavy Duty All-Purpose
  Cleaner A4
Yagash A8

Continental Labs, Inc.

Big "G" A4
Blue Granulated Sewer And
  Drain Solvent L1
Boilerite-100 .... G6
Conquer A1
Orange Deodorant C1
Red Hot L1

Continental Oil Company

All Season Motor Oil SAE
  10W-30 .... H2
All Season Motor Oil SAE
  10W-40 .... H2
Bentone Grease No. 1 .... H2
Bentone Grease No. 2 .... H2
Conoco "C" Redind Oil 33 .... H2
Conoco "C" Redind Oil 45 .... H2
Conoco "C" Redind Oil 68 .... H2
Conoco Heat Transfer Oil H2
Conoco Polar Start DN-600
  Fluid H2
Conoco Polar Start DN-600
  Gear Oil H2
Conoco Polar Start DN-600
  Grease H2
Conoco 2105 Oil SAE 140 .... H2
Conoco 2105 Oil SAE 80 .... H2
Conoco 2105 Oil SAE 90 .... H2
Conolith M Grease No. 1 .... H2
Conolith M Grease No. 2 .... H2
Conoplex Grease "C" No. 1 ...... H2
Dectol Oil 11 R&O H2
Dectol Oil 116 R&O H2
Dectol Oil 140 R&O H2
Dectol Oil 15 R&O H2
Dectol Oil 21 R&O H2
Dectol Oil 33 R&O H2
Dectol Oil 51 R&O H2
Dectol Oil 76 R&O H2
Dectol Oil 92 R&O H2
E.P. Conolith Grease No. 0 ...... H2
E.P. Conolith Grease No. 00 ..... H2
E.P. Conolith Grease No. 1 ...... H2
E.P. Conolith Greast No. 2 ...... H2
El Mar Oil 20-20W, 30, 40 ...... H2
Gear Oil 100 .... H2
Gear Oil 114 .... H2
Gear Oil 136 .... H2
Gear Oil 160 .... H2
Gear Oil 220 .... H2
Gear Oil 50 .... H2
Gear Oil 72 .... H2
Gear Oil 84 .... H2
GP Dectol Oil 140 R&O H2

Continental Organization Of/ Distributor Enterprises Inc.

Code 1005-1 Liquid Hand Dish
   Wash A1
Code 1090-1
   Institl.Liq.Drn.Opnr.&Mntnr L1
Code 330-X50 Machine Dish
   Wash Compound A1
Code 335-5 Inst. Deep Fat
   Fryer Cleaner.......................... A8
Code 341-X50 Supreme Low
   Suds Lndry Det. B1
Code 345-X10 Concentrated
   Dry Bleach................................ B1
Code 911 Heavy Duty Machine
   Scrub. Conc......................... A1
Code 915 General Purpose
   Cleaner................................... A1
Code 920 Heavy Duty General
   Purpose Clnr A1
Code 925-1 Ammoniated All
   Purpose Clnr. A4
Code 940-1 Bowl Cleaner.......... C2
Code 945 Alkaline Dgrsr &
   Oven Cln Conc. A8
Code 950 Extra Heavy Duty
   Degreaser A4,A8

Continental Research Corp.

Aluma Kleen................................ A3
Dura Solve A................................ A4
Dura Solve XX........................... A1
Dura Solve 70............................ A4

Continental Research Corporation

Aero-Glove Skin Shield Cream . E4
Aero-Grease................................ H2
C.R. Concentrate........................ A1
Caf-Fog Insecticide F1
Caf-Sect F2
Clean Guard................................ A4
Clear-All..................................... C1
Clor-Aid C1
Complete Non-Selective
   Surface Disinf. C2
CR-10... A1
CR-2... G6
CR-4... G6
Enzyme-70.................................... L2
Fleet Degreaser K1
Foam-Ox....................................... A3
Grab II.. E4
Graf.. H2
HD-10... A1
In-Mist F2
In-Mist-B F2
Jel-M.D... K1
Kitchen Kleen.............................. A1
Kitchen Magic A1
Lana High E1
Luster-B Liquid Polish................ A7
Micronism 100% Liq. Live
   Micro Organisms L2
Mosquito Larvaecide C1
No-Ox... C1
Non-Dust...................................... A1
Oven-Care A8
Power-Plus P1
Pro-Tec... C1
Process-Lube 210......................... H1
Safety-X K1
San-Mint C1
Sani-Bright Porcelain Cleaner C2
Sila-Slip H1
Soft N' Pink A1
Special Ox Acid........................... A3
Spray All....................................... C2
Super Foam.................................. C1
Super M.D. K1

Super 100...................................... A1
Super-Pent................................... H2
Surgi-Clean.................................. D1
Tar-Go... C1
Tar-Jel-"B".................................. K1
Trap L1
Tri-Mist C2
Trust C1
X-Algae D1
X-It Emulsion Bowl Cleaner...... C2
Xtra 932....................................... A1

Continental Research Corporation

Can-San C1
Emulsifate.................................... C1
Glow... C2

Contractors Chemical Corp.

#63 Degreaser A1
Floor Soap A1
Germicide RC-3........................... D2

Contrast Maintenance Chemicals, Inc.

Ammoniated Oven Cleaner A8
Classic Stainless Steel Cleaner ... A7
CMC Blitz.................................... F1
CMC Concentrate A4
CMC Kleen-Safe Bowl/Tile/
   Porcelain Clnr.......................... C2
CMC-3.. F1
CMC-5.. F1
Emulsi-Foam A1
Exo-Therm Sulfuric Acid Drain
   Opener L1
Saf-T-Solv K1
Scour Creme A6
Sila Lube H2
Spray Away F2
Spray N Wipe.............................. C1
Spring Aire Disinfectant
   Deodorant C1
Stainless Steel & Metal Polish A7
Stat... D1
Stat Disinfectant Cleaner &
   Deodorizer A4
Super Brite A3

Control Chemical Company

Concrete Cleaner........................ A4
Control Kleen A4
Detarnish A3
Metal Brite A3
No. 380 .... A2
Oven Brite A1
Power Chlor A2
Power Foam A1
Power Kleen A1
Power Suds A1
Purg Drain Pipe and Sewer
   Opener L1
PF-20.. A1
Release... A2
Rinse Brite................................... A1
Silver Brite A1
Stain Go A1
Super Chlor A2
Super Suds A1
360.. A2

Control Chemical Corporation

Ace Pipe Thread Sealer:............ H2
Activate.. F1
Blue Sky C1
D-Kloro.. C1
Dart.. A1
Decimate L2
Di-Kil... F2
Dish Kleen A1
Dry Pipe...................................... P1

Floor Ward A4
Go-Go A4
Grease-Go Waterless Hand
Cleaner.................................... E4
Hot Stuff.................................... A8
Ko Shan...................................... E1
Milk Stone Remover.................. A3
Munizyme.................................. L2
Neutrex...................................... C1
Neutron C1
Obliterate................................... L2
Open Cessme............................. L1
Punch.. A1
Q-Guard D1
Q-Trol....................................... D1
R. I. P. Insecticide Spray........... F2
Refresh C1
Roach & Bug Killer F2
Roach-Go................................... F2
Steam-O-Clean A1
Take-Off.................................... A4
Tee-Off...................................... A4
Tower Treat............................... G7
TNT... A4
Vanquish.................................... C1
White Cloud............................... C1

Control Chemical Enterprises
Lub Silicone Lubricant Spray.... H1

Conwood Corporation
Hot Shot Fly Bait....................... F2
Hot Shot Fly&Mosquito Insect
Killer...................................... F1
Universal Spray Insect Killer..... F1

Cook Chemical Company
CCC Cleaner Degreaser Extra
Heavy Duty A1
CCC Lemon Disinfectant........... C2
CCC Pine Odor Disinfectant...... C1

Cook's Industrial Lubricants Inc.
Cup Grease #2 .... H2

Cooke Laboratory Products
Cooke Ant Barrier...................... F2

Cooper U.S.A., Incorporated
Cooper Farm Disinfectant.......... C1
Cooper Saponified Cresylic
Solution.................................. C2

Copeco/ Custom Lubricants
Copeco Alco-AA Food
Machinery Grease H1
Copecote ACFG-1HTW H1
Copecote ACFG-1W H1
Copecote ACFG-2W H1
Copecote ACFG-3W H1
Copecote FO-250W.................... H1
Copecote FO-30W...................... H1

Copper Brite, Inc.
Roach Prufe F2

Coral Chemical Company
A-227 .... A3
Blu-Tide..................................... A1
Brite Clene Add A3
Brite Clene 25 .... A3
Brite Clene 30 .... A3
C.C. 403..................................... A2
C-Tide.. A4
Chlor-Clene 15........................... A2,Q1
Chlor-Clene 420......................... A1
Chloro-San D2,Q4
Clene A-21 .... A3
Clene Aid A1

Clene All 4 .... A1
Clene All 55 .... A1
Clene MGW................................ A1
Clene S.S.................................... A1
Clene SRO A1
Clene 1001 .... A2
Clene 24..................................... A1
Clene 27L................................... A2
Clene 308...,.............................. E1
Clene 315................................... A2
Clene 50..................................... A2
Clene 500................................... A2
Clene 509................................... A2
Clene 520................................... A2
Clene 57..................................... A2
Clene 777................................... A2
Clor-All A1
Cloral... A1
Concrete Brightener................... A3
Concrete Brightener #2.............. A3
Cor Cal #1................................. A1
Cor Cide D F2
Cor Jell...................................... A1
Cor Pop 03................................. C3
Cor Rinse 307 .... A2
Cor Saf 18 .... A1,G2
Cor Sheen 92 .... A1
Cor Steem 00 .... A2
Cor Steem 10 .... A2
Cor Steem 11 .... A2
Cor Steem 30 .... A1
Cor Steem 90 .... A2
Cor Vap 90 .... A1
Coral San No. 20........................ D1
Coralite 01.................................. A1
Coralite 02.................................. A1
Coralite 100................................ A4
Coralite 200................................ A4
Cordefom 92 .... Q5
Cordefom 93 .... Q5
Coredge 30 .... A2
Corem 909.................................. K1
Corem 92.................................... K1
Corem 93.................................... K1
Corfom 300 .... A1
Corfom 32 .... A1
Corfom 34 .... A1
Corfom 35 .... A1
Corfom 91 .... A1
Corfom 94 .... A1
Corlub 01.................................... H1
Corlub 80.................................... H2
Corlub 91.................................... H2
Corpol 34 .... A6
Corsheen Plus A1
Corsheen 333.............................. A2
Corsheen 908.............................. A1
Corsheen 909.............................. A1
Cortak 301 .... A2
Cortak 304 .... A2
Cortak 90.................................... A2
Cortak 96.................................... A2
Cortak 98.................................... A2
Corvin 00.................................... A3
Corvin 100.................................. A3
Corvin 11.................................... A3
Corvin 14.................................... A3
Corvin 17.................................... A3
Corvin 97.................................... A3
Descale 202 .... A3
Descale 90.................................. A3
E.C. Ultimate............................. A1,Q1
Egg Clene FMC A1,Q1
Egg Clene Plus A1,Q1
Flash Clene Super A2
Flash Clene 23 .... A2
Flash Clene 26 .... A2
Flash Clene 99 .... A3
Flash 23L................................... A2
Foam Clene 10............................ A2

GF 844....................................... A1
HSC... N2
Ibex C-51................................... G6
Lan-O-Lite C2
Li Cor CS A4
Li Cor 21 .... A2
Liqua-Steam 5............................ A1
Odor Con 5 .... C1
P. W. 88..................................... N4
PC 300....................................... A2
Sekt-I-Cide NT F1
Sewerex LF................................ L1
Sewerex 3................................... L1
Soil Solv 2000 .... K1
Spray Clene 6 .... A2
Spray Cor 41.............................. A2
Spray 30 .... A1
Spray 35 .... A1
Spray 90 .... A1
Super Clene................................ A2
Surcoat 36 .... A3
Surcoat 96 .... A3
Tank Clene 3.............................. C1
V-Clene N4
Vat Clene 50 .... A2

Corcoran Chemical Products, Inc.
Glass & All-Purpose Cleaner C1
Porcelain & Bowl Cleaner.......... C2
The Greaser H2

Cornell Chemical & Equipment Co., Inc.
Pyrenone Fog Spray F1
S Gard Space Spray.................... F1
WO-25 Rodenticide.................... F3

Correlated Products, Inc.
Bowlshine Bowl & Porcelain
Cleaner.................................... C2
Bowlshine X............................... C2
Contact Point Cleaner................. K2
Correlsept Disinfectant-Cleaner . D1
Creme-Clenz C2
CPI Krome Kleen Porcelain
Cleaner.................................... C1
CPI Residual Roach Liquid F2
Dirtsaway A1
Disinfectant-Sanitizer D2
Drain Cleaner L1
Dry-Tef\Spray Lubricant H2
End-Con P1
Foamy.. A4
Freen Mint Disinfectant............. C1
Germicidal Rinse D2
Grimeaway................................. C2
Help A4
Kleer Out L1
Liquid Satin Soap...................... E1
Liquid Steam Cleaning
Compound A1
Lotionized Glass &
Dishwashing Compound A1
Pine Odor Disinfectant C1
Spray K A4
Staphasept Hospital
Disinfectant Doed.................... C2

Corrosion Control Services
C-500... G6
CH-100 .... G6
CH-150 .... G6
CL-500....................................... G6
CS-760....................................... G6
HMP.. G6
PHP-20 .... G6
PHP-50 .... G6
SLT-10....................................... G6
SLT-12....................................... G7

84

Twice 98.. A1

Coventry Products Inc.
Alka-Klor LD............................. A1

Cowboy Chemical Company
Custer's Last Charge Sewer
   Cleaner.. L1
Hi Sierra.. C2
One Shot K1
Rio Pecos K1
Saddle Slick H2

Coyne Chemical Company
Coneen-Trol................................... F1
Insec-Trol...................................... F1
Super-Trol...................................... F1
Sure-Trol.. F1

CoDel Corporation
BTC New Power P1
Chain & Cable H2
Code-L A1
Code-T... H2
Code-32 Safety Solvent K1
Code-36 Contact Cleaner............ K2
Grease Cutter................................ A1
H.D.-300.. A1
Insul-8 .... P1
Invisible Glove E4
M.D.-200 .... A1
Moly-Dri H2
Ovec Ammoniated Oven
   Cleaner.. A8
Punch.. A1
Ridex.. E4
Sani-Lube H1
Stainless Silicone Spray H1
Thermo Klene................................ A1
Wyte-Ease H2

Craft Sanitary Specialties Corp.
All-Purpose Cleaner.................... A4
Craft Sprayklene.......................... C1
Germicidal Cleaner D1
Grease-Klene................................ A1

Crain Chemical Company
#1381 .... A2
Beyond A4
Boiler Compound G7
Con-O-Chlor.................................. D2
Con-O-Phene.................................. C2
Con-O-San...................................... A4
Con-O-Tox 2 .... F2
Concentrate.................................... A1
Crain Silicone Spray Food
   Grade H1
Dish-Brite A1
Dual... F1
DFC-40 Deep Fat Fryer
   Cleaner.. A2
Fogger No. 76 Super Fogging
   Insect Kille................................ F2
G-Go Hospital Disinfectant
   Deodorant C2
II-320 .... A1
Kleen-Eze....................................... A4
Lana.. E1
Mr. "C" Disinfectant
   Deodorant C1
New Glo.. C2
Nock-Out.. L1
P.S.S.-35 .... A7
Pluck-A-Chick N1
Power Plus Prilled Drain
   Opener.. L1
Quam 25 .... D1
Quam 75 .... D2

Robust Hand Cleaner.................. E4
RCR-235.. A3
SS-305 .... H2
X-Out C1

Crane Industrial Chemicals/ A Division of Larsal Inc.
Beulah.. C1
Good Bye Liquid Drain Opener L1
Olympic The Champion A4
Rebrite C2
Remox... A3
Selectric K1
Smooth N Creamy E1

Craver Supply Company, Inc.
Craco Hand Soap E1

Creative Chem.
Solvall.. A1

Creative Chemicals, Inc.
Cling-N-Clean Bowl &
   Porcelain Cleaner C2
Gush L1
Ram Liquid L1

Creative Industries
De-Zolv Sulfuric Acid Drain
   Opener....................................... L1
Emerald Ready To Use Liquid
   Hand Soap................................. E1
Evergreen Pine-Odor
   Disinfectant C1
Fact Concentrate All-Purpose
   Cleaner....................................... A4
Flash Cleaner-Degreaser............. A1
Zing Clinging Bowl Cleaner C2

Creed Laboratories
Creed Concentrate A1
CL-66.. A4
Duro Power A1

Crepaco, Incorporated
Airlube... H2
Compressorlube No. 30AF........ H2
Crepaco 280 .... A1
Crepaco 609 .... A1
Crepaco 626 .... A1
Crepaco 627 .... A1
Crepaco 648 .... A2
Crepaco 676 .... A2
Crepaco 703 .... A3
Crepaco 715 .... A3
Crepaco 759 .... D1
Crepaco 764 .... A3
Crepaco 778 .... H2
Homolube....................................... H2
Hydrauliclube H2

Cresset Chemical Company
Cresset Coral Hand Cleaner...... E4
Cresset Green Hand Cleaner..... E4

Crest Research Products
Basic... K1
Control Concentrate.................... C1
Crest Concentrate......................... A1
Goliath... K1
Herculox.. A3
Luxur.. E1

Crestfield Research
Luxur Liquid Hand Soap........... E1

Crestline Chemical Company
Fat-Away A1

Criter Chemco, Inc.
"Chem-Foam" A1
Big Stripper C1
Criter Chemco C-1 .... A1
Criter Chemco C-2 .... A1
Dish Suds "I" A1
Dish Suds "II" A1
Freezer Cleaner A5
Industrial Clean 1 .... A1
Industrial Clean 2 .... A1
RS-100 .... A1
Solve-It A4
Sudsy Ammonia C1
Super "A" A4,A8
Super "A" -100 .... A4,A8

Critzas Industries, Inc.
Goop Hand Cleaner E4

Crockett Chemical Company
C-211 .... A4

Crosby Chemicals, Inc.
Crosby 68-F P1

Crossco Manufacturing Corp.
Carbex C1
Glissen A1
Sanisation C2
Speed-Spray C1
Versatal C1

Crossland Laboratories, Inc.
Ivy Rose E1
Mercury A1
Mirochem A4
Naturair C1

Crouch Chemical
Stainless Steel & Metall Polish ... A7

Crouch Laboratories, Inc.
Crouch Drain Cleaner L2

Crouch Supply Company, Inc.
Acid-Klean A3
APC Cleaner A1
BD-150 .... A1
Cromal A1
Crouch Cleaner-Degreaser
  Concentrate A4,C1,A8
Crouch's Cir-Spray 341 Cleaner A1
Crouch's Coral Mist E1
Crouch's Hi-Lo Acid Cleaner ... A3
Crouch's ID-125 .... D2
Crouch's Pecel-It N4
CIP Acid Cleaner A3
CIP Acid Cleaner (concentrate) A3
D Foam Q5
D'Terge A1
DW-21 .... A1
Egg Washing Compound A1,Q1
Fluff A1
GPC Cleaner A1
Heavy Duty Super 85 .... A1
Improved Crouch's ID-05-R D2
Insect Spray F1
Insecticide 150 .... F1
Kanex A1,N2,G2
Kap Cleaner A1
Kleen-N-Brite A3
L.C.D. A1,Q1
No. 147 Heavy Duty Alkaline
  Cleaner A2
Petrolatum Spray H1
Petrolatum Spray Heavy Duty .. H1
Q-10 .... D1
Sani-Quat D2
Sanitizer-Bactericide D1

Ship Shape A1
Ship Shape-NC A1
Sodium Hypochlorite Solution ... D2,Q4,E3,N4
Spray Gel H1
Super Flakes A2
Super 85 .... A1,Q1
SP-14 .... A3
TCW Cleaner A4
Wegg-DF Q1
3222 Crouch Industrial Floor
  Clnr. A4
802 Centra-Matic Liquid Lube ... H2
888 Alkali A2

Crown Chemical Company
Amine Condensate Treatment ... G6
Colloid Boiler Water Treatment G6
Colloid/Chelate BWT G6
Compound Ca-3 .... G6
Compound S-4 .... G6
Diasyn-7 .... A1
Formula 600 .... G6
Hydrax 2 .... G6
Hydrax-4 .... G6
Liquid Drain Cleaner L1
Liquid Hand Soap E1
LHS-40 .... E1
Ox-Trol DC-P G6
Ox-Trol LQ-2 .... G6
PCA Boiler Water Treatment ... G6
SWEOS G6

Crown Chemical Company, Inc.
Foam-Up A1
USDAL E1

Crown Chemical Inc.
Pride Powder A1
PHC-71 .... A1
PHC-72 .... A1
Striplcene A4
Striplcene-TX C1
Sunbrite Laundry Bleach B1
1004 .... A2
15 Floor & General Cleaner A2
444 .... A1

Crown Chemicals
".5% Coated Warfarin" F3
A B Insecticide F2
C-20 Chlordane Oil Miscible
  Concentrate F2
Check Pest Food Plnt, Mill &
  Dairy Rm Sp. F1
Check Pest Indust. Aer. Insect.
  Concnt. F1
Checkfyre 207 .... P1
Code 888 Special Odor Control . C1
Code 999 Smoke Odor Control . C1
Crown Pyrenone Dairy & Food
  Plant Spray F1
Cygon 2-E F2
Dairy and Stable Spray F1
Diazinon 4E Insecticide F2
Diazinon 4S Insecticide F2
Felton No. 7 .... C1
Flea Spray F2
Food Plant & Dairyroom Insect
  Bomb F1
House & Garden Double
  Action Bug Killer F2
Ice Melting Pellets C1
Industrial Aerosol Insecticide ... F1
Industrial Aerosol Insecticide
  Gold F1
Lemon Disinfectant C1
Lindane Emulsifiable
  Concentrate F2
Malathion 57 .... F2
Mill & Food Plant Spray F1

New Purge II F1
New Purge III F2
Pest-Rid Home & Farm Bomb .. F2
Pest-Rid Pelleted Bait F3
Prolin Anticoagulant
  Concentrate F3
Prolin Meal Bait For Rats And
  Mice F3
Prolin Pelleted Rat Bait F3
PY-Fly Spray F1
PY-11 Emulsifiable Concentrate F1
Rat And Mouse Killer With
  Diphacinone F3
Rat Pucks F3
Roach & Ant Spray F2
Super Malvex F2
Synergized Pyrethrum
  Concentrate #3610 .... F1
Syntox Insecticide 2.50% F1
Syntox Insecticide 3.0% E.C. ... F1
Total F1
Vapona Spray Solution F2
4 lb. Chlordane Emulsion
  Concentrate F2
8 lb. Chlordane Emulsion
  Concentrate F2

Crown Industrial Products Co.
Food Grade Silicone Lubricant
  (Bulk) H1
Food-Gd Pure Silicone Lubrcnt
  #8036 .... H1
6035 Food Safe Lubricant H1

Crown Paper
All Purpose Neutral Cleaner A4
Checker-8 .... E1
Heavy Duty Cleaner-Degreaser . A4
Heavy Duty Stripper A4

Crown Research & Chemical
Corporation
Advanced CL A1
Advanced SH A2
Al-Safe A1
Al-Safe C A1
Al-Safe H A1
AC 410 .... A3
Brite A1
C 301 .... A1
Carburetor Cleaner K1
CP 401 .... A1
Glass and Mirror Cleaner C1
Honey-Do A1
Honey-Solve A4
HC-307 .... E1
HTC 301 .... A2
Kleen-Kit A1
LSC A1
MHC A2
PSC 50 .... A1
Solv-Safe K1
SHC A1
XC 301 .... A4
500 A A4
500 T A1

Crown Sanitary Chemicals Co.
Deep-A-Clean A1
Pow-A A4
Release A4
Sock C1

Crown Supply, Incorporated
Gleam A1

Crown Zellerbach Corporation
Lotion Mastr E1
Lotionmastr Cleanser E1

CO-80 .... A2
CO-900 .... A4
Super CO-900 .... A4

Curtis Industries, Inc.
Super De-Greez NB.................... A1

Curtis Noll Corporation
Custodi-All Strip Kleen No.
4401 .... A4
Selex-250 .... H1
White, Food Machinery Grease. H1

Custodial Supply Company
Meat Room Degrsr-Gen.
Purpose Clnr. A1
Quat Germicidal Cleaner D2

Custodial Supply Company
Clean Up A1
Foam-Away A1

Custom Blend of California
Custom Blend 301 .... F1
Custom Blend 402 .... F1

Custom Built Lubricants Co.
C-B All Purpose Gear Lube
SAE 140 .... H2
C-B All Purpose Gear Lube
SAE 80 .... H2
C-B All Purpose Gear Lube
SAE 90 .... H2
C-B AW Green Lubricant H2
C-B AW Green Lubricant No.
0.. H2
C-B Heavy Duty Motor Oil
SAE 10 .... H2
C-B Heavy Duty Motor Oil
SAE 20 .... H2
C-B Heavy Duty Motor Oil
SAE 30 .... H2
C-B Hydraulic Oil R&O Light .. H2
C-B Hydraulic Oil R&O
Medium H2
C-B Mineral Gear Lub SAE
140 .... H2
C-B Mineral Gear Lub SAE 80. H2
C-B Mineral Gear Lub SAE 90. H2
C-B Non Detergent Motor Oil
SAE 10 .... H2
C-B Non Detergent Motor Oil
SAE 20 .... H2
C-B Non Detergent Motor Oil
SAE 30 .... H2
C-B Unv Gr Lub Mil-L-2105B
SAE 140 .... H2
C-B Unv Gr Lub Mil-L-2105B
SAE 80 .... H2
C-B Unv Gr Lub Mil-L-2105B
SAE 90 .... H2
C-B White Bentone Lubricant ... H2
C-B White FM,Lubricant Code
12249 .... H2
Petrolith Lub Code No. 11106... H2
Petrolith Lub Code No. 11201... H2
Petrolith Lub Code No. 11301... H2
Realfilm AW Lubricant H2
8000 Improved Spraymatic
Gear Film H2

Custom Chemical Company
Boiler Compound M G7
Chain-Eze BB H2
Cleaner 5106 .... A2
Cleaner 5920 .... A1
Dish Wash #1 .... A1
Drain Cleaner L1
Floor Wash #1 .... A1

Foam Control HH....................... A1
Formula A-1 .... A1
Formula A-11.............................. A1
Formula A-3 .... C1
Formula A-4 .... A1
Formula A-6 .... A1
Formula A-7 .... A4
Formula A-8 .... A1
Formula AH-1 .... A1,Q1
Formula AH-2 .... A1
Formula AH-3 .... A1
Formula AH-4 .... A1
Formula AH-5 .... A1
Formula AH-6 .... A1
Formula AL-1 .... A1
Formula B-1 .... A2
Formula B-2 .... A2
Formula B-3 .... A1
Formula B-4 .... A2
Formula B-5 .... A2
Formula B-6 .... A2
Formula B-7 .... A1
Formula BH-1 .... A2
Formula C-1 .... A3
Formula C-11.............................. A1
Formula C-2 .... A3
Formula DQ-1 .... A4
Formula DQ-8 .... A4
Formula E-1 .... A6
Formula H-1 .... Q1
Formula V-551............................ A2
Formula 333 .... K1
Grout Cleaner P A3
Hand Wash FF C2
Ice Melt R C1
Oven Cleaner A8
Pan Wash #1 .... A1
Rinse Dry #1............................... P1

Custom Chemical Company
Caustic Soda Beads A1
Cleaner Car Clean A1
Cleaner Chlor Plus A1
Cleaner LD 25 .... B1
Cleaner LDC L1
Cleaner 215A A1
Cleaner 302C.............................. A2
Cleaner 304C.............................. A2
Cleaner 412C.............................. A2
401A.. A1
402A.. A1
403A.. A1
404A.. A1

Custom Chemical Corp.
Hard Surface Cleaner and
Degreaser A1

Custom Chemicals, Incorporated
Bowl Cleaner C2
Brite Degreaser Super Conc
Multi-Purp Cl.......................... A1
Brite Lac-Stone Milk Stone
Remover A3
Brite Neutra-Clean A1
Brite Pink Lotion Hand Soap..... E1
Brite Stainless Steel Plsh & Surf
Presrv.. A7
Liquid Drain Opener................... L1
Sani-Phen 10-X D1

Custom Chemicides Inc.
Rodeth Blocks F3

Custom Engineering
Brawnee....................................... A4,A8
Custom Cannery Special............. A3
Custom Concrete Cleaner........... C1
Custom Descalor A3

Dalcoene P-22	G6	
Domalco SW	G7	
Oxy-Purge SP	G6	
Volex	G6	

Dalco Chemical Company/ Division of Dalco Corporation

Concentrate	F1
Concentrate "C"	F1
Daleo Chlorez 11	F2
Dalco Concentrate D	F1
Daleo Heavy Duty Detergent Disinfectant	A4
Linrez	F2
Malrez	F2
Mill Spray	F1
Roach Spray	F1

Dalco Corporation

C-10 Wash Powder	A1
Chloroderm Antiseptic Hand Soap	E1
Dal-Phene	A4
Dalco #225	F1
Daleo 3610	F1
Dalco 603	C1
Daleo 603 Delux Liquid Cleaner	A4
Disperse	A4,A8
Electric Dishwash	A1
Grand Slam	A4
KP Grill and Oven Cleaner	A4,A8
Pink Ramco Suds	A1
Ramco Suds	A1
Sani-Rinse	D2,E3
Sani-Rinse Two	D2
Soft Hand Cleaner	E1
Starkleen	A4
Super-Sheen	A7
Surgisan Antiseptic Hand Soap	E1
Window Cleaner Concentrate	C1
X-109 Super Cleaner	A4
X-609 Concentrate	A4
X-99 Special Cleaner	A1
100 Cleaner	C1
501 Special Cleaner	A4

Dalco Industries, Ltd.

Dalco Kleen DK 70	A4
Dalco-Solve DS 50	K1

Dalton & Son Inc.

Duke Extra Heavy Duty Cleaner-Degreaser	A1
Lemon Disinfectant	C2
Mint Disinfectant	C2
Oven and Griddle Cleaner	A8
Pine Odor Disinfectant	C1

Dameron Enterprises, Incorporated

Vi-Lan DE	A1
Vilan Bact-Clean #8100	A4
Vilan Skin Cleanser	E4

Damkoehler Chemical & Paper Co., Inc.

Marvelle Acid Clnr-Milkstone Remvr	A3
Marvelle Dairy Cleanser	A1

Damon Chemical Company, Inc.

All Out	L1
Boiler-Rite Boiler Compound No. 9	G6
Bug-Dize	F1
Concrete Fl. & Drveway Clnr. No. 93	A4
D-C Disinfectant	D1
D-C Multipurpose Cleaner	A4

Da-Brite Hog Scald	N2	
Da-Brite No. 225	A1	
Da-Brite No. 350	A1	
Da-Kleen #2	A1	
Daco Rapid De-Scaler	A3	
Daco Stainless Shine	A7	
Daco-San	C2	
Disinfectant D-C Cleaner	A4	
Fogg-R	F1	
FF Cleaner	A2	
Germicidal D.C.	D1	
Heat-Wave	A5	
Lark	A1	
Mim-Rid	A3	
Poly Strip	A1	
Smokehouse Cleaner #2	A2	
Ster-O-Kem No. 10	D1	
Ster-O-Kem No. 26 Disinfectant	D2	
Synthosol D	A4	
SF-77	A4	
SPD	L1	
Winterfect	C2	
Zymogen UC	L2	

Damsco Chemical Company

Auto Industrial Cleaner & Degreaser	A1
Emulsifiable Degreaser	K1
Hand Soap	E1
Hard Surface Cleaner	A4
Heavy Duty Steam Cleaner	A1
Safety Solvent	K2
Synthetic Cleaner	A4

Dan-Sales Company

F-239	A1

Dana Chemicals

Compound DL 20	G6
Compound DP 01	G6
Compound DP 06	G6

Danbury Chemical Products Corp.

Formula P.C. 6-11 Protective Shield Clnr	A7
Formula 202 Mint Disinfectant	C2
Formula 616 Fortified Pine Odor Disinf.	C1
Formula 660	E1
Formula 700 All Purpose Cleaner	A1
Formula 77X	A1
Liquid Live Micro Organisms	L2

Dane Sales And Service

Dane 100-D	A1
Dane 115-DP	A1
Dane 220-DP	A1
105D	A4
115D	A1
120D	C1
500S	C1

Darling & Company

Tri-Dar 33/1 Curing Compound	P1
Tri-Dar 66/1 Anti-Spall Coatiing	P1

Darmex Industrial Corporation

Darmex ACO 10	H2
Darmex ACO 1040	H2
Darmex NTG	H1
Darmex NTG-1	H1
Darmex NTO 30	H1
Darmex NTO 50	H1
Darmex NTO 90	H1
Darmex 1050	H2
Darmex 123 HT	H2

Darmex 9140	H2
DX-123 Grease	H2
NTO 10	H1

Daron International Corporation

All-Purpose	A1
Alum-Klean	A3
Ammo-Klean	A4
Anco	P1
Batman	P1
Big-Kleen	B1
C&C Lube Lubricant & Protective Shield	H2
Cleancrete	A4
Clear View	C1
D-Grease	C1
Daron 501	F1
Dissolve-All	L1
DI-295	K2
DI-437	A3
Electro Safe	K2
Electro-Clean	K2
Electro-Clean	C1
Enz-A-Way	L1
Grime-Away	E1
Handi Wash	E1
Hot Kleen	A2
Hy-Glide Non-Flam. Lub. & Release Agent	H1
Lemon-Clean	C1
Liquid Stripp	P1
Magic	A3
Mr. Muscle	L1
Open-Up	L1
Oven Clean	A2
Power-Klean	A4
Rust-Away	A3
Stripp	P1
Triple Clean	A4
Volt Try	H2

Darrill Industries

Standard Detergent No. 100	A1
Standard Detergent No. 101	A1
Standard Detergent No. 103	A2
Standard Detergent No. 104	A1
Standard Detergent No. 105	A3

Dart Research, Inc.

Auto Strip	A4
Bol-Power Non Acid Bowl Cleaner	C2
Breakthrough	F2
Brute Force	A4
Checker's Aid	A4
Cling	K1
Dar-Ox	A3
Darco 868	C2
Darco 871	C2
Darlectric	K1
Darzyme 250	L2
Defender	D2
Deo-Kleen	C1
Dry-Lon Teflon Lubricant Dry Spray	H2
Dyna-Pow Sulfuric Drain Opener	L1
DG-11 Shaker	C1
Hand Kleen	E4
Hand-Aid	E4
Handy Waterless Hand Cleaner	E4
Johnny On The Spot Toilet Bowl Cleaner	C2
Maintain	A4
Micro-Staph	D2
Nu-Shine	A7
Reflect	A7
Scooter	A4
Septene Conc. Bowl Cleaner	C2
Strip All	A4

Super Brute...............................A4,A8
Thermadart..................................L1
Thermasolv..................................L1
Turnzit Plus.................................H2
Versatile......................................A1

Dartmouth Chemical
Dura Solv.....................................K1

Datek, Inc.
Clean-Sweep................................E1
Concentrated Odor and Grease
  Control.....................................C1
Dasol Heavy Duty Water Sol.
  Saf. Solvt.................................A1
Deoxidizing Chemical Cleaner...A3
Detergent Concentrate................A1
Grease Gun..................................H2
Jell Degreaser and Tar
  Remover...................................K1
Power Foam.................................C1
Safety Solvent.............................K1
Silicone Spray.............................H1
Tri-Jel...L1

Datex, Incorporated
Datex Super Enzymes for
  Sewerage Sys...........................L2

Davey-Fitch Company, The
Chlorey..A1
DF-150...A1
DF-160...H2
Flo-Kleen.....................................A1
Kleen Haus..................................A2

Davidson, C. L. Chemical Inc.
Crown Fas-Foam..........................A1
Crown Ice Go...............................A5
Hi-Ho Lotionized Hand Cleaner E1

Davies-Young Company, The
Buckeye Acid Bright....................A3
Buckeye Blockade.......................F2
Buckeye Blue............................A4,C2
Buckeye Blue Frost......................C2
Buckeye Bol-Clean......................C2
Buckeye Bug-M...........................F2
Buckeye Confidence....................C1
Buckeye Dy-Chlor.......................D1
Buckeye Dy-Phase Bowl
  Cleaner....................................C2
Buckeye Dyodyne........................D1
Buckeye Dysh..............................A1
Buckeye EDC...............................A4
Buckeye Gard..............................D1
Buckeye Glad-Latb'r....................C1
Buckeye Glass & Window
  Cleaner....................................C1
Buckeye Glint..............................A7
Buckeye Legion...........................F1
Buckeye Liberate.........................A1
Buckeye Liquid Hand Soap
  5000...E1
Buckeye Maintenol......................A4
Buckeye Natraspray Hello..........C1
Buckeye Ram...............................A1
Buckeye Roadstar G.P. Steam
  Cleaner....................................A4
Buckeye Roadstar H.D. Steam
  Cleaner....................................A2
Buckeye Sani-Q...........................D2
Buckeye Sanicare........................C1
Buckeye Sanicare Lemon Quat..A4,C1
Buckeye Sanicare Mint Quat......A4,C1
Buckeye Sanicare Pine Quat......A4,C1
Buckeye Sanicare Quat-128......A4,C1
Buckeye Sanicare Quat-256......A4,C1
Buckeye Sanicare Quat-64.........A4,C1

Buckeye Sparkle..........................C2
Buckeye Super Gard....................A4
Buckeye Swath.............................A8
Buckeye Two-O-One....................A4
Buckeye XL-100...........................A4
EMC-1..K1
Formula E-051.............................A4
Formula E-107.............................E1
Formula E-638.............................A1
Formula 053.................................A2
Formula 10../................................A4
Formula 104.................................E1
Formula 141.................................E1
Formula 256.................................A1
Formula 26...................................A3
Formula 265.................................A1
Formula 275.................................A1
Formula 421.................................A4
Formula 422.................................A4
Formula 440.................................A4
Formula 54...................................A5
Formula 558.................................L1
Formula 607.................................A1
Formula 630.................................G7
Formula 631.................................G6
Formula 652.................................A1
Formula 76...................................G6
Formula 91...................................A4
Formula 981.................................N1
Frontier B.S.D..............................D1
Frontier B-Wash...........................A2
Frontier C-P.................................A2
Frontier Chlor-Press....................A1
Frontier Chlor-San 12.................D2
Frontier Circle 8..........................A1
Frontier Circo..............................A1
Frontier Cleans Away..................A3
Frontier EEC-1.............................A3
Frontier Fair................................A1
Frontier Flash..............................A1
Frontier Foam Add......................A2
Frontier Foam Chlor...................A1
Frontier Hello..............................A4
Frontier Hot Stuff........................L1
Frontier Hy Press.........................A1
Frontier Hy-Lite...........................A4
Frontier Hy-Test 85.....................A2
Frontier HS-5000..........................E1
Frontier HTC...............................A1
Frontier I-O-San..........................D2
Frontier IPC-85............................A1
Frontier Kon-Kleen......................A4
Frontier Liquid Glow...................A1
Frontier Phos Acid......................A3
Frontier Phos 40..........................A3
Frontier Sanitizer Q....................D2
Frontier See Suds........................B1
Frontier Super Power..................A1
Frontier Supreme........................A1
Gem 40...E1
Marvel...Q1
PDC 5734.....................................A1
Syntho Soap................................H2

Davies, D. W. & Co., Inc.
"Aqua-Treat #900".......................G2
"Aqua-Treat 400".........................G2
"Aqua-Treat 800".........................G2
"Big-Joe".....................................E4
"Boiler Kleen" No. 600................G6
"Boiler-Kleen" No. S-100............G6
"Cento-USO"................................A1
"Coco Soap 40"...........................E1
"Dairy-Caustic"...........................A2
"Dairy-Kleen".............................A1
"Dr. Jones"..................................C2
"Egg-Shine".................................Q1
"Feather-Degreaser"....................P1
"Foam-Kleen"..............................A1
"Glass-Shine".............................A3

"Klear All"...................................L1
"Kleen-Hands"............................E1
"Kleen-Kote"................................A7
"No-Pit 100".................................G6
"No-Pit 200".................................G6
"No-Pit".......................................G6
"Oxy-Out 100".............................G6
"Pink Magic" Cleansing Creme. E4
"Pipe-Kleen"...............................A1
"Sani-Rinse"................................D2
"Speed-Kleen".............................A3
"Super-Khlor" Bleach.................P1
Acid-Kleen...................................A3
Algae-Guard.................................G7
Boiler Kleen-500..........................G6
Concrete Cleaner.........................A1
D.C. 99...A4
D.W. Concentrate.........................A4
Davies "Chain-Lube"...................H2
Davies "Sani-Kill".......................F1
Davies "Scale-Out".....................G6
Davies "Ultra Kill"
  Concentrate..............................F1
Davies "Ultra Kill" Fogging
  Insect Spray.............................F1
Davies Disinfectant......................C1
Davies Fogging Insect Killer.....F2
Davies Grease-Solv 50................A2
Davies Hospital Type
  Deodorant Disinf.....................C2
Davies Hot Shot...........................L1
Davies Industrial Penetrant........H2
Davies.Klor-Kleen.......................A1
Davies-"KGC".............................A8
Drain Kleen.................................L1
Easy-Glide...................................H2
Fabulous......................................A1
Foam Ban.................................A1,G7
Gentle Magic...............................A1
Green Magic.................................A1
Green Magic Hand Cleaner........E1
L-15 Liquid Hand Soap...............E1
L-20 Liquid Hand Soap...............E1
L-40 Liquid Hand Soap...............E1
Machine Dishwashing
  Compound................................A1
Machine Dshwashing Comp.
  Chlorntd..................................A1
Magic Glass.................................C1
No Freeze.....................................A5
No Scale.......................................G6
No-Touch.....................................A1
Polysiloxane.................................H1
Porcelain Cleaner........................C2
Scale-Ban....................................G6
Smoke House Cleaner No. 120..A1
Solvo-Kleen.................................A4
Stop-Rust 100..............................G6
Suds-EE..A1
Surgical Liquid Hand Soap.........E1
Triple Kleen.................................A4
Ultra-Khlor..................................A1
Ultra-Kleen..................................A1
White Tiger Institutional Chlor.
  Clnsr.......................................A6

Davis Manufacturing Co., Inc.
705 Liquid Live Micro
  Organism..................................L2
Buggon..F2
Converta Concentrate..................A4
CJ-975 Heavy Duty Cleaner.......A1
D & G Detergent..........................A1
D and W Conc. Multi-
  Purp.Neut.Clnr........................A1
Da-Co...Q3
Da-Co Chlorinated Dairy
  Cleaner....................................Q4
Da-Co E-Z Sol Bowl Cleaner......C2
Da-Co Glass Lustre.....................C1

90

Daxchem A Division of Oxford Chemicals

BC-184 .... A3
BC-195 .... A3
BC-84 .... C2
BWC-28 .... A2
Daxchem AMC-51 .... A3
Daxchem BC-217 .... C2
Daxchem BC-95 .... L1
Daxchem BWC-39 .... A2
Daxchem CCO-17 .... A4
Daxchem Devil Dog A1
Daxchem DC-206 .... H2
Daxchem DC-84 .... K1
Daxchem DRC-17 .... A1
Daxchem FC-17 .... A4
Daxchem FCD-39 .... D1
Daxchem FCD-40 .... A4
Daxchem GCP-117 .... A1
Daxchem GCP-62 .... A1
Daxchem GCP-84 .... A1
Daxchem HDW-17 .... A1
Daxchem ISC-417 .... F2
Daxchem Jet A4
Daxchem LCL-51 .... H2
Daxchem LD-262 .... A1
Daxchem LD-51 .... A1
Daxchem LDT-128 .... B1
Daxchem LDT-139 .... B1
Daxchem LDW-117 .... A1
Daxchem LGC-117 .... A1
Daxchem LGC-62 .... A1
Daxchem LHS-39 .... E1
Daxchem LMR-128 .... A3
Daxchem LMR-39 .... A3
Daxchem LMR-51 .... A3
Daxchem LRS-39 .... A3
Daxchem LS-106 .... D1
Daxchem LS-62 .... D2,E3
Daxchem OM-6340 .... H1
Daxchem PHD-173 .... A4
Daxchem PHD-206 .... A4
Daxchem PLB-62 .... B1
Daxchem PSD-17 .... C1
Daxchem RC-17 .... G7
Daxchem SC-84 .... A2
Daxchem SC-95 .... A1
Daxchem SCH-84 .... A2
Daxchem SD-28 .... A1
Daxchem SF-17 .... L1
Daxchem SF-39 .... L1
Daxchem SHD-62 .... A2
Daxchem SMC-28 .... C1
Daxchem TL-17 .... N3,P1
Daxchem Wave A1
DLC-17 .... A1
FC-273 .... A4
FC-62 .... A1
GC-28 .... C1
ISC-351 .... F1
Jubilee B1
LCW-17 .... A1
LCW-73 .... A1
LD-95 .... A1
LDT-117 .... B1
LDW-17 .... A1
LDW-84 .... A1
LGC-106 .... A1
LGC-51 .... A1
LHS-17 .... E1
LMR-40 .... A3
LS-17 .... D2
PD-17 .... C1
PHD-73 .... C1
PL-73 .... L2
PL-95 .... L2
SD-51 .... C1
TYC-140 .... A2
TYC-84 .... A2

Daycon Products Co., Inc.

Deo-Drane C2
Fresh'n II A1
Hy-Q D1
Insecticide F2

Dayton Chemical Corporation

"Spray Foam" A4
Blue Blaze Concentrate Cleaner A1
Blue J Heavy Duty Cln, Dgrsr,
  & Stripper A1
Clean Brite Heavy Duty
  Cleaner A4
Easy Clean Glass Cleaner C1
Germicidal Disinfectant Clnr
  Deo. Fngcd D1
Liquid Steam Cleaning
  Compound A1
Mint Disinfectant C1
Pine Odor Disinfectant C1

De-Oxx, Incorporated

Deo-Roach & Ant Killer F2

Dean Scientific's

Deans Quatdean D1

Dearborn Chemical (U.S.)/ Chemed Corporation

Aqua-Serv B-202 Pulverized
  BWT G6
Aqua-Serv B-205 Pulverized
  BWT G6
Aqua-Serv B-208 Pulverized
  BWT G6,N2,G2
Aqua-Serv B-212 .... G6
Aqua-Serv B-214 .... G7
Aqua-Serv B-215 .... G6
Aqua-Serv B-216 Powdered
  BWT G6
Aqua-Serv B-222 .... G6
Aqua-Serv B-224 .... G6
Aqua-Serv B-226 .... G6
Aqua-Serv B-227 .... G6
Aqua-Serv B-266 Pulverized
  BWT G6
Aqua-Serv B-267 .... G6
Aqua-Serv B-268 .... G6
Aqua-Serv B-269 .... G6
Aqua-Serv BC-45 .... G7
Aqua-Serv BC-88 .... G7
Aqua-Serv C-312 .... G3
Aqua-Serv C-319 .... G7,G2
Aqua-Serv DSR A3
Aqua-Serv LSR A3
Aqua-Serve B-217 .... G6
Aqua-Serve B-270 .... G6
Aquafloc 401 .... G1
Aquafloc 408 .... G1
Aquafloc 409 .... G1
Aquafloc 411 .... G1
Aquafloc 421 .... G1
Dearborn Alkatrol G6
Dearborn Aqua-Serv B-211 ... G6
Dearborn Aqua-Serv B-271 ... G6
Dearborn Hydrazine G6
Dearborn Turbine Defoulant G6
Dearborn Water-Tec BC-88 ...... G7
Dearborn 1010 .... G6
Dearborn 1015 .... G6
Dearborn 1020 .... G6
Dearborn 1035 .... G6
Dearborn 1040 .... G6
Dearborn 150 .... G6
Dearborn 152 .... G6
Dearborn 155 .... G6
Dearborn 156 .... G6
Dearborn 157 .... G6
Dearborn 169 .... G6

Dearborn 17 .... G6
Dearborn 19 .... A1
Dearborn 200 .... G3
Dearborn 201 .... G6
Dearborn 203 AF G6
Dearborn 206 .... G6
Dearborn 229 .... G6
Dearborn 240 .... G6
Dearborn 241 .... G6
Dearborn 244 .... G6
Dearborn 250 .... G6
Dearborn 253 AF G6
Dearborn 255 .... G6
Dearborn 258 .... G6
Dearborn 291 .... G6
Dearborn 354 .... G6
Dearborn 502 .... G6
Dearborn 519 .... G7
Dearborn 521 .... G7
Dearborn 527 .... G7
Dearborn 541 .... G7
Dearborn 601 .... G6
Dearborn 603 .... G6
Dearborn 61 .... G6
Dearborn 63 .... G6
Dearborn 652 .... G6
Dearborn 653 .... G6
Dearborn 659 .... G6
Dearborn 659 LPA G6
Dearborn 66 .... G6
Dearborn 693 .... G6
Dearborn 694 .... G6
Dearborn 740 .... G5
Dearborn 751 .... G7
Dearborn 81 .... G6,G2
Dearborn 840 .... G1
Dearborn 841 .... G6,N2,G2
Dearborn 846 .... G1,N2
Dearborn 851 .... G2
Dearborn 863 Cooling Water
    Treatment G7,L1
Dearfloc 4943 .... G1
Dearmeen G6
Dearox 4870 .... G6
Dearox 4872 .... G6
Dearsol 105 .... A1
Dearsol 144 .... A3
Dearsol 45 .... G7
Dearsol 92 .... K2
Deartrol 4760 .... G6
Deartrol 4770 .... G6
Deartrol 4780 .... G6
Deartrol 4842 .... G6
Deartrol 4844 .... G6
Ecomeen G6
Endcor 4600 .... G6,G2
Endcor 4602 .... G2
Endcor 4689 .... G1
Endcor 4700 .... G7
Endcorfilm 4960 .... G6
Endcormeen 4910 .... G6
Endcormeen 4920 .... G6
Endcormeen 4930 .... G6
Exfoam 4860 .... G6
Klar-Aid 21 .... G1
Lime-Treet G7
Poly Tec BU G6
Polycarb G6
Polymate 160 .... G6
Polymate 661 .... G6
Polymate 662 .... G6
Polyphos 67 .... G2
Polytec BA G6
Polytec BB G6
Polytec BC G6,N2,G1
Polytec BD G6
Polytec BE G6
Polytec BF G7
Polytec BI G6
Polytec BJ G6

Polytec BL G6
Polytec BO G6
Polytec BQ G6
Polytec BS G6
Polytec BT G6
Polytec CD G7
Polytec CE G7
Polytec CI G7
Polytec CJ G1
Polytec CK G7,G1
Polytec DC G7
Polytec DR A3
Polytec LC G7
Polytec LR A3
Polytec RB G6
Polytec RC P1
Polytec RD G6
Polytec RE G6
Polytrol G6
Resin Cleaner 223 .... P1
Scale-Cleen A3
Sodium Sulfite-Decharacterized. G6
Super Filmeen 14. .... G6
Water-Tec RB-262 .... G6
Water-Tec RB-285 .... G6
Water-Tec RB-290 .... G6
Water-Tec RC-390 .... G7,G2
Water-Tec RD-96 .... A3
Water-Tec RS-480 .... G6
Water-Tec RS-491 .... G6
Zeoquest 102 .... P1

Decker's, Incorporated

Decker's Kills-It II D2
Deo-Drane C2
Impact A3
Kills-It Sanitizer & Disinfectant. D2
New Generation Foam Cleaner. A1
77 New Type Chemical Cleaner A1

Decra Cleaning Supply Company

Decra Low Foam Cleaner-
    Degreaser A4

Deeco Incorporated

WA-100 .... P1

Deep Valley Chemical Company

Deep Clean Dry A1
Deep Clean Liquid A1
Deep Clean Liquid Plus............. A4
Deep Cleaner Concentrate A4
Deep Down' Sewer Cleaner L1
Deep Duty Floor Cleaner C1
Deep Foam A1
Deep Valley Dehmer................. A3
Deep Valley Laundry
    Detergent............................. B1
Liquid Hand Soap E1
Liquid Hand Soap Plus.............. E1

Deere & Company

Detergent Additive.................... A1
Germicidal Sanitizer.................. D1
Spray Detergent Number 1........ A1
Spray Detergent Number 2........ A1
Spray Detergent Number 3........ A1

Deezeto Lubricants, Incorporated

FM Grease (AA)........................ H1
FM Grease EP........................... H1
FM Grease Hi-Temp No. 2......... H1

Del Chemical Corporation

Ban-Ice C1
Command A4
D-Tarer K1
Del 10-11 .... A4
Han-D....................................... E4

Hi-Glow A1
Invisa-Shield............................. P1
N-Zymes Formula STD-1000 .... L2
N-Zymes Formula STD-1000
    Sewage Digest. L2
Ox-Off...................................... C1
Prolong..................................... G3
Steameze................................... A1

Del Vel Chemical Company

Formula 1030 .... A4,A8
Khlor-Kleen A1
MSR.. A4

Delaware Valley Water Treatment Corp.

NJ-Antifoam G7
NJ-B 100................................... G6
NJ-B 103M G6
NJ-B 104................................... G6
NJ-B 110F................................. G6
NJ-B 115................................... G6
NJ-B 116................................... G7
NJ-B 117................................... G7
NJ-C 201 .... G7
NJ-C 221 .... G6
NJ-C 230 .... A3

Delco Chemical Products

Delco 2010 .... A3
Delco 2013 .... A2
Delco 2014 .... L1
Delco 2020 .... A2
Delco 2030 .... A2
Delco 2055 .... A1
Delco 2062 .... A1
Delco 2150 .... A1

Delcrest Foods

Acid Cleaner No. 1 .... A3
Chain Lubricant........................ H2
Chlorinated General Purpose
    Cleaner................................ A1
Defoamer N1
Double Strength Acid................ A3
Heavy Duty Cleaner No. 20 ...... A2
Hi Alkali Cleaner A2
Iodine Sanitizer D2,E3
Liquid Alkali No. 10................. A2
Liquid All Purpose Cleaner A1
Liquid Bactericide.................... D2
Liquid Caustic Additive No. 40. A1
Liquid CIP Acid No. 30............ A3
Liquid CIP No. 28 .... A1
Powdered Chlorinated All
    Purpose Cleaner................... A1
Powdered CIP No. 25 .... A1

Delma Chemical Company

Del 40 .... A1

Delta Chemical Laboratories

Lightning.................................. A4
Super Jet................................... A4
XK-100 .... A4

Delta Foremost Chemical Corporation

Foremost F-0114 Fortigel.......... A1
Foremost F-0137 Detergent A1
Foremost F-0176 Super XXX
    Fortigel............................... A1
Foremost F-0249 Compound...... M1,M3
Foremost F-0256 Loob A1,H2
Foremost F-0379 XHD
    Laundry Detergent............... B1
Foremost F-733 Foam Additive A1
Foremost F-749 Cleaner for
    Equip. A1
Foremost F-769 Rust
    Preventative H1

92

WK Window Kleener C1

DeWitt Chemical Company

All Surface Aerosol C2
Aluminum Cleaner &
  Brightener A3
Anti-Pollution Bowl Cleaner C2
Auto Mist Air Sanitizer G7
Auto Mist Insect Killer F1
Auto Mist Odor Neutralizer C1
Bol-Aid A3
Ceramic Tile Cleaner C2
Chain Lube H2
Chlor A1,Q1
Con-Clean A4
Concentrated Glass Cleaner C1
D.P. 111 .... A2
D.P. 113 .... A1
D.P. 114 .... A1
D.P. 115 .... A3
D.P. 116 .... D1
D.P. 117 .... H2
D-13 .... A1
D-13-V A1
D-21 Concentrate A1
D-21 Concentrate-V A1
D-23 .... A4
D-27 Insecticide F1
D-48 Powdered Hand Cleaner ... E4
DeWitt A-D C2
DeWitt AK C2
DeWitt Bowl Brite C2
DeWitt Caress C2
DeWitt Chlorine Bleach B1
DeWitt Chloroterg D1
DeWitt D-3A F1
DeWitt Dry Moly Lube H2
DeWitt DW Lubricant H1
DeWitt Extra E1
DeWitt Fly Bait F2
DeWitt Foam-A-Way A1
DeWitt Formula G C1
DeWitt Formula 2056-S F1
DeWitt Formula 225 .... A4
DeWitt Formula 24 .... C1
DeWitt Formula 376
  Chlorinated Steamer A1
DeWitt Formula 68 .... C1
DeWitt Formula 91 .... A4
DeWitt Formula 91-A D1
DeWitt FF Sanitizing Hand
  Cleaner E2
DeWitt Grease Ban C1
DeWitt Lemon Fresh A4
DeWitt Lube H2
DeWitt LFD A1
DeWitt Micro 2 Insecticide F1
DeWitt Micro-3 Insecticide F1
DeWitt Moisture Gard P1
DeWitt Moly (Aerosol) H2
DeWitt Monitor C1
DeWitt No. 4155 .... B1
DeWitt No. 7 .... E1
DeWitt Ranger A8
DeWitt Rust Remover A3
DeWitt Sanitary Spray
  Lubricant USP H1
DeWitt Scram Liquid Steam
  Cleaner A1
DeWitt Scrub Up E1
DeWitt Snap E4
DeWitt Soft E1
DeWitt SP G7
DeWitt Touch E1
DeWitt Ultra E1
DeWitt Wrap-Up C1
DeWitt 1D78 .... A5
DeWitt 16D75 .... G6
DeWitt 19D74 .... A8
DeWitt 2D73 .... E1

DeWitt 21D65 .... C3
DeWitt 21D75 .... A1
DeWitt 22D69 .... K1
DeWitt 29-D-73 .... A1
DeWitt 4-D F2
DeWitt 4D74 .... H1
DeWitt 7D78 .... A1
DeWittco 15 .... A1
DeWittco 8 .... A2
DeWittco 9 .... A2
Foamer A1
Formula 135 .... C1
Formula 150 .... K1
Formula 16D59-V A1
Formula 200 .... K1
Formula 34 Car Wash A1
Formula 344 .... G7
Formula 375 .... A1,B1
Formula 5458 .... A1
Formula 81-V A4,C1
Garbage Truck Maintainer C1
Glass Cleaner A1
Glaze C1
Grout Cleaner C2
GTC L1
H-59 No-Pine A4
Hand Lotion E4
Heavy Duty Cleaner A6
Ice Melt C1
Kream Kleaner A6
L-51 Cleaner A2
Liquid Ice Melting Compound .. C1
Liquidizer Soap A4
LDS L1
LMC A1
Micro D C1
Mr Plumber C1
No-Pine Cement Cleaner A1
Open Gear Lubricant H2
Pine Disinfectant C1
Pinky E1
Release C1
Stainless Steel Polish A7
Steam Cleaner A2
Stove Aid Stove & Oven
  Cleaner A1
Strip Fast Steam Cleaner A2
Super Ice Melt C1
Super Machine Dishwash
  Compound A1
Super Sil H1
Super Wit-O-Bead Concentrate . A1
Tile and Grout Cleaner C2
Wit-D-Germ C2
Wit-Deo C1
Wit-Trol 2-S L1
Wit-Wonder A4
Wit-Wonder-V A4
Witt-Aid D-13 .... A1
Witt-Aid D-21 .... A1
Witt-Aid Foamer Concentrate A1
Witt-Aid LMC A1
Witt-Aid No. 7 .... E1
Witt-Aid No. 99 .... A1
Witt-Aid Wonder A4
Witt-Aid 13D61 .... N2,N3
Witt-Aid 15D78 .... A2
Witt-Aid 19D71 .... A2
Witt-Aid 22D65 .... A3
Witt-Aid 22D67 .... A1
Witt-Aid 225 .... A4
Witt-Aid 35D72 .... E2
Witt-Aid 375 .... A1
Witt-Aid 376 .... A1
Witt-Aid 38D62 .... A1
Witt-Aid 40D70 .... H1
Witt-Aid 53D69 .... B1
Witt-Aid 54D69 .... A1
Witt-Aid 9D68 .... D2,E3
Witt-All Kleen A1

Witt-Amine D1,Q3
Witt-AC C2
Witt-Check Antiseptic Liq
  Hand Soap E1
Witt-Check D-13 .... A1
Witt-Check D-21 .... A1
Witt-Check Defoaming
  Eggwash Q1
Witt-Check Feather Softener N1
Witt-Check Floor Cleaner A4
Witt-Check Foam Depressant A1,Q5
Witt-Check LMC A1,Q1
Witt-Check No. 1 Space Spray .. F1
Witt-Check No. 3 .... F2
Witt-Check No. 4 .... A4
Witt-Check No. 99 .... A1
Witt-Check 10D61 .... A3
Witt-Check 12D63 Bact.
  Eggwash Q1,Q4
Witt-Check 22D67 .... A1
Witt-Check 225 .... A4
Witt-Check 31D70 .... A1
Witt-Check 35D72 .... E2
Witt-Check 375 .... A1
Witt-Check 376 .... A1
Witt-Check 4-D F2
Witt-Check 40D70 .... H1
Witt-Check 51D65 .... A4
Witt-Check 9D68 .... D2,E3
Witt-Check 9D71 .... A1
Witt-Check 91 .... A4
Witt-D-Fend A4
Witt-Dis-Solvit L1
Witt-Haste A1
Witt-I-Deel A6
Witt-Mitt E4
Witt-O-Green C1
Witt-O-Kreem E4
Witt-O-Matic Deodorant C2
Witt-O-Matic Insecticide F1
Witt-O-Safe Utility Cleaner C2
Witt-O-San C1
Witt-O-Shield P1
Witt-O-Stop C1
Witt-O-Suds E1
Witt-O-Zyme B1
Witt-Odo-Kill C1
Witt-Oil-Sorb J1
Witt-Perform D1
Witt-Pona F2
Witt-Punch E4
Witt-Rid F1
Witt-Save A4
Witt-Sparkle Aerosol C2
Witt-Trol 1-F L1
Witt-Wash A1
Witt-Wonder-V A8
Wittamine A D2,Q3
Wittco No. 28 .... H2
Wittco No. 30 .... A4
Wittlon H1
Wittrend A1
Wittrol Extra L1
X-8956 .... N2
X-9156 Chemical Cleaner A3
X-9356 .... N3
1D60 .... A3
1D62 .... A1
1D71 .... A1
10D61 .... A1
10D67 Foam Depressant A1,Q5
1010 (No-Pine) A4
1160 .... C1
12D63 .... Q1
13D60 .... A4
13D61 .... N2,N3
14D62 .... A1
15D64 .... A2
16D59 .... A1
16D73 Chlorinated Egg Wash ... Q1

17D65 .... Q2
18D62 .... A2
18D71 .... M1
21D59 .... A2
21D69 .... A1
22D65 .... A3
23D62 .... A2
23D70 Cooking Pot &
    Machinery Clnr...................... A1
24D69 .... A1
24D70 Concrete Floor Cleaner.. A4
25D64 .... A1
25D70 Equipment Cleaner......... A1
26D70 Chlorinated Cleaner A1
28D61 .... A2
28D68 .... A1
28D69 .... A1
33 Boiler Compound G6
38D62 .... A1
38D70 .... A1
39D67 .... Q1
39D70 .... A1
4-D F2
44D70 Egg Wash Q1
45D69 Liquid Cleaner................ A1
48D70 .... A1
49D67 .... A2
49D70 .... A1
5D74 .... A1
50D67 .... A1
51D65 Freezer Locker Cleaner.. A4
53D69 .... B1
54D69 .... A1
55D71 .... A2
6D59 .... N1
6D66 .... A1
7D64 .... N2
75.. D2,Q6,E3
75D70 .... C2
8D59 .... A1
80.. D1,Q1,Q6
8056 Packing House Cleaner.. A1
81 All Surface Cleaner.............. A1
8652...................................... A2
9D61 .... A2
9D67 .... A1
9D68 .... D2,E3
99.. A1

DeYoung Chemical Company
DC-88 Cleaner Degreaser A4

Diagraph-Bradley Industries, Inc.
Quick Spray General Purpose
    Silicone H1
Quik-Spray Food Grade
    Silicone H1

Diamond Chemical & Contracting Company
All Purpose Cleaner
    Concentrate A4
Big "D" A1
Boilerite 100.......................... G6
Cocoanut Oil Hand Soap......... E1
Dual Power-43........................ D1
Klean-14 .... A1
Lemon Power 20..................... A4
Liquid Drain Opener................ L1
Safety Solvent........................ K1
Tile and Porcelain Cleaner C2

Diamond Chemical Co.
1080...................................... A1

Diamond Chemical Co., Inc.
Diamond Freezer Kleener.......... A5
Spray-Fog F1

Diamond Chemical Company
General Purpose Spra #2.......... F1
Spectrum D1
Super 175............................... A1

Diamond Janitorial Supply Company
Patent Non-Butyl Emulsifier A1
Patent Plus A1

Diamond Products, Inc.
Super Enzymes for Sewerage
    Systems L2

Diamond Shamrock Corporation
Caustic Soda Beaded................ A2,G6,L1,N2,N3
Caustic Soda, Flake................. A2,G6,L1,N2,N3
Caustic Soda, Granular A2,G6,L1,N2,N3
Caustic Soda, Liquid 50%......... A2
Chelated Alkali Superkel A2
Clipper Cleaner....................... B1
Compounder's Caustic Soda...... A2,G6,L1,N2,N3
CH-100 .... A2
General Maintenance Cleaner ... C1
Hardnox Alkali A2
Hi-Lite B1
Hi-Ratio Silicate A2,B1
Hi-Test Alkali #3 .... A2
Kin-Klor B1
Liquid Superkel Concentrate A1
Liquid Superkel 100 .... A2
Liquid Superkel 1000 .... A2
Liquid Superkel 150 .... A2
Liquid Superkel 1500 .... A2
Liquid Superkel 200 .... A2
Liquid Superkel 2000 .... A2
Liquid Superkel 250 .... A2
Liquid Superkel 300 .... A2
Liquid Superkel 350 .... A2
Liquid Superkel 400 .... A2
Liquid Superkel 500 .... A2
Liquid Superkel 600 .... A2
Liquid Superkel 700 .... A2
Liquid Superkel 800 .... A2
Liquid Superkel 900 .... A2
Maxi-Det B1
Orthodet A2,B1
Ortholate A1
Paralate A2
Paralate S A2
Sanuril 115 .... G7
Soda Ash A1,G6
Sodium Meta., Anhyd/All
    Grades A1,Q1,N1,G6,N4,N2
Sodium Meta., Pentahyd/All
    Grades A1,N1,G6,N4,N2,N3
Sodium Orthosilicate/
    Anhydrous.......................... A2,B1
Special Alkali #45.................... A2
Special Alkali #60.................... A2
Special Alkali #70.................... A2
Wondaco B1

Diamond-Rite Products Company
Do-All Creme Cleanser C2
Drain Solvent......................... L1
Formula X-100........................ A4
Hospital Disinfectant
    Deodorant.......................... C2
Lotionized Detergent A4
Porcelain Cleaner C2
Sani-Clean D1,Q3
Sta Brite Bowl Cleanse............ C2

Diatomic Chemical Company
Super Sorb J1

Dickler Chemical Laboratories, Inc.
All Surface Cleaner.................. A1
Concentrated Floor Conditioner C1

Hand Soap Liquid E1
Hand Soap Lotion.................... E1
Heavy Duty Bowl Cleaner......... C2
Heavy Duty Steam Cleaner
    Liquid A2
Liquid Drain Opener &
    Maintainer L1
Oven & Grill Cleaner................ A8
Pink Lotion A1
Porcelain and Bowl Cleaner
    Protector C2
Power Cleaner A4
Quat-Klean-Clnr-Disnf-Deo-
    Fngcd-Vired........................ D1
Rinse Disinfectant-Sanitizer
    Deodorizer.......................... D2
Stainless Steel Cleaner & Polish. A7
Surface Scum Remover A4
Wind-O-Shine C1

Dietary Products
D-Line All Brite C1
D-Line Foam Oven Cleaner....... A8
D-Line Oven Cleaner
    Foam.Carbon Solvntzr........... A8

Dilco Chemicals
Boilerite 100 .... G6
Steamrite 200 .... G6

Dill, J. J. Company, The
Aerosol Insecticide.................. F1
Dill Rat-Kill Paraffin Rat
    Blocks F3
Formula B30-150 Hvy Dty
    Pyren. Mill Fog F1
Formula B30-300 Hvy Dty
    Pyren. Mill Fog F1
Formula B30-70 Hvy Dty Pyr
    Mill Fog............................. F1
Formula B50-250 Ex. HD
    Super Mill Fog F1
Formula C-10 Pyrenone Fly
    Spray................................ F1
Formula C15 Hvy Dty Pyr.
    Mill Spray.......................... F1
Formula M50 Pyr Fogging
    Insecticide.......................... F1
Formula V-5 Sugar Fly Bait F2
Formula 163 Rat Kill............... F3
Formula 66 Mal-Fog
    Concentrate F2
Pyrethrin Ems......................... F1
Pyrethrin OS........................... F1
Pyrethrin Roach Spray
    Concentrate F1
Tri-Mal Residl Form.78
    Roach&Ant Spray................. F2
Tri-Ron Residl Form.75
    Roach&Ant Spray................. F2
Tri-Zin Residl Form.77
    Roach&Mill Spray................. F2
Tri-Zin Residual Formula 81..... F2

Diluent, Incorporated
Big Dog Concrete Cleaner......... A4
Brightline Special A2
Deluxe Machine Dishwashing
    Comp. A1
Dynamo Detergent................... A1
Extra Heavy Duty Concrete
    Clnr A4
Golden Gator Concrete Cleaner C1
Golden Magic Cleaner A4
Grime Scat A1
Iodo-175 Concentrated Iodine... D2
Liquid Hand Soap 15................ E1
Liquid Hand Soap 20................ E1
Liquid Hand Soap 40................ E1
Liquid Steam Cleaner A2

Multi-Blue A1
New Formula Pink Cleaner A4
No. 80 Space Spray.................... F1
Pine Odor Disinfectant C1
Pine Oil Disinfectant Coef. 5 ..... C1
Pink Satin E1
Premium Fly & Roach Spray..... F1
Quaternary Amm. Germicide
    Conc. D1
Sewer Solvent............................ L1
Steam Cleaning Compound A2
SP-1105...................................... A1
SP-182.. A1
SP-25.. A1
Wax Scat A1
X-150 Chlor. Machine Dish
    Wash A1
360 Milkstone Remover A3

Dipco Inc.

FM 100 Lubricating Grease H1
FM 109 Lubricating Grease No.
    1. .... H1
FM 109 Lubricating Grease No.
    2. .... H1
FM 109 Lubricating Oil No. 00. H1
FM 619 Lubricating Oil No. 1... H1
FM 619 Lubricating Oil No. 2... H1
FM 619 Lubricating Oil No. 3... H1
FM 619 Lubricating Oil No. 4... H1
FM 619 Lubricating Oil No. 5... H1
FM 619 Lubricating Oil No. 6... H1
FM 619 Lubricating Oil No. 7... H1
FM 619 Lubricating Oil No. 8... H1
FM 619 Lubricating Oil No. 8A H1
FM 754 Lubricating Oil No. 1... H1
FM 754 Lubricating Oil No. 2... H1
FM 754 Lubricating Oil No. 3... H1
FM 754 Lubricating Oil No. 4... H1
FM 754 Lubricating Oil No. 5... H1
FM 754 Lubricating Oil No. 6... H1
FM 754 Lubricating Oil No. 7... H1
FM 754 Lubricating Oil No. 8... H1
FM 754 Lubricating Oil No. 8A H1
111 Wateresist CA #1................ H2
111 Wateresist CA #2................ H2
116 GP Lube CA #1 .... H2
116 GP Lube CA #2 .... H2

Dishmate

All Purpose Laboratory
    Cleaner.................................. A1
Bake Pan Cleaner A1
Chlorinated Machine Dish
    Washing Cmpd. A1
Deep Fat Fryer Cleaner A2
DM #295.................................... G7
DM #466.................................... G6
DML #400.................................. G2
DML #422................................... G7
DML #430................................... G6
DML #455................................... G6
Food Plant Cleaner A1
Hychlor A1
Packing House Cleaner.............. A1
Quaternary Ammonium
    Germicide Conctd. D1
Sewer Pipe Cleaner.................... L1

Distribution Affiliates H. K. Porter Company, Inc.

da Medicated Skin Cream E4
da Medicated Skin Lotion E4
da Protective Cream Solvent-
    Resist. Type........................... E4
da-30 .... E4
da-61 .... E1
da-72 .... E1

Distributors Processing, Inc.

Micro-Aid.............................. A4,L1

District Supply, Incorporated

Certified Controlled Lo Foam
    Clnr A4

Diversey Chemicals Div. of The Diversey Corporation

Accel.. A1,E1
Add-2... A2
Asid-O-Phy A3,G7
Bac-Stop..................................... D2,Q6
Bril-Tak A2,L1
Bru-Spra #1 .... A2
Bulk-Et A1
Can Nu A1
Chain-Eze................................... A1,E1,H2
Cleaner Disinfectant.................. D1
CDC.. A1
CST.. A1
D-Phene...................................... C1
D-Stone....................................... A1
Dairy Deelite A3
Dairy Divoluxe........................... A1
Dairy Fortify.............................. A1
Dairy Gleemate A1
Dairy Korral A1
Dairy Liquasol A2
Dairy Shur-Brite A2,L1
Dairy Shur-Klor A2
Dairy Shurtrak 1 .... H2
Deogen Plus................................ A1
Di-Foam...................................... A1
Dibac D2,Q4,N4
Dican XX A3
Dicoloid FF................................. A3,Q1,Q3
Dicolube SL................................ A1,H2
Diflex.. A1
Diflex W..................................... A1,N4
Dilac A3
Diokem.. A1
Diopreen..................................... A1
Dipak A1
Disinfectant Deodorizer............. C2
Diton A A1,Q1
Diton B.. A1,Q1
Diverfoam Plus........................... A1
Diversey Dairy Shurtrak 2........ H2
Diversey Dairy Super Dilac...... A3
Diversey Drain Cleaner.............. L1
Diversey No. 1 .... A3
Diversey Scale Remover A3
Diverside K.................................. F1
Diverside KS............................... F1
Diversol CX Arodyne................. D1,Q1,Q3
Divogleam A2,L1
Divoklens E1
Divoklor..................................... A1,Q1,Q3
Dual-Dip P1
DS-9-333...................................... A3
Eggs-Act Q1,Q5
Eggs-Pel Q5
Elect.. A1,Q1
Elite A2
Everite... A3
F-120... A1,Q1,Q3
F-237... H2
F-262... A2
Glo-Tak A2
Hand Cleaner/Sanitizer.............. E2
Heavy Duty Acid Cleaner A3
Hog Scald Compound................. N1,N2
Improve....................................... A2
Key A1
Kold-Temp.................................. A1,H2
Kontrex...................................... A1
Liqua-Brite A2
Liqua-Brite V.............................. A2

Liqua-Brite X A2
Liqua-Flex A2
Liquid Aluminux A2,N4
Liquid Diokem............................ A1
Liquid Protex.............................. A1
Liquid PMMC............................. A1
Liquid Scalite............................. G2
LD-2... H2
No. 19 .... A1
No. 909 .... A1
Oven Cleaner A8
Pan-O-Mite................................. A1
Peel-Away.................................... N4
Perfon A3,A6
Protex A1
Pyron A1
PMMC... A1
PX 1636 .... A1
PX-1655 .... H1
PX-1690...................................... A4
Q.E.D. 11 .... A1,Q1
Quaternary Cleaner Disinfectant A4
Relion A2,N4
Result Plus A1,Q1
Saf-Sol D1,Q3,N4
Scalite... G2
Schlitz Lube #400...................... H2
Selcon Formula 117 .... A2,L1
Selcon Formula 149 .... A2,L1
Selcon No. 1 .... A3
Selcon No. 10............................. A1
Selcon No. 2 .... A1,Q1,Q3
Selcon No. 3 .... A1
Selcon No. 4 .... A1
Selcon No. 6 .... A2
Selcon No. 7 .... A3,Q1,Q2
Selcon No. 8 .... A2
Selcon No. 9 .... A2
Selcon 100 .... A1,E1,H2
Selcon 101 .... H2
Selcon 103 .... A2
Selcon 119 .... A2
Shur-Lube M H2
Shurlube LDH............................. H2
Shurlube LDM............................ H2
Shurlube LDS.............................. H2
Shurlube LDS-B.......................... H2
Shurlube Pet............................... H2
Shurlube XH............................... A1,H2
Shurspray A2
Spec Tak No. 50......................... A2,N4
Spec-Tak No. 2........................... A2,N4
Spec-Tak No. 3........................... A2,N4
Spec-Tak 1000............................ A2,L1
Spectrum D2
Spray Insecticide F1
Stainless Steel Cleaner And
    Polish A7
Stream... A1
Super Foam-Brite A1
Super Takeoff A4
Sure Kill Brand.......................... F1
SCA-658 .... Q5
SCA-659 .... A1,H2
SCA-661 .... H2
SCA-675 .... A3
Tik-Et A1
Tower.. A1
Tripe Wash Compound.............. N3
Tripel.. A3
TIG A1,E1
Udder-Bac D2,Q6
UD-100.. D1
Whirl... A2
Whirl Way B1
Whirlaway Booster B1
Whirlaway Break......................... B1
Whirlaway Chloro-Eight B1
Whirlaway Dry Bleach............... B1
Whirlaway Fabric Softener B2

Drewtrol-2501.............................G6
Drewtrol-3500.............................G6
Drewtrol-3501.............................G6
Drewtrol-4500.............................G6
Drewtrol-4501.............................G6
Drewtrol-5500.............................G6
Drewtrol-5501.............................G6
Drewtrol-6500.............................G6
Drewtrol-6501.............................G5
Drewtrol-7500.............................G6
Drewtrol-7501.............................G6
Drewtrol-8500.............................G6
Drewtrol-9500.............................G6
Ion Exchange Resin Cleaner.......P1
Organic AL...................................G6
Organic-ACAF-FG.......................G6
Organic-U14L..............................G6
OCS-M...G6
OST-P..G6
P-55..G7
PHR...G7
Saf-Acid (Indicating)...................A3
Saf-Acid G Indicating.................A3
Saf-Acid G Non-Indicating........A3
Saf-Acid Non-Indicating.............A3
Steamfilm-FG..............................G6
SLCC-35.......................................G6

Dri-Rite Company, The
Dri-Rite..J1

Dri-Slide, Incorporated
Dri-Slide Molybdenum
    Disulfide Base H2

Drum Chemical Corp.
Glisten Pan Cleaner A1

Drummond American Corporation
Assure Hospital Disinfectant
    Deodorant...............................C2
Best Effort Low Foam Cleaner.A4
Cavalcade Acidulade Foam
    Clean r....................................A3
Dash Creme Cleanser
    Concentrate............................C2
Hand-Some Lotionized Liquid
    Soap E1
Hay-Maker Fogging Insect
    Killer (Aer.) F2
Head-Way Silicone Lubricant
    Spray.......................................H1
Hold-Fast Emulsion Cleaner......C1
Key-Note Heavy Duty Cleaner.A4
Lunge Bowl Cleaner
    Disinfectant............................C2
Mandate Glycol Air Sanitzr &
    Surf.Disnf................................C1
Mine-Field Roach & Ant Killer.F2
Mine-Field Roach & Ant Killer
    (Aerosol)..................................F2
Mobilize Steam Cleaner A2
On DutyHospital Germicidal
    Detergent................................C1
On Your Mark Vandal Mark
    Remover..................................K1
Open & Shut Nut & Bolt
    Lsnr&Rust Pntrnt H2
Patina Glass Cleaner C1
Pipe Down Prilled Drain
    Opener.....................................L1
Pounce Bowl Sanitizer................C2
Proud Waterless Hand Cleaner..E4
Quest Disinfectant Cleaner
    Deodorant...............................C1
Rectify Deoxidizer And
    Cleaner.....................................A3
Sentinel Coconut Liquid Hand
    Soap E1

Sequester Detergent
    Disinfectant Cleaner..............A4
Sprint Multi-Purpose Cleaner....A1,E1
Status Germicidal Foam
    Cleaner....................................C1
Strive Non-Acid Bowl Cleaner..C2
Tight Spot Dry Spray
    Lubricant.................................H2
Touch And Go Multi-Use
    Cleaner/Degreaser C1

Dryden Oil Company, Inc.
Drydene B&R Bearing
    Lubricant #2............................H2
Drydene Bench Oil Light............H1
Drydene Bottle Capper & Filler
    Oil H1
Drydene Dalze Bench Oil H1
Drydene Divider Oil....................H1
Drydene Food Machine
    Lubricant.................................H1
Drydene Lid Oil #350.................H1
Drydene Oil Absorbent J1
Drydene Press Oil 150 ....H1

Du-Rite Chemical Company, Inc.
#505 Cleaner................................A2
#506 Cleaner................................A2

Dub-L-Kleen Chemical Corporation
D.K. No. 88 Truckleen................A1
Dub-L-Enzyme 'A'.......................L2
Dub-L-Enzyme 'B'.......................L2
Dub-L-Enzyme 'C'.......................L2
Dub-L-Enzyme 'D' L2
Dub-L-Flo....................................C1
Dub-L-Kleen 44...........................C1
Dub-L-Power...............................A1

Dugger Products Company, Inc.
HC-50 ....A1
HD-100 Concrete Cleaner C1
HD-50 Deep Vat Cleaner............A2
XD-30 Active A4
XD-37 Heavy Duty
    Concentrate............................A4
XD-40 Big Mack A1

Duke Industries, Inc.
Coco-Glyco Hand Soap..............E1
Disinfectant Cleaner....................A4,C1
Disinfectant-Sanitizer-
    Deodorant D2
Duke All Purpose Cleaner A4
Duke Chlorinated Deluxe
    Machine Dishwash..................A1
Duke Heavy Duty Degreaser A4,A8
Duke Powdered Pot & Pan
    Cleaner.....................................A1
Duke-Kleen Powdered Hand
    Soap C2
Foam Away A4
Lemon Scented Disinfectant A4
M.S.A.P..A4
PHF Fog & Contact Spray F1
Wax Stripper................................A4

Dumont Sales Company
#20 Conveyor Lubricant.............H2
#30 Conveyor Lubricant.............H2
Act 2...A1
Acto Degreaser/Cleaner..............A1
Du-O-Tex......................................A4
Dumont Empac Cleaner A4,A8
Dumont Super Cleaner A1
Force...A4
Milo A1

Dunrite, Incorporated
All Purpose Concentrate A4

Avancie D1
Blue Tint Toilet Bowl Cleaner ..C2
C-It-Go...L1
Creme Cleanser............................C2
D-400..D1
De-Greaser Detergent.................A1
Disinfectant Spray C1
Glass Cleaner...............................C1
Glass Cleaner (Aerosol)..............C1
Heavy Duty Concentrate A4
Insect Spray F1
Le Du Deodorant........................C2
Lime Solvent................................A3
Liquid Hand Soap E1
Liquid Steam Cleaner A1
Mechanics Hand Cleaner............E4
Pine Cleaner................................C1
Sani-Shine Dsnfctnt Bathroom
    Clnr C2
Spring Blossom............................C1
Toilet Bowl Cleaner....................C2
Toilet Bowl Deodorant...............C2
Wall Block Deodorants C2
Wax Stripper................................A4

Duo Chem Labs Inc.
Dizolv..L1

Dura-Chem Inc.
Insta-Kleen A4

Dura-Chem, Inc.
Concrete Cleaner..........................C1
Dura-Mint Disinfectant...............C2
Dura-Wash Detergent A1
Oven & Griddle Cleaner.............A8

Durable Products Inc.
Odor Control Granules #100.....C1
Odor Control Granules #200.....C1
Odor Control Granules #300 W C1

Durall Manufacturing Company, Inc.
Dura-Kleen A3
Dura-Sol Concrete Floor
    Degreaser.................................C1
Metal Prep A3

Durstan Ltd.
Mistone..A7

Durvet, Inc.
Iodine-Cleaner Sanitizer..............D1

Duryea Products Company, Inc.
Durco Foam Quick A1
Durco Freezer & Locker
    Cleaner.....................................A5
Durco Steam Clean A1
Durco 8 Degreaser & Cleaner ...A4
Durco 8 Heavy Duty Cleaner.....A4
Durcophene 100 ....A4
Durcophene 128 ....A4
Durcoquest...................................A4

Dutch Glo Chemical Company
Aqua-Glo Degreaser Heavy
    Duty Conc................................A1
Coconut Oil Hand Soap..............E1
Hand Dishwashing Detergent ...A1
Lemon Disinfectant......................C2
Lemon Scented All Purpose
    Cleaner.....................................A4
Lime Away De-Liming
    Compound................................A3
Mint Disinfectant.........................C2
Pine Odor Disinfectant C1

101

TSL-6-FT..................................A1
TSL-6A-FT................................A1
TVC-25-LT...............................A1
Ultra-Pure................................A1
Vac-Eze....................................A1
Wash-Up...................................E1
Wax Away................................A4
WC-5000.................................G7
WC-5020.................................G6
WC-5050.................................G6
WC-5060.................................G7
WC-5070.................................G7
X-E-Cute (Aerosol)..................F1
X-E-Cute (Bulk).......................F2
X-Pel.......................................G7
X-Perge...................................G7
X-980......................................A1
Zdu...K1
Zephyr-oid...........................A1,H2
Zolv..K1
100..A3
211.......................................A3,B2
306..N4
422M.......................................A1

DuChemCo Products Company

Acid Cleaner B.........................A3
All Purpose Cleaner..................A1
Ambeco Chlorinated Cleaner.....A1
Ambeco Drychlorine Shroud
    Bleach..................................B1
Concrete Cleaner......................A4
Dri Acid Cleaner.......................A3
Electrical Solvent......................K2
Gelled Acid Cleaner..................A3
Granulated Detergent................A1
Heavy Duty Cleaner..................A2
Heavy Duty Foam Cleaner.........A1
Ice Melting Compound..............C1
Liquid Acid Cleaner..................A3
Liquid Detergent.......................A1
Liquid Foam Cleaner.................A2
Liquid Hand Cleaner.................E1
Liquid Hand Soap A..................E1
Liquid Heavy Duty Cleaner.......A1
Safety Bleach............................B1
Scald Compound.......................N2
Sewer Cleaner..........................L1
Shroud Detergent......................B1
Smokehouse Cleaner.................A2
Tree Clean................................A2
Tripe Denuding Compound A...N3
Tripe Denuding Compound B...N3
Tripe Wash 10C.........................N3
Tripe Wash 5C..........................N3
Trolly And Metal Cleaner..........A2

DuCor Chemical Corporation

Beats All Insecticide..................F2

DuPont de Nemours, E.I. & Co.

GBS (Globular Sodium
    Bisulfate).............................A3
Hydrochloric Acid.....................A3
Hydroxyacetic Acid..................A3
Perone 35.................................N3
Sodium Metasilicate..............A1,N3
Sulfamic Acid...........................A3

Dymon, Inc.

Action.....................................E1
Anti-Seize................................H2
Aqua-Solv................................C1
Artic Clean..............................A5
Baysect....................................F2
Clean-N-Shine..........................C1
Deodorant Granules...................C1
Double Action Concentrate.......F2
DC-1250...................................A4
Electrical & Mechanical Helper.H2

Expel.......................................F2
Industrial Formula Hand Soap...E1
Liquid Live Bacteria..................L2
Lotionized Hand Soap...............E1
Mechanic's Friend.....................H2
Power Foam.............................C1
Sanitary Lube...........................H1
Silicone Spray..........................H1
Sta-N-Less................................A7
Stay-N-Remove........................C1
Syntha Mist.............................F1
Tri-Jell...................................L2
True Grit..............................C1,A8

Dyna Systems/ A Div. of Partsmaster, Inc.

Big Moly.................................H2
Clean N' Dry............................H1
CB-15.....................................H1
Dura-Lube Silicone Lubricant...H1
Dyna-Pack...............................P1
Dyna-Purge.............................L1
FP-116....................................H1
Giv-Way..................................H2
Grease-B-Gone.........................E1
HYD 10/30 Multi-Grade
    Hydraulic Oil.......................H2
Link-Lube................................H2
Mo-Ban...................................K2
MB-10 Moly Grease..................H2
MD-113...................................H2
PN-105....................................H2
PR-120....................................C3
Results.....................................F1
Sta-Ded...................................F2
Tac-N-Hold Belt Conditioner....P1
Teflon Tape.............................P1

Dyna-Chem Laboratories

Chem CC 11.............................K2
Food Safe................................H1
Glove Skin Shield Cream...........E4
Lightning All Purpose Cleaner..C1
S. S. Pol..................................A7
Solv..K1
Spray-Fog................................F1
Super Lube Silicone Spray.........H1
Tef Lube..................................H2

Dyna-Mist Chemical Company, Inc.

Dyna-Mist All Purpose Cleaner.A1
HP-9 High Pressure Boiler
    Feed Comp...........................G6
Quik Drain Snake......................L1

Dynamic Chemicals

Big "D"....................................A1
Brigade....................................C1
Centurion.................................A4
Chain & Cable Lubricant...........H2
Clean Kote...............................C1
Derma Clean Concentrated
    Liquid Skin Cl......................E1
Duo Quat.................................D1
Dura Cide Residual Insecticide
    w/Baygon.............................F2
Dyna Cide...............................F1
Dyna IC-150.............................A1
Dyna Zyme..............................L2
Dynamo Granular Sewer and
    Drain Solvent.......................L1
Grip..P1
Hydro Sol................................A1
Hydrochem 100........................G6
Hydrochem 200........................G6
Hydrochem 2000......................G7
Hydrochem 300........................G7
Hydrochem 400........................G7
Oxidex....................................A3

OC-227 Concentrate.................. C1
Safety Solvent............................ K1
Sani Dyne.................................... D2

Dynamic Chemicals International

Bottle Wash.................................. A2
Formula 201L A1
Formula 231 .... A1
Formula 233 Steam Cleaner A2
Formula 239 .... A1
Formula 269 .... A1
Formula 630 .... A1
Tank 71 Heavy Duty Vat
  Cleaner.................................... A2

Dynamic Supply, Inc.

Jewel.. A1
Lemonee...................................... A4
Liquid Super "C" A4
Neutral Cleaner A4
Super Z A4

Dynasurf Chemical Corporation

CS-117 Cleaner A4
Dynabol....................................... C2
Dynaphene................................... A4
Dynasept...................................... A4
Industrial Degreaser................... A1
Lotion-Creme Hand Soap Pink.. E1
Odorless Cleaner A1
Oven Cleaner C-447 .... A8
Total Cleaner A4

Dynatech Inc.

Action.. C2
Dyna-Sect Super Fly & Roach
  Spray....................................... F1
Dyna-Sol A4,A8
Esteam A4
Kitchen Klean A4,A8
Kleenz-All C1
Micro-Shield C1
Quatrex Germicidal Detergent... A4
Sani-Guard Hospital
  Disinfectant & Deo................ C1
Sani-Lube H1
Smooth Waterless Hand
  Cleaner.................................... E4
Super Solv................................A4,A8

Dynatek

D-34 Safety Solvent K1
Do All C1
Dyna-Soar A1
DB 50 Contact Cleaner............... K2
Kill'M Insect Spray..................... F2
Kleen E4
Lithlube....................................... H2
Oven Cleaner A8
Possit-Grip P1
Sani-Lube H1
Shield E4
Steel Shine Stainless Steel Metal
  Polish A7
T-Tape H2
TFE T-Lube H2

DyChem International, Inc.

"Report" A1
Ampsol K2
Amstrip.. C1
Big Ben.. A4
Big Ben II A4
Blizzard....................................... B1
Brite A3
Brite & Shiney A3
Brite & Shiney III A3
BLC 300 .... B1
Cleansit A1

Cool-Steam.................................. A1
CC 217... A1
CDS-320....................................... D2
DC 303 .... A1
Elbow Grease A4
Floor Brite A4
GP-117... A1
Hands-Down................................ E1
HD Steam A1
Jack Flash A1
Krush... A4
Lion-X D2
Maximum..................................... A1
Quick Drain L1
Ruf Rig.. A1
Scale Safe A3
Scale X A3
Scale-XX A3
Score.. E1
Take 5.. A4

E

E & Z Chemicals, Inc.

Blue Butyl Cleaner...................A4,A8

E H B Laboratories, Incorporated

EHB Formula 102 .... A2
EHB Formula 12.......................... A1
EHB Formula 1332...................... A1
EHB Formula 1332B.................... A1
EHB Formula 178 .... A1
EHB Formula 277 .... A1
EHB Formula 280 .... G5
EHB Formula 30 AC................... A1
EHB Formula 32.......................... A1
EHB Formula 36.......................... A1
EHB Formula 44.......................... A2
EHB Formula 46.......................... A1
EHB Formula 50.......................... A1
EHB Premier E1

E S P Sales

Cleaner Degreaser A1
No. 216 Multi-Purpose Cleaner/
  Dgrsr A4
No. 229 All Purpose Cleaner A1

E T C Chemical Company Div. of ETC Sales Corporation

ETC #105 .... A4
ETC #115 .... C2
ETC #120 .... A4
ETC #125 .... A4
ETC #140 .... E1
ETC #145 .... C2
ETC #160 .... E1
ETC #175 .... C1
ETC #185 .... H2
ETC #190 .... C1
ETC #200 .... E4
ETC #205 .... A4
ETC #210 .... E1
ETC #215 .... A3
ETC #230 .... A4
ETC #235 .... A3
ETC #250 .... F1
ETC #255 .... A4
ETC #260 .... C1

E.D.S. Chemicals & Coatings Co.

Below Zero................................... A5
E.D.S. Quaternary Cleaner......... D1
Hot Stuff..................................A4,A8

E.P.G. Chemicals

#88 Scale Cleaner A3

A-108 H Boiler Water
  Treatment Q6
T-2 Cooling Tower Treatment .. Q7

E-Mac Dairy Brush Company

Silicone Spray.............................. H1

E-Quality Chemical Co., Inc.

Foam-Eze A1

E-Z Janitor Supply Co.

E-Z Foam-A-Way....................... A1

Eagle Chemical Company

#48 Cleaner C1
ADE.. L1
Bi Faze K1
Bright... C1
Caustic-Jell A1
Clean It-All Purpose Cleaner A1
Cling Degreaser........................... K1
Coil & Fin Renovator A3
Concentrate A1
Concrete Cleaner A4
D.I.G. F2
Deodorant Nuggets..................... C1
Essence of Almond Waterless
  Hand Cleaner E1
Fast Oven And Grill Cleaner...... A8
Finest A1
Floor Cleaner A1
Foam Tex A1
Glisten C2
Heavy Duty Steam Cleaner
  #5730 .... A2
Heavy Duty Steam Cleaner
  #5773 .... A1
Hi Quat D2
Hog Scald.................................... N2
Hook and Trolley Cleaner.......... A2
Industrial Water Treatment........ G7
Insect-X....................................... F1
Institutional Insecticide.............. F1
Lemon Hand Soap E1
Neutra-Clean A1
Non-Residual Space Spray F1
Paint and Varnish Remover C3
Pinex C1
Pink Concentrate A1
Pot N'Pans A1
Power Cleaner #1143 .... A1
Power Crystals C1
Quadrizyme.................................. L2
Ram Rod L1
S.A.I.D. II.................................... D2
Solvall... B1
Steam Cleaner 1170..................... A4
Strip C1
Super Tuff Concrete Cleaner A4
1.2.3. Concentrated Powd.
  Granules A4
1001... A4

Eagle Chemicals, Inc.

Cleaner Car Clean....................... A1
Cleaner Caustic Soda Beads A2
Cleaner Chlor Plus...................... A1
Cleaner Chlor Plus Two.............. A1
Cleaner Chlor Plus 100................ A1
Cleaner Conditioner FM............. G2
Cleaner Defoamer S.................... A1
Cleaner Deodorant AR............... C1
Cleaner Jelling Agent A1
Cleaner LD 25 .... B1
Cleaner LDC................................ L1
Cleaner Rust Stripper C A2
Cleaner Strip Add A1
Cleaner 126 AE A2
Cleaner 215A A1

Economy Chemical Company

Blue Lightning A1
Bowl Cleaner C2
Drain Pipe Opener L1
W.C. Concentrate A1

Economy Compounds, Incorporated

#EC-21 Safety Solvent K1
Chlorinated Cleaner A1
Econo-Chlor D1
Economy #10 Edible Lubri. Oil H1
Economy Blitz A1
Economy EC-100 Solvent K2
Economy Fantastic...................... A4
Economy Kem-Suds A4
Economy Liquid Detergent A1
Economy MSR Cleaner.............. A3
Economy Pan Wash A1
Economy-115 Evaporator-Tube
Clnr A2
Economy-124 Smokehouse-Clnr A2
Economy-125-Smokehouse-Clnr A2
Economy-14 Floor-Cleaner A2
Economy-141 Floor-Cleaner A2
Economy-142 Floor-Cleaner A2
Economy-143 Floor-Cleaner A1
Economy-15 Liquid Hand Soap. E1
Economy-150 Liquid General
Clnr A1
Economy-34 General-Cleaner A1
Economy-35 General-Cleaner ... A2
Economy-45 General-Cleaner ... A1
Economy-46 General-Cleaner ... A1
Economy-50 Trolley-Cleaner..... A2
Economy-55 Tripe Clean............ N3
Economy-70 Steam-Cleaner A1
Economy-71 Steam-Cleaner A1
Economy-73 Steam-Cleaner A1
Economy-82 Trolley-Cleaner..... A2
Economy-84 Trolley-Cleaner ... A3
Economy-88 Laundry
Compound B1
Economy-95-PF Laundry
Comp. B1
Economy-96 Laundry
Compound A1,B1
Economy-97 Laundry
Compound A1,B1
EC 94 Compound A1
Pride Concentrate....................... A4
Pride Powder............................... A1
PHC-71 .... A1
PHC-72 .... A1
Sewer Cleaner............................. L1
Striptclene TX C1
Sunbrite Laundry Bleach........... B1
1004... A2
15 Floor & General Cleaner...... A2
444.. A2
925 Concentrate........................... A1
950 Concentrate........................... A1
99 Compound A1

Economy Lubricating Company, Inc.

Boiler Colloids G6
Boiler Crystalloids...................... G6
Economy Boiler Colloids 99 ...... G6

Economy Products Company, Inc.

Car Wash Soap............................ A1
Ciodrin Insect. Emulsif.
Concntrt F2
Ciodrin-Vapona Insecticide........ F2
Econ Acid Clean A3
Econ Clean #100.......................... A1
Econ Clean #200.......................... A2
Econ Dry Acid Clean................... A3
Econ Laundry Clean.................... A1
Pyrenone Aerosol Spray............. F1

Rat-On F3
Ratou-Lyn F3
Ridz.. F3
Septic Tank Cleaner.................... L2
Vapona Insect. Emulsif.
Concentrate F2
Vapona Insecticide Resin Strip.. F2
57% Malathion Emulsif.
Concentrate F2

Economy Services & Sales Company

B-103 .... G6
B-105 .... G6
B-106 .... G6
B-107 .... G6
B-109 .... G6
B-415... G6
B-421... G6
B-709 .... G6
B-710 .... G6
B-711 .... G6
B-712 .... G6
B-713 .... G6
B-719 .... G7
B-720 .... G7
B-721 .... G7
B-722 .... G7
B-723 .... G7
B-809 .... G6
B-810 .... G6
B-811 .... G6
B-812 .... G6
B-813 .... G6
B-909 .... G6
B-910 .... G6
B-911 .... G6
B-912 .... G6
B-913 .... G6
BCO-20.. G7
C-300... G6
EFA-12.. G7
ENF-18.. G7
L-250... G7
L-300... G6
LFA-255...................................... G7
LFA-255-2................................... G7
LFA-517...................................... G7
MA-1000...................................... G7
MA-103.. G7
MA-103 F G7
MA-105.. G6
MA-110.. G6
MA-115.. G6
MA-130.. G7
P-250... G6
S-400Q G6
500-S G6
501... G6
510... G7
511... G7
515... G7
516... G6
517... G6
525... G6
560... G7
560-A G6
570... G7
700-A G7
700-AHB G7
700-D... G7
700-E... G7

Edco Chemical Company, Inc.

A-Sept.. C2
Blast Insect Spray....................... F1
Distene Disinfectant D1
Egg Wash..................................... Q1
ED 20 Cleaner A4
General Purpose Cleaner............ A1
Germolene.................................... D1

107

Shur-Sheen A1,Q1
Silicone Spray H1
Smoke House Cleaner A2
Super Cleaner A1,Q1
Trolley Cleaner A2
U.S.P. Liq. Petroleum Spray H1

Emerson Chemical Products Co.
Emco Super Clean A1
Emsol 133 .... A4

Emery Industries, Inc.
Emery 2880B H2
Emery 2881A H2
Emery 2890A H2
Emery 2990A H2
Emery 2991B High
　Temperature Chain Lub.......... H2
Emery 2992 Rot Vane/Screw
　Type Comprs Fl H2
Emery 2993 Reciprocating
　Comprs Fluid H2
Emery 2997A H2

Emil Asch Inc.
"Q-10" D2
Dermatone Liquid Hand Soap ... E1
EACO Suds Liquid Detergent... A1
Surfacetone Formula 17............. A1

Emkay Chemical Company
Emkapol PO-18 .... A1

Empak Industries, Inc.
Code 2382 Food Mach. Grease.. H1

Empire Chemical Company
Aero Fog.................................... F1
Alu Cleaner................................ A1
B.P.R. Comp.............................. A1
Bio Soap Hand Powder C2
Bowl Kleen C2
Comp 3 .... A2
Comp 4 .... A1
Drainz.. L1
Economy Pine-O-Lene C1
Emco Fly Bait F2
F.P.S. Compound B1
Fleet.. D1
Gran-Noxo C2
Hy-Ge-No C2
Jaxo Cleanser A6
Latherine Liquid Soap Conc. E1
Marvex A4
Mill-O-Spray.............................. F1
Mill-O-Spray Mulsifiable F2
Mill-O-Spray X........................... F1
Nu-Gleem A1
Oxidoff...................................... A3
Repelocide................................. F1
Roachkil F1
Slayz F2
Ster-O-Lene............................... D1,E3
Sudsbrite................................... A4
Super-X Cleaner A4
SC-10 .... C2
T.S.P. A1
Thermo-Cide.............................. F1
U-B-9 Cleaner A4
Velvet Cleaner Lotion A4
Velvet Liquid Soap E1

Empire-Chemical Company
Supreme Cleaner A1

Empire International
Acterge...................................... A4
Aqua-Em 75 .... G7
Aqua-Em 82 .... G7

Aqua-EM D150 .... C1
Aqua-EM Z90............................. L2
Aqua-EM 10............................... G6
Aqua-EM 112............................. A3
Aqua-EM 117............................. A3
Aqua-EM 15............................... G7
Aqua-EM 20 .... G6
Aqua-EM 35 .... G6
Aqua-EM 50 .... G7
Aquatex`.................................... P1
Bactrol...................................... C1
Buto-Pip F1
Buto-Pip Concentrate................. F1
Buto-Pip Contact Insecticide..... F1
Capsulate P1
Coilene...................................... A1
Coilmatic P1
Crown Cleaner A4
De-Lite A4
Dermasurge................................ E1
Derust A3
Di-Lec K1
Di-Lec Aerosol........................... K1
Disperse K1
Dustex....................................... P1
Dynasol A4,A8
Electrosolv K1
Electrosolv (Aerosol).................. K1
Elox A3
Emcide...................................... F1
Emdee....................................... A1
Emfog F1
Engeen...................................... K1
Fluffy Suds................................ A1
FPC No. 44................................ A4
Glist Bouquet C1
Glist Bouquet Aerosol C1
Greasolv L1
Handez...................................... E4
Hep A4
Humect...................................... P1
Hygex A4
Kleershield C1
Lan-O-Coat E4
Lavon A4
Lavosan C2
Leader....................................... E1
Lectro-Safe................................ K2
Lectronic................................... H2
Lectronic (Aerosol)..................... H2
Liquasheen A1
Lubem DM 100 .... H2
Lubem FG 110 .... H1
Lubem FG 30 .... H1
Lubem SL 120 .... H2
Odorsolv.................................... C1
Panstrip..................................... K1
Penetronic H2
Penetronic (Aerosol).................. H2
Pineo C1
Pire-O-Shine.............................. A1
Pow-R-Jet X10 .... A1
Power Foam 3100 .... A1
Re-Zist F2
Remolex A3
Rust-Go A3
S.S.T. (Aerosol)......................... A7
S.S.T. (Bulk) A7
Saf-T-Steam.............................. A1
San-A-Can C1
Sani-Cide D2
Scale-Away A3
Scale-O-Sol A3
Scale-Rid A3
Scum-Go C2
Sew-R-Kleen II L1
Silube H1
Spectrum A4
Stripsall C3
Stripsall (Aerosol)...................... C3

109

Tempox.................................L1
Terge....................................B1
Thawz...................................C1
Tileen...................................C2
Trax.....................................P1
UP.......................................A4
X-It......................................A3
Zymo....................................L2

Empire Laboratories See Empire International

Emsco Chemical Company
Spray-Away..........................A2

Emsco Industries, Inc.
Crete Kleen...........................A2
Kleenrite Lemon Scent Stnlss.
  Steel Plsh............................A7

Emulso Corporation, The
Bowl Cleaner.........................C2
Cleans It Cleaner...................A1
Emco "a"..............................F1
Emco-e-................................F2
Glenzene...............................A1
Grease Gobbler......................A1
Hexzene No. 2.......................D1
Liquid Hand Soap..................E1
Nine Ninety Nine...................A1

Endico Laboratories, Inc.
Enden A................................A1
Endene..................................A4
Foam'N-Kleen.......................A1
Freeza-Kleen.........................A5
Freezer & Locker Cleaner #466 A5
Powdered Dish Wash
  Detergent No. 812................A1
Sani-Kleen.............................D1

Endura Chemical Corp.
Insta-Brite............................A1

Enerco Corporation, The
Enerco 204............................G7
Enerco 327............................A3
Enerco 328............................G7
Enerco 413............................G6
Enerco 414............................G6
Enerco 415............................G7
Enerco 416............................G6
Enerco 417............................G6
Enerco 418............................G6
Enerco 419............................G6
Enerco 420............................G6
Enerco 4204..........................G7
Enerco 4327..........................A3
Enerco 4328..........................G7
Enerco 4413..........................G6
Enerco 4414..........................G6
Enerco 4416..........................G6
Enerco 4417..........................G6
Enerco 4418..........................G6
Enerco 4419..........................G6
Enerco 4420..........................G6
Enerco 443............................G6
Enerco 4443..........................G6
Enerco 4445..........................G6
Enerco 445............................G6
Enerco 4541..........................G6
Enerco 4553..........................G6
Enerco 4624..........................G6
Enerco 4625..........................G6
Enerco 4662..........................G6
Enerco 4663..........................G6
Enerco 4664..........................G6
Enerco 4701..........................G1
Enerco 4702..........................G1

Enerco 4729...........................G1
Enerco 4732..........................P1
Enerco 5204..........................G7
Enerco 5327..........................A3
Enerco 5328..........................G7
Enerco 541............................G6
Enerco 5413..........................G6
Enerco 5414..........................G6
Enerco 5416..........................G6
Enerco 5417..........................G6
Enerco 5418..........................G6
Enerco 5419..........................G6
Enerco 5420..........................G6
Enerco 5443..........................G6
Enerco 5445..........................G6
Enerco 5450..........................G6
Enerco 553............................G6
Enerco 5541..........................G6
Enerco 5553..........................G6
Enerco 5571..........................G6
Enerco 5624..........................G6
Enerco 5625..........................G6
Enerco 5662..........................G6
Enerco 5663..........................G6
Enerco 5664..........................G6
Enerco 5701..........................G1
Enerco 5702..........................G1
Enerco 571............................G6
Enerco 5729..........................G1
Enerco 5732..........................P1
Enerco 624............................G6
Enerco 625............................G6
Enerco 662............................G6
Enerco 663............................G6
Enerco 664............................G6
Enerco 701............................G2
Enerco 702............................G2
Enerco 729............................G1
Enerco 732............................P1

Eng-Skell Company
Esco Mighty Concentrated
  Detergent.............................A4

Engelhard Minerals & Chemicals Div.
Auto-Dri...............................J1
Poise....................................J1
Sol-Speedi-Dri.......................J1

Engineered Chemicals of Florida, Inc.
Cigo Alk. B-I.........................G6
Cigo Liq. Ox-Go-II................G6
Cigo Phos-T..........................G6
Cigodyne D...........................D2
CIGO Boiler Water Treatment
  BWT-R..............................G6
CIGO Boiler Water Treatment
  BWT-RC............................G6
CIGO Boiler Water Treatment
  BWT-RCS...........................G6
CIGO Boiler Water Treatment
  BWT-SW............................G6
CIGO Boiler Water Treatment
  BWT-SWC..........................G6
CIGO Boiler Water Treatment
  BWT-SWC-S........................G7
CIGO Boiler Water Treatment
  Organic-C...........................G6
CIGO Boiler Water Treatment
  Sulfite-M.............................G6
CIGO 111..............................Q1
CIGO 112..............................Q1

Engineered Lubricants
Enclean MC-II.......................A4
Enlubol Anti-Wear Hydraulic
  Fluid 210.............................H2
Enlubol Anti-Wear Hydraulic
  Fluid 340.............................H2

Enlubol Anti-Wear Hydraulic
  Fluid 500.............................H2
Enlubol ACO-495-BB.............H2
Enlubol ACO-745-BB.............H2
Enlubol EP-BB Lubricant 10....H2
Enlubol EP-BB Lubricant 140...H2
Enlubol EP-BB Lubricant 20.....H2
Enlubol EP-BB Lubricant 30,....H2
Enlubol EP-BB Lubricant 40.....H2
Enlubol EP-BB Lubricant 50.....H2
Enlubol FGG-BB Grease..........H2
Enlubol FGL-OCL-10 Grease....H2
Enlubol FGL-OCL-25 Grease....H2
Enlubol FGL-OCL-375.............H2
Enlubol FGL-OCL-44..............H2
Enlubol FGL-OCL-95..............H2
Enlubol FGO Oil 539..............H1
Enlubol FGO Oil 927..............H1
Enlubol Grease AA-1..............H2
Enlubol Grease AA-2..............H1
Enlubol T-88 Grease...............H2
Enlubol 500 Grease No. 0.........H2
Enlubol 500 Grease No. 1.........H2
Enlubol 500 Grease No. 2.........H2

Engineering Chemical Services, Inc.
L-101A.................................G6
L-117A.................................G7
L-148...................................G6
L-149...................................G6
L-153...................................G6
L-158...................................G6
P 270A.................................G6
P-209...................................G6
P-212A.................................G6
P-214...................................G6
P-226A.................................G6
P-252...................................G6

Enpro, Incorporated
Blue Blaze.............................A4
Crack Down..........................A1
Environ Citrus Fragrance.........A4
Envo....................................D1
Power-R-Plus........................A4

Ensign Products Company
Ensign #577 Lubricall..............H2
Ensign 595 Food Grade Lube....H1
318 Versatol..........................H2

Entech Systems Corp.
Entech Fog-5.........................F1
Prentox Residual Spray............F2
Resmethrin Insect Spray..........F1

Enterprise Chemical and Paper Corp.
Enco PC...............................A4

Enterprise Mill Soap Works
Special Cleaner......................A1

Enterprise Paper and Chemical Co., Inc.
Enco Kleen Mist.....................A1
Entech II...............................A1
K-M-C-5 Liquid Drain Solvent..L1
K-M-C-6 Insecticide...............F2
KMC 1.................................A1
KMC-2.................................A1

Enterprise Products Company
'Bye Insecticide Spray No. 38....F2
'Bye Residual Insecticide Spray. F2
Big-Brute No. 202..................A1
Boot Hill Insect Spray.............F2
Boot Hill Roach & Fly Spray....F1
Boot Hill Vaporizing Spray
  Concentrated........................F1

Dynafloc 271 .... P1
Dynafloc 373 .... P1
Dynafloc 375 .... P1
Dynafloc 377 .... P1
Dynafloc 73 .... P1
Dynafloc 80 .... P1
Dynafloc 81 .... P1
DynaFloc 175 .... L1
DynaFloc 176 .... L1
DynaFloc 273 .... L1
DynaFloc 72 .... L1
DynaFloc 78 .... L1
DynaFresh 700 .... C1
DynaFresh 701 .... C1
DynaFresh 702 .... C1
DynaFresh 704 .... C1
DynaFresh 705 .... C1

Environmental Sales, Inc.
Esco-Foamy Cleaner Degreaser A1
Esco-Sanisep Sanitizer
   Disinfectant D2
Esco-Sept II C2
Esco-Sol E1

Environmental Services, Inc.
E.S.I. Premium Multi-Cleaner.... A4

Environmental Services, Inc.
ECA Fogging Concentrate
   5628 .... F1

Environmental Standards, Inc.
Enviro Lift A4
Enviro-Sept D1
Enviro-Sept II C2

Enzyme Industries of the USA, Inc.
Grease-G0 .... A1,L1
Sanzyme A1,L1

Epic Chemical Co.
Easy Rider H2
Smooth Sailing H2

Epic Chemical Sales Corp.
All Purpose Cleaner
   Concentrate A1
Antiseptic Liquid Hand Lotion.. E1
Antiseptic Liquid Hand Soap..... E1
Antiseptic Liquid Hand Soap
   Concentrate E1
Breakout A1
Breakthrough A1
Cleaner-Degreaser-Wax
   Stripper A1
CP 00 .... A1
CP-1 Odorless Sanitizing
   Agency D1,Q3
CP-115 .... H2
CP-12 .... A1
CP-12-NJ A1
CP-131 .... A4
CP-150 .... A1
CP-21 .... A2
CP-21-NJ A2
CP-26 .... A1
CP-26-NJ A1
CP-29 .... A1
CP-29-NJ A1
CP-31 .... A4
CP-31-NJ A4
CP-33 NJ A1
CP-35 .... A1
CP-37 .... A1
CP-37-NJ A1
CP-41 .... A1
CP-41-NJ A1
CP-55 .... Q2

CP-57 .... A3
CP-57-NJ A3
CP-59 .... A3,Q2
CP-59-NJ A3
CP-6 Dtrgent-Santzr Iodophor
   Type D2,Q6
CP-7 .... D2
CP-8 Conctrd Liq. Acid
   Sanitizer A3,D1
Detergent Cleaner-Disinfectant.. D1
Disinfectant Cleaner A4
Epic Astro Equipment Cleaner.. A4,A8
Epic Conveyor Lube
   Concentrate H2
Epic Feather Softener N1
Epic K-66 .... A2
Epic Liq. Conv. Lube No. 1 ...... H2
Epic Liq. Conv. Lube No. 2 ...... H2
Epic's Skid Rid A4
EP-1 .... A1
EP-106 .... A3
EP-110 .... A1,Q1
EP-110-NJ A1
EP-111 .... D1
EP-116 .... A2
EP-122 .... A2
EP-122-NJ A2
EP-125 .... A2
EP-125-NJ A2
EP-128 .... A1
EP-128-NJ A1
EP-133 .... A2
EP-133-NJ A2
EP-140 .... A2
EP-140-NJ A2
EP-142 .... D1
EP-150 .... A2
EP-150-NJ A2
EP-16 .... A1
EP-16-NJ A1
EP-19 .... A3
EP-2 .... A1
EP-2-NJ A1
EP-20 .... A1
EP-20-NJ A1
EP-22 .... C2
EP-22-NJ C2
EP-23 .... A1
EP-25 .... L1
EP-25-NJ L1
EP-29 .... A1
EP-29-NJ A1
EP-33 .... A2
EP-33-NJ A2
EP-37 .... A1
EP-37-NJ A1
EP-39 .... A1,Q1
EP-39-NJ A1
EP-4 .... A1
EP-4-NJ A1
EP-42 .... A1
EP-42-NJ A1
EP-44 .... A1
EP-49 .... N2
EP-5 .... A1
EP-5-NJ A1
EP-53 .... N2
EP-54 .... N2
EP-54-NJ N2
EP-60 .... A1
EP-60-NJ A1
EP-65 .... A1
EP-65-NJ A1
EP-73 .... A2
EP-73-NJ A2
EP-74 .... A2
EP-74-NJ A2
EP-76 .... A1
EP-76-NJ A1
EP-81 .... A1

EP-81-NJ A1
EP-83 .... A1
EP-83-NJ A1
EP-84 .... A1
EP-84-NJ A1
EP-87 .... A1
EP-87-NJ A1
EP-89 .... A2
EP-89-NJ A2
EP-91 .... A2,Q1
EP-91-NJ A1
EP-93 .... A2
EP-93-NJ A2
EP-98 .... A2
EP-98-NJ A2
EP-99 .... A2
EP-99-NJ A2
Fortified Odor Disinfectant C1
Heavy Duty Breakthrough A1
Liquid Bactericide D2
Liquid Conveyor Lube H2
Liquid Hand Soap E1
Mint Disinfectant C2
Oven & Grill Cleaner A8
Pine Odor Disinfectant C1
Pine Oil Disinfectant C1
Quarry Tile Cleaner A4

Epic Chemicals, Incorporated
pH 7+ Pink Lotion Hand
   Cleaner E2
Alkakleen Floor Cleaner A4
All Purpose Lime and Scale
   Solvent A3
Ama-Clean C1
AK-Shun Melt C1
Bar Foam Glass Washing
   Powder A1
Bar Foam Liquid Detergent A1
Blue Magic Sudsing Detergent .. A1
Breach Chlorinated Detergent A1
Chlorinated Liq. Machine
   Dishwash. Comp. A1
Choice Powdered Detergent A1
Coronet Oxygen Bleach B1
Coronet White Laundry Deter. . B1
Dandy Soft Pink Lotion
   Detergent A4
Del-Mar Synthetic Detergent A1
E-Z Suds A1
End-Germ Concntrtd Disinf.
   Clnr. D1
Epic Concentrated Liq. Steam
   Clng. Cmpd. A1
Epic Z Machine Dishwashing
   Compound A1
Episuds Laundry Detergent
   (Powdered) B1
Episuds Liquid Laundry
   Detergent B1
Epitame Liquid Laundry Sour... B1
Epitrol Fabric Softener B2
Epitrol Liquid Fabric Softener . B2
Fairway Pine Oil C1
Fury Car Washing Compound... A1
Fury Instant Spray Cleaner A4
Garage Floor Cleaner C1
Germicidal Q-Lab Spray
   Cleaner C1
Glide Concentrated Liquid
   Cleaner A1
Grade "A" Liquid Hand Soap ... E1
Guard-All C1
Happy Hands Powdered Hand
   Soap C2
Liquid Bingo Laundry
   Detergent B1
Multi-Purpose Cleaner A1
Pride Of The Kitchen Liquid
   Detergent A1

112

Super Butyl Cleaner.....................A4
Super Neutral Cleaner A1
Super Power #2 ....A4
Super Solv #2............................A4,A8
Tempered Drain Opener............L1

Erbrich Products Company

Epco Bowl Cleaner......................C2
Epco Quick Clean Toilet Bowl
  Cleaner...................................C2
SX-3 ....D2,G4,B1
SX-5 ....G4,B1

Ergon Chemical Products/ Div. of Quality Lamp And Supply Inc.

All-Kleen......................................A1
Degree-Sor...................................A4
Dishwashing Lotion....................A1
Drane-Free...................................L1
Hi-Pro-Steam A2
Lather..E1
Met-L-Brite..................................A7
Oven-Grill...............................,......K1
Windo-Kleen...............................C1

Erickson Chemical Company

Amine Condensate Treatment
  #72PF....................................G6
AS-17...P1
Boiler Water Treatment #164....G6
Boiler Water Treatment #278....G6
Boiler Water Treatment #295....G6
Boiler Water Treatment #713....G6
Boiler Water Treatment #852....G6
Boiler Water Treatment #854....G6
Boiler Water Treatment #863....G6
Boiler Water Treatment #864....G6
Boiler Water Treatment #865....G6
Boiler Water Treatment #869....G6
Boiler Water Treatment #875....G6
Boiler Water Treatment #876....G6
Boiler Water Treatment #877....G6
Boiler Water Treatment #878....G6
Boiler Water Treatment #890....G6
Boiler Water Treatment
  Formula-888 ....G6
Cooling Water Treatment #250.G7
Cooling Water Treatment #252.G7
Cooling Water Treatment #307.G7
Cooling Water Treatment #310.G7
Polyphosphate Treatment #187.G2
Polyphosphate Treatment #192.G2
SH-16 ....P1
Water Treatment #180 ....G6
Water Treatment #195 ....G2
Water Treatment #303 ....G7
Water Treatment #664 ....G6
Water Treatment #72 ....G6
Water Treatment #857 ....G6
Water Treatment #870 ....G6
Water Treatment E-12 ....A3
Zeo Clean....................................P1
Zeo Clean 199C P1
Zeo Clean 199F P1

Erny Supply Company

Bowl Cleaner C2
Esco C-20....................................A1
Esco 435 ....A1
Esco 480 ....A1
Esco 806 ....A3
Heavy Duty Cleaner A4
Hypersteam 143 ....A2
Liquid Hand Soap E1
Pine Scrub Soap .'........................C1

Espar Chemical Corp.

Espar-Kleen A4,A8
Pink Velvet Hand Soap..............E1

Essential Chemicals Corp.

Chloroderm Antis. Liq. Hand
  Soap 20%................................E1
Ecco Quat 10 ....D1
PM A619 Liquid Hand
  Dishwash A1
PM Concentrated Bowl Clnr. C2
PM Drain Pipe Opener..............L1
PM Freezer Cleaner....................A5
PM Mild Prcln Grout
  Tile&Bowl Clnr. C1
PM Tile,Porcelain,&Toilet
  Bowl Clnr...............................C2
PM 103K Window Cleaner........C1
PM 1502 Super Garage Floor
  Clnr..C1
PM 18T3-Trust B1
PM 20X Hvy Dty Mach.
  Dshwsh Powder A1
PM 2026 NPN A4
PM 2032 Low Suds Cleaner A4
PM 2070 Visco Pine Clinging
  Cleaner...................................C1
PM 2072 Liquid Laundry
  Detergent................................B1
PM 2089 Ecco Solvent Cleaner.C1
PM 2095 Xtra Duty Wax
  Stripper..................................A4
PM 2300 Econo-Strip..................A4
PM 2304 Ltnzd Liq. Hand
  Clnr. RTU...............................E1
PM 2319 Hvy Dty
  Clnr.Degrsr&Wax Strppr........A4
PM 2323 Heavy Duty Machine
  Dshwsh. Pwdr. A1
PM 2328 Econo-Cleaner.............A4
PM 2343 Concentrated Cleaner-
  Degreaser A4
PM 2346 Stnlss Steel, Porcelain
  & Enl Cl C2
PM 2405 Non-Acid Bowl
  Cleaner...................................C2
PM 2546 All Purpose Cleaner
  & Degreaser C1,A8
PM 2701 Quick Suds
  Degreasing Cleaner A4
PM 2705 Industrial Heavy Duty
  Cleaner...................................A1
PM 294 D-Limer A3
PM 299 Delimer&Cement
  Cleaner...................................A3
PM 308 Lotionized Liq. Hand
  Dshwsh...................................A1
PM 325 Concentrated Liq.
  Hand Soap E1
PM 33E Laundry Detergent B1
PM 337 Sof-tutch Pink Lotion
  Hand Clnr. E1
PM 338C Liq. Hand Soap E1
PM 340C Liq. Hand Soap
  Concentrate E1
PM 35E Laundry Detergent B1
PM 354 Hvy Dty
  Clnr.Degrsr.&Wax Strppr.......C1
PM 358 Quick Suds Degreasing
  Cleaner...................................A4
PM 359 ....A1
PM 360 ....A1
PM 387 Supreme Powdered
  Cleaner...................................A4
PM 393 ....A1
PM 394 ....A1
PM 395 Mechanics' Hand
  Cleaner...................................E4
PM 400-1 Superbase All Purp.
  Clnr..A4
PM 401-1 Superbase Wax
  Strppr&Degrsr.........................A4
PM 403 Liq. Parts Degreaser.....C1
PM 501 Economy Cleaner A4

PM 504 Xtra Active
    Concentrate A4
PM 508 Hand Dishwash
    Powder A1
PM 581 All Purpose Cleaner A4
PM 6X Prep Floor Conditioner . A4
PM 601 Liquid Steam Cleaner ... A4.
PM 601 Low Suds Wax Stripper A4.
PM 603P Xtra Dty Wax
    Strppr(Hgh Actv).............. C1
PM 6140 Liquid Hand
    Dishwash A4
PM 830 Industrial Heavy Duty
    Cleaner A1
PM 9020R Antisp. Liq. Hand
    Soap 20% E1
PM 9040R Antisp. Liq. Hand
    Soap 40% E1
PM 9066 Pink Lotnzd Antisp.
    Hand Soap E1
PM 9067 Pink Lotnzd Antisp.
    Liq Hand Soa E1
Quat Rinse D2,E3
Quat 42 .... D1
Trust Antiseptic Hand Soap...... E1
2710 Trust Lotion Hand Soap..... E1
329 Trust Lotion Hand Soap...... E1

Etasol Distributing Company
Etasol Industrial Cleaner A1

Eton-Colby Chemical Company
Caustic Soda-Cluconate
    Solution A2

Etsol Products, Inc.
All Purpose Solvent Cleaner C1
Etsol 733......................... A4

Eureka Laboratories, Inc.
Con-Mist A4
Heavy Duty General Purpose
    Cleaner A4
Liqui-Drain L1
No. 200 Super Butyl Cleaner A4
Sana-Clene Concentrate........... C1
Total-Clean A1
Zip-100 All Purpose Cleaner A1
201 Heavy Duty Cleaner A1

Everall Products Inc.
EP 360 Aluminum Cmplx Food
    Mach.Grs."AA" H1

Evergreen Chemical Company
Degreaser Non-Butyl A1

Excel Chemical Company
Excel Toilet Bowl Cleaner C2
No Fume L1
X-L Concentrate.................. A1
X-L Foam A1

Excel Chemical Company, Inc.
Add A1
Aluminum Chiller Cleaner A1
Blast A1
Blitz A1
Block-Aid L1
Contract......................... F2
CCP-100 .... A1
CL-4600 .... A1,Q1
Excel 3DVF....................... D1
EL-350............................ A4
Food Processors Insecticide...... F1
Formula 960 .... A1
G-4400 .... A1,Q1
G-480 .... A1
Genteel A1

Gleam A1
Jet Clean A1
Lectri-Sol K1
Liquid Hand Soap E1
NT................................ A4
Odo-Go............................ C1
Phos-Clene A3
Pine-O-Dis....................... C1
Regal............................. C2
Seabreeze......................... C1
Stuff............................. A1
Super Bowl C2
SPI F1
Thunderbolt Automotive K1
Triple K.......................... A1
Windo Brite....................... C1
XL-200............................ A1
Z-2100............................ A4

Excel-Mineral Company
Absorbs-It........................ J1
Clean Dri......................... J1
Quick-Sorb J1
Safe-T-Sorb J1

Excelsior Varnish & Chemicals, Inc.
Odor Conteractant Water
    Soluble R-664-W C1
Odor Counteractant Oil Soluble
    R-665-O C1
Odor Counteractant Surface
    Treat. R-667-.................. C1
Powdered Drain Opener Fast
    Acting V 252................... L1
Powdered Drain Opener Slow
    Acting V 253................... L1
Rack Wash 601 .... A2

Excetra Chemical Company
All Purpose Cleaner A2
Boilerite 100 .... G6
Steamrite 200 .... G6

Exsel Industries
Multi Power Cleaner
    Degreaser-Stripper A4

Exsl Chemical Company
Acid Cleaner BN................... A3
Awake A4
BW-38............................. A2
Cal Can Cleaner RC-5 .... A2
Cal Can Cleaner 1 .... A2
Cal Can Cleaner 2 .... A2
Cal Can Cleaner 4 .... A1
China Dip......................... A1
Chlor MDL-7721 .... A1
E-Z Kleen A4
Exslite #10 BOC A2
Exslite 'Purg' Drain Cleaner L1
Exslite BHG-762................... A1
Exslite Chlor Cal A1
Exslite Condor Chlor............. A1
Exslite Controlled Suds Shroud
    Cl............................ A1,B1
Exslite CCL Special A1
Exslite D-Ice..................... C1
Exslite E-13...................... A1
Exslite E-14...................... A1
Exslite Enzadet B1
Exslite F-417..................... A1
Exslite F-430 .... A2
Exslite Fluff A1
Exslite Foam Up................... A1
Exslite H & T 110 .... A2
Exslite Hand Soap E1
Exslite Hi Glo #10............... A1
Exslite HD AL Spray Cleaner... A2
Exslite HD-44 .... A2

Exslite Lash...................... A2
Exslite Lor Leen Chlorinated A1
Exslite Lube #1 .... H2
Exslite Lube #3 .... H2
Exslite Lube #7 .... H2
Exslite LF-EMR.................... A3
Exslite LG Neutralizer B1
Exslite LHD Chlor A2
Exslite LST 461 .... A4
Exslite LST 465 .... A1
Exslite M-X-17.................... A2
Exslite Monarch A1
Exslite Multi 51 .... A1
Exslite MAC 2001 .... A1
Exslite Neutrakleen No. 1........ A1
Exslite Neutrakleen No. 2........ A1
Exslite Neutrakleen No. 3........ A1
Exslite No Foam Poultry Scald. N1
Exslite Powdered Hand Soap.... C2
Exslite Prefer A1
Exslite Prefer No. 2.............. A1
Exslite PA........................ A3
Exslite Rapid Rinse A1
Exslite Special Chlor............. A1
Exslite Sulfamic Acid............ A3
Exslite Super Kleen A1
Exslite Tripc 55 .... N3
Exslite TM Cleaner A6
Exslite V-Kleen A1
Exslite WOW-75.................... A2
Exslite 2-DFF Cleaner A2
Exslite 2-OFC A1,Q1
Exslite 2-OFHD A2
Exslite 2-SMC.................... A2
Exslite 2-SMC #40 .... A2
Exslite 2050E A1,Q1
Exslite 3004 .... A2
Exslite 4004F A1
Exslite 4033 .... A1
Exslite 438 HD A2,B1
Exslite 4460 .... A1
Exslite 500 .... A2
Exslite 526 .... A1
Exslite 712 .... A2
Exslite 7130 .... A2
Exslite 7270 .... A1
Exslite 7530 .... A2
Exslite 7535-CC A1
Exslite 955 .... A2
Exsol-Chlor D1
Floor Cleaner BN A1,B1
Formula 1155 .... C1
Formula 7858 CW A1
Golden A1
Hi-White Bleach B1
Misty Fabric Conditioner B1
MDW A2
Rosette........................... A1
Sil-Dip A3
Silver Dip........................ A1
The Blue Max A1
The Cleaning Gel A1
The Cleaning Gem A4
Ultra-Peel N4

Extraction Systems, Inc.
Heavy Duty Degreaser 302........ A4

Exxon Company, U.S.A.
Alert M 275....................... H2
Andok M 275....................... H2
Andok 260 .... H2
Bayol 72 .... H1
Bayol 90 .... H1
Bayol 92 .... H1
Beacon Q 2........................ H2
Carum 330......................... H1
Coray 100......................... H2
Coray 15.......................... H2
Coray 150......................... H2

114

Vársol 5 .... K1
Zerice 22 .... H2
Zerice 42 .... H2
Zerice 46 .... H2
Zerice 68 .... H2
2921 Technical White Oil H1
2935 White Oil H1

Ezzy Clean-Up Supplies & Equip.
Grease Gone A4,A8

F

F & H Chemicals
Acid Detergent A3
Acigen 2012 .... A3
Formula No. 100 .... A1
Formula No. 127 .... A2
Lime Scale Remover A3
Sanitize D1

F & H Food Equipment Co.
C.I.P. Acidic Scale Remover A3
Chlorinated Egg Washing
   Compound Q1
Chlorinated General Cleaner A1
Chlorinated Granular Cleaner A2
Heavy Duty Acid Detergent
   Cleaner A3
Heavy Duty Liquid Detergent ... A1
Pressure Washing Compound A1
Synthetic Conveyor Lubricant ... H2

F E C Industries
Aqua-Clean A1
Lana Suds A1
N.B.C. Non-Butyl Cleaner-
   Degreaser A1

F F Industrial Service & Supplies, Inc.
Bacteriasol A4
Deluxe Machine Dshwshng
   Compound A1
Dynamo A1
Extra Heavy Duty Concrete
   Cleaner A4
Golden Gator Concrete Cleaner C1
Golden Magic Cleaner A4
Grime Scat A4
Hydro Sol C2
Industrial Cleaner A4
Iodo-175 .... D2
Liquid Hand Soap 40 .... E1
Liquid Steam Cleaner A2
Multi Blue A1
No. 600 Hog Scald N2
No. 80 Space Spray F1
Pine Odor Disinfectant C1
Pine Oil Disinfectant C1
Pink Cleaner A4
Premium Fly & Roach Spray F1
Quick Drain L1
Safe A1
Sewer Solvent L1
Steam Cleaning Compound A2
Wax Scat A4
100 Plus Drain Pipe Opener L1

F M C Corporation
Defoamer 16 .... Q5
Disodium Phosphate G6,N2,G1
Freshgard 4A Q1
Hexaphos G6,G1
Hydrogen Peroxide 35% HP N3
Monosodium Phosphate A3,G1

Phosphoric Acid A3
Potassium Tripolyphosphate A1
Soda Ash Dense 160 .... N1,N2,N3,G1
Soda Ash Grade 100 .... N1,G6,N2,N3, G1
Sodium Acid Pyrophosphate N2
Sodium Sesquicarbonate N1,G1
Sodium Tripolyphosphate A1,G6,N2,G1
Tetrapotassium Pyrophosphate
   Granular A1
Tetrasodium Pyrophosphate A1,G6,N2,G1
Trisodium Phosphate A1,G6,N2,N3, G1

F R M Chem. Incorporated
Cir-Clean A1
Gain S.V.P. A3
Super 8 .... A1

Facet Enterprises, Inc.
Deep Clene Cream Hand
   Cleaner E4

Fairbank Corporation, The
Facilite Giant Wash A2,N4

Fairfield American Corporation
Aqueous Food Plant Pyrn. Fog.
   Insect F1
Automatic Sequential·
   Pressurized Spray F2
Compactor and Kitchen
   Insecticide F1
Drione Insecticide F2
Drione Insecticide Spray F2
Food Plant Fogging Insecticide. F1
General Purpose 0.25-0.25
   Insect Killer F1
Industrial Aqueous Pressurized
   2.0-0.4 .... F1
Industrial Aqueous Pressurized
   4.0-0.5 .... F1
Industrial Pressurized Spray
   2.0.-0.4 .... F1
Mosquito Spray Concentrate F1
Multi-Purpose Pyrenone Insect-
   Conc. F1
Pressurized Spray Multi-Use
   Insecticide F1
Pyrenone Dairy and Food Plant
   Aerosol F1
Pyrenone Diazinon Dual Use
   E.C. F2
Pyrenone Diazinon E.C. F2
Pyrenone Diazinon Residual
   Spray F2
Pyrenone Double-A Plus Fly
   Spray F1
Pyrenone Durshan Dual Use
   E.C. F2
Pyrenone Dursban Roach &
   Ant Spray F2
Pyrenone Food Plant Aerosol
   Insecticide F1
Pyrenone Food Plant Fogging
   Insecticide F1
Pyrenone Gen. Purp. Aqueous
   Insecticide F1
Pyrenone General Purpose
   Household Spray F1
Pyrenone General Purpose
   Spray F1
Pyrenone Indust. Spray
   Emulsfbl. Conc. F1
Pyrenone Malathion Mosquito
   Fog. Insect. F2
Pyrenone Malathion Residual
   Spray F2
Pyrenone Mill Spray F1

Pyrenone Mill Spray 2-0.2 Oil
Type.. F1
Pyrenone Mosquito Fogging
Spray...................................... F1
Pyrenone Multi-Purpose
Insecticide.............................. F1
Pyrenone One Shot Hi-Pressure
Fogger F1
Pyrenone PCO Roach
Concentrate F1
Pyrenone Space Spray F1
Pyrenone 1-0.2 Food Plant
Spray...................................... F1
Pyrenone 25 5-M.A.G.C............ F2
Roach & Ant Liquid Spray........ F2
Roach & Ant Spray Aqueous F2
Roach & Ant Spray Pressurized F2
Roach & Ant Spray Pressurized
w/Diazinon F2
Roach And Cricket Bait F2
Synthrin Industrial Preszd.
Spray 0.50............................. F1
Synthrin Industrial Spray 0.25 ... F1
Tetralate Multi-Purpose
Insecticide E.C........................ F1

Fairfield International Corp.

Enviro Blast Solid C1
Enviro Solv Odor Control
Liquid #316 .... L1
Enviro Solv Oven & Grill
Cleaner................................... A8
Enviro-Bac L2
Enviro-Bac MD L2
Enviro-Blast L1
Enviro-Clean A4
Enviro-Solv-E............................ K1
Enviro-Solve A1
Poly Solv................................... L1
Quat Disinfectant...................... D1
Shure-Scent C1

Falco Chemical Company

Falclean #1 .... Q1
Power-Plus Concentrated
Cleaner................................... A1
Quick Spray H1

Famous Lubricants, Incorporated

Fasol A-180 White Oil............... H1
Fasol A-200 Certified H1
Fasol A-340 Certified H1
Fasol A-70 White Oil................. H1
Food Machinery Lubricant H1
Food Machinery Lubricant No.
0... H1
Orange FM Lubricant No. 0....... H1
Orange FM Lubricant No. 2....... H1

Far Best Corporation

No. 6575 43 Cleaner.................... A1
7062 Eazy Kleen........................ A4
7184 Liquid Universal Cleaner... A4

Farewell, W. K., Company

Concentrate 45 .... A1
P T L Conquest A1
PTL Foam-Aid A1
PTL Impact A1
PTL Kutter Plus......................... A4
PTL Renew A3
PTL Smokehouse Cleaner A2
PTL Sparkle A1
PTL Super-Kleen A1
PTL Trolley Cleaner A2
PTL Vanish A2
PTL Vantage Plus....................... N3

Farmers Chemical & Equipment Co.

Industrial Cleaner A4

Truck Wash #4 .... A1

Farmers Union Central Exchange, Inc.

Spray-Fog F2

Farmland Industries, Inc.

Co-Op BRB Grease..................... H2
Co-Op Dexron Auto. Transm.
Fluid B-11043......................... H2
Co-Op Diesel Engine Oil SAE
10W.. H2
Co-Op Diesel Engine Oil SAE
20-20W................................... H2
Co-Op Diesel Engine Oil SAE
30.. H2
Co-Op Diesel Engine Oil SAE
40.. H2
Co-Op Hydrol No. 150............... H2
Co-Op Hydrol No. 315............... H2
Co-Op Hydrol No. 465............... H2
Co-Op Natrl & LP Gas Eng.
Oil SAE 10W.......................... H2
Co-Op Natrl & LP Gas Eng.
Oil SAE 20............................. H2
Co-Op Natrl & LP Gas Eng.
Oil SAE 30............................. H2
Co-Op Natrl & LP Gas Eng.
Oil SAE 40............................. H2
Co-Op Premium Light Gun
Grease..................................... H2
Co-Op Super H.T.B. Hydraulic
Oil H2
Double Circle Motor Oil SAE
10W.. H2
Double Circle Motor Oil SAE
20-20W................................... H2
Double Circle Motor Oil SAE
30.. H2
Double Circle Motor Oil SAE
40.. H2
Double Circle Motor Oil SAE
50.. H2
Indol Oil No. 10 .... H2
Indol Oil No. 2 .... H2
Indol Oil No. 3 .... H2
Indol Oil No. 35 .... K1
Indol Oil No. 4 .... H2
Indol Oil No. 5 .... H2
Indol Oil No. 7 .... H2
Indol Oil No. 8 .... H2
Indol Oil No. 9 .... H2
Lith-Gard 2 .... H2
Super Miscol White Packer Oil . H1
Universal Gear Lube SAE 140 .. H2
Universal Gear Lube SAE 80.... H2
Universal Gear Lube SAE 90.... H2

Farnam Companies

Kill'Em F2
Lure'Em P1

Farris Chemical Company, Inc.

Alka-Aid................................... G6
ACSC-BD.................................. G7
BAF-10...................................... G7
BAF-15...................................... G7
Co-Jel #200................................ G6
Co-Jel S-201.............................. G6
Co-Jel S-202.............................. G6
Co-Jel S-203.............................. G6
Co-Jell 600 .... G6
Cond-Trol G7
Ex-19.. G7
No-Ox....................................... G6
No-Ox-L.................................... G6
No-Scale #400 .... G6
No-Scale #410 .... G6
No-Scale #420 .... G6
No-Scale #440 .... G6
No-Scale #444 .... G6

No-Scale #460 .... G6
No-Scale #485 .... G6
Poly-Sur G6
Poly-Trol LH............................. G6
Scale-Trol L.............................. G6
SC-46 A G6
SC-46-B G6
SC-46-C..................................... G6
Vap Adjust................................ G6
Vap Trol.................................... G6
Void-Ox.................................... G6

Fast Products

Power Cleaner A4

Faultless Starch/Bon Ami Company

Bon Ami Cleaning Cake............. A6
Bon Ami Cleaning Powder......... A6
Bon Ami Polishing Cleanser A6
Kleen King For Copper &
Stnls Steel A6

Federal Chemical Co., Inc.

Arab Industrial Strength
Aerosol F1
Arab Power Pak with SBP
1382 Resmethrin F1

Federal International Chemicals

"Solv-Kwik" K1
Acidulade A3
Advantage 100,......................... A4
Advantage 128........................... A4
Advantage 256........................... A4
At-Last A1
Blue Concentrate Detergent
Cleaner................................... A4
Capture A4
Capture Neutral Fragrance D1
Clean-All A4,E1,A8
Conquest II Citrus Fragrance A4
Conquest II Neutral Fragrance.. D1
Conquest 256 Lemon Fragrance A4
Conquest 256 Neutral
Fragrance D1
Dyno-Might A4,A8
Fedphene 100............................. A4
Fedphene 128............................. A4
Foam-Kwik A4
Grease-Go C1
Hi Pressure Cleaner.................. A4
Ice-Cold.................................... A5
Low Foam Butyl Cleaner.......... A4
Low Foam Detergent A4
Mark VIII Plus.......................... A4
Mark VIII Spra Cleaner A4
Municipal Sewer Cleaner L1
Power-Solv K1
Steam-Kleen Liq. Steam
Cleaner................................... A1
Sub-Zero................................... A5
Super Con-Treat A3
Task Force Industrial
Degreaser A4
Wash Off Remover C3

Federated Chemicals

Federated (Floral Fragrance)
Disinf. Cln A4
Federated Cln/Disinf/Deod
(Citrus Frag.) A4
Federated Cln/Disinf/Deod
(Phenol) A4
Federated Neutral Fragrance.... D1
Foaming Cleaner A1
Freezer/Locker Cleaner A5
Meat Plant Steam Cleaner......... A1

Feedwaters, Incorporated

Algor G6

960...A1

First U.S. Chemical Company
BWT-1..G6
BWT-2..G6
First U.S. BWT-1.........................G7
First U.S. CHWT-1......................G7
First U.S. CWT-S1.......................G7
First U.S. CWT-S2.......................G7
First U.S. CWT-1.........................G7
First U.S. Microbiocide 15.........G7

Fischer Chemco, Incorporated
All Purpose Liquid......................A1
All Purpose Powder....................A1
Blast No. 1 Liquid.......................A4
Blast No. 1 Powder.....................A1
Milk Stone Remover Liquid.......A3
Milk Stone Remover Powder A3
Pan Cleaner.................................A1
Super Blast A2

Fischer-Lang & Company, Inc.
F-L Liquid Hand Soap E1
Last Quat Germicidal Cleaner ...D1
Liquid Heavy Duty Detergent...A1
Recon Conditioner-Cleaner A1
Syn-X...A1

Fish, A. J. Oven Company
Oven Lubricant H2

Fiske Brothers Refining Co.
Evap Oil......................................P1
Lubriplate Aero..........................H2
Lubriplate Air Tool Lubricant...H2
Lubriplate Auto. Trans. Fld
  Type "A"Suffix"A".................H2
Lubriplate Auto-Guard
  (Aerosol)..................................H2
Lubriplate Auto-Guard (Bulk)...H2
Lubriplate Auto/Marine-Lube
  A H2
Lubriplate AC-O H2
Lubriplate AC-1 ....H2
Lubriplate AC-2 ....H2
Lubriplate AC-2A H2
Lubriplate AC-3 ....H2
Lubriplate AC-4 ....H2
Lubriplate APG-75......................H2
Lubriplate Ball Bearing..............H2
Lubriplate Chain & Cable Fluid H2
Lubriplate Chain & Cable
  Fluid(Aerosol)..........................H2
Lubriplate DS-ES........................H2
Lubriplate FC-50 ....H1
Lubriplate FMO 350 Spray........H1
Lubriplate FMO-350 ....H1
Lubriplate FMO-350L H1
Lubriplate FMO-500 ....H1
Lubriplate FMO-500 CCH H1
Lubriplate FMO-500L H1
Lubriplate FMO-85.....................H1
Lubriplate FMO-85L...................H1
Lubriplate FMO-900....................H1
Lubriplate High Temp H2
Lubriplate HO-0 ....H2
Lubriplate HO-00 ....H2
Lubriplate HO-1 ....H2
Lubriplate HO-2 ....H2
Lubriplate HO-2A H2
Lubriplate HO-3 ....H2
Lubriplate HO-4 ....H2
Lubriplate HO-5 ....H2
Lubriplate Low Pour Hydraulic
  Oil H2
Lubriplate Low Pour HDS
  Motor Oil SAE 30...................H2
Lubriplate Low Pour HDS
  Motor Oil SAE10W H2

Lubriplate Low Pour HDS
  Motor Oil 20-20W H2
Lubriplate Low Temp H2
Lubriplate Marine-Guard
  (Aerosol)..................................H2
Lubriplate Marine-Guard (Bulk) H2
Lubriplate Mist Oil....................H2
Lubriplate Multi-Lube A H2
Lubriplate Multi-Lube A-1.........H2
Lubriplate MH Oil H1
Lubriplate MO-Lith No. 2...........H2
Lubriplate No. APG-140.............H2
Lubriplate No. APG-250.............H2
Lubriplate No. APG-80...............H2
Lubriplate No. APG-90...............H2
Lubriplate No. 0..........................H2
Lubriplate No. 1..........................H2
Lubriplate No. 102 1/2 ....H2
Lubriplate No. 105 ....H2
Lubriplate No. 105V H2
Lubriplate No. 107 ....H2
Lubriplate No. 110 ....H2
Lubriplate No. 115 ....H2
Lubriplate No. 1200-2 ....H2
Lubriplate No. 130-A..................H2
Lubriplate No. 130-AA...............H2
Lubriplate No. 130-AAA............H2
Lubriplate No. 2..........................H2
Lubriplate No. 3 ....H2
Lubriplate No. 3-V......................H2
Lubriplate No. 4..........................H2
Lubriplate No. 5555 ....H2
Lubriplate No. 630-A..................H2
Lubriplate No. 630-AA...............H2
Lubriplate No. 630-AAA............H2
Lubriplate No. 630-2...................H2
Lubriplate No. 630-2 Special......H2
Lubriplate No. 70 ....H2
Lubriplate No. 707 ....H2
Lubriplate No. 78 Oil..................H2
Lubriplate No. 8 ....H2
Lubriplate No. 914 Oven Chain
  Lubricant H1
Lubriplate No. 930-A..................H2
Lubriplate No. 930-AA...............H2
Lubriplate No. 930-AA-S H2
Lubriplate No. 930-AAA............H2
Lubriplate No. 930-AAA-S.........H2
Lubriplate No. 930-2...................H2
Lubriplate No. 930-2-S................H2
Lubriplate Oven Chain
  Lubricant H2
Lubriplate PM-500 ....H1
Lubriplate Solvent......................H2
Lubriplate Super FML-O...........H1
Lubriplate Super FML-1 ....H1
Lubriplate Super FML-2 ....H1
Lubriplate Super FML-2
  (Aerosol)..................................H1
Lubriplate Super FML-3 ....H1
Lubriplate Super GPO Motor
  Oil SAE 10W...........................H2
Lubriplate Super GPO Motor
  Oil SAE 30...............................H2
Lubriplate Super GPO Motor
  Oil SAE 40...............................H2
Lubriplate Super HDS Motor
  Oil SAE 10W...........................H2
Lubriplate Super HDS Motor
  Oil SAE 30...............................H2
Lubriplate Super HDS Motor
  Oil SAE 40...............................H2
Lubriplate Super HDS Motor
  Oil SAE 50...............................H2
Lubriplate Super HDS Motor
  Oil SAE20-20W........................H2
Lubriplate STO-3 ....H1
Lubriplate Wheel Bearing H2
Lubriplate 184.............................H2
Magic Soluble Oil.......................H2

117

Magic Soluble Oil Special H3
Stamping Oil 7 .... P1
35 Soluble Oil H3

Fisons Corporation
Ficam W F2

Fitch Dustdown Company, The
All American Heavy Duty
  Butyl Cleaner A1
Break-Down A1
Fido B-838.................................... A4
Fido Brand P-28 Insect Spray.... F1
Fido Brand Steric A1
L-50 Fido Insect Spray F1
Pink Lotion Hand Soap E1
Satin-Wash Hand Soap E1

Five Flags Chemical Co.
Bleach Cleanser A6
Block Out C1
Blue Mist A1
Five Ways A4
Freeze-R-Kleen............................ A5
Glass Cleaner C1
Insect Spray F1
Pipe Coat, ... P1
Pro-Degreaser A1
Rose-Bowl C2
Skin Shield Cream E4
Stainless Steel and Metal Polish. A7
Supre-Solve A3

Flag Janitorial Supply
Great Stuff A1

Fleetwood Chemical Co.
Acc-U-Solv #19 .... A4

Flexabar Corporation
Jelly De Rust A3
Liquid Germicidal Cleaner A4
Odor Control 1600 .... C1
Remove All A4

Flexo Chemical Company
Flexo-Terg A4

Flintkote Co./U.S. Lime Div.
Flintkote Hydrated Lime............. N3

Flo-Cem
Pynammo C1
Pynclene C1
Super D-G..................................... A1
Versa-Clene A4
Water Soluable Deod Conc
  Neut Bubble Gum.................... C1

Flo-Kem, Incorporated
All Purpose Cleaner A1
All Purpose Cleaner
  Concentrated A4
All Purpose Liquid 1093 .... A1
Bottle Wash 0437......................... A2
Circulation Cleaner 0873 .... A2
Coconut Liquid Hand Soap
  15% E1
Cooker Cleaner 1027.................... A2
Drain Opener L1
Dry Chlorine Bleach.................... A1,B1
Flo-Away Cleaner A4
Flo-Dine Disinfectant.................. D2
Flo-Kem 9-90 Conc. Fogging
  Insecticide F1
Flo-Kem 9-95
  Ant&RoachSpyResid w/
  Dursban F2
FloKem Acidet.............................. A3,Q2

FloKem Defoamer Q5
FloKem Egg Wash........................ Q1
FloKem Flo-AC A3
FloKem Flo-Con A4
FloKem Flo-Crete......................... A1
FloKem Flo-Kleen 57 .... A2
FloKem Flo-Klor A1
FloKem Flobac.............................. A2
FloKem Flobrite............................ A2
FloKem Flodet A1
FloKem Flomatic A1
FloKem Flopower.......................... A2
FloKem Flosan A3
FloKem Flosheen A1
FloKem Flosheen HD A2
FloKem Floterg............................. A1
FloKem Flothru A1
FloKem Foamzall.......................... A1
FloKem Fos Det 31 .... A3
FloKem Fos-Klor.......................... A1
FloKem Fosbrite A3
FloKem FA-21 .... A3
FloKem FGC A1
FloKem FK 204 .... A1
FloKem FK 913 .... A3
FloKem GC-20 .... A1
FloKem Kleener 96....................... A2
FloKem Klorshine D A1,Q1
FloKem Laundry Compound....... B1
FloKem MC-100............................ A1
FloKem PC-61 .... A1
FloKem PC-95............................... A1
FloKem Spray Wipe 90 .... A1
FloKem Spraykleen X A2
FloKem TC#1................................ A2
FloKem Veg Peel.......................... N4
Freezer Locker Cleaner............... A5
FKL Lotion Soap.......................... E4
General Purpose Cleaner 0690... A1
Glass Cleaner Concentrate C1
Heavy Duty Degreaser................. A4
Io-Dis Disinfectant D2
Liquid Acid Cleaner 1032............ A3
Liquid Conveyor Chain Lube.... H2
Liquid Steam Cleaner 0931 .... A1
Low Suds Cleaner A4
Maraud Industrial Spray............. F1
Milkstone Remover 0539 .... A3
Neutral Cleaner C1
Oven Cleaner 1141 .... A8
Paste Conveyor Chain Lube H2
Peel Additive 0927 .... N4
Quaternary Germicidal Cleaner. A4
Smoke House Cleaner A2
Spot Remover 0492...................... G5
Spray & Wipe Cleaner-
  Degreaser A1
Triple-2 Germicidal Cleaner....... A4
Water Base Degreaser A1
358 Glass Cleaner (Ready to
  Use) C1
999 Odrls Concntrtd Insecticide
  Spray F1

Floor Care Supply
Break Thru A1

Floor Systems, Incorporated
Trouble Spots A4

Florasan Corporation
Contact Cleaner........................... K2
No Sweat....................................... P1
No. 25 Insect Spray..................... F1
Sani-Lube H1
Silicone Spray H1

Florida Paper Company
APC Bowl Cleaner And
  Sanitizer.................................. C2

APC Germicidal Utility
  Cleaner..................................... C1
APC Glass Cleaner C1
APC TB Disinfectant
  Deodorant C2
APC Waterless Hand Cleaner.... E4

Floridin Company
Cal-Flor Dry................................. J1
Flor-Kleen J1
Florco J1
Florco-X....................................... J1
Florex LVM J1
Formula 44................................... J1

Flow Laboratories, Inc./ Environmental Cultures Division
Cultured Plmbr Grs,Trap&Drn
  Line Cl...................................... L2
Cultured Plumb Environ Odor
  Contl C1
DBC Plus L2

Fluoramics, Inc.
Chem-8 .... H2

Flushing Plastics Corporation
MAE 13C2AB Power Wash
  Detergent A1
MAE 13C2111 Power Wash....... A1
Steam Cleaning Machine
  Additive.................................... A1

Foam Spray Industries
"Champion Supreme" A4
"Champion" A1

Foam-Pak Company
Foam-Pak General Purpose
  Degreaser A4
Foam-Pak Germicidal Cleaner... D1
Foam-Pak Safety Hard Water
  Scale Remover A3

Foley Janitor Supply & Service Co.
Acid Bright A3
Excel A1

Ford Tractor Operations/ Ford Motor Company
BTC-350 Liquid Detergent
  Additive.................................... A1
BTC-710 Detergent...................... A1
BTC-730 Caustic Free
  Detergent................................. A1

Foremost Maintenance Supply Co. Inc.
#510 Fore-Lite Multi-Purpose
  Cleaner..................................... A1
#530 Lo-Foam Auto Scrub........ A4
#540 Slugger A4
#560 Spray & Clean.................... A1
#605 Co Co Liquid Hand Soap. E1

Foresight Of Omaha, Inc.
F-101.. A1
F-40+ Cleaner-Degreaser A4
F-804 Heavy Duty Industrial
  Cleaner..................................... A4

Formula Marketing Co.
C-4 Steam&Pressure Clng
  Compound A1

Formulated Products, Inc.
100 Smoke House Cleaner.......... A2
101 Smoke House Cleaner.......... A2

Foss, M. L. Inc.

Foss-Cream Sanitary Lubricant . H1
Release Silicone Mold Release ... H1

Foster Chemicals Inc.

Bravo C1
Chain Lube #69........................... H2
Chain Lube #70........................... H2
Chelaclean BW A2
Chelaclean HD-5 .... A2
Chelaclean P-67 .... A2
Chlelaclean KC............................ A2
Emulso... A1
Fosteracid 100.............................. A3
Fostracide.................................... A4
Fostracide EX............................... A4
Fostracide HD A4
FP A1
FP-10 .... A1
FP-15 .... A1
FP-18 .... A1
FP-20 .... E1
FP-35 .... A1
FP-38 .... A4
FP-45 .... C1
FP-70 .... C2
Machine-Matic BF A1
Non-CC K1
Wint Mint II C1

Foster Company

Respond Degreaser A1

Four Star Chemical

"FSL" Food Equipment
   Lubricant................................. H1
"118" Fluorocarbon Lubricant... H2
"910" Contact Cleaner K2
Double Duty 32............................ A4
Grip P1
Powr-Thru Drain Opener.......... L1
SS-201 .... K1
Vy-Ser-Eze................................... H2
421 Silicone Spray H1

Fowler Sales & Distributing Co.

Formula 3 Squared Chlor. Gen.
   Purp. Cl. A1
Husky... A2

Foxco

Foam-Away A1

Framar Industrial Products Inc.

Spray Fog F1

Franchise Services, Inc.

FSI Ammoniated All Purpose
   Cleaner.................................... A4
FSI Liquid Hand Dish Wash A1
FSI Machine Dish Wash
   Compound................................ A1
Institutional Liq Drn Pipe Opnr
   & Maintn L1

Franciscan Forge Mfg. Co.

Aqua-Jet All Purpose
   Degreaser Clnr......................... A1
Aqua-Jet Liquid Steam A1

Frank C. Mendes Supply Co.

Conquer....................................... A1

Frank Industries, Inc.

Fran-Chem A1

Frank Miller & Sons, Inc.

Concentrated Inhibited
   Chemical Clnr.......................... A3

Condensate Line Conditioner..... G6
Deo-Gran C1
Disinfectant-Sanitizer-
   Deodorizer Conc...................... D2
Emulsifiable Spray Concentrate. F1
Germicidal Cleaner A4
Heavy-Duty Floor Cleaner C1
Hospital Type (Mint) Disinf. &
   Clnr... C1
Insecticide Concentrate............... F1
Liquid Steam Cleaning
   Compud. H.D.Form. A1
Lotionized Hand Soap E1
Lubricant-Coolant H2
Mira-Pel A7
Mr. Melt C1
Multiple Purpose Boiler &
   Feedwtr. Treat......................... G6
Paint Stripper.............................. C3
Powdered Hand Soap C2
Virucide....................................... D1

Franklyn Sales Company, Inc.

Quick-Release A1

Franlynn, Incorporated

Citrosene 10 .... C1
Formula 12 FF 12 .... A1
Formula 28 Mr. Steam FF 28 ... A1
Lynnpol No. 96 Concentrate...... A1

Freco Chemical Co.

Offense... D2
The Bomb A1

Fred H. Williams Co.

Textile Grade Dry Silicone
   Spray....................................... H1

Fred's Chemical & Boiler Service

Chemical No. 240S...................... G6

Frederick Gumm Chemical Co., Inc.

Stero 24 .... A1

Fredricks Manufacturing Corp.

Lubri-Cone Spray........................ H1
Safti-Flex Lubri-Spray H1

Freedman, S. & Sons, Inc.

Liquid Hand Soap E1
SFS All Purpose Spray Cleaner A4
SFS Grill & Oven Cleaner A8
SFS Heavy Duty Cleaner........... A1
SFS Liquid Hand Dishwash...... A1
SFS Lotionized Liquid Hand
   Dish Wash A1
SFS 301 Hvy Dty Stripper
   Degreaser A4
SFS 40 All Purpose
   Concentrate Cleaner................ A4
SFS 64 Disinf. Cl. Citrus
   Fragrance A4

Freeport Manufacturing Company

Freeport Waterless Hand Cln.
   Knuckles Up E4

Freitas Enterprises

Emulisifer Cleaner Degreaser A1
New Non Butyl Degreaser......... A1

Fremar Corporation

Lemon Disinfectant..................... C2

Fremont Industries, Inc.

#101... A3
#101X... A3
#109... A1

#11	A1	
#110	A1	
#12 Acid Cleaner	A3	
#14A	A6	
#16	A1	
#160	A1	
#160XX	A1	
#180 New	D1	
#190F	A1	
#2014 Food Vat Cleaner	A2	
#2020 Brightener	A3,A6	
#2025 Acid Cleaner	A3	
#2026 Acid Cleaner	A3	
#203HD	A3	
#215	A1	
#22	A1	
#220	A1	
#227	A1	
#227M General Cleaner	A1	
#227PF General Cleaner	A1	
#23	A1	
#23D	A4	
#23S	A4	
#2300 Detergent-Sanitizer	D1	
#2301 Detergent-Sanitizer	D1	
#2302 Detergent-Sanitizer Conc	D1	
#2303 Detergent-Sanitizer	D1	
#237	A1	
#24	A2	
#25	A2	
#26	A3	
#3015	A2	
#302	A1	
#3024F Alkaline Rust Remover	A1	
#3036 Food Vat Cleaner	A2	
#304F	E1	
#3045 General Maintenance Cleaner	A1	
#3045PF General Maintenance Cleaner	A1	
#305F	A3	
#3061 General Cleaner	A1	
#3066 General Maintenance Cleaner	A1	
#307	A1	
#3074	L1	
#310 Fremo-Chlor	D1	
#311	A1	
#322	A2	
#324	A2	
#336 Food Vat Cleaner	A2	
#34	A2	
#34F	A4	
#360	A4	
#4	A1	
#4 Special Pot & Pan Cleaner	A1	
#40	A2	
#4006 Utensil Cleaner	A1	
#4022 Cleaner Detergent	A1	
#4032 H.D. Liquid Cleaner	A1	
#4072	A1	
#4073	A1	
#4080 General Purpose Detergent	A1	
#4081	A1	
#41	A1	
#41A	A1	
#42HD	A2	
#42S	A2	
#4908	A1	
#50	A2	
#51H	A1	
#52PW	A1	
#7	A1	
#7A	A1	
#7B	A1	
#70HD	A1	
#707 Alkaline Cleaner	A1	
#71X	A1	

#90 Boiler Water Treatment	G6
#90C Boiler Water Treatment	G6
#900 Boiler Water Treatment	G6
#901 Boiler Water Treatment	G6
#902 Boiler Water Treatment	G6
#91	G6
#91C Boiler Water Treatment	G6
#910N	A1
#911	G6
#911C Water Treatment	G6
#911HD Boiler Water Treatment	G6
#912	G2
#913	G6
#92	G6
#92P Boiler Water Treatment	G6
#922	G2
#923	G6
#924	G6
#930	G6
#931	G6
#932	G2
#94	G6
#9400C	G6
#9401	C1
#9402	C1
#9403	C1
#9404	C1
#942	G2
#95 Boiler Water Treatment	G6
#953 Boiler Water Treatment	G6
#96 Boiler Water Treatment	G6
#960 Collodial Oxygen Scavenger	G6
#963 Boiler Water Treatment	G6
#98 Boiler Water Treatment	G6
#98C Colloidal Treatment	G6
#98P Colloidal Treatment	G6
#9901	G6
#9904 Municipal Water Treat.	G3
Fremont 2012HW Water Soluble Lubricant	H2
Fremont 2015 Low Foam Acid Cleaner	A3
Fremont 2306 Detergent Sanitizer	D1
Fremont 2307 Sanitizer	D1
Fremont 250	A2
Fremont 752 Low Temp. Cleaner	A1
Fremont 753 Low Temp. Cleaner	A1
Fremont 8020 Machine Utensil Detergent	A1
Fremont 825 Organic Drain And Line Clnr.	L2
Fremont 826 Bacteria-Enzyme Digester	L1
Fremont 8505 Steam Line Treatment	G6
Fremont 8511 Steam Line Treatment	G6
Fremont 8515 Steam Line Treatment	G6
Fremont 8517 Steam Line Treatment	G6
Fremont 8530 Steam Line Treatment	G6
Fremont 8536 Steam Line Treatment	G6
Fremont 8540 Steam Line Treatment	G6
Fremont 8542 Steam Line Treatment	G6
Fremont 8555 Steam Line Treatment	G6
Fremont 8561 Steam Line Treatment	G6
Fremont 8565 Steam Line Treatment	G6
Fremont 8567 Steam Line Treatment	G6

Fremont 921 Water Conditioner	G2
Fremont 955 Boiler Water Treatment	G6
Fremont 9931 Sanitizer	D2
Fremont 9932 Disinfectant	D2
Fremont 9933 Disinfectant	D2
Fremont 9954	G6
Fremont 9961 Liquid Oxygen Scavenger	G6
Fremont 9962	G6
Fremont 9965 Flocculent	G1
Fremont 9966 Flocculent	G1
Fremont 9980 Liquid Caustic	G6

Frick Company

Frick No. 5 Oil	H2
Frick Oil No. 2A	H2
Refrigerating Machine Oil No. 3	H2
Refrigeration Machine Oil No. 4	H2

Fricke & Peters Paper Company, Inc.

FP 320 All Purpose Cleaner/ Degreaser	A1

Fried Industries, Inc.

Stericlean Granular	A1,Q1
Stericlean Liquid	A1,Q1

Friendly Chemical Company

Banshee	K2
Chava Industrial Solvent	A1
Fas-Cut Cleaner & Descaling Agent	A3
Foil Demoist. Compd. f/Elec. Equip.	P1
Foil Demoist. Compd. f/Elec. Equip.(Aer)	P1
Friendlube 100	H2
Friendlube 100 Extreme Pressure Gear Oil	H2
Friendlube 200	H2
Friendlube 300	H2
Friendly 9900 Cooling Water Treatment	G7
Frisk	E1
FVS-7 Heavy Duty Stripper	K3
Herald Concntd. All-Purpose Cleaner	A1
L-88 High Temp. Hvy Dty. Lubricant	H2
Mable	H2
Maxi-Flex	H2
Muscle	H2
Plow Concentrated Liquid Drain Opener	L1
Poly-Slip	H1
Proto-Shield	C1
Steam Chisel	A1
Will-Open	L1

Friendly Systems/ Division of NCH Corporation

Accept	H2
Armor Guard	A1
Assist	A3
Avenger	A1
Breakthru	A8
BP-1	C2
Charade	C1
Cide-Way	F1
Clenforize	A4
D.O.A.	F2
FAS-Hand	E1
Geartrain	H2
Ice-Melting Pellets	C1
Let-Go	H2
Lineman	L1

Moist-Out (Aerosol) K2
Moist-Out (Bulk) K2
Move-On K1
Naturalizer Biological
　Treatment System L2
No-Scrub A8
Pine-Point C1
Pinion H2
Problem-Solver K1
PC-400 (Aerosol) C1ᵇ
Ranger Food-grade White
　Lubricant H1
Sani-Breeze C1
Sani-Breeze (Aerosol) C1
Sewer-Magic L1
Sharpshooter F2
Shear-Glo C1
Single-Step A1
Steam Chisel A1
Stride H2
Strike-Out F1
Submit A1
Super Strip K3
Surround SAE 80W-90 .... H2
Surround SAE 85W-140 .... H2
Trail-Blazer F2

Frito-Lay, Incorporated
F.L. Acid Cleaner A3
F.L. Caustic Cleaner A2
F.L. Equipment Cleaner A1
F.L. Floor and Wall Cleaner A4
F.L. Gel for Acid Spray P1

Fritzsche Dodge & Olcott, Inc.
Animal Product Deodorizer
　47100 .... C1

Frosty Acres Brands, Inc.
Heavy Duty Machine Scrubber
　· Concentrate A1
1005-1 Liquid Hand Dish Wash. A1
1090-1 Inst. Liquid Drn Pipe
　Opnr & Mntr L1
330-X50 Machine Dish Wash
　Compound A1
335 Institutional Deep Fat
　Fryer Cleaner A8
341-X50 Supreme Low Suds
　Laundry Det. B1
345-X10 Concentrated Dry
　Bleach B1
915 General Purpose Cleaner A1
920 Heavy Duty General
　Purpose Cleaner........................ A1
925-1 Ammoniated All Purpose
　Cleaner A4
940-1 Bowl Cleaner C2
945 Alkaline Degreaser & Oven
　Cln Conc A8
950 Extra Heavy Duty
　Degreaser A4,A8
955-1 Ready-To-Use Glass
　Spray.. C1

Fruchtman Floor Supply
Command Duz-All A1

Fuld-Stalfort, Incorporated
Brevity Blue Liquid Scouring
　Creme A6
Cherry Blossom Frangrance
　Insect. Spray F2
Comp Clnr. Sanitizer Disinfnt.
　Deodorant C1
CTC C2
D-San 1330................................... D1
D'Part Roach & Fly Spray F1
D'Part Vaporizing Spray........... F1

E-Z-Kleen A1
Electrical Parts Cleaner K2
Enviro-Sep C1
F-S-78D Disinfectant Cleaner ... D2
F-1000... D2
Forty Percent Hand Soap........... E1
Fuld Professional Strength
　Insect Killer F1
Fuld-O-Solv A4
Gold Wrench Penetrating Oil H2
HD4 Problem Stripper &
　Heavy Duty Cln. C1
I-O-Teen Adjusted D1,Q6
Jamaica Insecticide Spray F2
Just 1 Defoamer A1
Just 1 Emulsifiable All-Purp
　Insct. Conc C1
Just 10 Cleaner Disinfectant A4
Kinetic Safety Elec.Mtr.&Pts.
　Degrsr.. K2
Like Magic Ammoniated
　Window & Hard Surf.............. C1
Like Magic Window Glass
　Cleaner C1
Liminate Liquid Residual
　Insect... F2
Liminox Liquid Residual Insect. F2
Mint-Tergent................................ A4
Open Gear & Wire Lube H2
Out L1
P-19 Disinfectant Pine Odor C1
Pink Cleaner Concentrate
　(Odorless) A1
Pink Tergent Cleaner.................. A1
Reem Drain Opener &
　Maintainer L1
S'Gone Germicidal Clnr. and
　Deodorant C1
Scum-Go 12 .... A4
Silicone Lubricant H1
Solarbrite...................................... A3
Solarine Metal Polish C1
Spring Time Insect. Preszd.
　Spray II F2
Stainless Steel Cleaner A7
Super Butyl Blue A4
Super Staze A1
Switer Bowl Cleaner................... C2
Triad A4
Veltone Antibacterial Hand
　Soap E1
Wildcat A1
15% Liquid Hand Soap E1

Fuller Brush Company, The
Formula 937 Liquid Steam
　Clnr.. A1
Formula 938 Water Soluble
　Deg. Sol..................................... C1
Formula 939 Strippng&Degrsng
　Cl... A4
Fullclean A1
Fullguard D1
Fullguard Germicidal Cleaner .. D2
Fullphene II A4
Fullpower Cleaner/Degreaser ... A4
Fullsan A4,C1
Fullsparkle Cleaner A4
Fullsparkle Window Cleaner...... C2
Industrial Insect Spray F1
Liquid Bowl Cleaner................... C2
Super Fullphene A4
3 Way Bowl & Bathroom
　Cleaner C2

Fuller, H. B., Company Monarch Chemicals Division
Monacid 250................................. A3
Monarch #12 .... A4
Monarch #3 .... A1

Monarch Acid Clean.................. A3
Monarch Acid Rust Remover.... A3
Monarch Add Chlor D2,Q4
Monarch Boot Wash D1
Monarch Bottleshine A2
Monarch BW-5 .... A2
Monarch BW-90 .... A2
Monarch C.I.P. Acid A3
Monarch C-S............................... D2
Monarch Chain Clean................ H2
Monarch Chlor Brite.................. A1
Monarch Circ Brite A2
Monarch Cots-O-Dine A3
Monarch Crest A1,Q1
Monarch Crevat A1
Monarch Dart........................... A2,B1
Monarch Degreaser..................... A1
Monarch Egg Wash Compound Q1
Monarch Experimental 600 .... Q1
Monarch Foam-X... .... A1
Monarch Germicidal Hand
　Scrub.. E2
Monarch GPC #10 .... A1
Monarch H.D. 325 .... A2
Monarch H.S.-Chain Lube H2
Monarch Heavy Duty Floor
　Cleaner A4
Monarch I-Bac............................. D2
Monarch Industrial Laundry
　Detergent................................... B1
Monarch K3-Chain Lube........... H2
Monarch K6-Chain Lube........... H1
Monarch Laundry Bleach
　Powder B1
Monarch Laundry Sour.............. B1
Monarch Liquid Sep-Ko............ A1
Monarch Liquid Super Ream..... A1
Monarch Low Foam Iodophor
　Germicidal Det........................ D2,Q6
Monarch M-9-B A1
Monarch Mon-O-Dine D2
Monarch Mon-O-San A1
Monarch Monacid A3
Monarch Monalum A1
Monarch Monoklor Liquid........ D2,Q4
Monarch Monoklor Power......... D1,Q4
Monarch MP-2 Acid................... A3
Monarch Oxygen Bleach B1
Monarch ORP-1 Boiler Water
　Additive..................................... G6
Monarch Peri B1
Monarch Phosphate Rust
　Inhibitor.................................... P1
Monarch Prime A1
Monarch Quat 1000.,.................. D2
Monarch Renu Acid A1
Monarch Sep-Ko A1
Monarch Sep-Ko BTC................ Q1
Monarch Sewer Cleaner L1
Monarch Shell White.................. Q1
Monarch Silicate Rust Inhibitor. P1
Monarch Solide D1
Monarch Special Caustic
　Additive..................................... A3
Monarch Spin A2
Monarch Super Can-O................ A1
Monarch Super Jet...................... A1
Monarch Super Kabon................ D1
Monarch Super Ream A1
Monarch Super 200..................... A3
Monarch Super-R........................ A1
Monarch ST-1 Steam Line
　Treatment.................................. G6
Monarch Teem A1,E1
Monarch Water Conditioner A1,G6,G2,G5
Monarch WT-12 Water
　Conditioner................................ G7,G2,G5
Monarch 10BTL.......................... G6
Monarch 11 .... A1
Monarch 1300 .... A1

All Purpose Laboratory
Cleaner A1
Bake Pan Cleaner A1
Blue Pressure Spray Wash A1
Borax Hand Soap C2
Chlorinated In Place Dairy
Cleaner A1
Chlorinated Mach. Dish
Washing Compound A1
Deep Fat Fry Cleaner A2
Deluxe Concrete Cleaner A4
Dermaterg 100 A A1
Derusting Cold Vat Compound . A2
Derusting Hot Vat Compound... A8
DC-20 Pressure Spray Wash ... A1
DC-40 Automatic Spray Wash .. A1
Foaming Scouring Cleanser/
bleach A6
Food Plant Cleaner A1
Hand Dshwshng Compound
No. 1 .... A1
Heavy Duty All Purpose
Cleaner A1
Heavy Duty Steam Cleaner A2
Hot Vat Compound A2
Hunter's Deep Fat Fryer
Cleaner A2
Hychlor A1
HPH Pink Automatic Spray
Wash A1
Ice & Snow Melt No.1 .... C1
Low Suds Laundry Compound . B1
Low Suds Laundry Compound
No. 2 .... B1
Machine Dish Compound No. 1 A1
Medium Duty Steam Cleaner A1
Medium Suds Laundry
Compound B1
Neutro Vapor Steam Cleaning
Comp. A1
Non Chlorine Bleach B1
Non Suds Detergent A1
Packing House Cleaner A1
Pinklor D1
San-I-Soft B2
Sewer Pipe Cleaner L1
Special Concrete Cleaner A4
Truck & Car Wash -Brown ... A1
Truck & Car Wash -1 .... A1
Truck & Car Wash -2 .... A1
Truck & Rig Spray Wash A1
Washing Powder A1
White Pressure Spray Wash A1
Yellow Hand Dish Wash A1
10% Chlorine Bleach B1
16% Chlorine Bleach B1
20% Chlorine Bleach B1

Gem, Incorporated

Aerocos Ant & Roach Spry w/
.25% SBP-1382 .... F1
Aerocos Crawl. Inset Kill.w/
.25%SBP-1382 .... F1
Aerocos Disinfectant Cleaner &
Deodz. A4
Aerocos Disinfectant Deo. &
Air Fresh. C1
Aerocos Disinfectant
Deodorant C1
Aerocos Double Action
Residual Insect. F2
Aerocos Fly & Mosq. Spray w/
.30% D-Trans F1
Aerocos Flying Insect Kill.w/
.25%D-Trans. F1
Aerocos Food Grade Silicone
Lubricant H1
Aerocos Glass Cleaner. C1
Aerocos Utility Cleaner C1

Gemini Chemical Corporation

All Purpose Cleaner A1
Descaler & Delimer A3
Emulsion Degreaser C1
Gemini Aluminum Cleaner &
Brightener A3
Gemini Liquid Steam Cleaner ... A1
Safety Solvent #2 .... K1

Gemini Chemicals, Incorporated

Blue Suds A1
Bo Pan A1
Glide Bars H2
Star Kleen Plus 393 .... A3
Star Shine 211 .... A1
Star Suds A1
Star Terg A4
Super Cid 396 .... A3
217 Jet Off A4
218 Strip Off A2
300 Glide A1
317 Luster Plus A1
320 De Lux A1
323 Super Chlor A1
324 All Klor A1
326 Blast Off A2
328 Star AC A1
329 Star HD A2
332 Star Chlor A1
380 Supreme Performance A1
381 D-Terg A1
394 Star Kleen A3
397 Cid A3
519 Chem Terg A1
537 Raise C1

Gemini Distributing Company

Biox 200 .... D1,G1

Gene Labs

Activated Alkaline Powdered
Drain Opener L1
Aerosol Disinfectant C1
All-Purpose Non Flammable
Safety Solvent K1
Belt Dressing P1
Big-B 2 .... A4,A8
Bigestant 250 .... L2
Bio-Kleen A1
CD D2
De-Grease K1
Disinfectant Sanitizer
Deodorizer D2
Drain-Away L1
Frigid Cut H2
Graphite H2
Jell C1
Lemon Sewer Sweetener L1
Liquid Hand Soap E1
Mint Deodorant C1
Multi-Purpose Penetrant H2
Odor-Ban C1
Paint and Varnish Remover ... C3
Penetrating Oil H2
Polysiloxane Release Agent... H1
Silicone Lube H1
Spray-Fog F1
Stainless Steel Cleaner A7
Stericide D1

Gene Smith Chemical Co., Inc.

SS-18 Liquid A1
SS-18 Powder A1

General American Oil Company

Gold Shield 100 .... H1
Gold Shld All Purp. Lith
Grease Grd 1 .... H2
Gold Shld All Purp. Lith
Grease Grd 2 .... H2

Gold Shld All Purp. Lith

Grease.Grd 3 .... H2
Gold Shld All Wthr Gr Oil
162P/80/90/140 .... H2
Gold Shld Compressor Oil H2
Gold Shld Ice Mach Oil Wax
Free 4025 hvy H2
Gold Shld Ice Mach Oil Wax
Free 4025 lgt H2
Gold Shld Ice Mach Oil Wax
Free 4025 med H2
Gold Shld Moly Bentone
Grease AP Grd 1 .... H2
Gold Shld Moly Bentone
Grease AP Grd 2 .... H2
Gold Shld Moly Bentone
Grease AP Grd 3 .... H2
Gold Shld OCT 500
Compressor Oil H2
Gold Shld SR-1 Gear Oil ... H2
Gold Shld Wheel Bearing
Grease S H2
Gold Shld White Litho Grease
Grade 1 .... H2
Gold Shld White Litho Grease
Grade 2 .... H2
Gold Shld White Litho Grease
Grade 3 .... H2

General Bionomics Corporation

Aqualite A1
AquaBond G2

General Chemical, Inc.

Break Away A1
Gen Chem Formula 700 .... A1
Len-Phene C1
Myra-Cyn C1

General Distributors, Inc.

Heavy Duty Non-Butyl
Degreaser A1

General Drug & Chemical Corp.

10% Sodium Hypochlorite
Solution D2

General Electric Company

AF-10 .... Q5
AF-71 .... Q5
AF-72 .... Q5
M&R 10 .... P1
M&R 30 .... P1
RTV-102 White P1
RTV-103 Black P1
RTV-106 Red P1
RTV-108 Translucent P1
RTV-109 Aluminum P1
RTV-112 White Pourable P1
RTV-116 Red Self-Leveling P1
RTV-118 Translucent P1
Silicone Emulsion AF-75 .... Q5

General Environmental Science Co

LLMO Liquid Live Micro-
Organisms L2

General Import-Export Company

Par-C-Lay Multi-Purpose
Cleaner A1

General Industrial Chemical Corp.

Workhorse Heavy Duty
Cleaner-Degreaser A4,A8

General Industrial Supply Co.

GI-4 Liquid Compound Stm &
Prs Clnr A1

FS Bleach B1,P1
FS Cleanser A6
G-P Chlor D2,Q2,Q4
Top Clor D2

Georgia-Tennessee Mining & Chemical Co.
Inst-A-Sorb Super J1
Instant-Dri J1
Whiz-Zorb J1

Ger Chem Industries
Extra Strength Hvy Dty Water
    Soluble Cl A1

Gerald McManus Chemical Co.
McManus Hand Liquid Soap E1

Gerard Supply Co.
Gerard I-O-Teen Disinfectant
    Adjusted D2

Gessco Sales Division
Formula 101 .... G6
Formula 102 .... G6
Formula 150 .... G6
Formula 151 .... G6
Formula 152 .... G1
Formula 154 .... G1
Formula 50 .... G6
Formula 52 .... G6
Formula 54 .... G6
Formula 60 .... G6
Gessco Basameen N G6
Gessco 105-X G6
Gessco 106 .... G6
Gessco 145 .... G7
Gessco 208 .... G7
Gessco 208-B G7
Gessco 209 .... G7
Gessco 210 .... G7
Gessco 40-D G6
Gessco 551 .... G6
Gessco 90 .... G6
Gessco 92 .... G7
Navy Special Formula G6

Getty Refining & Marketing Co.
Petroleum Insoluble Lube 1641 .. H2

Geyer-New York Equipment & Chemical Co.
Perfection Power Cleaner A1

Giant Oak Industries, Inc.
Medadine E4

Gibney, R. B. & Sons, Inc.
Big "G" Detergent Cleaner-
    Disinfectant D1
3-D All Purpose Clnr, Dgrsr,
    Wax Strip A1

Gibralter Chemical Corp.
Arctic Clean A5
Chemical Degreaser No. 1 .... K1
Extra Strength Heavy Duty
    Water Sol. Cl. A1
Foam Kleen A1
Gibralter Bacterial-Enzyme L2
Gibralter Clean It-All Purpose
    Cleaner A1
Gibralter Odor Control
    Granules C1
Super Kleen A4

Gibson Products Co.
Gib-Foam A1

Gil Chemical Company
C-S 26 .... A1
Chlorit D1
D-97 .... C1
Formula One Hundred D1
GL-601 Boilerized G6
GM22 .... A1
Iodex D2
MS-77 .... A3
OP88 .... A4
Pride A1
TC-60 .... A2
880A L1

Gilbert Chemical Corporation
Gil-Chem D1
PVL Germicidal Cleanser A4
Super Mov-It A4

Gilbert, G. T. Company, Inc.
All Purpose Cleaner &
    Degreaser 410 .... A4
All Purpose Cleaner 1500 .... A4
All Purpose Neutral Cleaner
    415 .... A4
Blue Car Wash 1517 .... A1
Car Wash 1537 .... A1
Chlorinated Machine Dish
    Wash 1503 .... A1
Concrete Cleaner 1506 .... A4
Disinfectant Deodorizer
    Cleaner A4,C1
Economy Hand Soap 135 .... E1
Extra Heavy Duty Steam
    Cleaner 1511 .... A2
Extra Heavy Duty Truck Wash
    1525 .... A1
Fogging Insecticide F1
Heavy Duty Cleaner &
    Degreaser A1
Heavy Duty Cleaner &
    Degreaser 412 .... A4
Heavy Duty Liquid Steam
    Cleaner 420 .... A1
Lemon Odor Disinfectant A4,C1
Machine Dish Wash 1504 .... A1
Odorless Disinfectant-
    Deodorant-Santzr. D2
Pine Odor Disinfectant C1
Pink Liquid Dishwash 306 .... A1
Pure Coconut Hand Soap 130 ... E1
Residual Insecticide F2
Sewer Pipe Cleaner 1508 .... L1
Space Spray Insecticide F1
Steam Cleaner Non-Caustic
    1510 .... A1
Steam Cleaner 419 .... A4
Super Solv A4,A8
Truck & Car Wash 1515 .... A1

Gillis Chemicals, Inc.
G-Brite Lemon Scented Stnlss.
    Steel Clnr A7
G222 .... A1

Gilmer Industries, Inc.
Gil A Super Chlor D1
Gil Acid Sanitizer LF A3,D1
Gil Beads A1
Gil Better Egg Wash Q1
Gil Bigger Dog A1
Gil Blitz A4
Gil Boil Away A2
Gil Bowl Cleaner C2
Gil Brite Wash B1
Gil BG Cleaner A2
Gil BV-20A A3
Gil Can-Tort A1
Gil Caustic A2,G6,L1,N2,
                                                      N3

Gil Caustic Plus A2
Gil Chlo G.P. A1
Gil Chlor Hi-Suds A1
Gil Chlor Kleen A1
Gil Chlor Machine Wash............. A1
Gil Chlor Suds A1
Gil Chlor Suds Heavy Duty A1
Gil Chlor-W A1
Gil Chlorinated Cleaner................ A1
Gil Clean A1
Gil Conc. G.P. Powder A1
Gil Concrete Cleaner A4
Gil Deep Clean A2
Gil Defoam A1,A1
Gil Dehair N2,N3
Gil Dermal Kare E1
Gil Econo Clean A1
Gil Egg Wash Q1
Gil F.P. Liquid Cleaner A1
Gil Feather Off-LF N1
Gil Feather-Off............................. N1
Gil Foam-Ade A1
Gil G.P. Liquid A1
Gil Grime-Away A2
Gil H.D. Steam Powder A2
Gil Heavy Duty Cleaner A2
Gil Heavy Duty CIP Cleaner A1
Gil Hifoam Acid............................ A3
Gil High Suds A1
Gil Hook And Trolley................... A2
Gil HD Cleaner A2
Gil Jel A1
Gil L.D.O.C. L1
Gil Liquid Chlor 15 .... D2
Gil Liquid CIP A1
Gil Liquid Power A1
Gil Liquid Spray A1
Gil Liquid Steam A1
Gil Metal Strip.............................. A2
Gil Metal-Brite.............................. A3
Gil O-Dine D2,E3
Gil Organo-Chlor D1
Gil P.R. A3
Gil Phene A4
Gil Phene Plus............................... A4
Gil Poultry Scalder N1
Gil Powdered CIP A1
Gil Powdered CIP And Spray
    Cleaner.................................. A1,Q1
Gil PR Acid A3
Gil Quick Clean............................ A4
Gil San..................................... D2,Q3
Gil Shackle Klean and Lube A1,H2
Gil Soft Suds A1
Gil Speedy Cleaner
    Concentrate A1
Gil Speedy II A1
Gil Spray Cleaner.......................... A1
Gil Spray Off................................. A4
Gil Strip-Away A1
Gil Sud-Z-Wash............................. A1
Gil Super Speedy A1
Gil Tray Wash A1,Q1
Gil Tripe-Brite.......................... N2,N3
Gil Truck & Car Wash A1
Gil Vat Cleaner A2
Gil Wildcat A1
Gil 333 .... L1
Gil-Bacti Derm Hand Cleaner
    Lotion E1
Gil-Super Insecticide
    Concentrate F1
Residual Roach Spray................... F2

Ginn Chemical Company
Big "G" A1

Glasby Maintenance Supply Company
"Satin Glo" Stnlss Steel Cl &
    Polsh Conc A7

Glazer Chemical Corporation
C-25 Ice & Snow Melter C1
C-26 Laundry Soap Powder........ B1
C-3 All Purpose Cleaner.............. A1
C-39 Interior Exterior Cleaner... A1
G-40 Skale Remover A3

Glissen Chemical Co., Inc.
Nu-Foam A1

Global Laboratories, Inc.
Bac-Off E1
C I P Acid A3
Do All-P H.D. A2
Freedom Silicone Spray H1
FPL-22 Food Equipment
    Lubricant H1
Global #81 .... A2
Global 2020 .... A4
Global-Plus................................... A1
Grab Belt Dressing....................... P1
Heavy Duty Degreaser A4
Insure Battery Terminal Clnr &
    Protect....................................... P1
Link Chain & Cable Lubricant .. H2
Muscle A4
Non Chlor P-2 .... A1
Pro Remove 1 .... A3
Pro-Remove................................... A3
Protective Hand Cream E4
Release H2
Super Degreaser A4
Super Muscle A4
Switch Contact Cleaner................ K2
Total L-1 .... A1
Total L-2 H.D. A2

Globe Chemical Company, Inc.
#95.. C2
A-161 Sof T Suds A1
A-162 Ezy Suds A1
A-230 Chlornate A1
A-231 Brillant A1
A-235 Accent A1
Accent A1
Alto Pine A1
ABC A-290 Po Pac Ho Cleaner A1
B.G. Spray F2
B.W.T. 100 .... G6
B.W.T. 300 .... G7
Barrier Creme E4
Black Jack C1
C. F. R. A3
Chain Lube.................................... H2
Clear View Glass & Window
    Cleaner...................................... C1
Cream-O-Septic E1
Double-Power................................ D1
Dri-Kleen E1
Dyne Germicide Concentrate D2
Economy Electro Kleen
    Solvent...................................... K2
Ezy Steam Cleaner A2
Fast Break A1
Formula A-R A1
Formula 1231-EW Q1
G-12 .... A4
Germ Purge D1
Getz-All Concrete Cleaner............ A4
Globan C1
Globe Aftawash A1
Globe Alkatron A1
Globe Atackit A1
Globe Eggcondition 90
Globe Eggcondition No. 2 .... Q1
Globe Flushox A2
Globe Liquid Chain Lube H2
Globe Metastrip A2
Globe Multikon A1

Globe P-H-C Formula II........... A1
Globe Power Flush....................... A1
Globe Scog Remover................. N2
Globe Smack Formula II............. A1
Globe Soakit A3
Globe Tripecure N3
Globe Zymes L2
Globe-San..................................... D1
Globe-San Formula 21................. D1
Globe-Sol C2
Globecide F2
Globolene A1
GL Concentrate............................ A4
GM Concentrate........................... A4
GM Concentrate-Odorless A1
Heavy Duty Steam Cleaner P.... A2
Heavy Duty Steam Cleaner
    Powder A2
Hospital Disinfectant
    Deodorant C2
Hotcide F1
Hydroclo C2
Johnny Brite.................................. C2
Kleen-O Brite................................ A1
Kleer View C1
Kut-Zit Drain Opener.................. L1
L-S Cleaner C1
Lab-Kleen..................................... C1
Liquid Hand Soap E1
M-157 Yellow Cat C1
Magic Kleen.................................. C1
Malathion Fly Bait with DDVP F2
Mark IV Cleaner A4
Odor Solve C1
Open Gear Lubricant H2
Out-Side F1
P.H. Cleaner................................. C2
Panacea A1
Pesto Spray F2
Pine-O-Septic C1
Pyrenone Insecticide F1
R-140 Stainless Steel Cleaner &
    Polish C1
Red Hot Sewer Solvent................ L1
Remuv-Ox A3
Sani-Phene 10-X D1
Sewer Solvent............................... L1
Starbrite.. A7
Super 500 Thermal Fog Spray... F1
Thermal Fog Roach Spray........... F1
Thry-Gly C1
Triple 'T' A1
Vapo Mist..................................... F2
Waterless Hand Cleaner............... E4
30% Malathion in Oil F2
50% Malathion Concentrate........ F2

Globe Services
Chemi-Jel...................................... C1
Floor Syn A4
Rapid-Clean................................... C1
123 Spray Cleaner A4
15% Liquid Hand Soap E1
20% Liquid Hand Soap E1
40% Concentrate Liq. Hand
    Soap E1

Go-Jo Industries, Inc.
Go-Jo Antiseptic Lotion Hand
    Cleaner...................................... E1
Go-Jo Hand Cl w/Deep Scrub
    Fine Italn Pu E4
Go-Jo Original Form. Hand
    Clur E4

Go-Pro, Inc.
Blue Butyl Cleaner A4,A8

Gold Cross Products, Inc.
Servis Liquid Hand Soap............ E1

Servis Liquid Scrub Detergent .. A4

Gold Kist, Incorporated

Acid Brite.. A3
Acid Brite I..................................... A3
Acid Brite II.................................... A3
Acid Cleaner................................... A3
Alkaline Powder............................. A1
All Purpose Clnr. Concentrate... A1
Caustic Cleaner.............................. A2
Cleaner... A1
Concrete Cleaner........................... A1
CIP Cleaner A1
Deodorant-Stripper C1
Egg Wash................................. A1,Q1
Feather Softener N1
Feather Softener and Defoamer. N1
Gold Kist B-102 Boiler
    Treatment................................... G6
Gold Kist B-104 Boiler
    Treatment................................... G6
Gold Kist B-105 Boiler
    Treatment................................... G6
Gold Kist B-106 Chelated
    Boiler Treat............................... G6
Gold Kist B-200............................. G6
Gold Kist B-300............................. G6
Gold Kist B-301............................. G6
Gold Kist B-303 Chelated
    Boiler Treat............................... G6
Gold Kist B-310 Chelated
    Boiler Treat............................... G6
Gold Kist Defoamer G6
Gold Kist NTA G6
GPC.. A1
GPC Powdered No. 1..................... A1
GPC Powdered No. 2..................... A1
Hi-Chlor D1,Q1,Q3
Hi-Foam A1
Liqui-Foam A1
Liquid Caustic Concentrate........ A2
Liquid Detergent A1
Low Chlor A1
Powdered Caustic Cleaner I A.. A2
Powdered Caustic Cleaner II A. A2
Powdered Egg Wash Q1
Processor Residual Sanitizer D1
TT Cleaner..................................... A1

Golden Bear Division Witco Chemical Corporation

Golden Bear's Cd 9914 White
    Min. Oil H1

Golden Circle Chemicals, Inc.

FZ-3.. A1
Mighty Mike Concentrate A1

Golden Glades Chemicals

Concrete Cleaner........................... A4
Liquid Hand Soap E1

Golden Products, Incorporated

LDC A1
Super 10 All Purpose Cleaner.... A4

Golden State Chemical Co.

GSC-77 .... A4
GSC-88 .... A1

Good News Distribution Center

Dishware And Utensil Cleaner .. A1

Goodwill Chemical Co., Inc.

Golden Glow All Purpose
    Cleaner....................................... A1
Golden Glow Cesspool&Septic
    Tank Cl....................................... L1

Golden Glow Drainpipe
    Solvent Liq................................. L1
Golden Glow Drainpipe
    Solvent(Flk) L1
Golden Glow Hand Cleaner C2
Golden Glow Hog Scour N2
Golden Glow Liquid Detergent A1
Golden Glow Safety Cleaner A1
Golden Glow Sewage
    Conditioner................................ L1
Golden Glow Smoke House
    Cleaner....................................... A2
Golden Glow Soap Beads A1
Golden Glow Speedy Steam
    Jenny... A2
Golden Glow Sudzy Ammonia.. C2
Golden Glow 515 Comp. Soap
    Beads.. A1

Goorland & Mann

Gem C-Bright A1

Gordon Johnson Company

SC-3 .... A3
SC-8 .... A1
SC-9 .... A1

Gordon's Supply, Inc.

Brute Liquid Organic Digester .. L1

Gordy Salt Company, Inc.

Evaporated Salt P1
Evaporated Salt B1
PHG Evaporated Salt................... P1
PHG Evaporated Salt................... B1

Government Chemical Products Corp.

Pink Lightning Concentrate A1

Grace-Lee Products, Inc.

All Purpose Compound A2
Aquamatic A1
Aquamatic 2B A1
Blue Wall Wash A1
Chain Lube.................................... H2
Controlled Suds B1
Controlled Suds with Bleach B1
Cream Cleanser.............................. C2
D.C. Cleaner A1
Dairy & Vat Cleaner A3
Dubl Duti Cleaner A4
E-M Hand Soap............................. C2
Empress Hand Soap E4
Extra Hvy Dty Liq Steam
    Cleaner....................................... A2
F-165 .... A3
F-199 .... A1
F-209 .... A3
F-210 .... A3
F-211 .... A3
F-212 .... A2
F-263 .... A1
F-266 .... A3
F-283 .... A3
F-295 .... Q5
F-348 Acid Cleaner...................... A3
F-360 Safety Cleaner................ A4,A8
Fect... A1
Formula N.F................................... A1
Formula XP-101 Steam Cleaner A2
Formula 140................................... A1
Formula 16..................................... A1
Formula 300................................... A1
Formula 400................................... A1
Formula 400-SA Utility Cleaner A4
Formula 9....................................... A1
G/L CarWash................................. A1
G/L Cleaner A6
G/L Detergent A1

G/L Liquid Cleanser A4
G/L Syntrate................................. A4
G/L 24 .... A2
G/L 62 .... A2
Glo Suds... A1
H.D.Hand Soap C2
Heavy Duty Liquid Steam
    Cleaner....................................... A2
Heavy Duty Steam
    Cleaner(Powder)........................ A2
Heavy Grease Detergent A1
Heavy Weight................................ A1
Knock Out A4
Lab Formula A1
Lab Formula 4 .... D1
Lid-128 .... A1
Lime Clean..................................... A3
Liquid All Purp A4
Liquid All Purp (Low Suds)...... A1
Lustre Stainless Steel Cleaner &
    Polish.. A7
LD Hand Soap C2
MD Hand Soap C2
No. 45 Soap.................................... A1
Ph D Surgical Hand Soap........... E1
Pink Pine Crystals (Odorless).... A4
Pink Velvet Lotion Hand Soap . E1
Plastic Dish Compound A1
Plus One... A3
Reet Bowl Cleaner C2
Sanitizer................................... D2,E3
Scale Remover................................ A3
Sewer Cleaner SC-36.................... L1
Skin Lotion E4
Special Hard Floor Cleaner A4
Special T-D.................................... A1
Spray Away A1
Spray Wash 100 .... A1
Standard Liquid Steam Cleaner. A2
Standard Steam Cleaner.............. A2
Synetic Suds A1
Tripl Duti...................................... A4
Tripl-Duti Low Suds A4
Waterless Hand Cleaner............. E4
XP 223 Liquid Detergent A1
133 Suds A1
225-Kleener(Odorless).................. A4
3-Way... A4
3-Way PF A4
611 Degreaser A4

Graco Products, Incorporated

Electro-Solv-A Solvent Cleaner
    & De-Grsr.................................. K1
Graco Pine-Odo............................ C1
Remov-All Concentrate............... C1

Graco, Incorporated

Dirt-tergent A1
Green Dirt-tergent A1

Grand Union Company

Sani-Clean Plus Neutral
    Fragrance D1

Gray Distributing Company

Custom Foamee.............................. A1

Gray Supply Company

Gold Strike Kwik-Foam............. A1
Gold Strike-Break-Down........... A1

Graymills Corp.

Super Agitene K1

Great Atlantic & Pacific Tea Co., Inc., The

A&P Supermarket Floor
    Cleaner....................................... A4

Great Lakes Biochemical Co., Inc.
Meta-Phos G6,G2
Micro-Zyme L2

Great Lakes Chemical Corporation
BromiCide G7
Meth-O-Gas F4

Great Lakes Corporation
Brilliant Green Dye #2 .... M3
C.L.D. Liquid Denaturant.......... M1
Solvent #690............................. K1

Great Lakes Distributing, Inc.
Spray Foam............................... A1

Great Western Supply Co.
G. W. Super Grelease A1

Greater Mountain Chemical Co., Inc.
GMC Nu-Blu A1
GMC 127.................................. A2

Green Mountain Research Corp.
Bio-Terge Concentrated All
  Purpose Clnr. A1
D.F.C..................................... A4
Electro-Solve Safety Solvent K1
Greasex Jell Degreaser K1
ODXL C1
Pentra Jell H2
Tri-Solvex A1

Green's, Inc.
"Extra" A1
Atlas Creme Cleanser C2
Big "G" A1
Blast Off K1
Boiler Treat............................. G7
Bulldozer................................ C2
Culture................................... L2
Custom Maid A1
De-Grease A1
Dizolv A4
Doz A1
Flush Out Organic Digester....... L1
Lemon Poli-Steel Stnlss Stl
  Clnr&Plsh A7
Liqui-Grease H2
Mash E1
Muscle A1
No. 333 Insecticide.................. F1
Nuthin'................................... A4
Oven Cleaner A8
OX A3
P & E Bowl Detergent,
  Deodorant & Disinf................ C2
Polisteel A7
Skin It Kleen........................... A8
Spray-Ez................................. C1
Steam-O A1
Super 444................................ F1
Wildcat A4,A8

Greene-Douglas Maintenance Industries, Inc.
G-D 1000................................ A1
G-D 1050................................ A1
G-D 1099................................ A1
Superac A1

Greene, E & Company, Inc.
Greenetree All-Purpose Cleaner
  #560................................... A4
Greenetree Coconut Oil Liq
  Hand Soap #99..................... E1
Greentree Hvy Dty Stripper Cl.
  Degrsr................................. A1

Greenwood Chemical Co., Inc.
Greenwood-100-C C1
Improve A1
Klor-Klene A1
Liquid Mildsuds........................ A1
Sanityze 72 .... D1
Shine-C A1
Sudsy-C B1

Greenwood Chemical Sales, Inc.
Glo-Bowl................................. C2

Greer Laboratories, Inc.
Alltreat 300 .... G7
Big "G" A1
Dual-27 .... D1

Griffin Bros., Incorporated
Antiseptic Liq. Soap 40%
  Concentrate E1
Auto-Magic A1
Blu Blazes C2
Bowl San Emulsion.................... C2
Brand-Y C1
Cleaner Disinfectant A4
Diamond Liquid Soap................. E1
Disinfectant Cleaner A4
Formula 100 .... A4
Formula-56 Sanitizer D1
Fresh-Ette C1
Glass Cleaner C1
Golden Pinol............................ C1
Greasol K1
Griff-N-Off.............................. A1
Griffin's Fasfoam A4
GSS One Swipe Stainless Steel
  Cln Polish A7
Heavy Duty. Ammon. Glass
  Cleaner.............................. C1
Heavy Duty Griffin-Suds C1
Hypine-8 .... C1
Jolt A4
Liqua-Suds A1
Locaten Sanitizer...................... D1
Non-Scents C1
Odor-Ban C1
Opal 40 Liquid Soap E1
Phen-All 7 Cleaner.................... A4
Red-Hot Stripper C1
Ritenow Cleaner C2
Spring E4
Var Brite Cleaner C1
20% Antiseptic Liquid Soap E1

Griffith & Company, Inc.
Sterii-Kleen A4

Griffith Company, Inc., The
Hy-Det Liquid A1
Hydro Flow A1
Hydrosteam (Green Label).......... A1
Hydrosteam (Red Label) A1
Hydrosteam A.P. Powdered
  Compnd. A1
Hydrosteam No. 50 .... A1
Power Wash Liquid
  Concentrate A1
Powerwash P............................ A1

Griffith Laboratories, Inc., The
Erado..................................... D1
Klenzall A1

Growth Marketing Corporation
APC-23 All Purpose Cleaner A1
ASV-33 Disinfectant Deodorant C1
BCL-25 Bowl Cleaner C2
Controlled Situation Metered
  Insect. Aer.......................... F2

CCR-13 Coil Conditioner A3
DOR-14.................................. L1
DOR-15A................................ L1
DOR-15S................................ L1
DSD-33 .... A4
DSF-33 .... A4
DSP-33 .... C1
DSQ-33 Germicide.................... D1
E-Z Kleen C1
ELD-13 Electric Motor
  Degreaser K2
ELD-13NF Electric Motor
  Degreaser K1
END-13 Engine And Parts
  Degreaser K1
Formula 8-B Boiler Treatment...G6
Formula 8-BL Boiler And Line
  Treatment G6
FPL-11 Food Equipment
  Lubricant H1
HSP-23 Hand Soap E1
HSP-235.................................. E1
HTC-31.................................. A2
OCG-55 Odor Control............... C1
OXC-13 Oxidizing Cleaner........ A3
Sol-10.................................... A1
Sol-20 Cleaner......................... A1
Springtime C1
SCL-15 .... A1
SOL-15 Solvent A1
TRR-13G K1

Growth Marketing Corporation of Nebraska
APC-23 All Purpose Cleaner.....A1
BCL-25 Bowl Cleaner C2
CCR-13 Coil Conditioner A3
DOR-15 S Drain Opener........... L1
DSF-33.................................. A4
DSP-33 .... C1
ELD-13 .... K2
ELD-13NF Electric Motor
  Degreaser K1
END-13 Engine and Parts
  Degreaser K1
Formula 8 B Boiler Treatment...G6
Formula 8 BL Boiler & Line
  Treatment G6
Gromark SIP-11 Spray Silicone
  Release............................... H1
HSP-23 .... E1
OCG-55 .... C1
OXC-13 Oxidizing Cleaner........ A3
SCL-15 .... A1
SOL-15 .... A1
TRR-13G K1

Gruss Industries, Incorporated
Compound 10........................... A1
Compound 99........................... A4

Guarantee Chemical Company
Auto Scrub.............................. C1
Boot Hill................................. F3
Brite A1
Cherry Gran C1
Cling C2
D Germ D1
Doz/It A1
Get Zit L1
Kitch-Solv A4,A8
Look Out................................. C1
Medi-Creme E1
Mighty A3
Odoban C1
Protect G6
Royal Flush............................. L1
Safe Walk C1
Safety Solvent.......................... K1

Shopmate A4

Guarantee Exterminating Co.
Supertox Brand Food Plant
  Fog. Insect. F1

Guarantee Exterminators, Inc.
Guarantee's Insect Fogger F1

Guardian Chemical & Supply, Inc.
Guard Ammoniated Wax Strip .. A4
Guard Germicidal Cleaner A4
Guard Pressure Cleaner A4
H.D. Caustic Cleaner A2
H.D. For Grease Removal
  Cleaner C1
H.D. High Caustic P.S. Cleaner A2
H.D. Steam Cleaner A1
Heavy Duty Cleaner A4
Non Caustic Chlorinated
  Cleaner A1
Non Caustic Cleaner A1
Non Caustic Processing Equip.
  Cleaner A1

Guardian Chemical Company
Atox F1
Chickodine D1,Q3
Concentrated Liquid Toilet
  Soap E1
Conta A2
Disolvit L1
EKTA A2
Germicide D1
Guardex A4
H.D. Cleaner A1
Idocide D2
Liquid Toilet Soap E1
Nuthin' A4
PHC Cleaner A1
Release H1
Skin-It-Kleen A8
Squirt C1
Task A4
7-C A1

Guardian Chemical Specialties Corp.
Aqua Kleen A4
Artic Kleen A5
Assure F A4
Assure N D1
Command 30 .... D1
Command 40 .... A4
Design 100 .... A4
Design 200 .... A4
Design 300 .... A4
Foam OP A1
Grease-B-Gone A4
Pack Kleen A1
Saf-T-Solv K1
Ultra 300 .... D1
Ultra 400 .... A4

Guardian Company Inc.
Formula 400 Industrial
  Antiseptic Soap E1
Our Eight Formula 111GC A1

Guardian Industrial Products, Inc.
Etch-Crete A4
Gard-Solv A1

Guardian IPCO, Incorporated
#550 Alkaline Drain Cleaner L1
#551 Acid Drain Cleaner L1
Guardian-Ipco #1 .... G6
Guardian-Ipco #1-T G6
Guardian-Ipco #101 Threshold
  Treatment G2

Guardian-Ipco #102 Threshold
  Treatment G6
Guardian-Ipco #170 .... G7
Guardian-Ipco #2 .... G6
Guardian-Ipco #2-S G6
Guardian-Ipco #21 .... G6
Guardian-Ipco #3 .... G6
Guardian-Ipco #30 .... G6
Guardian-Ipco #370 Boiler
  Treatment G6
Guardian-Ipco #400 Boiler
  Treatment G6
Guardian-Ipco #43 Steamline
  Treatment G6
Guardian-Ipco #44 Steamline
  Treatment G6
Guardian-Ipco #45 Steamline
  Treatment G6
Guardian-Ipco #46 Steamline
  Treatment G7
Guardian-Ipco #47 Steamline
  Treatment G7
Guardian-Ipco #6 On-Line
  Boiler Cleaner G7
Guardian-Ipco #7 .... G7
Guardian-Ipco #99 Rust
  Remover G7
Vulcan #88 .... G7
Vulcan #89 .... G7

Guest Paper Company
GP Lemon Disinfectant C2
GP Pine Odor Disinfectant C1
Pink Lotion A1

Gulf Atlantic Chemicals, Inc.
GAC G6
GAC-681 .... G6

Gulf Chemical Supply Co., Inc.
"Charlie" Heavy Duty Cleaner .. A4,A8

Gulf Coast Laboratories, Inc.
Everklear G2
Hydroklear G2
Hydroklear Plus G2
Liqui-Phos 400 .... G2
Liqui-Phos 500 .... G2
Tri-Lux G2

Gulf Oil Corporation
Automatic Transmission Fl.
  Type F H2
Gulf Automatic Transmission
  Fl.Dexron II........................... H2
Gulf Cut Soluble Oil H2
Gulf E.P. Lubricant HD100 .... H2
Gulf E.P. Lubricant HD150 .... H2
Gulf E.P. Lubricant HD220 .... H2
Gulf E.P. Lubricant HD32 .... H2
Gulf E.P. Lubricant HD320 .... H2
Gulf E.P. Lubricant HD460 .... H2
Gulf E.P. Lubricant HD68 .... H2
Gulf E.P. Lubricant HD680 .... H2
Gulf Endurance 14 .... H2
Gulf Endurance 150 .... H2
Gulf Endurance 19 .... H2
Gulf Endurance 9 .... H2
Gulf Eskimo C H2
Gulf Harmony 115 .... H2
Gulf Harmony 150 .... H2
Gulf Harmony 150 AW H2
Gulf Harmony 22 .... H2
Gulf Harmony 220 .... H2
Gulf Harmony 32 .... H2
Gulf Harmony 32 AW H2
Gulf Harmony 46 .... H2
Gulf Harmony 46 AW H2
Gulf Harmony 68 .... H2

Gulf Harmony 68 AW H2
Gulf Harmony 68 E.P. H2
Gulf Harmony 90 .... H2
Gulf Legion 100 .... H2
Gulf Legion 15 .... H2
Gulf Legion 150 .... H2
Gulf Legion 19 .... H2
Gulf Legion 220 .... H2
Gulf Legion 32 .... H2
Gulf Legion 39 .... H2
Gulf Legion 46 .... H2
Gulf Legion 58 .... H2
Gulf Legion 97 .... H2
Gulf Mineral Seal Oil H2
Gulf Multi-Purpose Gear
  Lubrant. 85W/140 .... H2
Gulf Multi-Purpose Gear
  Lubricant 80W/90 .... H2
Gulf Paramount 100 .... H2
Gulf Paramount 150 .... H2
Gulf Paramount 22 .... H2
Gulf Paramount 32 .... H2
Gulf Paramount 46 .... H2
Gulf Paramount 68 .... H2
Gulf Plastic Petroleum B H2
Gulf Plastic Petroleum E H2
Gulf Precision Grease No.1 .... H2
Gulf Precision Grease No.2 .... H2
Gulf Precision Grease No.3 .... H2
Gulf Security 100 .... H2
Gulf Security 100 AW H2
Gulf Security 115 .... H2
Gulf Security 180M H2
Gulf Security 20 .... H2
Gulf Security 32 .... H2
Gulf Security 32 AW H2
Gulf Security 320M H2
Gulf Security 46 .... H2
Gulf Security 46 AW H2
Gulf Security 460M H2
Gulf Security 68 .... H2
Gulf Security 68 AW H2
Gulf Security 788 .... H2
Gulf Senate 320 D H2
Gulf Senate 375 .... H2
Gulf Senate 400 D H2
Gulf Senate 460 .... H2
Gulf Senate 680 .... H2
Gulf Seneca 100 .... H2
Gulf Seneca 150 .... H2
Gulf Seneca 180 .... H2
Gulf Seneca 22 .... H2
Gulf Seneca 32 .... H2
Gulf Seneca 46 .... H2
Gulf Seneca 68 .... H2
Gulf Seneca 77 .... H2
Gulf Stoddard Solvent K1
Gulf Super Duty Motor Oil
  10W H2
Gulf Super Duty Motor Oil
  20W/20 .... H2
Gulf Super Duty Motor Oil 30 .. H2
Gulf Super Duty Motor Oil 40 .. H2
Gulf Supreme Grease No. 0 .... H2
Gulf Supreme Grease No. 1 .... H2
Gulf Supreme Grease No. 2 .... H2
Gulf Supreme Grease No. 3 .... H2
Gulf Supreme Grease No. 4 .... H2
Gulf Synfluid White Oil 2cs H1
Gulf Synfluid White Oil 4cs H1
Gulf Synfluid White Oil 6cs H1
Gulf Transmission Oil 140 .... H2
Gulf Transmission Oil 250 .... H2
Gulf Transmission Oil 90 .... H2
Gulfcrest 32 .... H2
Gulfcrown Grease 0 .... H2
Gulfcrown Grease 1 .... H1
Gulfcrown Grease 2 .... H2
Gulfcrown Grease 3 .... H?
Gulflex Moly .... -

129

Gulflex Poly.................................H2
Gulflex-Al H2
Gulflube Motor Oil H.D. 10W ..H2
Gulflube Motor Oil H.D. 20W/
20...H2
Gulflube Motor Oil H.D. 30 .....H2
Gulflube Motor Oil H.D. 40 .....H2
Gulflube Motor Oil H.D. 50 .....H2
Gulflube Motor Oil X.H.D.
10W.......................................H2
Gulflube Motor Oil X.H.D. 30 .. H2
Gulflube Motor Oil X.H.D. 40 .. H2
Gulflube Motor Oil X.H.D. 50 .. H2
Gulflube Motor Oil 10W............H2
Gulflube Motor Oil 20W/20 ......H2
Gulflube Motor Oil 30................H2
Gulflube Motor Oil 40................H2
Gulflube Motor Oil 50................H2
Gulflube MOtor Oil X.H.D.
20W/20..................................H2
Gulfpride Multi-G-Extended
Drain (For I).........................H2
Gulfpride Multi-G-Extended
Drain(For II).........................H2
Gulfpride Single-G 10W............H2
Gulfpride Single-G 20W/20 ......H2
Gulfpride Single-G 30................H2
Gulfpride Single-G 40................H2
Gulfspray Ant Roach Killer
Formula 17............................F2
Harmony 150D H2
High Temperature Grease H2

Gulf State Chemical & Supply Co.
Lemon Disinfectant No. 13 ....C1

Gulf States Paper Corporation
E Z Complete All Purpose
Cleaner..................................A1
E-Z Big Power A1
E-Z Coconut Oil Hand Soap E1
E-Z Insect Killer F1
E-Z Power Clean A1

Gustone Products, Inc.
Latha-Leaf................................E1

Guyard, Inc.
Formula G-510 ....A4

H

H & E Chemicals, Inc.
H & E 143 Power Wash Liquid. A1
H & E 161 Heavy Duty Steam
Cleaner..................................A1
H & E 191 Water Soluble
Degreaser...............................A4
H & E 404N All Purpose
Neutral Conc. Cl. A1
Silicone Release (Bulk)..............H2

H & H Janitorial Supply
Spontaneous Industrial Cleaner..A1

H & W Sales Company
Degrease-All A1

H F M Inc.
No Solvents Degreaser A1

H M K Distributors
HMK Degreaser-Cleaner............A1

H. & W. Chemicals, Incorporated
Lustre General Cleaner MC
No. 1....................................A1
Lustre Granular A1
Lustre Liquid Acid Cleaner A3
Lustre N-550.............................C1
Lustre 900 ....A2
Lustre 990-X A2

H.E.K. Chemical Products
Formula 140..............................A1
Formula 140L............................A1
Formula 360..............................A1
Formula 360L............................A4

H-O-H Chemicals, Inc.
B-615.......................................G6
B-622.......................................G6
B-629.......................................G6
B-665.......................................G6
B-669.......................................G7
B-702.......................................G6
B-703S.....................................G6
B-707.......................................G6
C-311.......................................P1
Pur-Flo 987L L1
Pur-Flo 991L L1
Pur-Flo 993 ....L1
Pur-Flo 995 ....L1
SC-20.......................................G6
SC-32.......................................G6

Ha-Co
D.C.D. Concentrate A1

Haag Laboratories, Incorporated
Chemi-Jel Floor Cleaner C1
Conveyor Lubricant (Greasless) H2
CP 00......................................A1
Floor Syn. Concentrate A4
Foamanol Hand Dish Wash
Liquid A1
GLD A4
Hi-Ho Lotionized Hand Cleaner E1
L C All Purpose Industrial Cln
& Dgrsr A1
Liquid Scrub Soap....................C1
Mapp Concentrate A1
Medi-Soap E1
Pi-No-Germ C1
QAT 1000................................D2
QAT 160..................................D1
QAT 450..................................D2
QAT 900..................................D1
Rapid-Clean..............................C1
Rapid-Clean Plus C1
Remove-It A4
Scrub-It....................................A1
Super-Concentrate A1
Zephyr-Brite Neutral Floor
Clnsr....................................A4
Zephyr-Phene C1
123 Spray Cleaner A4
15% Liquid Hand Soap E1
20% Liquid Hand Soap E1
40% Concentrate Liquid Hand
Soap.....................................E1
45% Vegetable Oil Jelly Soap ... A1,H2
55% Conveyor Lubricant...........H2
65% All Coconut Oil Jelly
Soap.............................A1,H2
65% Conveyor Lubricant...........A1,H2
65% Vegetable Oil Jelly Soap ... A1,H2
80% Vegetable Oil Jelly Soap ... A1,H2

Hach Chemical Company
RoVer Rust Remover & Resin
Bed Cleaner...........................P1

Hadco Corporation
Ab-Solv A1
Aci-Det....................................A3
All-Kleen Heavy Duty A4
All-Kleen Klor..........................D1
Blue Butyl Cleaner A4
Bowl Kleen C2
Bug-Doom................................F1
Butylful Blue............................A4
Chlorinated All Purpose
Cleaner..................................A1
Concrete Floor Cleaner A1
Correct Germicidal Detergent ... D1
Cove-Kleen Soil Penetrant Gel
Odorless................................C1
Drain Pipe Opener A1
Electrolite KLM 0001 PF A1
Extra Hi Concentrate A3
Fog-O-Death F1
Food Processors Cleaner A3
Glitter Liquid A1
HST Kitchen Degreaser A8
Liqui-Cide No. 100 Insecticide .. F1
Liqui-Cide No. 50.....................F1
Liquid Drain Pipe Opener.........L1
Liquid Hand/Face Soap
Concentrate E1
Liquid Hand/Face Soap
Dispenser Ready....................E1
Liquid Hand/Face Soap
Number 20.............................E1
Liquid Hand/Face Soap
Number 30.............................E1
Luxury Bathroom Cleaner.........C2
LST Kitchen Degreaser A1
No. 21 Concentrate A1
Quat-Amo D2
Rac IV A4,A8
Super Ab-Solv A2
Super One A1
Super Steam Concentrate A1
X-Static....................................A4
X-Stream A1
511 Heavy Duty Cleaner-
Degreaser...............................A4,A8
606 General Purpose Detergent. A1
809 General Purpose Detergent. A8

Halco Engineering Company
IMP...A3
Rat Nix Blox F3

Hale Sanitary Supply Co.
H.D.C. A1

Hallbro Chemical Company
Alkali/No. 2..............................A2
Liq. C.I.P. Cleaner A2
No. 1 General Cleaner...............A1
No. 12 H.T.S.T. Cleaner............A2
No. 2 C.I.P. Cleaner A1
No. 4 General Cleaner...............A1
No. 5 Acid Cleaner A3
No. 5 Tank Wash A1
No. 6 Acid Cleaner A3
No. 7 C.I.P. Cleaner A2
3 General Cleaner......................A1

Hallemite/ Lehn & Fink Industrial Products Division
Epoxy Rok/Epoxy Hrdnr 1/
Epoxy Resin 2........................P1
Food Area Degreaser A4
Halco Stripper C3
Hallemite Epxy Flrng/Hrnr 1/
Resin 2..................................P1
Por-rok Cement P1
Por-rok Grout...........................P1

130

Halt Products
HP 10..................................... A1
HP 12..................................... A1
HP-151 Spray Cleaner-
  Degreaser A4
HP-153 Detergent Concentrate.. A4
HP-155 Quaternery Cleaner A4
HP-203-Paint & Varnish
  Stripper C3
HP-402-Freezer & Locker
  Cleaner A5
HP-404 Safety Solvent
  Degreaser K1
HP-405 Water Soluble Dgrsrng.
  Solvent............................... C1
HP-406 Liquid Steam Cleaner.... A1

Hamblet & Hayes Company
Tek Klean GCB......................... A2

Hammond Laboratories, Inc.
Aqua-Solv Concentrate.............. A1
Fomacid.................................. C2
HL Concentrate........................ A4
Stat-256.................................. D1

Hammons Products, Incorporated
Hammons Waterless Skin
  Cleanser E4
Pink Lotion Skin Cleanser.......... E1
Shimmering Lotion Skin
  Cleanser E1

Hampshire Chemical Division
Hamp-Ene 100 .... G6

Hampton Janitorial & Paper Co.
Coliseum Brand Blue Cleaner A4

Hamway, D.S. Company
27-B...................................... B1

Hanco Manufacturing Co., Inc.
#575 Insect Spray F1
"200".................................... A4
Bowl Sanitizer......................... C2
Chain Lube 101........................ H2
Chain Lube 202........................ H2
Electro................................... K1
Grout Cleaner.......................... C2
Gum Remover........................... C1
Han-O-Lan "A" E4
Han-O-Lan Cleansing................ E4
Hano Mint............................... C2
Hanozone C1
Hi Purity Contact Cleaner.......... K2
Hospital Disinfectant
  Deodorant C2
HC Concentrate........................ A4
Lathe..................................... E1
New Power Battery Terminal
  Clnr & Protec P1
Seal-Ease Teflon-Tape H2
Silicone Spray.......................... H2
Skin Guard.............................. E4
Star Heavy Duty Cleaner............ A4
Sur-Grip Belt Dressing P1
SC-245................................... A1
SC-250................................... A1
Towercide................................ G7
270 Paint Remover.................... C3
500 Drain Treatment.................. L1

**Hanco Products Division/ Britt Tech
  Corporation**
Hanco Germicidal Sanitizer D1
Hanco No. 240 Detergent........... A1
Hanco No. 700 Detergent........... A1

Hanco 880 Germicidal
  Detergent............................. D1
No. 735 Detergent..................... A1
750 Special Purpose Detergent .. A1

Hancock Industries Inc.
Hancock 75T............................ H1

Handi-Clean Products, Inc.
All Purpose Cleaner
  Concentrated A4
Almost Heaven Space Spray
  and Deodorant C1
Bacteriasol Germicidal Cleaner.. A4
Big Dog Concrete Cleaner.......... A4
Clean-It................................. A1
Concentrated Bowl Sanitizer...... C2
Double-D Disinfectant-
  Deodorant C1
Drain Pipe Opener &
  Maintainer L1
Extra Heavy Duty Concrete
  Cleaner................................ A4
Formula H4C Industrial Liquid
  Deodorant C1
Formula 955-T.......................... C1
Fresh Cherry Blossom Space
  Spray Deo. C1
General Grant Cleaner............... C1
General Lee Multi-Purpose
  Cleaner & Dgsr...................... A4
General Purpose Low Sudsing
  Cleaner................................ A4
H-100 Mystery Cleaner.............. A4
Handi-"G"........................... A4,A8
Handi-Cide Hospital
  Disinfectant Deo.................... C2
Handi-Way Instant Spray
  Cleaner................................ C1
Insecticide for Crawling Insects F2
Liquid Creme Cleanser
  Concentrate C2
Liquid Hand Soap 20%.............. E1
Lotionized Hand Soap E1
NuAir All Purp. Liq Deo. and
  Air Frshnr C1
Pine Odor Disinfectant C1
Pine Oil Disinfectant Coefi #5... C1
Premium Fly & Roach Spray...... F1
Prometheus Alkaline Cleaner..... A2
Sewer Solvent.......................... L1
Steam Cleaning Compound A2
Stripper Degreaser.................... A4
SST Stainless Steel And Metal
  Polish A7
X-150 Chlorinated Machine
  Dishwash A1

Handicap Agency
Tom Savage Perfumed
  Deodorant Blox C2

Handyman
Concentrated All Purpose
  Cleaner-Dgrs......................... A1

Hanson Maintenance Systems
Eliminate A1

Hanson, R G Company, Inc.
AM 16 Multi-Purpose Cleaner
  Concentrate A1
FPC...................................... A1
GC64 Germicidal Cleaner D1
HD 36.................................... A4
HD 690................................... A4
Super 14................................. A1
VHD 400................................. A2
X-OX A3

Hantover Incorporated
Non-Slip-Absorbent 50000......... J1

Harbor Chemical & Engineering Corp.
#125 Boiler Water Treatment.... G6

Hardi Chemicals
Deep Six................................. L1
Hardi-Power A4
Lemon-Aide E1
Solvital.................................. A4

Hardy Salt Company
K.D. Granulated Salt................. P1
Lake Maid Ex Crse Salt Crystl
  Kiln Dried P1
Lake Maid Kiln Dried Coarse
  Salt Crystl P1
Lake Maid Kiln Dried Feed
  Mix Salt P1
Lake Maid Kiln Dried Hay &
  Stock Salt C1
Lake Maid Salt Crystals Kiln
  Dried.................................. P1
Rock Salt Crystals.................... P1
Snow White Crystals P1
Zeo-Gram............................... P1
Zeo-Tabs P1

**Harley Chemicals/ Div. of Concord
  Chemical Co., Inc.**
Activated Pine Type
  Disinfectant C1
Banish It Q A4
Bar Q Disinfectant Sanitizer
  Deodorant D2
Con-Sen A1
Creamedic E1
Dishtergent............................. A1
Foam Up A1
Green Satin Glove C2
Harco Tincture of Green Soap .. E1
Harco 162B Conctd Liq Stm
  Clng Cmpd A1
Harley F.F. Anti-Bacterial
  Hand Soap............................ E2
Industrial Creamedic E1
Lemonee 8 Disinfectant C2
Ov-N-Kleen............................. A8
Pine Jelly Soap C1
Pine Oil Disinfectant Coefi 5 .... C1
Pine Scrub Soap C1
Pink Lotion Hand Soap E1
Rinz-Free................................ A4
Scourge.................................. C1
Winta-Dis Disinfectant.............. C1
15% Liquid Hand Soap E1
3-D All Purpose, Hvy Dty
  Industrial Clnr A1
40% Liquid Hand Soap E1

Harlou Products Corporation
Formula BH-5 for Boilers.......... G6

Harper Brush Works
Formula #730 .... A4
Formula #731 .... A4
Formula #732 .... C1
Formula #770 .... A1

**Harrill Chemical and Paper Supply,
  Inc.**
Big Spike Cleaner Stripper &
  Degreaser A4
Heavy Duty Steam & Pressure
  Clng. Cmpd........................... A1

Harris Corporation
Guard.................................... C2

X-33 .... K1

Harris Janitor Supply Co.
Conquest Plus A4

Harris Janitor Supply Co., Inc.
Clear-Pane C1
Code 1000 Phosp. Free Clnr H-
13 .... A4
D/Zolv A1
H-X-24 Bleach Cleanser A6

Harrison Bros. Janitorial Supply Co.
Emerald Liquid Soap E1
Harco Jolt 45 .... A4
Liquid Hand Soap E1
Pinol No. 5 .... C1
Super Cleaner #3 .... C1

Harry Alter Co., Inc., The
Allclean A4
Coil Clean C1
Metalclean A4
Scale-Clean A3
Spray N'Wipe C1

Harry Miller Corporation
Reversol 492 .... L1

Hart Chemical Co.
Hart Formula 715 .... A1
HS-15 Residual Spray
Containing Bagon F2
HS-27 Disinfectant Cleaner D1
Pine Odor Disinfectant Coef. 5.. C1

Hart Laboratories
Chain Belt Lubricant Type 1 ..... H2

Hartford Chemical Corporation
#227 Super Neutral Cleaner A1
Acid Free Ice Melter C1
Enzymes L2
Heavy Duty Cleaner A4
Safety Solvent Cleaner &
Degreaser K2
Super Neutral Cleaner A1
Tempered Drain Opener L1

Hartford Chemical Corporation
Foam-Kleen 22 .... A1

Harvard Supply
Pearl Liquid Toilet Soap E1

Havatampa Corporation
Blast Off A4
Coco-Kleen Hand Soap A1
Emerald Pine C1
En-Gard Waterless Hand
Cleaner E4
Enviro Kleen C2
Hava-Crete C1
Hava-Dine D1
Knock Out F1
Steel Glo A7

Haviland Products Company
Acid Cleaner #138 .... A3
Acid Cleaner #139 .... A3
Aluminum Cleaner FG A1
B-5-S Egg Cleaner Q1,Q3
Chlorisol Q4
Cleaner #124 BFG A2
Concrete Cleaner C1
Havasan CS D1
Havasan LB-12 .... D2
Havasol LS K1
Milkstone & Lime Remover A3

Packing House Cleaner A1
Pintex C1
Powdered Delimer A3
Purgit L1
Purple Sparkle A1
Rust And Scale Remover #132. A3
S.S.S. J1
Swish A4
Ultra Clean D A1

Hawk Industries, Inc.
Du-All A4
HD-77 .... A1
Lift C1
Pine-Air C1
Poly Scour Power A1
Power Pac Concentrate A1
Scour Powder A4

Haymar Sanitary Supply
Greasaway A1
Super Greasaway A1

Haynes Manufacturing Co., The
Haynes USP Liquid Petroleum
Spray H1
Lubri-Film H1
Lubri-Film Spray H1

Health-Aids, Inc.
Chem-Rel MD A1

Heat-Power Engineering Co., Inc.
#120 Boiler Feedwater
Treatment G6
#122 Boiler Feedwater
Treatment G6
#124 Boiler Feedwater
Treatment G6
#190 Boiler Feedwater
Treatment G6
#220 Boiler Feedwater
Treatment G6
#410 Boiler Feedwater
Treatment G6
#930 Resin Cleaner P1
Potable Water Treatment #730. G1

Heather Products International, Inc.
MacDuff Heavy Duty Cleaner .. A4

Hedgetree, Inc.
LOCC Liquid Organic Cleaning
Compound A1
Tub And Tile Cleaner A3

Heinrich Fischer Company
490 Immaculene A4

Heisler-Green Chemical Co.
Boiler Water Softener G6
Boiler Water Treatment G6
HG 454 .... G6
HG 455 .... G6
HG 6 .... G6
HG 7 .... G6
HG-C270 Water Treatment G7
Scale Solve #9 .... G7

Heljo Inc.
Heljo 700 .... A1

Heller B. & Company
No-0130 Grip J1
Ozo Washing Powder A1
Tru-White Washing Powder A1

Hemisphere Chemical Corp.
Mitey-Kleen A4

Henry Chemical Industries, Inc.
No-Scent Deodorant Nuggets.... C1

Herbert Chemical Company, The
C.K. Chlorinated Cleaner A1
C.K. General Purpose Cleaner... A1
C.K. 62 Cleaner A1
Premier Boiler Water Treat.
151 .... G6
Premier Boiler Water Treat.
199 .... G6
Premier Boiler Water Treat.
236 .... G6
Premier Boiler Water Treat.
236V G6
Premier Boiler Water Treat.
237VS G6
Premier Boiler Water Treat.
238 .... G6
Premier Boiler Water Treat.
250 .... G6
Premier Boiler Water Treat.
403 .... G6
Premier Boiler Water Treat.
403S G6
Premier Boiler Water Treat.
403V G6
Premier Boiler Water Treat.
409 .... G6
Premier Boiler Water Treat.
414 .... G6
Premier Boiler Water Treat.
414H G6
Premier Boiler Water Treat.
414V G6
Premier Boiler Water Treat.
415 .... G6
Premier Boiler Water Treat.
555 .... G6
Premier Boiler Water
Treatment No. 460 .... G6
Premier Boiler Water
Treatment No. 606 .... G6
Premier BH-40 Detergent Clnr.
& Degrs. A4
Premier Cooling Tower Treat.
86 .... G7
Premier Egg Cleaner 2 .... Q1
Premier Inhibitor Solution 518.. G6,G1
Premier Liquid Cleaner
Concentrate A1,E1
Premier P-200 Inhib.
Propln.Glycol Conc. P1
Premier Steam & Condensate
Treat. No. 1 .... G6
Premier Steam & Condensate
Treat. No. 2 .... G6
Premier Steam & Condensate
Treat. No. 3 .... G6
Premier Sure Fire Boiler Wtr
Treat. 155 .... G6
Premier Tank Cleaner Acid A3
Premier Water Treatment 380.... G1
Premier 425 Dry Inhibitor G6,G1

Herbert Webb Co., Inc.
Silicone Spray (Aerosol) H1

Hercules Chemical Co., Inc.
Dark Cutting Oil P1

Hercules Packing Corporation
Food Machine Grease Style
No. 97 .... H1

Hercules, Incorporated
Antifoulant Compound AF 501. G7
Antifoulant Compound AF 502. G7
Antifoulant Compound AF 504. G7

132

Antifoulant Compound AF 506 . G7
Antifoulant Compound AF 510 . G7
Antifoulant Compound AF 511 . G7
Antifoulant Compound AF 512 . G7
Antifoulant Compound AF 513 . G7
Antifoulant Compound AF 545 . G7
Boiler Treat. Chemical BL 205.. G6
Boiler Treat. Chemical BL 217.. G6
Boiler Treat. Chemical BL 227.. G6
Boiler Treat. Chemical BL 234.. G6
Boiler Treat. Chemical BL 236.. G6
Boiler Treat. Chemical BL 241.. G6
Boiler Treat. Chemical BL 249.. G6
Boiler Treat. Chemical BL 250.. G6
Boiler Treat. Chemical BL 255.. G6
Boiler Treat. Chemical BL 261.. G6
Boiler Treat. Chemical BL 265.. G6
Boiler Treat. Chemical BL 273.. G6
Boiler Treat. Chemical BL 279.. G6
Boiler Treat. Chemical BL 284.. G6
Boiler Treat. Chemical BL 290.. G6
Boiler Treat. Chemical BL 291.. G6
Boiler Treat. Chemical BL 294.. G6
Boiler Treat. Chemical BL 295.. G6
Boiler Treat. Chemical BL 296.. G6
Brisgo II P1
BL 351 .... G6
BL 352 .... G6
BL 354 .... G6
BL 357 .... G6
Corrosion Inhibitor CR 404 .... G7
Corrosion Inhibitor CR 408 .... G7
Corrosion Inhibitor CR 430 .... G2
Hercules SP 944 .... G6
Microbiocidal Comp MB 102..... G7
Microbiocidal Comp MB 120..... G7
Microbiocidal Comp MB 121..... G7
Microbiological Comp MB 100.. G7
Microbiological Comp MB 103.. G7
Microbiological Comp MB 108.. G7
Microbiological Comp MB 111.. G7
Microbiological Comp MB 118.. G7
Specialty Product SP 975 .... A3
Specialty Product SP 976 .... A3

Herculite Products, Inc.
Hercon Insectape With
  Chlorpyrifos F2

Herculite Protective Fabrics CorporationA Subsidiary of Health-Chem Corporation
Insectape.. F2

Hereford Janitor Supply Inc.
Mack's Foambrite........................... A3

Heritage House Supplies LTD.
Heritage Foam A1

Heritage Manufacturing Company
Betsy Ross Pink Lotion Hand
  Soap E1
Punch Oven & Grill Cleaner...... A8
Sparkle... C1
Steam-Away.................................... A1
Terrific All Purpose Cleaner....... A1
Tri-Hydrol...................................... A1

Hermetic Chemical Laboratories, Inc.
Activated Alkaline Powdered
  Drain Opener L1
All Purpose Non Flammable
  Safety Solvent K1
Apollo.. D1
Belt Dressing................................. P1
Centurion....................................... E4
D-A2.. C1
Digestant 250 .... L2

Euripedes...................................... D2
Graphite Fast Dry Lube............ H2
Harvest Moon A7
Hermes' Spray C1
Lemon Sewer Sweetener........... L1
Mirv 2....................................... A4,A8
Multi-Purpose Penetrant............ H2
Odin A1
Paint and Varnish Remover C3
Precision Instrument & Elet.
  Pts. Cl................................... K2
Slick Silicone Lube H1
Slipper H1
Spray Fog F1
Zeus D2

Herco International Industries, Inc.
Big Mac Concentrate A1
BT 100 .... G6
BT 300 .... G7
Clean Up A1
De Ox A3
Grimex....................................... A4
GC Digester............................... L1
Hizyme L2
HC 5000 .... L1
Insta Ox A3
Kaboom L1
Klenze.. C1
Koil Brite A4,P1
New Hands E1
Odorout...................................... L1
Red Streak.................................. L1
Rout A1
Steamline A2
TC-NC Concentrate................... G7

Hertron, Inc.
Economite 40 .... A4

Hess & Clark/Division of Rhodia Inc.
Carnebon "500" D1,G4

Hewitt Soap Company, Inc., The
#6200 Synthetic Detergent A1
#6688 Green Bar Soap H2

Hexcel/specialty chemicals
F.O. 253 Rust & Stain Remover A3
F.O. 328-C Heavy Duty
  Concrete Degreaser.................. A1
F.O. 385-A Detergent-Sanitizer-
  Cntrodrnt............................... A4
F.O. 389 Phos-Free A4
F.O. 497 Heavy Duty Steam
  Cleaner.................................. A1
F.O. 500 Special-Tee Solvent.... K1
F.O. 506 E-Mulse K1
F.O. 537 Heavy Grease And
  Oxide Remover........................ A4
F.O. 545 Freezer-Cleaner A5
F.O.-328-B Heavy Duty
  Concrete Degreaser.................. A4
FO 446 Heavy Duty Alkaline
  Degreaser A4
FO 464:...................................... A1
FO 534.. A1
FO 552.. A1
FO 554.. A4

Hi-Brett Chemical Co., Inc.
Boom C1
Drain Solve A L1
Drain-Solv.................................. L1
Formula No. 184 .... C2
Formula No. 8712...................... D1
Formula X-115A......................... A4
Formula 0-22............................. C1
Formula 0-33............................. A1

Formula 132 .... K1
Formula 185 .... C2
Formula 1881 .... D2
Formula 220.............................. A3
Formula 220X A3
Formula 89................................ A1
Hook Cleaner............................ A2
Jamboree A1
Liquid Hand Soap E1
Liquid Steam Cleaner Formula
  #88...................................... A1
Marathon 790............................ A1
Odor Ban.................................. C1
Presto Parts Cleaner................. K1
Safety Solvent B........................ K1
Safety Solvent C........................ K2
Sealer and Hardener for
  Concrete Floors...................... P1
Spartan All Purpose Cleaner...... A1
Sweet-Air C1
Whiz A1
Whiz Special A1

Hi-Lite Chemicals Co. Inc.
Apex Germicidal Detergent A4
Belle-Air Germicidal Detergent. D1
Formula BC-15 .... G6
Formula CCS-60......................... G7
Formula H-43............................. G6
Formula OL-45........................... G6
Formula OS-45........................... G6
Formula P-80.............................. G6
Formula RT-60 .... G6
Formula 35.................................. G6
So Soft Coconut Oil Liquid
  Hand Soap............................. E1

Hi-Valley Chemicals, Inc.
Acid Pipeline Cleaner A3
Boiler Treatment G6

Hiawatha Chef Supply, Inc.
Automatic Glass Wash............... D2,E3
Chlorinated Mech Dshwsh
  Cmpd Pressurized A1
Chlorinated Mechanical
  Dishwash Compound A1
Concentrated Glass Cleaner A7
Grease Remover......................... A8
Heavy Duty Chlorinated Mech.
  Dshwsh Cmpd A1
Heavy Duty Concrete Cleaner .. A2
Heavy Duty Floor Cleaner A4
Laundry Liquid Builder.............. B1
Liquid All Purpose Cleaner A1
Liquid Laundry Detergent B1
Liquid Laundry Soft B2
Low Foam Lime Remover.......... A3
Multi-Purpose Cleaner A1
Sanitizer.................................... D2

Hicks-Denver Co.
Safety Cleaner........................... A4

High Valley Products, Inc.
Boiler Treatment G7
Extra Heavy Duty Meat
  Packers Cleaner...................... A1
Meat Packers General Cleaner... A2

Highland Chemical Corp.
Hi Degrease #1004..................... A1
Hi-Clean #103............................ A4

Highpoint Marketing Corp., Inc.
H.P. 201 Floor Finish
  Maintainer Neu. Cl................. A1
H.P. 234 Wax & Floor Finish
  Stripper.................................. A4

H.P. 239 Ammoniated Wax Stripper A4
H.P. 502 All Purpose Concentrated Clnr. A4
H.P.512 NeutCoco.OilSoapFort.w/ Triclosan B2
HP-101 Degreaser Cleaner A4,A8
HP-112 Degreaser Cleaner Wax Stripper A1
HP-145 General Type Detergent Clnr Disnf D1
HP-156 Hospital Detergent Cleaner-Disnf D2
HP-167 Conc Det San Fngcd Disnf Deodz D1
HP-178 Butyl Cleaner Degreaser A4
HP-189 Concentrate A1
HP-212 Scum Remover and Renovator A1
HP-223 Foam Degreaser and Cleaner A1
HP-245 Heavy Duty Concrete Cleaner A4
HP-267 Hand Cleaner Lotion w/Septi-Chlor E1
HP-323 Heavy Duty Solvent Cleaner C1
HP-423 Steam & Pressure Cleaner A1

Higley Chemical Co.
Santroll A4

Hild-Chem Division
A-1 Hi-Lite A1
Foam-Rite A1
Hild All Purpose Cleaner A1
Hild Chlorinated Cleaner A1
Hild Control A4
Hild Control Neutral D1
Hild H. D. Caustic Cleaner A2
Hild H. D. Cleaner C1
Hild H. D. Liquid Cleaner A1
Hild Lubricant H2
Hild No. 64 Germicide-C A4
Hild No. 64 Germicide-N D1
Hild P. E. Cleaner A1
Hild Steam Cleaner A1
Hild Super Chaser A4
88 Neutral Cleaner A1

Hilex Division, Hunt Chemicals, Inc.
Hi-Lex Bleach D2,Q4
Hi-Lex 6-40 .... D2,Q4

Hill Manufacturing Company, Inc.
#1147 Master Foam C1
Alpha-Dyne D1
Anti-Foam H-1739 .... A1
Bee Wasp And Hornet Killer F2
Beta-Phene A4
Black Disinfectant(Phenol Coefficient 5) C1
Clean-All A4
Conveyor Lubricant H2
CK-341 .... A2
Du-More A4
End-Sweat P1
Foamy Bowl Cleaner C2
Food Grade Phosphoric Acid A3
Formula -381 .... A1
Formula 707 Instant Spray Cleaner C1
H.D. 23 Special A1
H-D-45 .... A1
Hi-Bowl C2
Hi-Kleen A1
Hilco #88 .... A1

Hilco Bane F2
Hilco No. 74 Roach And Insect Spray F2
Hilco San D1,Q3
Hilco-Lube H2
Hilco-Pride C1
Hilco-Rid F2
Hilco-Sec F1
Hilco-Suds A1
Hilco-Tox F1
Hillco No. 119 .... A1
Hillco No. 120 .... A1
Hillco No. 20 .... D1
Hillco No. 224 .... A1
Hillco No. 30 .... A2
Hy-Pine C1
Kreme-Kleen A6
Launder-Rite B1
Liquid Hand Soap #200 .... E1
LC-65 Liquid Steam Cleaner A2
Mint Disinfectant C1
Mr. Foam A1
No. 1033 Equipment Cleaner A1
No. 1208 Steam Cleaning Compound A2
No. 121 Keep-Kleen A1
No. 134 Ilas A4,A8
No. 1366 Cherry Masking Compound C1
No. 1372 Hi-Lustre A3
No. 1397 Hilcotrol C1
No. 1402 Foaming Cleaner A4
No. 142 Scale-Solvent C2
No. 1430 Hill's Dry Bleach B1
No. 147 Shane C2
No. 1477 Mighty Red A1
No. 1480 Tuffy E1
No. 1481 Foamer A1
No. 1482 Additive A1
No. 1568 Foaming Acid Cleaner A3
No. 157 Pyneco Disinfectant C1
No. 159 Lemon-Cide C1
No. 1678 Mad Dog C1
No. 168 Shackle Cleaner A3
No. 197 Surgisep E1
No. 200 Liquid Hand Soap F2
No. 212 Fly Bait F2
No. 26 Sewer Solvent L1
No. 304 Rust Stripper A2
No. 306 Aluminum Chiller Cleaner A1
No. 31 Flo-Thru Sewer Solvent L1
No. 574 Feather Penetrant N1
No. 766 Compound A1
No. 85 Hilcotrol N3
No. 909 Blue Flash A1
No. 959 Hilco-Solv C2
O.T. 155 .... C1
Odorless Lotion-Kleen E1
Odorless Pink Lotion Kleen-Up . E1
OT-155 Deodorant C1
Power L1
R.P.D. No. 255 .... C1
Sanitary Spray Lubricant, U.S.P. H1
Silicone Spray H1
Super Con "96" C1
Super-Chlor 84 .... D2
SC-1 Steam Cleaning Compound A2
SC-2 Steam Cleaning Compound A2
SC-3 Steam Stripper Compound A2
Vapo-Mist "35" F1
Vapo-Mist 500-D F2
WS-1291 Water Soluble Safety Solvent A4,K1
WS-309 Water Soluble Safety Solvent A4

101 Drain Lax L1
1518 Sep-Guard E1
1733 Track & Bearing Lubricant H2
85 Hilcotrol C1

Hillcrest Products Inc.
HC-232 Heavy Duty Cleaner A2
HC-235 S Heavy Duty Cleaner . A2
HC-43-CS Chlorinated Heavy Duty Cleaner A2
HP-10 Liquid Acid Cleaner A3
HP-20 Heavy Duty Cleaner A2
HP-21 Bottle Wash Alkali A2
HP-22 Trolley Cleaner A2
HP-29 Heavy Duty Cleaner A2
HP-30 General Cleaner A1
HP-31 General Cleaner A1
HP-40 Chlorinated CIP Cleaner A2
HP-90 Foam Booster A1
Smokehouse Cleaner #1 .... A1

Hillside Brush & Chemical Co.
Hillco Creme Cleanser A6
Hillco H-555 .... A4
Hillco H-717 .... A4
Hillco Jet Degreaser A4
Hillco Power Solvent A1
Hillco Special Mill Spray F1

Hillyard Chemical Company
Best-All A4
Briten-Zit A4
Clean-O-Lite D1
Clean-O-Lite II D2
CSP Cleaner A3
Dishware And Utensil Cleaner .. A1
Foam Cleaner #22 .... A1
Foam Cleaner #9 .... A1
H-101 .... D2
Hil-Phene A4
Industrial Degreaser/Cleaner A4
No. 115 Drain-Lax C1
Pro-Line Disinfectant Cleaner .. A4
Pro-Line Double Quat Disinfectant Clnr. D2
Pro-Line General Purpose Detergent A4
Protective Hand Cleaner E4
Q.T. D1
Re-Juv-Nal A4
Regular Hilco Cleaner A4
Super Grease Buster A1
Synthetic Cleaner #153 .... A4
Top Clean A4
Triple Strength Hilco Cleaner ... A4

Hilson Manufacturing Co.
HPC 1001 .... A1

Hilton-Davis Chemical Co., The
Cyncal 80% D2
Roccal 11 50% D2
Roccal 50% Technical D2

Hoch Company
Anti-Foam N2
AV-240 Tripe Cleaner N3
Calcan Wash A1
Chloro-San D1
Concentrated Liquid Rust Stripper A3
Cooler Cleaner A4
Diphacinone Rodenticide Concentrate F3
Duke Super Floor Cleaner A2
DBT Cleaner A3
F-C No. 100 .... A4
First Wash Tripe Cleaner N3

134

Hotsy Equipment Company

Super X Compound A1

Hotsy Pacific

Hotsy Super X Two A1
Hotsy 4461 .... A4
Hotsy 7270 .... A4
Super X A1
The Gem A4

Hou-Tex Specialty Company

Big Boy All Purpose Cleaner A1
Clear and Bright C1

Houghton, E.F. & Company

Rust Veto A-2 .... P1
Rust Veto A-2 .... P1
Rust Veto 344 .... P1
Rust Veto-110-D P1
Sta-Put 204 .... H2
Sta-Put 350 .... H2

House of Automotive Equipment

All Purpose Concentrate A1

Houston Paper Company, Inc.

Houston Duo A4
Houston 200 H.D. Alkaline
  Clnr. A2
Houston 300 Acid Cleaner A3

Howard Johnson Company

Acid Cleaner 500 .... A3
Aluminum Cleaner A1
Caustic Cleaner 100 .... A2
Caustic Cleaner 200 .... A2
Caustic Cleaner 300 .... A2
Chlorinated Circulation Cleaner
  500 .... A1
Iodine Sanitizer D2
Laundry Bleach B1
Laundry Detergent B1
Liquid Chlorinating Product D2,G1,E3
Liquid Water Softener B2
Manuel Cleaner 800 .... A1
Quaternary Sanitizer D2
Smoke House Cleaner A2
Terrifik C1

Howco Supply Inc.

Formula 77 Heavy Duty
  Cleaner Conc. A1

Howe Chemical Company

Cleaner L-130 .... A1
Cleaner P-101 .... A1
Cleaner P-103 .... A1
Cleaner P-112 .... A2
Cleaner P-114A A1
Cleaner P-117 .... A1
Cleaner P-120 .... A2
M.S.R. Acid Cleaner A3
M.S.R. Acid Concentrate A3

Hub Chemical Company

Acid Quat Dairy And Food
  Industry Clnr D2
Advance 100 .... A4
Advance 128 .... A4
Advance 256 .... A4
All-Purpose Non Flammable
  Safety Solvent K1
Belt Dressing P1
Bio-Kleen A1
Blazes A4
Charge A4
Concentrated Inhibited Muriatic
  Acid Cl A3

Conquer II A4
Conquer N D1
Conquer 256 L A4
Conquer 256 N D1
De-Grease K1
Digestant 250 .... L2
Disinfectant Sanitizer
  Deodorizer D2
Fluoro Carbon H2
Foams Off A1
Frigid A5
Frigid Cut H2
Graphite Fast Dry Lube H2
Hub Deo-Gran C1
Hub Deodorizer C1
Hub General Purp. Liquid Clnr. A1
Hub Germicidal Cleaner D1
Hub Heavy Duty Floor Cleaner C1
Hub Liq. Solvent Non-Flam.
  Clnr. A4
Hub Liquid Ice Melt C1
Hub Low Pressure Boiler
  Comp. G6
Hub Organic Acid Detergent A3
Hub Powdered Cleaning Comp. A1
Hub Rat Bait F3
Hub Three-Gly C1
Hub Window Glass Cleaner
  Conc. C1
Jell C1
Lemon Sewer Sweetener L1
Liquid Hand Soap E1
Mint Deodorant C1
Multi-Purpose Penetrant H2
Non-Acid Lime and Scale
  Remover A1
Odor-Ban Odor Control C1
Odor-Ban Odor Control with
  Mask C1
Paint and Varnish Remover C3
Penetrating Oil H2
Polysiloxane Release Agent H1
Precision Instrument &
  Electronic Pts Cl K2
Protector A4
Protector N D1
Ready-To-Use Non-Flammable
  Solvent Dgrsr K1
Rust and Corrosion
  Preventative H2
Shower Room Cleaner A3
Silicone Lube H1
Spray-Fog F1
Stainless Steel Cleaner And
  Polish A7
Super Power A4
Super Solv #2 .... A4,A8
Waterless Hand Cleaner E4
Waterless Hand Cleaner
  (Aerosol) E4

Hub States Corporation

Blue Ribbon Insect Spray F2
D.D.V.P. 2#Vapona F2
Di-Tox E P1
Duracide F2
Dursban 4E Emulsifiable
  Insecticide F2
Fog or Spray F1
Fogging Formula II F1
Hub States #147 .... F3
Hub States Rodent Blocks F3
Lindane 20% F2
Pyrethrins Concentrate F1
Residual Spray F2
Tower of Power F1
Trust F1
45% Chlordane-45 .... F2
50% Malathion Spray F2

Hubbard-Hall Chemical Co., The

H-H Lusterbrite L-141 .... A1
H-H Lusterbrite L-142 .... A1
H-H Lusterbrite L-144 .... A1

Huber Janitor Supplies, Inc.

Beaver Beads A1
Beaver Clean C A1
Beaver-Soil Toilet Bowl
  Cleaner C2
Don't Bug Me Insecticide
  Spray-Fog. F1
H-833 Liquid Steam Cleaner A1
H-855 Spray Cleaner A4
Huber Dis-Solves-It A1
Huber-Phene A4
Kill-M F1
Liquid Drain Pipe Opnr &
  Maintainer L1
Mint-O-Green Disinfectant C2
Pine Odor Disinfectant C1
Pow'r Cleaner-Degreaser #75 ... A1
Power-X L1
Pyr-O-Kil Insecticide F1
Sani-Dis Rinse D2
Super Huber Pink Dish-Lustre .. A1
40% Liquid Hand Soap E1

Huge' Company, Inc., The

"Excelcide" Insect. Aer Bomb
  Insect Fog. F1
Excelcide A-OK Residual F2
Excelcide Aerosol Insecticide F1
Excelcide C.M.S. Residual
  Conc. #2. .... F2
Excelcide C.M.S. Spra No. 3 ..... F1
Excelcide C+C Residl. Spra
  Cont. Baygon F2
Excelcide Canary Seed
  Rodenticide F3
Excelcide Cold Fog F1
Excelcide Cunilate Mold
  Inhibitor P1
Excelcide CS Fog No. 40 .... F1
Excelcide Diazinon 4E F2
Excelcide Dursban 2E
  Insecticide F2
Excelcide Dyna-Fog Contact
  Spra F1
Excelcide Dyna-Fog Contact
  Spra Special F1
Excelcide Enzyme Treatment ... L2
Excelcide Excelfume F5
Excelcide Fly Spra F1
Excelcide Fly Spra-Formula G .. F1
Excelcide Fogging Concentrate
  5628 .... F1
Excelcide Fogging Concentrate
  7192 .... F1
Excelcide Fogging Concentrate
  7211 .... F1
Excelcide Fogging Formula
  7207 .... F1
Excelcide General Purpose
  Spra No. 2 .... F1
Excelcide Hydrosol Rodent
  Bait F3
Excelcide Indstrl Aerosol Insect
  Bomb F2
Excelcide Kill-Kote F2
Excelcide Malathion
  Concentrate F2
Excelcide Micro-Encap. Rat &
  Mouse Bait F3
Excelcide Micromist F2
Excelcide Microspra F2
Excelcide Mill Spra F1
Excelcide Mold Control D1,G1
Excelcide Outside Residual F2

Hydro-Pure 600 .... G7

Hydrochem Corporation
All Purpose Cleaner.................. A1
Boiler Water Treatment EOS G6
Hy-Amine................................... G6
Hy-Chelate BWT G6
Hydrotex AC2 .... G6
Hydrotex F8............................... G6
Hydrotex HC3 .... G6
Hydrotex HC7 .... G6
Hydrotex 09C G6
Liquid Drain Cleaner................. L1
Liquid Soap E1
Ox-Scav L G6

Hydrotech Chemical Corporation
BWT 600 .... G6
BWT 610 FG.............................. G6
BWT 620 FG.............................. G6
BWT 623 FG.............................. G6
BWT 630.................................... G6
BWT 655.................................... G6
BWT 685.................................... G6
BWT 690.................................... G6

Hydrotex Industries
Deluxe No. 100 Non-Stain
   Textile Oil H2
Deluxe No. 216 "Improved" H2
Deluxe No. 217 "Improved" H2
Deluxe No. 218 "Improved" H2
Deluxe No. 219 "Improved" H2
Deluxe No. 302-U...................... H1
Deluxe No. 507-R...................... H1
Deluxe No. 528 Sp Hgh Spd
   Chn Lub w/Moly H2
Deluxe No. 529 High Spd
   Chain Lube.......................... H2
Deluxe No. 529Sp. Hgh Spd
   Chain Lube.......................... H2
Deluxe No. 670 Copr Gear
   Cote..................................... H2
Deluxe No. 711.......................... H2
Deluxe No. 712 S.P. H2
Deluxe No. 713.......................... H2
Deluxe No. 800-U...................... H1
Deluxe 1040............................... K1
Golden Kleenz............................ A1
Lubrakleen H2
No. 602 Perma-Temp w/Moly
   Medum H2
Powerkleen H2

Hydrotreat Engineering/ Div. of Hammond Laboratories, Inc.
BT-LS.. G6
RL-F... G6
Steam Trol-HH.......................... G6

Hygienic Sanitation Co., Inc.
Fly & Mosquito Spray w/.30%
   D-Trans F1

Hygin Sanitary Supply Co.
Safe-T-Clean Non Butyl
   Cleaner................................. A1

Hygrade Food Products Corp.
Hygrade Heavy Duty Liquid
   Cleaner................................. A2

Hysan Corporation
A.M. Concentrate...................... A1,E1
A-Plus.. C1
Aero-Bol Bowl Cleaner C2
Aerodet...................................... A4
Ambush Insect Killer F2
Aqua-Sect Insecticide................ F1

B-U Foam Cleaner C1
Bafix Germicidal Spray & Wipe
   Cleaner................................. C1
Bergamot................................... C1
Big Blow F2
Big Fyte..................................... C2
Blue Glass Cleaner C1
Blue Grotto............................... C1
Bowlex....................................... C2
C.I.K. Roach'N Ant Killer......... F2
C-Spra Disinfectant Deodorant.. C1
Can-I-San................................... C2
Cherry Plok! Insect Killer F1
Citrone Disinfectant Deodorant. C1
Clean & Neat E1
Clean Mint Disinfectant
   Deodorant C1
Cleaner for Deep Fry Kettles A2
Clear-Thru................................. C1
Clinch Hand Cleaner.................. E4
Concentrated Insecticide F1
Concentrated Liquid HandSoap. E1
Concept C2
D.T. Heavy Duty Cleaner......... C1
Dek Concrete Cleaner C1
Dirtex.. A1
Disan D1,Q3
Disinfectant #7 Hospital Disinf.
   Deod..................................... C2
Disinfectant #8 .... C1
Do-Rite...................................... L1
Early Mist C1
Emes Open Gear Lubricant
   Aerosol H2
F.S. 200 Sanitary Spray
   Lubricant H1
Foaming Cleanser....................... A6
Futron 25................................... D1,Q3
Fyte Hospital Germ. Detergent. C1
Fyte 13 Hospital Disinfectant..... C1
FIK-20 Insect Killer.................. F2
Gly... C1
H-28 Cleaner A4
Hint Of Mint C1
Hot Rod L1
Hy-Clear..................................... C1
Hy-Fog....................................... F2
Hy-Grip...................................... P1
Hy-Kil.. F1
Hy-O-Dine D1,Q6
Hy-O-Lan Super Duty Soap C2
Hy-Od-Abate C1
Hy-Slurp.................................... J1
Hy-Tef.. H2
Hy-Treat Shower Suds E4
Hyaction Improved Oven
   Cleaner................................. A8
Hycide G7
Hycreme..................................... A4
Hydene A4
Hydrang Prilled Drain Opener .. L1
Hyhex F1
Hypine 7..................................... C1
Hyron Soothing Soap C2
Hys-Solv.................................... K1
Hysect.. F1
Hyshine...................................... A7
Hysil Silicone Spray Aerosol H2
Hysilan Barrier Creme............... E4
Hyso-Base Emul. Degreaser...... K1
Hyspray 77................................. C1
Hyzex... C1
HP-673....................................... G6
HP-88 Sewer Solvent................. L1
HY-T Metered Insecticide
   Aerosol F2
I.S. Sewer Solvent...................... L1
K.O. Dose Insect Killer............. F1
Klop... F1
KO-40 Insect Killer................... F1

Lemon Hyshine A7
Lemon 20 Disinfectant
   Deodorant C2
Liqui-Klenz C2
Liquify C1
Lo-Sope E1
Ludene C2
Master-Kleer L1
Medicide C2
Mintene C2
MIM Silicone Lubricant H1
N.R. Electronic Solvent
   Degreaser K2
New Super Smooth Cleansing
   Creme E4
No Fume L1
Number One Hosp. Disinf
   Deod Aerosol....................... C1
NAB H2
Opalium Cleaner A1
OB-40 Insect Killer F1
Pearl Creme A4
Pine Oil Disinfectant Coef. 3 ..... C1
Pine Oil Disinfectant Coef. 5 ..... C1
Plok! Insect Killer F1
Poly-Zag.................................... C2
Porcena C2
Pow.. L1
PS 75 Insect Killer F1
Q.B. Hand Cleaner E4
Saniwash-FF E2
Shear Gel C1
Slash.. A3
Sleek Germicidal Cleaner &
   Deodorant C1
Smite.. F1
Smite 25..................................... F1
Smite 35..................................... F1
Smite 50..................................... F1
Smooth Waterless Hand
   Cleaner................................. E4
Sno-Cleang C2
Snoap A1
Solvene Pink Powder.................. C1
Splendid..................................... A7
Spoox Residual Roach Liquid.... F2
Spray N' Wipe Multi-Use Clr.
   Degreaser C1
Sta-Brite.................................... A7
Sterizone.................................... C2
Super Hy-Kil Insect Killer F1
Super Hykil Liquid Insecticide .. F1
Super Hysanite A1
Super Syn Suds.......................... A1
Surgent E1
Swinger A4
Teffy H2
Thrice Steam Cleaner A2
Traffic Light C2
Vac.. A4
Vandalex.................................... K1
Victory Disinfectant 5 .... C1
Vigate #2 Air Sanitizer
   Deodorizer C1
Vigate #3 Air Sanitizer
   Deodorizer C1
Vigate #4 Air Sanitizer
   Deodorizer C1
Vigate #5 Air Sanitizer
   Deodorizer C1
Vigate #6 Air Sanitizer
   Deodorizer C1
Vigate #7 Air Sanitizer
   Deodorizer C1
Vigate #8 Air Sanitizer
   Deodorizer C1
Vigate #9 Air Sanitizer
   Deodorizer C1
Vigate One Air Sanitizer
   Deodorizer C1

VIP C1
Wax-Rid C1
Wint Mint C1
Wisp Of Spice C1
X-OX A3
15% Liquid Coco Hand Soap E1
361 Insect Killer F1
40 Plus E1
6209 Heavy Duty Cleaner A1

Hyte Engineering, Inc.
Hyte #214 .... G6
Hyte #550 .... G6
Hyte 611-L5 .... G2
Hyte 705 .... G7

I A C Chemical Company
Floor Cleaner A4
IAC Dish Detergent A1
Scrappy Hand Cleaner E1
Scrappy Industrial Pine Cleaner C1
Scrappy Instant Action Cleaner. A1
Scrappy Multi-Purpose Cleaner. A1
Scrappy Spray Cleaner A1
Scrappy Spry Clnr Steam Clng.
   Comp. A4

I B A, Incorporated
FC-900 Chlorinated Egg Wash .. Q1

I C I Americas Inc.
Talon Rodenticide Pellets F3

I M C Chemical Group, Inc.
Zorb-All J1

I M C Corporation
Big "G" A4
Boilerite-100 .... G6
Citrosene 10 .... C1
Fogging Spray Concentrate F1
Hydrolytic Enzyme Bacteria
   Complex L2
Lemon Tree C1
Power Concentrate C2
Red Hot L1
Whirlpool Drain Opener L1
6/49 Concentrate A4

I Mart
Innkare B-4 Heavy Duty All
   Purpose Clnr. A1
Innkare Emerald 15 .... E1
Innkare Emerald-40 .... E1
Innkare Golden-Glo A1
Innkare Klensade D1
Innkare Microbio-Cide C2
Innkare Odorless Disinfectant ... D2
Innkare Pine Prill C1
Innkare Sprite A8
Innkare Sulfuric Acid Liquid
   Drain Cln. L1
Innkare 2-Way Degreaser A4

I T T Building Services
ITT 700 Foam A1

I T T Gwaltney Inc.
CUC A1
DFC-STR-HTC A2
HA-FC A2
La-CGC, CIP A1

I W M Corporation
B-115 Boiler Water Treatment ... G6
B-116 Boiler Water Treatment .. G6
B-130F Boiler Water Treatment G6
B-131F Boiler Water Treatment G6
B-132F Boiler Water Treatment G6
B-141 Boiler Water Treatment ... G6
B-146 Boiler Water Treatment ... G6
B-150 Boiler Water Treatment ... G6
B-170 Boiler Water Treatment ... G6
B-171 Boiler Water Treatment ... G6
S-30 Condensate Treatment G6
S-36 Condensate Treatment G6
S-40 Condensate Treatment G6
S-46 Condensate Treatment G6

I W T Inc.
Concentrated Odor and Grease
   Control C1

I.V. Company, Inc.
Blue Sparkle A1
Super 771 Concentrate A1

Impac Chemical Products Inc.
Impac Heavy Duty Floor
   Cleaner A4

Impala Chem Labs., Inc.
Activated Alkaline Powdered
   Drain Opener L1
Bio-Kleen A1
Concrete Cleaner A4
CD D2
De-Grease K1
Digestant 250 .... L2
Drain-Away L1
Dual Chlor D1
Flash A4
Heavy Duty Soluble Indust Wtr
   Cl & Dgrs A1
Jell C1
Lemon Sewer Sweetener L1
Liquid Hand Soap E1
Magic A1
Meat Plant Cleaner No. 50 .... A1
Meat Plant Cleaner No. 70 .... A2
Mint Deodorant C1
Odor-Ban Odor Control C1
Odor-Ban Odor Control with
   Mask C1
Peak LF A3
Pentrating Oil H2
Perma-Clean A1
Scale & Film Remover A3
Spray Fog F1
Stainless Steel Cleaner A7
Steam Cleaner AD 1235 .... A2
Super Solv #2 .... A4,A8
Truck Wash Steam Cleaner
   Pressure Wash A1

Imperial Industries, Inc.
Amber-Glo A4
Deluxe A4
Dynasty Blue A4
Glisten C1
Golden Princess E1
J-Jax Concentrate A4
Lemonize C1
Mint-O C1
Pathocide D1
Pine-O C1
Pink Lemon-Aid A4
Pink Princess E1
Sky Power A4
Velvet E1

Imperial Oil & Grease Company
A-890 Heavy Synthetic
   Compressor Lub. H2
A-890 Light Synthetic
   Compressor Lub. H2
A-890 Medium Synthetic
   Compressor Lub. H2
A-930 High Temperature Chain
   Oil H2
A-931 Light Spindle Oil H2
A-940 High Temperature Chain
   Oil H2
A-942 Hydraulic Oil H2
A-943 Hydraulic Oil H2
A-945 Hydraulic Oil H2
Molub Alloy White M1D
   Grease 0 .... H2
Molub Alloy White M1D
   Grease 2 .... H2
Molub-Alloy 823-0 FM Grease .. H1
Molub-Alloy G.O. 690 M1D H2
Molub-Alloy G.P. Oil M1D H2
Molub-Alloy Go-L-M1D H2
Molub-Alloy Spindle Oil M1D .. H2
Molub-Alloy XTO Heavy H2
Molub-Alloy XTO Light H2
Molub-Alloy XTO Medium H2
Molub-Alloy XTO Super
   Heavy H2
Molub-Alloy 785 FM Grease H1
Molub-Alloy 785-0 FM Grease .. H1
Molub-Alloy 815 Grease H2
Molub-Alloy 823 FM Grease H1
Tribol 770 Circulating Oil H2
Tribol 771 Circulating Oil H2
Tribol 772 Circulating Oil H2
Tribol 773 Circulating Oil H2
Tribol 774 Circulating Oil H2
Tribol 775 Circulating Oil H2
Tribol 776 Circulating Oil H2
Tribol 779 Circulating Oil H2
White Mid Grease 1 .... H2

Imperial Paper Company
Elite A1

Imperial, Inc.
Aqueous Food Plnt Pyrne Fog
   Insect. F1
Drione Insecticide F2
Food Plant Fogging Insecticide. F1
Food Plant Spray F1
Pyrenone All-Purpose Spray F1
Pyrenone Super Spray
   Emulsifiable Conc. F1
Pyrenone 1-.2 Food Plant Spray F1
Rat & Mouse Killer F3

Imperial, Inc.
Silicone Lubricant H1
V Belt Conditioner P1

Inco Chemical Supply Co., Inc.
Egg Wash Q1
I.A.C. A3
Inco 360 Dish Brite A1
Inco 375 Blue Concrete Cleaner C1

Incon, Inc.
CC-Acid A3
CD-Acid A1
CH-I Cleaner A1
CH-Cleaner A2
CHCL-I Cleaner A1
CM Cleaner A2
CMCL Cleaner A1
Formula EW Q1
Formula EWCL Q1
GFM Cleaner A1

139

GFM-d Cleaner A1
GFMCL A1
GMH Cleaner.................................. A2
GMM Cleaner.................................. A1
Remove A1
Shurfoam A.................................... A1
Shurfoam E A1

Independence Chemical Co.
All Purpose Detergent................ A1
Bora-Soap...................................... C2
Con-Cleen....................................... C1
Deep Fat-Fryer Cleaner A2
Hi-Suds A1
I-Deen-2 Disinfectant D2
Indco Bleach.................................. B1
Indco Detergent K...................... A1,Q4
Indco LA 17 .... A1
Indco LA 19 .... A1
Indco LD 11 .... A4
Indco LD 12 .... A4
Indco LD-13 .... A1
Indco LF 12 .... D1
Indco LG-11 .... D2
Indco LM-10.................................. G6
Indco LM-11 .... G6
Indco LM-12.................................. G6
Indco LM-13 .... G6
Indco LM-14.................................. G6
Indco LM-15 .... G6
Indco LM-16.................................. G6
Indco LM-17.................................. G6
Indco LP 10 .... C2
Indco LR 10.................................... C1
Indco LR 11.................................... C1
Indco LR 12.................................... C1
Indco PA-31................................... A1
Indco PB 25 .... A2
Indco PB-24 .... A2
Indco PC-10................................... A3
Indco PC-11 .... A3
Indco LA 10 .... A1
Iudeo-LA 11 .... A1
Indco-LA 12 .... A1
Indco-LA 14 .... A1
Indco-LA 15 .... A1
Indco-LA 18 .... A1
Indco-LB 10.................................... A2
Indco-LB 13.................................... E1
Indco-LBB 10 .... H2
Indco-LBB 11 .... H2
Indco-LC 10.................................... A3
Indco-LC 11.................................... A3
Indco-LD 10.................................... A4
Indco-LQ 10................................... E1
Indco-LQ 11................................... E1
Indco-LQ 12................................... E1
Indco-Nancy Brand Laundry
  Detergent................................... B1
Indco-PA 10.................................... A1
Indco-PA 11.................................... A1
Indco-PA 12.................................... A1
Indco-PA 13.................................... A1
Indco-PA 14.................................... A1
Indco-PA 15.................................... A1
Indco-PA 16.................................... A1
Indco-PA 17.................................... A1
Indco-PA 19.................................... A1
Indco-PA 20.................................... A1
Indco-PA 21.................................... A1
Indco-PA 22.................................... A1
Indco-PA 23.................................... A1
Indco-PA 24.................................... A1
Indco-PA 26.................................... A1
Indco-PA 27.................................... A1
Indco-PA 28.................................... A1
Indco-PA 30.................................... A1
Indco-PB 11.................................... A1
Indco-PB 12.................................... A2

Indco-PB 13................................... A2
Indco-PB 14................................... A2
Indco-PB 15................................... A1
Indco-PB 16................................... A2
Indco-PB 17................................... A2
Indco-PB 18................................... A2
Indco-PB 19................................... A2
Indco-PB 20................................... A2
Indco-PB 21................................... A2
Indco-PB 22................................... A2
Indco-PB 23................................... A2
Indco-PO 10................................... L1
Indco-PX 10................................... N2
Independence Bowl Cleaner....... C2
Liberty Bell Beads........................ A1
Scotch Boy Liquid Handsoap E1
Sterine 100..................................... D2

Independent Chemical & Supply Co.
Impact... L1

Independent Chemical Corp.
All Purpose Cleaner A1
Can Wash A1

Independent Chemical Corporation
B.T. 1500 Concentrate G6
B.T. 300.. G7
S.T. 400 .... G7

Independent Dairy Equipment Affiliation
I.D.E.A. Chlorinated CIP
  Cleaner...................................... A1
I.D.E.A. Liquid Chlorinated
  CIP Cleaner A1
I.D.E.A. Liquid Sanitizer D2,B1

Indiana Chemical Corporation
S & W All Purpose Chlorinated
  Clnr... A1
S & W General Cleaner.............. A1
S & W Hook & Trolley Cleaner A2
S & W Laundry Detergent.......... A1
S & W Loaf Mold Cleaner......... A2
S & W Low Foam Detergent A1
S & W New Hook & Trolley
  Cleaner...................................... A2
S & W Powdered Acid............... A3
S & W Smokehouse Cleaner A2
S & W Special Acid Cleaner..... A3
S & W Tree & Hanger Cleaner . A2
S & W Tripe Cleaner N3
S & W Truck Cleaner A1
Tripe Cleaner #60....................... N3

Indianhead Manufacturing Co.
Cal-Chief Bug Killer F1
Fly Killer F1

Indusco Chemical Products
Product No. 580 Tug-N-Barge... A1

Industrial & Institutional Supply Co., Inc.
Foodlube H1
Insta-Cide...................................... F1
OVEC... A8
Sil-Ease.. H1
Skin Shield Cream....................... E4
Spots-Off C1
Stepol... A7

Industrial Bearing and Transmission Co.
Food Machinery Lubricant H1

Industrial Chemical Division of Leo Silfen
Formula 100 Mint Disinfectant.. C2
Formula 100 Mint Hosp. Germ
  Disinf. C2
Formula 101 Neutral Cleaner.... A4
Formula 101-S A4
Formula 101-X Floor
  Condtnr&Clnr.......................... A4
Formula 106 All-Purpose
  Cleaner...................................... A4
Formula 106-X All-Purpose
  Cleaner...................................... A4
Formula 11 Disinfectant-
  Cleaner...................................... A4
Formula 15..................................... D2
Formula 178 Freezer Cleaner.... A5
Formula 179 Freezer Cleaner.... A5
Formula 22 Pine Oil
  Disinfectant.............................. C1
Formula 220 Liquid Hand Soap. E1
Formula 44 N-B Degreaser
  Cleaner...................................... A4
Formula 47 Soap Scum
  Remover A1
Formula 50..................................... D2
Formula 77 Multi-Purpose
  Cleaner-Dgrsr. A4,A8
Formula 77-X Multi-Purp.
  Cl&Degrsr C1
Formula 801................................... L2
Formula 808................................... L2
Formula 81..................................... F2
Formula 91..................................... F1
Formula 91-X................................ F1
Formula 996 Hvy Dty Steam
  Clnr... A1
Formula-171 Foaming Cleaner.. A1
Hard Surface Cleaner.................. A4

Industrial Chemical
Success... A1

Industrial Chemical & Supply
Lift A4,A8
Task Master................................... A1

Industrial Chemical & Supply Co.
M.S.R. Acid Cleaner................... A3
Pelican Brand General Cleaner. A1
Pelican Brand Sanitizer Cleaner. D1
Pelican Brand Smokehouse
  Cleaner...................................... A2

Industrial Chemical Cleaner Company
"Power Plus".................................. C2

Industrial Chemical Co. of S.F. Inc.
Indco Acid Cleaner #103....... A3
Indco Action Alkali No. 54....... A2
Indco Bottle Cleaner D-20......... A2
Indco Cleaning Compound
  #104.. A4
Indco D-10-A................................. A1
Indco F.B. Cleaner...................... A2
Indco F.F.C. No. 40 Cleaner A1
Indco L.C.C. No. 1...................... A4
Indco MP-3 Cleaner.................... A4
Indco No. 46................................. A2
Indco No. 49................................. A2
Indco No. 5.................................... A1
Indco No. 50................................. A2
Indco No. 51................................. A2
Indco 43 Cleaner A1
Indo No.52 Cleaner..................... A2

Industrial Chemical Labs., Inc.
Aqua-Kil... F2
Black Boiler Compound G6

ICC MLC................................A5
ICC 207................................A4

Inter-County Farmers Co-op
Coop Brand Chlorinated
  Detergent........................Q1,Q3
Coop Brand Sanitizer.................Q1

Intercem
Food Plant Boiler Treatment.....G7
Heavy Duty Degreaser And
  Cleaner.............................A4,A8
Kill Sect 25-S...........................F1
Kill Sect-1...............................F1
Kill Sect-3...............................F1
Kill Sect-5 Food Plant Fogging
  Insect................................F1
No. 4 Resido-Sect
  Residl.Roach&AntLiqSpy......F2
Ultra-Suds...............................A1

Interchem
Mint Disinfectant......................C1

Interchem, Inc.
Chase Automatic Washing
  Machine Det. B1
Formula 700............................A4
Freedom.................................E1
Germa-Sep..............................D1
Remove Heavy-Duty Granular
  Floor Cleaner.....................C1

Interco Company
Interco 204 NP.........................A4

Interconti Chemical Corp.
Formula 924 Pressure Washer....A1
MA-11 Mild Steam Cleaner A1
Spin-Off..................................A1
X449 Heavy Duty Alkaline
  Emulsifier...........................A1

Intercontinental Chemical Corp.
Polycon Industrial Oil................H2

Intercontinental Chemical Corp.
Con-O- 1467 Peeleze.................N4
ICC Clean-N-San 234 ....C1
ICC Concentrated Spray
  Emulsifiable Conc................F1
ICC Descaler 166......................A3
ICC Dysodine 236....................D2
ICC Grill Dazzl........................A1
ICC Liquid Hand Soap..............E1
ICC Nu-Grili...........................A8
ICC Poly-Con FG-2 ....H1
ICC San-O-Clear 235................D1
ICC W 03................................G7
ICC W 08................................G6
ICC W 09................................G7
ICC W 12................................G7
ICC W 21................................G6
ICC W 24................................G6
ICC W 27................................G6
ICC W 28................................G6
ICC W 30................................G6
ICC W 32................................G6
ICC W 33................................G6
ICC W 34................................G6
ICC W 35................................G6
ICC W 37................................G6
ICC W 38................................G6
ICC W 40................................G6
ICC W 41................................G6
ICC W 46................................G6
ICC W 71................................A3
ICC W 72................................P1
ICC W 73................................A2

ICC W-60................................G7
ICC W-62................................G7
ICC W-64................................G7
ICC 101.................................A1
ICC 103.................................A1
ICC 105.................................A1
ICC 1061...............................A1
ICC 1070...............................A1
ICC 108.................................A1
ICC 1137...............................A1
ICC 1138...............................A1
ICC 1141...............................A1
ICC 1144...............................A1
ICC 115.................................A1
ICC 1150...............................A1
ICC 116.................................A1
ICC 117.................................A1
ICC 119.................................A2
ICC 120.................................A2
ICC 121.................................A1
ICC 123.................................A1
ICC 124.................................A1
ICC 125.................................A1,B1
ICC 126.................................A1
ICC 127.................................A1
ICC 128.................................A1
ICC 130.................................A1
ICC 131.................................A1
ICC 132.................................A1
ICC 133.................................A1
ICC 140.................................A1
ICC 141.................................A1
ICC 142.................................A1
ICC 1466 Ezzepeel...................A4,N4
ICC 1468 Ezzepeel...................B2
ICC 1470...............................C1,N4,G2
ICC 1471...............................C1,N4,G2
ICC 149.................................A1
ICC 150.................................A1
ICC 160.................................A1,B1
ICC 162.................................A1,B1
ICC 165.................................A3
ICC 169.................................A3
ICC 191.................................A1
ICC 192.................................A4,A8
ICC 194.................................A1
ICC 200.................................A1
ICC 204.................................A1
ICC 205.................................A2
ICC 205D...............................A2
ICC 207.................................A2
ICC 2077...............................A1,B1
ICC 2078...............................A2
ICC 208.................................A2
ICC 2103...............................A3
ICC 211.................................A1
ICC 212.................................N4,G2
ICC 213.................................N4,G2
ICC 217.................................A3
ICC 221.................................A1
ICC 230.................................D1
ICC 237.................................D1
ICC 237-A..............................D2
ICC 238.................................A3
ICC 241.................................G1
ICC 242.................................P1
ICC 243.................................G1
ICC 244.................................A3
ICC 245.................................G4
ICC 258.................................A3
ICC 262.................................H2
ICC 270.................................H2
ICC 291.................................A1
ICC 309.................................A1
ICC 310.................................A1
ICC 3101...............................A1
ICC 3104...............................A3
ICC 3105...............................A1
ICC 3106...............................A3
ICC 3108...............................A1

Isis Foods, Incorporated

Lee All Purpose Cleaner
  Concentrate A1
Lee Blue Glass Cleaner C1
Lee Bowl & Porcelain Cleaner .. C2
Lee Concentrated Hvy Dty
  Skin Cleanser E1
Lee Creme................................... A7
Lee Deep Fat Fry Cleaner A2
Lee Degreaser.............................. A4
Lee Disinfectant Cleaner A4
Lee Heavy Duty Cleaner A4
Lee Lemon Disinfectant
  Deodorant C2
Lee Liquid Coco Hand Soap E1
Lee Liquid Drain Solvent L1
Lee Porcelain, Tile, & Grout
  Cleaner.................................... C2
Lee Prilled Drain Opener.......... L1
Lee Regular Disinfectant
  Deodorant C2
Lee Roach'N Ant Killer............. F2
Lee Sanitizer D1
Lee Sanitizer Iodine Germicide . D1
Lee Sparkle Glass Cleaner C1
Lee Spray N' Wipe Cleaner
  Degreaser C1
Lee Stainless Steel Cleaner &
  Polish A7
Lee Triclosan Liquid Soap......... E1
Super Lee Insect Killer.............. F1

Iso-Chem Division of Sage Systems, Inc.

Lemon Scrub A1

J

J & B Associates/ Division of Lloyd Chemicals, Inc.

F-239....................................... A1

J & B Industries

CDD-10..................................... C1
Descaler #1................................. A3
DCS 50...................................... D1
DOS Synergist............................ F1
Husky.. A4
JB-101 .... G6
LD-2 .... A1
Steam Clean A1
Won-Der C1

J & D Enterprises

Rosey Red................................... A1

J & G Sales Corporation

Aquacide F1
Cocoanut Oil Hand Soap
  (Concentrate) E1
Dual-27 .... D1
Emulsion Bowl Cleaner.............. C2
Fogging Spray Concentrate F1
L-Tox Spray F1
Lectra Solv K2
Odo-Rout.................................... C1
Un-Block L1
Whomp......:................................ L1

J & J Chemical Company

Acid Cleaner and Brightener A3
Anti-Foam One............................ Q5
F-5 Chlorinated Cleaner............. A1
Food Plant Cleaner #1............... A1
Formula 272............................... Q1

Formula 372.............................. A1
J & J Blaze A1
J & J Booster Q1
J & J Egg Click Q1
J & J Foam................................ A1
J & J Formula 173...................... A2
J & J Hand Clean E1
J & J Hand Soap E1
J & J Nil Foam Q1
J & J Strip A1
J & J Suds A1
J & J Suds Plenty A1
Liquid MFP Cleaner................... A1
Lite Egg Wash........................... Q1
Powdered MFP Cleaner.............. A1
Scald Vat Defoamer................... N1

J & J Chemical Company

Chain & Cable Lubricant-
  Protectant H2
Contact Cleaner K2
Easy Task Degreaser A4
Flying Insect Concentrate F1
Residual Insect Spray................. F2
Safety Solvent K1
Super Blue Concentrate A4,A8
Wyte-Ease H2

J & J Chemicals, Incorporated

Blast L1
Chem-Spec General Purpose
  Dtrgnt A1
Heavy Duty Cleaner & Wax
  Stripper A4
Kwik Heavy Duty Detergent A1
Steam Machine Compound A1
White Floor Cleaning Powder... A4
White Powder Hand Soap.......... C2

J & J Rangel Distributor Co.

J-J 99....................................... L2

J & P Chemical Corporation

H.D. Steam Cleaner A4
Liquid Hand Soap E1
Minute Clean 480....................... A4
Neutral Cleaner 3 .... A4
Odor Control Solvent C1

J & S Custom Chemical

AC #18...................................... A3
C.I.P... A2
Chlor-Cleen................................ A1
Formula #42............................... A2
G.P.C.. A1
H.D. 205.................................... A2
Hy-Cleen A1
J & S-207 Quick Concrete
  Cleaner.................................... C1
J & S-749 40% Coconut Oil
  Liq. Hand Soap....................... E1

J & V Sales & Services, Inc.

Contact Cleaner K2
S S C Metal Polish A7
Touch-Up All Purpose Cleaner . C1

J Drewsan Company

Tackler Cleaner-Degreaser......... A1

J W F Inc.

Alive Bacteria/Enzyme L2

J. J. Enterprises

HP-10 Steam And Pressure
  Cleaner.................................... A1

Jack Flint & Son, Inc.

Magi-Clean A1

Magic All Purpose Steam........... A1
Magic Car and Truck Wash
  Powder........................... A1
Magic Coconut Oil Hand Soap.. E1
Magic Misty Alka-Solve
  Degreaser & Clnr. A4
Magic Pine Odor Disinfectant ... C1
Magic Safety Cleaning Solvent.. K1
Magic Suds............................. A4
Magic Super Hi Alkaline
  Cleaner............................. A4
Magic Task Master Surfact
  Strnthnd Conc.................... A4
Magic 614 Brute Steam Clnr
  H.D. Powder....................... A1
Super................................... A1

Jack Frye Sales & Service
Ease..................................... A4
Kickoff-P.............................. A4
Ripoff................................... A8
Sudsy................................... A1

Jack Huppert Company, Inc.
Bacto-Fun-Zyme..................... L2

Jacks Manufacturing Company
Anti-Foam............................. G6
Boiler Compound No. 2.......... G6
Bowlbryte.............................. C2
Cement Floor Cleaner C1
Cooling Tower Treatment........ G7
Drasol.................................. L1
Formula 50............................ G6
Formula 50 Type B................. G6
Grease Solvent....................... C1
Hot Water Tank Treatment..... G3
Jacks Coil Cleaner.................. A3
Non-Fuming Jacks Coil Cleaner A3
Porclean................................ C2
Powdered Boiler Comp. Form.
  100.................................. G6
Powdered Boiler Comp. Form.
  120.................................. G6
Special Acid Descaler.............. A3
Steam Generator Compound.... G6
WS Boiler Compound.............. G6

Jackson Paper Company
Concrete Cleaner,................... A1
Lemon Disinfectant No. 10....... C1
Liquid Hand Soap E1
Neutral Cleaner A4
Pine-Dis-5 Pine Oil Disinfectant C1

Jadeo Chemical, LTD.
"91" Industrial Cleaner A4

Jadco, Inc.
Bio-Clean.............................. L2

Jade Chemical Company
Jade H.D.-25 .... D1

Jade Chemicals, Incorporated
De-Grease A4
Quick Cleaner A1

Jaguar Chemical Corporation
Clean It-All Purpose Cleaner A1
Clean-It................................ A1
Concentrated Inhibited
  Chemical Cleaner A3
Concrete Cleaner.................... A4
Descaler & Delimer................. A3
Drain-Flo............................. L1
Emulsifiable Cleaner K1
Flo-Go No. 308 .... C1
General Purpose Liquid
  Cleaner............................. A1

Germicidal Cleaner C1
Hand Soap............................ E1
Heavy Duty Organic Drain
  Cleaner............................. L1
Ice Melting Compound............ C1
J. A. E. Agricultural Enzyme.... L2
J-T650.................................. A4
Jaguar Controlled Suds........... B1
Jaguar Germicidal Cleaner...... D1
Jaguar Lemon Hand Soap........ E1
Jaguar Odor Control Granules .. C1
Jaguar Sewer Solvent L1
Liqui-Drain........................... L1
Liquid Hand Soap 40%
  Concentrate E1
Liquid Solvent Non-Flammable
  Cleaner............................. A4
Neutral Cleaner A1
Organic Acid Detergent A3
Pine Odor Disinfectant C1
Quat Cleaner D1
Safety Solvent....................... K1
Super-Kleen A4
Three-Gly............................. C1
1001 Butyl Power Cleaner....... A4

Jalcar Industries
Red Dog,............................... C1

Jamard Company, The
Ovec Ammoniated Oven
  Cleaner............................. A8

James Austin Company
Austin's A-1 Bleach................. D1,B1

James Huggins & Son, Inc.
Cresyl Fluid C2
Hug-O-Dine.......................... D1
Kiltarol Disinfectant Coefi 6...... C2
Ortho-Solve........................... C1
Pine Odor Disinfectant Coefi 2.. C1
Pine Oil Disinfectant Coefi 5 .... C1

James Varley & Sons, Inc.
#200 Vaporizing Insecticide F1
#202 Select Hand Soap E1
#40 Liquid Soap..................... E1
#400 Insecticide F1
#404 Select Hand Soap E1
All American 32-Skidoo A4
Anti Staph............................ A4
Aqua Kill Insecticide F2
Aqua Kill Insecticide Fogger..... F2
Beta-Trol.............................. C2
Blue Label Concentrate Iodine.. D2
Bowl Creme C2
Clear Bowl Cleaner................. C2
Clinger................................. C2
Closet Deodorant For Indoor
  Toilets.............................. C2
Coal Tar Disinfectant Coefi 5.... C1
Coal Tar Disinfectant Coefi 6.... C1
Creme Cote A4
Crystal Clear A1
Deo-Sul Liquid Drain Pipe
  Opener............................. L1
Drain Pipe Cleaner................. L1
Dyna Lube H2
Dyna-Sol.............................. A4
Dysul................................... L1
Emulsyde.............................. A7
Emulsyde Ready For Use.......... A7
Gentle-Mild Bowl and
  Porcelain Cleaner C2
Glyco Mist............................ C1
H.W.C. Waterless Hand Clnr
  (Bulk) E4
H.W.C. Wtrlss Hand
  Clnr(Aerosol)..................... E4

Hospital Creyslic Disinfectant.... C1
J.V. Concentrate..................... A4
Lano Var Cream Lotion Hand
  Soap................................ E1
Lemon Scented Disinfectant C1
Med-I-San Anti-Bacterial Liq.
  Soap 1320 .... E1
Med-I-San Anti-Bacterial Liq.
  Soap 1340 .... E1
Mist 'N Wipe C1
New Disinfectant Cleaner A4
No-Ox.................................. C1
Non-Selective Surface
  Disinfectant C2
Pale Pine 8........................... C1
Pine Oil Disinfectant Phenol
  Coefi 5 .... C1
Pink Frost............................. E4
Porcelain Cleaner C2
Power-Pak Insect Killer F1
Pyn-A-Roma Disinfectant C1
Pyrethrum Type Fly Spray........ F1
Q-Tabs Sanitizing Tablets......... D1
Quaternary Ammon. Germ.
  Coefi #5 .... D1
Quaternary Ammon. Germ.
  Coefi #10.......................... D1
Quaternary Ammon. Germ.
  Coefi #20.......................... D1
Quaternary Ammon. Germ.
  Coefi #25.......................... D1
Quiet Please.......................... C2
Red Label Beta Trol C1
Special Synthetic Butyl Cleaner A4,A8
St. Louis Blue Window Cleaner A1
Sunny Suds........................... A1
Super Crystal Clear................. A1
SS-96 Liquid Detergent........... A1
Tincture of Green Soap............ E1
Var-Tar Remover.................... C1
Varco Resistant Roach Spray F2
Varco Supreme Carburetor
  Cleaner............................. K1
Varco Viscous Phosphoric
  Cleaner............................. C2
Varlectric Safety Solvent
  Degreaser K1
Varley Lemon Scented Non-
  Selective Disnf.................... C1
Varox Inhibited Chemical
  Cleaner............................. A3
White Label Concentrated
  Iodine.............................. D2
Winter-Phene........................ C1
20% Pure Coconut Oil Liq.
  Hand Soap......................... E1
40% Pure Cocoanut Oil Liq.
  Hand Soap......................... E1

James White Corporation
Dazie Disk............................ C1

Jan-San Supply Co.
Foam-Now A1
Grease Off............................. A1
Jan-A-Mite A1
Jan-San Spray Silicone Release.. H1
Super 263.............................. A1

Janco Chemical Supply, Inc.
Foam-Klean........................... A1
Jan-Quat.............................. A4
Jan-Quat 64 .... A4
Jan-Quat 64-N D1
Jan-Quat-N D1
Jan-Quat-256 .... A4
Janco Super Clean A4
Janco SJ-212 .... A4
Janco SP-005......................... A4
Janco WSD-66........................ C1

146

147

Jeffcool P-200 .... P1

Jefferson Food Products Corp.
Daisy Bleach C2,B1,P1

Jefferson Products, Inc.
Remuv-OX A3
Zip A4

Jefferson Service Company
J-54 .... F1
Sta-Brite A7

Jefferson Supply Company
Silicone Spray H1

Jen-Mor Products, Inc.
Floz P1
Pro-Teet F1
Rip A3
X-"Em" F1

Jersey Industrial Chemical Products Inc.
Boiler Water Treatment
   Compound G7
Scalite G6
Softite G6

Jersey Janitor Supply
Clover Industrial Cleaner #8 ..... A4
Clover Steam Kleen A1
Hospital D1

Jeryco Chemical & Supply Co., Inc.
#999 Liq. Boiler & Feed Water
   Treatment G6
Bombast A4,A8
Con-Mist A4
Control II D1
Control 111 .... A4
Control 562 .... A4
Edge-652 .... A4
Edge-821 .... A4
Fume-Free #500 Liquid Drain
   Opener L1
H Grab-6 .... D1
Jeryco #211 .... A1,E1
Jeryco #211 Concentrate A1
Jeryco #221 .... A4
Jeryco #245 .... A3
Jeryco #425 Residual Roach
   Liquid F2
Jeryco #430 Fly & Roach
   Spray F1
Jeryco #510 Clean-Sweep L1
Jeryco #511 HA Bowl
   Sanitizer C2
Jeryco #601 .... C1
Jeryco #605 .... F2
Jeryco #609 .... C2
Jeryco #615 .... H2
Jeryco #645 Solvent Degreaser . K1
Jeryco #647 Lemon Scented
   Disnft. Deo. C2
Jeryco #88 Sewer Solvent L1
Jeryco BK-57 Bug Killer F1
Jeryco No. 20 .... A2
Jeryco V1003 .... H1
Jeryco V101 .... L1
Jeryco V103 .... L1
Jeryco V104 .... C2
Jeryco V107 .... L2
Jeryco V111 .... C2
Jeryco V115 .... L1
Jeryco V206 .... A4
Jeryco V207 .... A4
Jeryco V208 .... A1
Jeryco V226 .... A4

Jeryco V227 .... A1
Jeryco V228 .... A5
Jeryco V407 .... E1
Jeryco V408 .... A3
Jeryco V412 .... C1
Jeryco V423 .... A8
Jeryco V440 .... C2
Jeryco V503 .... K1
Jeryco V533 .... A8
Jeryco V601 .... C1
Jeryco V834 .... L1
Jeryco V902 .... C1
Jeryco 641 .... H2
Klean-Quik A1
Nitro Clean 197 The Reserve
   Power Detrgt A4
Sana-Clene Concentrate 1000 ..... C1
Scrub-8 .... A4

Jesco Lubricants Co., Inc.
AC Track Roller H2
Ball & Roller Bearing No. 1 .... H2
Ball & Roller Bearing No. 2 .... H2
Helox A H2
Helox B H2
Lith-O-White No. 2 .... H2
Lithium-AM 1 .... H2
Lithi3m AM 2 .... H2
Lubri-Vis 1 .... H2
Realfilm AW H2
Realfilm Evapo H Gear Shield .. H2
Realfilm 3F H2
Water Pump H2
White Food Machnry Grease
   BB H2
3000 Mill. H2
3001 .... H2

Jet Chemical Inc.
Jet Spray Biodegradable
   Detergent A1

Jet Sales, Incorporated
All Purpose Cleaner A4
Blast Off A4
Clean A4
Jet Liquid Dishwash A1
Jet Sheen Stainless Steel Clrn.
   & Polish A7
Mark Ten D2

Jet Supply Company, Inc.
Aerojet A1
Algaecide C1
Ammoniated Oven Cleaner A8
Biocide C2
Cherry Blossom Insecticide
   Spray F1
Citrosene 5727 .... A1
Cocoanut Oil Hand Soap E1
Concrete Cleaner, Liquid A4
Deluxe Dish Wash A1
Derma-Jet Waterless Hand
   Cleaner E4
Drip-Dri P1
Dual-27 .... D1
Flora-Mint C2
Germisyl Coef. 10 .... A4
Glass Cleaner C1
GP 2000 .... A1
Hydroxade A1
Invisible Skin Shield Cream E4
Jet-ise A1
Jet-O-Dyne 20 .... D2
Jet-O-Dyne 30 .... D2
Jet-O-Lene C2
Jet-O-Pine C1
Jet-O-San D1,Q3
Jet-O-Sol C2
Jet-O-Steam A, Liquid A2

Jet-O-Strip C1
Jetco Jint Foam A1
Jetco-Cide Spray F1
Kleer Glass Cleaner C1
Lemon Disinfectant C2
Lemon Disinfectant Coef. 6 .... C2
Lemon Disinfectant No. 10 .... C1
Liquid Hand Soap E1
Liquid Steam Cleaner A2
Lo-Sudz A4
Mint Disinfectant C1
New Power Battery Terminal
   Cln & Protec. P1
O.L.Concentrate A1
Odor Gone C1
Odorless Concentrate Cleaner .. A1
Odorless Disinfectant C2
P.C.D.-10. .... A4
Para-Phyll. C2
Pine Oil Disinfectant C1
Pine Scrub Soap C1
Pine-Dis-5 .... C1
Poultry-San D1
Quik Clenz C1
R.I.S.-15 Residual Insecticide
   C/Baygon F2
Red Hot L1
Safety Hard Water Scale
   Remover A3
Safety Solvent K1
Sani-Lube Food Equipment
   Lubricant H1
Sewer Solvent L1
Shamrock 880 .... A4
Silicone Spray H1
Spotless Foaming Germicidal
   Surface Clnr C2
Steam Cleaner A4
Steri-Fog D1
Super J C1
Super Rig Wash A4
Sur-Grip Belt Dressing P1
Surgical Soap 40% C2
Task Force A1
Technical A.P.C. A4
Technical D.W. Base A4
U.N.C.L.E. Insecticide F1
Un-Stop Drain Opener L1
Victory Cleaner A4
W.C.-170. .... C1
W.C.-85. .... C1
1421-31 Jet Food Industry
   Spray F1
1421-31 Jet Jetcocide Non-Tox
   AA Fly Sp. F1
1421-69 Jet No. 39 Roach&Ant
   Spray F1
15% Liquid Hand Soap E1
20% Liquid Hand Soap E1
40% Liquid Hand Soap E1
6/49 Concentrate A4

Jet-Lube, Incorporated
Aluminum Complex Food
   Mach. Grease E.P. H1
Calcium No. 2 Food Machinery
   Grease H1
FM-W Special H1

Jett-Co Products, Inc.
Acid Cleaner B A3
Chlorinated Cleaner A1
Concrete Cleaner A4
Dri Acid Cleaner A3
Dry Chlorine Shroud Bleach B1
Granulated Detergent A1
Heavy Duty Cleaner A2
Liquid Acid Cleaner A3
Liquid Detergent A1
Liquid Hand Soap A E1

149

K

K & G Enterprises, Inc.
"Spotless" A1
C-Thru.. A7
Derma-Suds..................................... E1
Ezee Glide....................................... H1
Glisten A4
Grime Stopper A4
Irish Mist C1
Kwick Free H2
Kwick Klean.................................... K1
Liquid Grease H2
Multi-Klean A1
Multiple Bacteria L2
Power-House A4
Reely Klean C2
Rust Off C1
Sil-Lube H1
Sila-Guard G1
Speedy Spray A1
Spray Klean A1
Steam-Klean.................................... A1

K & R Distributing Co., Inc.
De-Gree-Sol Solvent Cleaner
Clnr & Dgrsr......................... A4,A8

K and M Products
Activ VIII A1
K and M Lemon Hand Soap...... E1

K C I Chemical Company
Carmel-Solvent A3
Klene Vat-50.................................... A2
Klener-E.. A2
Klor Klene A1

K D F Enterprises
Enterprise 8000............................... A1
KDF 298 .... D1
KDF 350 .... D1

K E B, Incorporated
K&S Heavy Duty Degreaser C1
K&S Kel Klen A1
K&S Phase One................................ C1

K H Supply Company
Goetz-All.. A1

K O Manufacturing, Inc.
1010 Power Plus............................. A4
211.. A1
217.. A4,A8
226.. A2
777.. A1

K R C Research Corporation
KRC No. 7 Cleaner....................... A4
Minus-Zero...................................... A5
Twink A4

K Yamada Distributors, Ltd.
Alki A1
Non Butyl Degreaser A1
Super D A1

K.E.W. Chemical Company
Q100 General Purpose Cleaner.. A4
Q200 Heavy Duty Cleaner/
Degreaser A4

K.L. Sales
K.L. Powdered Cleaning
Compound.............................. A1

K-Klean Chemical Company
Chelated Alkali #1 .... A2
Chlorinated C.I.P. Cleaner A1,Q1
Chlorinated Heavy Duty
Cleaner..................................... A2
Concentrated Acid A3
Conveyor Lubricant #2............. H2
Dry Bleach.................................. B1
Econo-Klean A2
Foam Booster.............................. A1
Foam Two A1
General Cleaner #1.................. A1,Q1
General Cleaner #2.................... A1
General Cleaner #3.................... A1
General Cleaner #3 SP.............. A1
Heavy Duty Cleaner A2
Heavy Duty Cleaner No. 2....... A2
Hot Tank Cleaner....................... A2
HD Spray Cleaner................... A1,Q1
K-Klean Cleaner-Degreaser A1
K-Klean Coco 40 Hand Soap..... E1
K-Klean Conveyor Lubricant H2
K-Klean Liquid Sanitizer........... D2
Light Duty Cleaner A1,Q1
Liquid Acid Cleaner A3
Lite & Brite................................ B1
Mean-Klean................................. A1
Powder Foam A1
Safety Bleach B1
Smokehouse Cleaner A2
Special C.I.P. Cleaner A1,Q1
Trolley Klean.............................. A2
Trolley Oil H1
Truck Wash A1

Kadison Laboratories, Inc.
Sparkleen #1................................ A1

Kalco Chemical Company
KC-20 Moisture Guard.............. H2
Remuv-Ox................................... A3

Kalco Sales Company/ Cleancraft Industries, Inc.
Quatcide A4

Kammson Industries, Inc.
Blue Supreme............................... A1
Industrial Creamedic-HD E1
Trans-100..................................... A4

Kamo Manufacturing Co., Inc.
Big K.. A1
BK-62.. A2
BST... A2
Chloro Clean............................... D1
Dart... A3
Emulsion Bowl Cleaner &
Disinfectant C2
Formula APK No. 2
Concentrate............................. A1
Germicidal Cleaner D1
Hi-Foamy.................................... A1
K-40 Liquid Hand Soap............. E1
K-42 Smokehouse Cleaner......... A2
K-44 Activated Sewer Solvent .. L1
K-99 Insecticide.......................... F1
Kamo-F.I.S. Insecticide F1
Kamo's Liquid BST..................... A1
KTC Tripe Cleaner N3
No Roma Concrete Cleaner A4
Remov-Ox.................................... A3
Steam Cleaner GP....................... A2
Vapo-Cide 400 .... F1

Kanco Tech., Inc.
Hotsy Super X.............................. A1

Kane Chemical Company
Liquid Household Insect
Spray #1.................................. F2

Kankakee Industrial Supply Co.
KM 550 Hy-Foam Degreasing
Cleaner..................................... A1
KM-505 Safety Solvent.............. K1

Kansas City Chemical Company
Super-Clo D2,Q4

Kansas Janitorial Supply
Neutral Concentrate.................... A1
Torol All Surface Cleaner.......... A1

Kanter, R. J. Company, The
Horseoline #2.............................. H1

Kapeo Industries, Incorporated
#10 Super Concentrate Cleaner. A1
#100 Super Odor
Counteractant.......................... C1
#22 Frozen Freezer Cleaner A4
#50 Hard Surface Cleaner......... A4
Antiseptic Liquid Hand Soap.... E1
Formula No. 60 .... A1
Formula NP 812.......................... D2
Formula 812................................. D1
Industrial Bowl Cleaner............. C2
Kapco Odor Counteractant
Cleaner..................................... A4
Liquid Hand Soap E1
No. 25 Germicidal Cleaner........ D1
No. 55 Hard Surface Cleaner.... A4
Safety Solvent 1.......................... K2
Safety Solvent 2.......................... K2
Sparkle Emulsion Bowl Cleaner C2
Super Clean-Up A1
Super No. 23 Cleaner &
Stripper.................................... A1
Super Shine-All Concentrated
Cleaner..................................... C1

Kar Products Inc.
65379 White Grease H2
65381 Hi-Temp Dry Lubricant
(Aerosol).................................. H2
65385 Penetrating Oil (Aerosol). H2

Kare Products, Inc.
Knock Out F1
Roach & Ant Killer..................... F2

Karen Janitorial Supply Co.
Miracle Might............................... A4

Karrousel Kemical Korporation, Inc.
"Food-Plant" Insecticide F1
"Food-Plant" Insecticide Super
Strength.................................... F1
"Grime-Away" A1
"Jell-Karre" K1
"Karre-Ox".................................... A3
"Karrousel Klean-Kitchen" A1
"Kontack Odor-Trol" C1
"Safety-Karre" K1
"Silli-Slide" H2
"Skummy".................................... K1
"Super-Degreaser" K1
"Trash Kan-Karre" C1
Gentle Stroke............................... E4
Grime-Away S.S............................ A1
Lub-It H2

Kay Chemical Company
Ban-Or A1
Ban-Or 50..................................... A1

Kelgraf Company, The/ A Div. of Anderson Oil&Chemical Co.,Inc.

1100CS	H2
1200BS	H2
1300MS	H2
1400 M	H2

Kellermeyer Chemical Company

#173 Liquid Hand Soap	E1
Cold Room Cleaner	A5
Foam-Away Cleaner	A4
Grease Lightning	A1
Kelco Concentrate	A4
Master Industrial Bowl Cleaner	C2
San-O-Wite Toilet Bowl Cleaner	C2
Swish	A4
T-O.L.C.-300 Dispersant	G6
T-425 Return Line.Treatment	G6
WG-5 Disinfectant	C2
WT-525 Boiler Water Treatment	G6

Kellogg-American, Inc.

Rotary Oil KA RO 200	H2
RE-100 Oil	H2

Kelso Company

Acid Drain Opener	L1
Bowl Cleaner Disinfectant	C2
Caustic Drain Opener	L1
Concentrated Multi-Use Cleaner	A1
Cream Cleanser	C2
Heavy Duty Cleaner	A4
Heavy Duty Disinfectant Cleaner	A4
Hospital Disinfectant Deodorant	C2
K-Ox Chemical Cleaner and Deoxidizer	A3
Lemon Odor Disinfectant Deodorant	C2
Liquid Lotion Hand Soap	E1
Lotionized Hand Dish Detergent	A4
Pine Oil 5 Disinfectant	C1
Porcelain and Bowl Cleaner	D1
Roach 'N Ant Spray(Aerosol)	F2
Roach-N-Ant Spray	F2
Safety Solvent Degreaser	K1
Silicone Spray	H2
Stainless Steel Polish	A7
40% Liquid Hand Soap Concentrate	E1
6209 Cleaner	A1

Kelton Company, The

Kel Steam 202	A1
Kelco Detergent	A1
Kelco Detergent 202	A1
Kelco Detergent 404	A1

Kem Manufacturing Corp.

A.R. 69	H1
Aquashield	P1
Beltack	P1
Bio-Strain	L1
Blast	A8
Blast	A4
Boiler Treat	G6
Boiler Treat 1200	G6
Check Spice	D1
Coil Clean	P1
Coilex	A1
Concrete Cleaner	A4
Control	B1
D-Mark	K1
Durakem EP 500	H1

Dust Bond	P1
Econocide	F1
Emuls-It	K1
Exsoil	B2
Ez Bait Weather Resistant Rat Killer	F3
Fabrite	B1
Faze	C2
Fogicide	F1
Fry Off	K1
FD-100	F1
Gentle Formula #4	A1
Germex	C1
Germex Aerosol	C1
Grout Gleem	C2
H.G. Concentrate	A4
Hands-Off	A4
HR 230	L1
Insta-Kleen	A3
Insulex	P1
K 901	K1
K.O.K.	A4,A8
Kem "77"	A4
Kem Kill B	F2
Kem Lube	H2
Kem-O-Lan	C2
Kemcide	F1
Kemelt	C1
Kemist Air Sanitizer Hospital Type	C1
Kemist Odor Killer	C1
Kemox	A3
Kemsect	F1
Kemsurge	E1
Kemterge	A4
Kemtreet Krust	A3
Kemtreet M.S.R.	G7
Kemtreet Rx 150	G7
Kemtreet SX-11	A3
Kemtreet 100	G6
Kemtreet 1120	A3
Kemitreet 200	G6
Kemtreet 201	G6
Kemtreet 203	G6
Kemtreet 300	G6
Kemtreet 450	G6
Kemtreet 456	G6
Kemtreet 800	G7
Kemtreet 820	G6
Kemtreet 850	G7
Kemtrex	A1
Kemtrol	C1
Kemtronic	H2
Kemtronic Aerosol	H2
Kemycin	E1
Kill	F2
Kleergard II	C1
Lactosolv	A3
Lathereeze	E1
Lectrasol	K1
Lectrasol Aerosol	K1
Lectrokem	K1
Liqui Lube	H2
Micro-Cellerate Formula 3	L2
Miro-Kem	A1
Neutradine	K1
New Formula 7-11	C1
New Vigilante	E4
Odortrol	C1
Penetrate	C1
Pine 17	C1
Porcel	C2
Quad	A1
Quikcide Concentrate Insecticide	F1
Quikcide Contact Insecticide	F1
Quikcide Insecticide	F1
Ren-O-Grout	C2
Ren-O-Grout Ready To Use	C1
Residex	H2

Rustless II A3
San-A-Lube H1
San-O-Van C1
Sanitrol D2
Shield E4
Shine Brite A1
Silikem H2
Skavenger L1
Spraytrol P1
Stain-Go A1
Stainless C2
Stapol (Bulk) A7
Stapol Aerosol A7
Steam All A1
Strip-N-Flush (Aerosol) C3
Strip-N'Flush C3
Super Foam 310 .... A1
Superterge C1
Thermakem L1
Viro-Phene C1
Zip A1
Zone Fog F1

Kem-Co Chemical, Inc.

Acid Quat Dairy and Food
    Industry Clnr D2
CD Cleaner Disinfectant Deodz
    Fungicide D2
Descaler&Delimer/
    Boil.,Swim.Pools,Etc. A3
Di-Verse A4,A8
Di-Verse Butyl Cleaner A4
Eliminator L1
Gentle Touch E4
Glass-N-Utility Cleaner C1
Grease Gun H2
Heavy Duty Stripper Cleaner &
    Degreaser A4
Hospital Foaming Cleaner C2
Kwik Kleen Concentrate A1,E1
NP Cleaner Disinfectant Deodz
    Fungicide D2
Safety Solvent Cleaner &
    Degreaser K1
Sani-Gleem C2
Spray-Fog F1
Surgi-Power E1
Wash-Up Liquid Coco Hand
    Soap E1
Wipe Out Multi-Use Cleaner
    Degreaser C1

Kemaloid Corporation, The

Kemaloid #AL-25 .... G6
Kemaloid #AL-40 .... G6
Kemaloid #OS-15 .... G7
Kemaloid #OS-30 .... G6
Kemaloid #PO-20 .... G7
Kemaloid #PO-35 .... G6
Kemaloid Dry Heavy Duty
    #201 .... G6
Kemaloid Liquid-HP #102 ...... G7
Kemaloid Liquid-LP #101 ...... G7
Kemaloid Steam Line
    Treatment #310 .... G6
Kemaloid Water Treatment G6
PO-23 .... G6

Kemco Chemical Company

Bingo A3
Bowl Brite C2
Chlor-Kleen A1
Flash A4
Glass Cleaner C1
Green Giant Hand Dshwshng
    Detergent A1
Hi-Foam A1
Invisible Skin Shield Cream E4
K-Wash A1
Kemco #1 .... A1

Kleen-Brite A4
Lite A1
No. 25 Insecticide F1
Oceana E1
Pink Lotion A1
Pro Fite A1
Quick Clean Oven Cleaner A8
Safety Electric Parts Cleaner ... K1
Spray and Wipe All Purpose
    Cleaner C1
Super Flash A4
SSC Stainless Steel Metal
    Polish A7
Valiant Hand Soap E1
Victory Cleaner-Disinfectant C2
Winda-Shine Glass Cleaner C1

Kemco Environmental Products Co.

Ammoniated Oven Cleaner A8
Chain & Cable Lubricant H2
K-26 Silicone Spray H1
KE-25 Insecticide F1
KE-32 Contact Cleaner K2
KE-40 .... K1
Lanolin Waterless Hand
    Cleaner E4
Sani-Lube Food Equipment
    Lubricant H1
Super Clean All Purpose
    Cleaner C1
Sur-Grip P1
TEF Sealer H2
Wyte-Ease Lithium Grease
    Lubricant H2

Kemex, Incorporated

Insect Spray F2
K 190 .... A4
K 25 .... A1
K 32 Grease Remover A1
K 34 .... A1
K 35 .... A1
K 350 .... E1
K 355 .... E1
K 570 .... A2
K 571 .... A1
K 580 .... A2
K-302 Liquid Dishwash A1
K-320 .... A1
K-33 .... A4
K-340 .... A1
K-36 .... A1
K-38 Grease Remover A1
K-380 .... E1
K-44 .... A4
K-48 Foam-Kleen A1
K-852 .... L1
K300 Liquid Dishwash A1
K881 Super Enzymes L2
Lemon Disinfectant No. 6 .... C1
Orange Bacteriasol Germicidal
    Cleaner C1
Premium Fly & Roach Spray F1

Kemicals, Incorporated

K-2AC A1

Kemin Industries, Incorporated

Kem San D1
Kem Wet P N1
Liquid Mold Curb P1

Kemtron Products, Inc.

CD Cleaner Disinfectant
    Deodorizer Fngcd D2
NP Cleaner Disinfectant
    Deodorizer Fngcd D2
Odor-Ban Odor Control C1
Ready-To-Use Non-Flammable
    Solvent Dgrsr K1

Stericide Disinfectant Defilmer
    Deo. Det. D1
Super Power #2 .... A4
Super Solv #2 .... A4,A8
Suspend Bio-Kleen A1

Ken-Ray Industrial Products Corp.

Power Instant Spray Cleaner A4

Kendall Co., The

Kendall Food Plant Fogging
    Insecticide F1

Kendall Refining Company/ Div. of Witco Chemical Corporation

All Oil Strat.Min. Oil Gear
    Lub. SAE 140 .... H2
Dual Action Heavy Duty
    Motor Oil H2
Kenlube C-931 .... H2
Kenlube L-412 .... H2
Kenlube L-421 .... H2
Kenlube L-425 .... H2
Kenlube M-612 .... H2
Kenlube M-621 .... H2
Kenlube S-825 .... H2
Kenlube S-831 .... H2
Kenoil 047 R&O EP Turbine
    &Hydraulic Oil H2
Kenoil 053 R&O EP Turbine
    &Hydraulic Oil H2
Kenoil 065 R&O EP Turbine
    &Hydraulic Oil H2
Kenoil 080 R&O EP Turbine
    &Hydraulic Oil H2
Three Star Gear Lubricant
    SAE 85 W-140 .... H2

Kensington Co. Inc., The

Germi-Kleen A4
Kold & Kleen A5
Multi Concentrate A4,A8

Kensington Products Co.

Kenpro Safety Clean 100 .... A1
Kenpro Safety Clean 200 .... A1

Kent Chemical Company

CD G6
CDX G6
KN G6
KNA G6
KNL G6
KNO G6
KRT G7
KS G6
KSL G6

Kenway Chemical Co. of Kentucky

Creme-A-San C2

Kernite/ Division of NCH Corporation

'K' Ey K1
'K' Go K1
'K' O. Sect F2
'K' Strip K1
'K'-Last H2
'K'-Seal P1
'K'-Steam A1
'K'an-'K'are A1
'K'era-Fog F1
'K'ernaire C1
'K'ernaire Aerosol C1
'K'il Mate F2
'K'on-Off A3
'K'oncentrate A1
'K'wick-Thaw C1
'K'Kwik-Pac'k' P1

152

888 S A1

Kerr-McGee Chemical Corp.
Trona Fine Granular Soda Ash . N3
Trona Granular Soda Ash N3

Key Chemical Co. Inc.
Auto Degreaser A1
Drain Flo................................... L1
Hard Surface Cleaner.................. A4
Super Heavy Duty All-Purpose
Cleaner................................... A4
Synthetic Cleaner A4

Key Chemical Company
Form. No.23HvyDty Wax
Remvr&Degr. A4

Key Chemical Company
Alive Bacteria/Enzyme LS-
1471 .... L2

Key Chemical Company
CD-95 .... G6
Keyco 150.................................. G6
Keyco 310.................................. G6
Keyloid 119............................... G6

Key Chemicals, Incorporated
Key-Chem 200 .... A4
Key-Chem 303-A....................... A3
Key-Chem 306 .... A4
Key-Chem 307 .... A1
Key-Chem 342 .... A1
Key-Chem 348 .... A1
Key-Chem 365 .... P1
Key-Chem 370 .... A1
Key-Chem 370A A1
Key-Chem 371 .... A3
Key-Chem 390NF A3
Key-Chem 503 .... D1
Key-Chem 503 A.................... A4,C1
Key-Chem 533 .... A4,A8
Key-Chem 760 .... A4
Key-Mix 100 .... A1
Key-Mix 107 .... A2
Key-Mix 111 .... A1
Key-Mix 113 .... A1
Key-Mix 117 .... A1
Key-Mix 125 .... A2
Key-Mix 131 .... A2
Key-Mix 139C............................ A1
Key-Mix 147 .... A1
Key-Mix 28 .... A3
Key-Mix 30 .... A1
Key-Mix 40 .... L1
Key-Mix 48 .... A1
Key-Mix 54 .... A1
Key-Mix 54H A1
Key-Mix 92 .... A2
Key-Mix 95 .... A1

Key Chemlab Corporation
Action....................................... C2
Attack....................................... F1
Bactrol Mint Odor Disinfectant . C2
Bactrol Pine Odor Disinfectant.. C1
Blast Off................................... A2
Breeze C2
Chem-Ox A3
Chem-Sil H2
D.D.D.. A4
Flow L1
Fogicide.................................... F2
Husky Creme Hand Clnr
(Aerosol)................................ E4
Key Kleen A1
Key Lube 201 .... H2
Key Lube 33 .... H1

Kleen (Aerosol) C1
Kleer.. C1
Power Off A2
Punch C2
Purge.. L1
Show Off................................... C2
Tef-Seal H2
X-L A4

Key Industrial Products, Inc.
Century..................................... A4
Domino...................................... A1

Key-Chem Industries
#15 Liquid Hand Soap E1
"P D Q" Lotionized Waterless
Hand Cleane........................... E4
Adios Insecticide F2
Concept C2
E.S.P... A4
E-Z Oven Grill & Cleaner A8
E-Z-Wrench H2
Enzymes L2
K-C Solv K1
Move-It Dry Silicone.................. H1
Nock-Out.................................. F2
Pro-Tek Moisture Guard H2
Refresh Lemon Scented............. C2
Royal Flush............................... L1
Safety Solvent Cleaner &
Degreaser K1
Swat F2
Swinger A4

Keystone Chemical & Supply Co.
Boiler Water Treatment............. G6

Keystone Division/Pennwalt Corp.
Condensed Oil 50 Medium H2
Cup Grease Medium H2
G.P. 20..................................... H2
G.P. 30..................................... H2
GG Medium............................... H1
K 600 .... H2
K.L.C. No. 6 .... H2
Keleo SAE 20 Motor Oil H2
Kelco SAE 30 Motor Oil H2
Kelco SAE 40 Motor Oil H2
Keycut 102 .... H2
KLC-2 .... H2
KLC-3 .... H2
KLC-4 .... H2
KLC-4A H2
KLC-432................................... H2
KLC-5 .... H2
KLC-543................................... H2
KLC-654A H2
KR-6 .... H2
KSL 222 .... H2
KSL-112 .... H2
KSL-114 .... H2
KSL-213 .... H2
KSL-214 .... H2
KSL-219 .... H2
KSL-220 .... H2
Moly 29 .... H2
Moly-81 LT. H2
Nevastane Heavy H1
Nevastane HT-1 .... H1
Nevastane Light H1
Nevastane Medium.................... H1
Nevastane SP Heavy.................. H1
Nevastane SP Light H1
Nevastane SP Medium............... H1
Nevastane SPS........................... H1
Nevastane SP6........................... H1
Nevastane 10............................. H1
Nevastane 20............................. H1
Nevastane 40............................. H1
Nevastane 5............................... H1

Nevastane 6 Light H1
Nevastane 6 Medium.................. H1
No. 1790 .... H2
No. 53-X Light H2
No. 81 Light C1
No. 81 X Light H2
No. 81 XX Light H2
No. 86 Light H2
No. 89 Medium Silicone H2
NP-500... H2
NP-750... H2
Pen Oil No. 1.............................. H2
PKN-1 .... H2
P3B H2
Silicone Release Agent &
   Lubricant H1
Velox 2062 .... H2
WG-A H2
WG-1 .... H2
WG-3 .... H2
1791... H2
2-K-6 .... H1
2K5 Medium H1
2K5X Light H1
49 Med H2
49-LT H2
5P6 .... H1
5P7 Heavy.................................. H1
5P7 Light H1
5P7 Medium H1
53-LT... H2
80 Light H2
80 XX Light................................ H2
81 EP Light H2
86 Med.. H2

Keystone Laboratories, Inc.
Keystone BT-642......................... G6
Keystone BT-652:........................ G6
Keystone BT-661-F G6
Keystone OR-70 .... G6
Keystone P-380............................ G6
Keystone PP-390 .... G5
Keystone SI-377 .... G6
Keystone SL-681 .... G6

Keystone Wiper & Supply Co.
Key-O-Kleen A4

Kilburn Chemical Company
Wind-O-Shine C1

Kilmer & Associates
K&A-300 .... A1

Kimbell Foods, Incorporated
Kalex Bleach................................D1,B1
101 Heavy Duty All Purpose
   Cleaner A1
305 Pine Odor Disinfectant C1
340 Lemon Disinfectant.............. C2
701 R-T-U Liquid Hand Soap.... E1

Kimco
Kemclean R-10 Concentrated
   Detergent................................. A4
Kemclean R-11 All Purpose
   Cleaner A4

Kimzey Chemical Company
Disburse A1
Kimquat 64-N D1

Kinetico Incorporated
K-Five G1

King Associates, LTD.
King Bath And Tile Cleaner...... A1
King Detergent Cleaner
   Disinfectant D1

King Liquid Hand Soap.............. E1
King Mint Disinfectant C2
King Oven & Grille Cleaner A8
King Pine Odor Disinfectant...... C1
King Pine Oil Disinfectant C1
King Pink Lotion Hand Soap E1
King Quik Oven & Grille
   Cleaner A8
King Super Cleaner..................... A4
King Super Strip A4
King Wax Stripper-Degreaser
   Cleaner A1

King Chemical Company, Inc.
King-Solv A1
Luster-King.................................. A4
Opalene Hand Soap..................... E1
Tri-Lite A1

King Chemical Company, Inc.
Kingco All-Purp Silcn Spray
   Food Grd H1
Kingco All-Purp Silcn Spray
   Food Grd 2% H1
Kingco All-Purp Silcn Spray
   Food Grd 3% H1
Kingco All-Purp Silcn Spray
   Food Grd 4% H1

Kinzua Environmental, Inc.
KE 109 .... A3
KE 114 .... A3
KE 147 .... A3
KE 22 .... A4
KE 31 .... A4
KE 31 .... A1
KE 89 .... A1
Power-Out.................................... L1
Power-Out Drain Opener L1

Kitten & Bear Chemicals, Inc.
Bear #9 Smoke Odor Control
   Concentrate P1
Bear 99 Smoke Odor Control P1
Concentrated Aerosol Purge 11
   Insect...................................... F1
Cygon 2-E Systemic Insecticide F2
Diazinon 4E................................. F2
Felton 77 Masking Agent C1
Industrial Aerosol Insecticide
   Conc.. F1
Kitten 88 Problem Odor
   Control.................................... C1
Kitten 88 Problem Odor
   Control Conc. C1
Malathion 57-E F2
N-1 Naled Fly Spray................... F2
PY-1 PY-Fly AA Grade Fly
   Spray....................................... F1
PY-3.. F1
Roach and Ant Spray F2
Sugar Fly Bait F2

Kjell Water Consultants, Inc.
C-10... G2
F-35... G2
K.W.C. Aqua Mag G2

Kleen Chemical Company, Inc.
All Kleen..................................... A1
All-Kleen Heavy Duty A1
Bowl Kleen #34 .... C2
D-505 Hand Dishwashing Det... A1
F-99... A1
G-707 Galvanized Cleaner......... A3
Hog Scald.................................... N2
Kleen Klor................................... D1
Kleen R-1000 .... C2
Kleen W-5.................................... A1

Kleen 01 .... E1
Kleen 02 .... E1
Kleen 11 .... A1
Kleen 22 .... A1
Kleen 23 .... A1
Kleen 24 .... A1
Kleen 25 .... A1
Liquid HD Alkaline Cleaner...... A1
P-101 .... A2
P-711... A1
WA-55 .... N1
1-4 Insecticide............................. F1
511-Plus Heavy Duty Cleaner-
   Degreaser A4,A8
88 Inhibited Acid Detergent A3

Kleen Kare Company
Big "G" A1
Cocoanut Oil Hand Soap
   (Concentrate) E1
Combine A4
Dual-27 Disinfectant Cleaner ... D1
Dyno Mite................................ A4,A8
DC-142 .... D1
Fogging Spray Concentrate F1
GPC-24 .... A4
KK Industrial Deodorant C1
Odo-Rout..................................... C1
R.I.S.-15 Residual Insecticide.. F2
Remuv-Ox A3
Steam Cleaner Liquid A2
Vita-Pine..................................... C1

Kleen-Rite
Big Jim Lemon Foam Germcdl
   Clnr & Disnft C2
Descaler....................................... C2
Kleen Coil Cooler Coil Cleaner. P1
Klog-Free Drain Opener L1
Laundry Detergent...................... B1
Lemon Disinfectant No. 6.......... C1
Liquid Hand Soap 20.................. E1
Liquid Steam Cleaner All Purp
   Heavy Duty A2
Low Phos Alkaline Detergent ... B1
Mighty Kleen Grease Remover. A1
Orange Bacteriasol Germicidal
   Cleaner C1
Premium Fly & Roach Spray..... F1
Pyrethroid Fly & Roach Spray.. F1
Residual Roach Spray................. F2
Safety Solvent............................. K1
Sparkle Kleen Liquid Dishwash A1
Stainless Steel Cleaner C1
Sweeping Compound (Red) Oil
   Base.. J1
X-150 Chlorinated Machine
   Dishwash A1

Kleenco Products, Inc.
Break-Thru................................... A4
Disinfectant Cleaner 140............ A4
Heavy Duty K-Suds..................... C1
Kleen-Off..................................... A1
Kleenzit A4
PleaScent C1

**Klein Kleen-All Company Div. of
Terminix International**
Blue Concentrate General
   Cleaner A4
Heavy Duty Cleaner &
   Degreaser T100........................ C1
Terminix Germicidal Cleaner.... D1

Klein Pest Control Company
Bug-X F1
Kilex F1
Rat Feast F3

Klenz-Klor Formula X-3 .... D1,Q1
Klenz-Scald Formula 2133 .... A1,N1
Klenz-Solv............................... A4,A8
Klenzmation Formula AC-30 ..... A3
Klenzyme L2
Kurb Formula HC-65................. G7
Laundry Detergent Super "S" ... B1
Liquid Cleaner 2......................... A1
Liquid Cleaner 3......................... A1
Liquid Corrosion Inhibitor 2575 A1
Liquid Egg Defoamer 2579........ Q5
Liquid Foam Booster And
   Cleaner................................. A1
Liquid Gen. Purpose Cleaner
   #2256 .... A1
Liquid Peeler N4
Liquid Sanitizer D2,Q4,N4,G1,
                                                       B1,E3
Liquid Smokehouse Cleaner....... A8
Liquid Smokehouse Cleaner
   2528 .... A4,A8 .
Liquid Sodium Hydroxide......... A2,L1,N4
Liquid-K Formula LC-300 .... A1
Low Foaming Liquid Cleaner ... A1
Low Temperature Lubricant...... A1,H2
Lube-Mor 2593 .... A1,H2
Lubri-Klenz.............................. A1,H2
Lubri-Klenz LF A1,H2
Mikroklene DF........................... D2,Q6,E3
Multi-Power Formula HC-9....... A1
M1-5 Formula M1-5.................... A1,Q1
Nu-Kleen................................... A3
OR-3 Acid Detergent Formula
   OR-3 .... A3
Phosphoric Acid......................... A3
Poultry Scald 2510 .... A1,N1
PL-190..................................... Q1
PL-3.. A3
Redi-Kleen A1
Retort Additive KX-2363 .... G5
Special Alkali Formula HC-41 ... A2
Speed Wash Formula LC-5........ A1
Spray Cleaner 2135 .... A1
Ster-Bac Formula KQ-12........... D2,Q3
Technical Oil 1 .... H1
Technical Oil 2 .... H1
Technical Oil 3 .... H1
Trichlor-O-Cide Formula XP-
   100.. D1
Tripe Bleach 2558....................... N3
Tripe Wash 2557........................ N3
Trolley Cleaner KX 2567 .... A2
Vega-Kleen Formula 1164...:.... N4
Vega-Kleen Formula 2437.......... A1,N4
Vital Pwdr Caustic Additive
   Form HC-52............................ A1
Vital 2 Formula 2446 .... A1
X-4 .... D2,Q4
XY-12 .... D2,Q4,N4,G4,
                                                       B1,E3
2100 Acid A3
2171 General Cleaner................. A1
2313 Liq. Alkaline Clnr Form.
   2313...................................... A4,Q1
2356 General Cleaner................. A1
2385 Plastic Wash...................... A1
2470 Conveyor Lubricant........... A1,H2
2485 Cleaner A8
2562.. A2
2564 Alkaline Cleaner................ A2

Klenzer Company, The

Heavy Duty A.P. Detergent....... A1
Heavy Duty Liq. Steam Clning
   Compd A1
Non-Caustic Liq. Steam Clning
   Compd A1

Klenzoid, Inc.

Kelox #10F................................ G6

Klenphos G2
Klenzamine G6
Klenzamine F1 .... G7
Klenzamine F2 .... G7

Klix Chemical Company, Inc.

All In One................................. D1
Bluebird.................................... A1
Boss... A4
Brawnee.................................... A4,A8
Chloro Clean............................. D1
Clean-CO 25.............................. A4
Concrete Cleaner....................... C1
Control B1
Control Plus.............................. B1
Do-It.. A4
Egg Wash #18-266..................... Q1
Extra Power Spray Insecticide .. F1
Foam-O-Gel A4
Freezer Locker Cleaner.............. A5
Germicidal Cleaner D1
Grease Trap Cleaner L1
Green Glass Cleaner C1
Grime Zapper E1
Insecticide Type 11 .... F1
Iodine Sanitizer......................... D1
K-100 .... C1
KLX Cleaner A1
Liquid Soap............................... E1
Liquid Suds Up.......................... A1
Logic B1
Med-I-Sept E1
Mighty Concentrated Detergent A4
Mill Spray F1
Phos-Clean A3
Pine Odor 5 Disinfectant C1
Powdered Laundry Bleach 16% B1
Power Pink Cleaner A4
Power Spray F1
Spray Clean............................... A4
Super Concentrated Control B1
Super-Q-Sanitizer....................... D1
Top-Flite A1

Knapp Chemical Company

Acid Descaler A3
Acid Detergent........................... A3
All Purpose Cleaner Granular ... A1
All Purpose Cleaner Liquid A1
Bottle Washing Compound A2
Bottle Washing Compound-
   Liquid A2
Chain Lubricant #1..................... H2
Chain Lubricant #2..................... H2
Chlorinated Machine Dshwshng
   Comp. A1
Egg Stain Remover..................... Q2
Egg Washing Compound............. Q1
Floor and Wall Cleaner.............. A1
Food Contamination Remover... A2
Knapp Fruit Wash...................... N4
Knapp Portable Restroom
   Deodorant-Clnr....................... C2
Liquid Detergent A1
Machine Dishwashing
   Compound A1
Peeling Compound N4
Pipeline Cleaner Heavy Duty..... A1
Pipeline Cleaner-Regular A1
Pot,Pan&Utensil Cleaner A1
Roach Control Insecticide
   Aerosol F2
Spra-Gon Drosophila Spray F1
Spra-Gon Roach and Fly Spray F1
Steam Cleaning Comp H.D.   ·
   Liquid A1
Steam Cleaning Comp Light
   Duty Liq.................................. A1
Steam Cleaning Comp Med
   Duty Powd.............................. A2

Steam Cleaning Comp Medium
Liq..A1
Steam Cleaning Compound
H.D. Powd..............................A2
Steam Clning Comp. Lgbt
Duty Powder A1

Knight Enterprises, Incorporated
AV 60 ....A4
Brite 400 ....A3
Brite 404 ....A3
BD 14 ....A1
Carb Off 400 ....A2
Carb Off 405 ....A2
Carb Off 44 ....A2
Cleen J...A2
Cleen 200.....................................A2
Cleen 70 AL A1
DJ 20 ....P1
DJ-20 ....P1
FLR.-Con A4
GLD A4
Hi-Ho Lotionized Hand Cleaner E1
HL Concentrate............................A1
HT-17 ....A1
Jel-Cleen......................................C1
JR-8..A3
Knight L-212................................C1
Knight 1 ....A1
Lub-29 ....H2
PH Reducer S...............................A3
Rapid Clean C1
Special 123 Spray Cleaner A4
Spra 16...C1
Super Concentrate........................A4
SC-55 ....A3
SRO 82 ....A2
T-21-L..C1
T-25-L..C1
VHL A1
Zephyr Brite A4
123 Spray Cleaner & Degrease..A4
15% Liquid Hand Soap E1
20% Liquid Hand Soap E1
200-B..A2
212..A1
40% Concentrate Liq. Hand
Soap E1
505..A1
506..A1
515..A1
770..A1
920..A1

Knight Oil Corporation
Knight's Grease Grabber E4
Knight's New Spray Nine A1
Knight's TB-X Hospital Spray
Disinf. C1

Knox Chemical Company
Canner's Special Insecticide F1
Diazinon 4S Insecticide F2
Fly Killer F2
Formula #111 ....F1
Formula DML F2
Formula MML..............................F2
Formula 311 Residual
Insecticide...............................F2
Hi-Pressure Aerosol Bomb F1
Knox Fly Spray Formula G..........F1
Knox Pyrenone Semi-
Concentrate..............................F1
Knox Rat & Mouse Cakes F3
Knox Special Contact Spray
Formula-D................................F1
Knox Special Mill Spray F1
Knoxit #304 ....A4
Malathion Insecticide 2...............F2

Malathion Insecticide 5...............F2
Mintene C2
M4F..F2
New Improved Prolin...................F3
Professional F.O.D. F2
Pyrenone A...................................F1
Pyrenone Concentrate C F1
Pyrenone O.T. 60-6 Insecticide..F1
Pyronyl Concentrate #3610.......F1
Residual "D" Insecticide F2
Residual Insecticide.....................F2
Tanner's Special Formula
Insecticide...............................F2
20% Liquid Hand Soap E1

Knox Oil of Texas, Inc.
Blue Spray Car Wash
Detergent..................................A1

Kobax Corp.
Solvent Cleaner & Degreaser.....K1

Koch Supplies, Incorporated
Hi-Voltage....................................A1
New "Old" Baldy Liquid Hog
Scald N2
Old Baldy Hog Scald...................N2
Smoke House And All Purpose
Cleaner.....................................A1
Tripurge.......................................N3

Kochem Inc./ Div. Power Engr'g & Supply
Formula CWT-41 ....G2
Formula H-211 Boiler Feed
Water Treat...............................G6
Formula HA-20 ....G6
Formula HB-13 Boiler Feed
Water Treat...............................G6
Formula HB-13A.........................G6
Formula HB-21 Boiler Feed
Water Treat...............................G6
Formula HCW-66........................G7
Formula HL-10 Boiler Feed
Water Treat...............................G6
Formula HL-10A.........................G6
Formula HL-18 Boiler Feed
Water Treat...............................G6
Formula LS-12 Boiler Feed
Water Treat...............................G6
Formula LS-20 Boiler Feed
Water Treat...............................G6
Formula RL-55 Boil. Corosn
Contrl Treat..............................G6
Formula RL-56 Boil. Corosn
Contrl Treat..............................G6
Formula SB-15 Boiler Feed
Water Treat...............................G6
Formula SB-19 Boiler Feed
Water Treat...............................G6
Formula SB-23 Boiler Feed
Water Treat...............................G6
Formula TW-250 Cooling
Water Treatment.......................G7
Formula TW-251 Cooling
Water Treatment.......................G7
Formula Z-930 Wtr Softener
Mineral Clnr.............................P1
Kochem AL-146 Algicide G7
Kochem AL-153 Algicide G7

Kohl's Food Stores
Liquid Hand Soap E1

Kohnstamm, H. & Company, Inc.
#996 Universal Detergent A1,B1
Blue Power Scouring Creme.......C2
Borochlor Dry Bleach B2
Borochlor Dry Bleach B1

Bowl Power C2
Bristol B2
Bug-Off Residual Insecticide......F2
Bug-Off Residual Insecticide
(Aer.) F2
Clear-Blue C1
Clenital Detergent B2
Colorsuds.....................................B1
Drytex B2
Eliminator Drain Opener.............L1
Ez-It Silicone Lubricant H1
Germ Spot....................................C1
H.K. Alkaline Cleaner A2
H.K. All Purpose Liquid
Cleaner.....................................A1
H.K. All-Purpose Liquid
Alkaline Cleaner A1
H.K. Bleach B2
H.K. Detergent Oil A1,B1
H.K. Dry Bleach B1
H.K. Manual Washing
Detergent..................................A1
H.K. Sodium Metasilicate...........B1
Hi-Chlor Dry Bleach B1
Impax A1,B1
Inex B1
Intersuds B1
Knock Out F1
Leveion Blue................................B2
Lime Solv A1,B1
Liquid Hand Soap E1
Medi-Cream E1
No. 1179 Detergent B1
Nuclor Dry Bleach.......................B1
Nuclor Extra Dry Bleach B2
Orange Fresh................................C1
Perlite Extra A2,B1
Perlite Regular A1,B1
Poly/Cot A4,B1
Pow-R-Clean...............................A4
Pow-R-Foam................................A1
Pow-R-Solv..................................K1
Ramrod...A4
Sani-Bowl C2
Sani-Phene Plus A4
Sani-Q..D1
Sani-Q-Extra D1
Simplex B1
Soft-Stat.......................................B2
Startex..B1
Steamer..A1
Super Sheen A1
Thundersuds B2
Thundersuds A4,B1
Trimite Detergent B2
Wheatex Starch B2

Kor-Chem, Inc.
AL-33 ....G6
Bact-Cide.....................................G7
Big 4 ....A1
Brute #175 ....A4
Concrete Cleaner.........................C1
Formula 828 On Line Descaler..G6
Formula 84 Dispersant................G6
Formula 92 Oxget........................G6
Formula 93....................................G6
Formula 95 SLT...........................G6
Formula 97 SL 11 ....G7
Formula 98 ST LN SS.................G6
Formula 99 Sperse AR................G7
Get's-It Pot&Pan Hnd Dish.Liq
Conc 181-L...............................A1
Heavy Duty Butyl Cleaner A4
Kleans-It Concentrate 174-L.......A4
Liquid Floor Scrubber Cleaner..A4
Odor Control 100 ....C1
Oven & Grill Cleaner..................A8
Re-Dox A3
Stainless Steel Cleaner A7

No. 540 Lotion Hand Cleaner.... E4
No. 9100 Laundry Detergent B1
No. 92 Laundry Detergent B1
Protective Cream (Water-
  Resistant) E4
Protective Cream PC-25
  (Solvent-Resist.) E4
Pum-Solv Antiseptic Hand
  Cleaner E4
Pure................................... C2
Royal Lotion Skin Cleanser E1
Safe-Guard C2
San-O-Clean C2
Sani-Anti Iodine Hand Cleaner .. E2
Summa Hand Cleaner C2
Supreme C2
550 Lotion Skin Cleaner E4

L

L & A Products-Northern Calif.
Lanco Formula 361 .... A1
Lanco-Chlor HD A2
LA Concentrate A4

L & A Products, Inc.
#20 General Cleaner A1
#44 .... A4
#46 .... D1
Formula 14 .... A1
Formula 14M A1
Formula 14PF A1
Formula 16 .... A1
Formula 36 .... A1
Formula 36L A4
Formula 40 .... A1
Formula 58 .... A3
Formula 64 .... A2
Formula 685 .... A1
Formula 690 .... A1
Formula 695 .... A2

L & J Chemical & Supply Co. Inc.
AC 555 .... A3
AC 666 .... A3
ASC 15 .... A1
CMC 25 .... A1
CMC 50 .... A1
FCP C1
GC 77 .... A1
GC 99 .... A1
HDC 101 .... A2
HDC 102 .... A2
HDC 103 .... A2
HDC 104 .... A2

L & P Paper Co.
Lots-of-Power A1

L & S Products
Liquid Dynamite A4

L K C Cleaning
Degreaser/Cleaner A1

L P S Research Laboratories, Inc.
LPS #1 .... H2
LPS #2 .... H2
LPS #3 .... H2
LPS Instant Contact Cleaner K2
LPS Instant Super Cleaner K2
LPS 100 ESA H2

L. B. Chemical Company, Inc.
L.B.D. #10 Sanitizer &
  Detergent D1

L.B.D. Algi-San 10 .... G7
L.B.D. Drain Opener&Sewer
  Solvnt L1
L.B.D. Hansom A1
L.B.D. Heavy Duty Concrete
  Clnr C1
L.B.D. Lite A3
L.B.D. Mark V11 Concentrate ... A1
L.B.D. Milky Bowl
  Clnr&Deodorizer C2
L.B.D. Pine Disinfectant C1
L.B.D. Spray Concentrate No.
  625 .... F1
L.B.D. Tip Top A1
L.B.D. Zippo Liq. Steamer
  Compnd A2
L.B.D. Zippo Pwdr Steamer
  Compnd A2

La-Mar Chemicals, Incorporated
Acid Etch A3
Acidulade 111 .... A3
ALC-77 Aluminum Cleaner A3
CF-28 Safety Solvent K1
CHD-77 Heavy Duty Cleaner ... A4
Dyno-Might II A8
DS-140 Disinfectant #10 .... D2
Floor Blast A4
FCF-28-29 Safety Solvent K1
FD #1 Neutral Fragrance D1
FD#1C Citrus Fragrance A1
FF 67 Foam Off....................... A1
FSC 800 Steam Cleaner A1
Grease A-Go-Go C1
HD 77 .... A4
IA-30 Inhibited Acid............... A3
Kennel Kleen II...................... A4
Kennel Kleen 1 Neutral
  Fragrance D1
Kwik Solv II.......................... K1
MW-82 Motor Cleaner C1
RS-102 Spray F1
Sparkle A8
Sub-Zero 40-Below A5
SC-65 .... L1
SC-800 Liq. Steam Jenny
  Compound A1
Wash Off Remover P1

Lab Automated Chemicals Div. Of American Chemmate Corp.
Actene-Z L1
Ampo Solv K1
Ampo-Solv K1
Applaud C2
Banish F2
Bi-Phase C1
Biomune.............................. D2,Q3
Bonifide A8
Broot................................ L1
Conetic A7
Corelium C2
D-Bac................................ C2
Delectrin............................ H2
Deo-Bol C1
Deo-Gen C1
Descend L1
Dish Wash Compound A1
Dispatch C1
Frigid-Kleen......................... A4
Glyde................................ A8
Hytox Chemical Cleaner A3
Industrial Bi-Solv K1
Industrial Steel Brite
  Concentrate A7
Lab Conquer A4
Lab Creme Lotion..................... E1
Lab Escort A4
Lab Flo-Kleen L1
Lab Glaspray C1

LaMar Chemical Service Inc.

Approve A2
Asset A2
Astro A2
Bost A2
Brite-Glo A1
Built Caustic A2
Deplete A2
Do More A3
Double A A3
ESE Silicone Spray H1
Gyro A3
Impact A2
Luster Glass Cleaner C1
Nu Brite A3
Oven Majic A8
Peel Aid N4
Results A3
Royal A2
Ultimate A2
Wow A3,B1

LaRosa Products, Inc.

AB-118 Acid And Alkaline
   Foamer A1
AC-114 Acid Cleaner A3
CA-113 Chlorinated Alkaline
   Cleaner A1
LHS-109 Liquid Hand Soap E1
LHS-117 Synthetic Hand Soap .. E1
MR-120 .... A1
SHC-119 Smoke House Cleaner A2
SLA-101 Shroud Laundry
   Alkali B1
SLB-110 Shroud Laundry
   Bleach B1
SLB-111 Shroud Laundry
   Bleach B1
SLD-100 Shroud Laundry
   Detergent B1
SLW-102 Shroud Laundry
   Surfactant B1
SS-108 Safety Solvent K2
TC-103 Trolley Cleaner A2
TC-104 Trolly Deruster A2
TN-116 Tripe Neutralizier N3
TO-105 Trolley Oil And
   Lubercant H1

Le Bro Chemical Company

LB 101 .... C1
LB 102A C1
LB 112 .... A7
LB 22 Heavy Duty Liquid
   Cleaner A4
LB 303 .... K2
LB 306 .... H2
LB 314 .... K1
LB 35 Degreaser A4
LB 505 .... E4
LB-30 Disinfectant Cleaner D1
Punch C2

Lea Chemicals, Inc.

Aide Hand Soap E1
AD-51 .... A3
Capri A1
Concrete Cleaner A3
Conveyor Lubricant No. 6 .... H2
D-Sperse G6
Dermaclean E2
Egg Wash Q1
Genie E1
Guardian D1
Jiffy Bowl Cleaner C2
Lea Algitrol G7
Lea AC-108 .... A4
Lea AC-216 .... A4
Lea D Rust And Shackle
   Cleaner A2

Lea Drain Opener L1
Lea Egg Sanitizer Q3
Lea EC-42 .... A1
Lea Foam Aid Cleaner A1
Lea HD-413 .... A1
Lea LC-15 .... A1
Lea Oven Cleaner A8
Lea Pine Scented Disinfectant ... C1
Lea Pot 'N Pan Cleaner A1
Lea Powdered Cleaner A1
Lea Quik Kill F1
Lea Tank & Pipe Line Cleaner .. A1
Lea Tile Bright Concentrate A3
Lea 1235 Disinfectant C1
Lea 305 .... D1
Lea 310 .... D1
Lea-Zyme L2
Leacide F1
Liquid Egg Destainer and
   Cleaner Q1,Q2
Liquid LC-17 .... A2
LC 100 Iodophor D1
LC 1231 Disinfectant - Cleaner . C1
LC 19 .... A1,E1
LC 38 .... A1
Orbit Hand Soap E1
Scale Remover A3
Steam-X A1
Super Scrub Hand Soap E1
Superb Heavy Duty Laundry
   Detergent A1,B1
Total E1
UP-404 .... A4
WD-7 .... C1

Leadership Products Company

All Purpose Heavy Duty
   Degreaser A4
All Purpose Wax Stripper A4
Liquid Hand Soap E1
Minute Clean A4
Steam Jenny Compound A1

Leahy-Wolf Company

Purity H1

LeahChem Industries, Inc.

Chlor-Clean 1 .... A1
Chlor-Clean 2 .... A1
H D Degreaser A4
Killtex-G-1 .... F1
Killtex-G-3 .... F1
L-C-40 .... A4
LC-50 .... A4
Stainless Steel Cleaner A3
Wetsteam A1

Lebanon Valley Paper Company

Dutch Valley Power Cleaner A4

Lee Chemical Corporation

Buster A1
LL-9 Fogging Concentrate F1
LP-85 .... A2

Lee Chemical, Inc.

Automatic Dishwash A1
Bio Clean 11 .... D1
Bio-Clean 2 .... D1
Bulls Eye Formula F A1
Bulls Eye Heavy Duty Cleaner . A1
Challenge Heavy Duty
   Degreaser A4,A8
Chlorinated Destainer A1
Dura-Suds Liquid Dishwash A1
Dura-Suds Lotionized
   Dishwash Detergent A1
Effecto Special Degreaser A4
Foamy Red Baron A1

159

Lime Be Gone A3
Liquid Hand Soap Ready to
  Use E1
Liquid Hand Soap 20% E1
Liquid Hand Soap 40% E1
Lotionized Hand Cleaner
  Concentrate E1
LC 12 All Purpose Cleaner A4
LC-30 General Purpose Cleaner A1
LC-4 All Purpose Powder A1
Mark Ten D2
Pink Dawn Hand Dishwash
  Liquid A1
Red Baron A1,A8
Restore 10 .... A1
Restore 60 .... A2
Result Chlor. Liquid Automatic
  Dishwash A2
Result Heavy Duty Bar Glass
  Wash A1
Result Heavy Duty Liquid
  Auto. Dishwash A2
S-S-C Super Suspension Cleaner A4
So Easy Cleaner & Polish A7
Tiffany Pink Bowl Cleaner C2
Tiffany Snow Bowl Cleaner C2
Velvet Touch Laundry
  Detergent B2
636 odorless Oven & Grill
  Cleaner A1

Lee Soap Co.
Zing Concentrate Detergent A1

Leeder Chemicals, Inc.
Improved Amm. G-1 Gen.
  Industrial Clnr A4
110-A A1
118-A A1
120-B A1
123-A A1
123-B A1
124-C A3
124-L H2
126-B Q1
129-C A3
144-D A3
147-D A3
150-D A3
167-R A1
175-R A1
176-V E1
184-Y D1
196-Y D2
199-Y D1,Q3
202-A A2
206-Y A1
208-A A1
209-Y A1
210-Y A1
211-A A2
214-Y A1
215-B A1
216-Y A1
217-A A1
219-B A2
219-Y Q1
219-Y Special Q1
226-C Q2
227-C Q2
232-Y Q1
234-D A3
237-D A3
244-Y Q1
272-V N1
315-A A2
317-A A2
321-A A2
324-A A2
329-M H1

331-M Q1
334-M Q1
368-G HD Alkaline Cleaner A2
383-V A4
385-V A2,N2
386-V A2,N3
395-V G-I Cleaner A4
397-V General Industrial
  Cleaner A4
400-Y Q1
402-Y Q1
402-Y Special Q1
403-Y Q1
404-Y Q1
405-Y Q1
407-A Q1
408-A A2
411-A A2

Lehrman, A.J. & Sons
Ajay Insect Spray Odorless F1
Ajay Medi Creme E1
Ajay Mint Disinfectant C2
Ajay Pine Oil Disinfectant C1
Ajay Spearmint C2
Ajay Staph-Out C1
Ajay Surgi-Mint Liq. Hand
  Soap E1
Ajay Wide Open L1
Dyna Kleen A1
Nitrolene Concentrate A1
Pink Magic A1

Leidy Chemicals Corporation
Leidy LA-12 .... A1
Leidy LA-14 .... A1
Leidy LA-15 .... A1
Leidy LB-10 .... A2
Leidy LC-10 .... A3
Leidy LC-11 .... A3
Leidy PA-10 .... A1
Leidy PA-11 .... A1
Leidy PA-14 .... A1
Leidy PA-16 .... A1
Leidy PA-17 .... A1
Leidy PA-19 .... A1
Leidy PA-20 .... A1
Leidy PA-24 .... A1
Leidy PA-26 .... A1
Leidy PA-28 .... A1
Leidy PB-12 .... A2
Leidy PB-13 .... A2
Leidy PB-14 .... A2
Leidy PB-17 .... A2
Leidy PB-18 .... A2
Leidy PB-19 .... A2
Leidy PB-20 .... A2
Leidy PO-10 .... L1

**Leland Chemical Division/ American
Bio-Synthetics Corp.**
"Smoky Joe" Smoke House
  Cleaner A1
Acid Cleaner A3
Ambio-Zyme Liquid L2
Ambio-Zyme Powder L2
Heavy Duty Cleaner 610 .... A1
Hog Scald #4 Slippery Jim N2
Hog Scald #5 Slippery Jim N2
Hook & Trolley Cleaner A3
Hook & Trolley Cleaner #14 ...... A2
Hook & Trolley Cleaner Liquid A3
L-711 Liquid Cleaner A1
Le-Brite HD B1
Le-Brite 10 .... B1
Le-Brite 421 .... B1
Le-Scald Liquid N2,N5
Le-Scald 3030 .... N2,N5
Liquid Hand Soap #2 .... E1

Liquid Hand Soap #3 .... E1
Liquid Hand Soap 15% E1
Liquid Hog Scald Slippery Jim . N2
Poultry Scald N1
PH-8 .... N3
Red Ball #21 Aluminum
  Cleaner A1
Red Ball No. 11 .... D1
Red Ball No. 31 .... A3
S.F. #11 Slippery Jim Hog
  Scald #3 .... N2
S.F. No. 11 Tripe Cleaner N3
S.F. Number 11-C A2
S.F. Number 11-XX A1
S.F. Number 43 .... A4
Slippery Jim Hog Scald N2
Special S.F. Number 11 .... A1
Steam Cleaner A1
Super Foam No. 126 .... A1
Tripe Cleaner No. 1005 .... A1
Tripe Cleaner 10-10 .... N3
Tripe Cleaner 11-00 .... N3

Lemajeur, W. W. Co.
E-Z Duz It Action Heavy Duty
  Cleaner C1
E-Z Duz It Automatic
  Dishwasher Det. A1
E-Z Duz It Bandage Antisp.
  Lotionzd Soap E1
E-Z Duz It Big Blow Super
  Fog.Insct.Kill F2
E-Z Duz It Blast Off A6
E-Z Duz It Cement
  Rejuvenator Floor Cl. C1
E-Z Duz It Cement
  Rejuvenator HD Flr Cl A2
E-Z Duz It Controlled Suds
  Laundry Det. B1
E-Z Duz It Cycle Four
  Disinfectant-Deo. C1
E-Z Duz It D.O.A. Roach
  Liquid F2
E-Z Duz It Deep Fatfry
  Cleaner A2
E-Z Duz It Degreaser A8
E-Z Duz It Equalizer A1
E-Z Duz It Flash A1
E-Z Duz It Hot Shot L1
E-Z Duz It Industrial Cleaner ... C2
E-Z Duz It Lime Remover
  Low Foam A3
E-Z Duz It Pro-Tect No. 1 .... D2
E-Z Duz It Pro-Tect No. 2 .... D2
E-Z Duz It Super Crud
  Remover G7
E-Z Duz It Super Scrubber
  Floor&Wall Cl. C1
E-Z Duz It Super Suds Powder A1
E-Z Duz It Zapper Germicidal
  Clnr & Deo. C1

Lendow
Breaker 1-9 .... A1

Lenox Laboratories, Inc.
Mercury A4

Leon Supply Company
Leon's Concentrate A4
Royal All Purpose Powder A1
Royal Aluminum Brightner A3
Royal AA Fly Spray F1
Royal DA Plus Fly Spray F1
Royal DA-30 Cleaner A1
Royal DA-30 Steam Cleaner A2
Royal DA-40 Concrete Cleaner. A4
Royal DA-40 Steam Cleaner A2
Royal Liquid Steam Cleaner#1 . A2
Royal Liquid Steam Cleaner#2 . A2

Obsol Heavy Duty Cleaner A1
Obsol Heavy Duty Liq.
   Concrete Clnr. A4
Odorless Concrete Cleaner A1
Porclean.................................... C2
Powdered Oleon A1
Ratex.. F3
Reddy Ice Machine Cleaner A3
Run Roach F1
Sanitiz D1
Scale King................................ A3
Scalex Scale Remover................ A3
Soilene A4
Solvol...................................... A1
Speedy Scale Remover A3
Super Bainicide........................ F1
Waterene Gems G7
Waterene 500 .... G7
Xtra Solve K2

Levenson Chemical Company

O'Kay Coconut Oil Liquid
   Hand Soap.............................. E1
O'Kay Diazinon-Pyrethrum
   Emlsf. Conc............................ F2
O'Kay Glass Cleaner.................. C1
O'Kay Hospital Disinfectant
   Deodorizer C2
O'Kay Levensoap Concentrated E1
O'Kay Pine 7 Sweet
   Disinfectant C1
O'Kay Surgent........................... E1
O'Kay Toilet Bowl Cleaner C2

Lever Brothers Company

Cold Water "All" A1
Exact Commercial Dishwashing
   Powder A1
Lever Profi So-Ponic Liq.
   Laundry Supply...................... B1
Lever Professional Conc. Dry
   Chlor. Bich............................ B1
Lever Professional LS1 Alkali ... B1
Lever Professional LS1 Blue
   Detergent.............................. B1
Lever Professional LS1
   Detergent.............................. B1
Lever Professional LS1 Fabric
   Softener B2
Lever Professional LS1 Liquid
   Bleach B1
Lever Professional Powdered
   Fabric Softr............................ B2
Lever Professional Scouring
   Cleanser A6
Lever Profsnl All Purp. Hard
   Surface Cl.............................. C1
Liquid Detergent DW-300 ... A1
Neptune Blue Synthetic
   Detergent.............................. A1
Neptune Deodorant Soap C2
Neptune Heavy Duty Liquid A1
OS-4 Built Detergent B1
OS-5 Built Detergent B1
Starlight Soap C2
Sunbeam Deodorant Soap C2
Sunlight Soap............................ C2
Wisk.. A1

Lewis Research Laboratories, Inc.

Drain-Aid L2

Lewis, A. M. Company

Heavy Duty Non Butyl Cleaner
   Degreaser A1

LeMasters' Janitor Supply Co.

Aseptic-Kleen A4
Germa-Phene L1

161

Task Force Cleaner &
   Maintainer A4

Libby Laboratories, Incorporated

ACD A1
Flor-Dozer A4
Hand Eze.................................. E1
Klene-Tec................................. A4
LQD Lightning Quick
   Detergent.............................. A4
Spray-Away A4
Strip-Tec C1
T & C Conditioner A4
15% Hand Soap......................... E1

Libby, McNeill & Libby

Libby Blue General Purpose
   Detergent.............................. A1
Libby Green Liquid Detergent.. A1
Libby Red Heavy Duty
   Detergent.............................. A2
Libby's Yellow Acid Detergent. A3
Liquid Blue General Purpose
   Detergent.............................. A1

Liberty Brand Inc.

Formula IND 100....................... K1
Formula IND 101....................... K1
Formula IND 102....................... K1
Formula IND 103....................... K1
Formula IND 105....................... K1
Formula IND 110....................... A3
Formula IND-129....................... A1
Formula IND-156....................... A3
Formula IND-160....................... A1
Formula MUN-361..................... C1
Formula MUN-373..................... L2

Liberty Chemical Corp.

Auto Kleen................................ A1
Combat A4
Glow.. A4
Grease Off................................ K1
Grime Free............................... A4
Irish Breeze.............................. C1
Maxi-Lube F/G H1
Power Plus............................... A1
Rust Buster............................... C1
Sila-Shield G1
Steam Off A1

Liberty Chemical Products Co.

Alkatone L G6
Amiline G6
Aqua Treat............................... G6
Aquacoll BWT G6
Aquakey BWT........................... G6
Aquaphos-P.............................. G6
Aquapol BWT G6
Aquatex G6
Chemicide SBA G7
Chemsoap Liquid Soap............... E1
Chemtort G5
Degresit A4
Drainfree L1
Dual Treat................................ G6
Lustra Soap E1
Minitkleen A4
Ox Scav-D G6
Ox Scav-S................................ G6
Ox-Scav-H................................ G6
Prestokleen A1
Steamex A1

Liberty Chemicals Ltd., Inc.

"Blast" Multi-Pur . Steam
   Cleaning Conc.p...................... A1
"Max" Liquid Steam Cleaning
   Concentrate A1

"Monster" Ultra Hot Sewer
Solvent L1
A-145 Disinfectant Cleaner A4
A-151 Pink Concentrate A1
A-152 Syntheso Neutral
Cleaner A1
A-161 Sof-T-Suds A1
A-162 Beaded Hand Dishwash
Compd. A1
A-230 Super Mach. Dishwash
Compd. A1
A-231 Prize Mach. Dishwash
Comp. A1
A-235 Deluxe Mach. Dishwash
Comp. A1
A-280 Egg Wash A1
Alpha 5000 .... A4,A8
B-165 Liquid Dishwash
Compound A2
B-194 White Stm. Pres. Cl.
Odrls. A2
B-195 Liq. Hvy Dty Stm Prssr
Clnr A2
B-196 Steam Pressure Cl. Med.
Duty A2
B-197 Extra HD St. Pressure
Cl. A2
Big "C" Concentrate A4,A8
D-155 Odorless Concrete
Cleaner A4
D-160 ADR Sudsing Cleaner A4
F-186 Chlorinated Krystals A1
F-187 Super Germicide D1
F-188 Germ I San. D2
G-100 Cocoanut Oil Hand Soap E1
G-105 Econo Hand Soap E1
G-110 Germicidal Hand Soap E1
Heavy Duty Form Liq. Stm
Cleaning Conc. A1
Inhibited Chemical Cl. &
Oxidation Rmvr. A3
L-120 Scal-Away Inhibited
Bowl Cl. C2
L-125 Blue Bowl Cleaner C2
L-130 Bowl CLeaner (Emulsion
Type) C2
L-135 Porcelain Cleaner C2
Liberty APC-27 .... A1
Liberty BD-85 .... C1
Liberty C-34 .... A4
Liberty C1P-150 .... A1
Liberty GC-42 .... A1
Liberty GP-280 .... A1
Liberty H.D. Alkalene A1
Liberty Lubricant TL
Concentrate H1
Liberty LD-515 .... A1
Liberty MC-81 .... A3
Liberty MPD-550 .... A1
Liberty S-311 .... A2
Liberty S-45 .... A1
Liberty Scalex A3
Liberty Superus A1
Liberty SH-26 .... A2
Liberty SHC-21 .... A2
Liberty SHC-32 .... A2
Liberty TC-38 .... A2
Liberty X-10 .... A1
M-157 H. D. Magic Concrete
Cl. C1
M-157 Magic Concrete Cleaner . C1
M-159 Econo Concrete Cleaner . C1
M-181 Pine Odor Disinfect.
Coef. 5 .... C1
N-211 Insecticide No. 111 .... F1
N-212 Space Spray No. 3 .... F1
O-215 ABC 50% Malathion
Emlsfbl.Conc. F2
Organic Acid Detergent A3
Po-PacHo Cleaner A4

Red Hot Sewer Solvent L1
Task Master E1

Liberty Labs
Super Bruce A1

Liberty Paper & Bag Company
Silicone Spray & Lubricant
#145 .... H1

Lico Chemicals, Inc.
Bruté A4
Catch 22 .... F1
Coolrite P1
Dielectric K1
Draintain L2
E-Lime-Inate A3
Excelerate L2
Expose A4
Lico Bacteriostatic Hand
Lotion Soap E1
Lico Blue A4
Lico Formulation 66-25 : D1
Lico Lemon Cleaner A4
Lico Liquid Hand Soap #1 .... E1
Lico Liquid Hand Soap #2 .... E1
Lico Liquid Hand Soap #3 .... E1
Lico Renew Conc. Emuls.
Bowl & Urnl. Cl C2
Lico Sanitizer-Cleaner 2150 ..... D2
Lico Stripper-Degreaser A4
Lico Super-X Stripper-
Degreaser A4
Lico 30 .... A4
Lico-Det A4
Lico-Suds A4
Liquithaw C1
Manicure E4
Noxout A3
Oven & Grill Cleaner A8
Perfloormance Plus A4
Pierce A4
Pink Lico-Suds A1
Prelude A4
Prevent C1
Pure K2
Recon P1
Revive A4,C1
Sapon A1
Slide'N Glide H1
Slugger B1
Solute K1
Stainless Brite A7
Torque P1

Lien Chemical Company
M828 Deodorant Concentrate C1
M8546 Solvent Action Cleaner .. C2
M917 Rodeth F3
S8515 Detergent Powder A6
140 Clienall A1
147 Clienware A1
152 Toilet Bowl Cleaner C2
154 Concentrated Nu Clien C1
159 Drain Pipe Opener L1
163 Washbrite A1
174 Grand Toilet Bowl Cleaner . C2
177 Grand A1
224 Scum Remover A1
227 Waterless Hand Cleaner E4
243 Shop & Garage Shock
Treat. C1
254 Clien Chloro D2
275 Waterless Hand Cleaner E4
305 Clien-Off A4
324 Clien-A-Matic A4
362 Royalien A4
403 Malathion Emulsion 50% F2
404 Fly Bait F2
410 Insect Spray F1

413 Vapocide F1
5001 Grimebuster Tile&Ceramic
Cl. A4
5002 Grimebuster Glass Clnr.
Conc. C1
540 Supreme-10. .... E1
592 Waste Treatment C1
806 Supreme 20 Liquid E1
816 N.P. Hand Soap E4
822 Sanitizer C2
8293 Deodorant Liquid C2
853 Magiclein A4

Life Soap & Chemical Co., Inc.
Acid B A3
Acid CA A3
Acid S. A3
Decharacterized Sodium Nitrite G5
Klean-A-Bric A3
Life Drain Cleaner L1
Life Super Foam. A1
LS #30 .... A1
LS #60 .... A1
LS #7 .... A1
New Acid B A3
Packers Detergent No. 25 .... A1
Sink A3
Soak Tank Cleaner A2
W.I.T. Steam Cleaner A1

Lightfoot Company, The
API A4
Laundri-Mate B1
LMI B1
15% Liquid Soap E1

Lighthouse Manufacturing Co., Inc.
A/C Coil Life I P1
Belt-It. P1
Big L C1
Big Red L1
Blast Off A4
Boiler Maker G7
Concentration A4
Flash Concrete-L. A2
Flash Concrete-P C1
Growl C2
Handee-Andee Waterless Hand
Cleaner E4
K.K. On Duty. A3
Kiss-Off F1
Kiss-Off Insecticide (Aerosol) ... F1
Kitchen Maid A1
Libby Lube H2
Lighthouse 100 .... G6
Lighthouse 1500 Concentrate ... G6
Lighthouse 200 .... G6
Lighthouse 300 .... G7
Lighthouse 400 .... G7
Lighthouse 924 .... G7
Lightox A4
Linda Gloveless Invis. Skin
Shield Crm. E4
Mi-T-Jac A4
Pine-O-Mine C1
Plug Away-I L2
Plug Away-M L2
Power Play D1
Scent-D-Garb. C1
Silcon Silicone Spray H1
Soft Touch A1
Super-L-Solvo (Aerosol) K2
Super-L-Solvo (Bulk). K1
Susie Stainless Metal Polish A7
Susie Stainless Cleaner And
Protector. C1
Terminal Man P1
V-Day A1
Vince Bombari F1
Zymatic L2

162

Bunny Bleach.................................B1
Formula "LSC" Liquid
   Concentrate..............................A1
Formula 2389.............................A1
Formula 711...............................A1
General Cleaner 1......................A1
General Cleaner 2......................A1
General Cleaner 3......................A1
Heavy Duty Liquid
   Concentrate..............................A2
Imperial General Cleaner...........A1
Keo-Break Detergent..................B1
Super Tetramex..........................B1

Livingston Laboratories
Heavy Duty Cleaner Degreaser.A1
Super Crest-O-San......................C2

Lockert Enterprises
Bulldog Commercial Clean Up
   Det...A1
Bulldog Floor Detergent............A4

Loctite Corp.
Locquic Primer T.......................P1
Loctite 242.................................P1
Loctite 271.................................P1
Pipe Sealant w/Teflon Catalog
   #92..P1
Retaining Compound 35.............P1
Retaining Compound 75.............P1

Lone Star Brush & Chemical Co.
APS Multi-Purpose Cleaner.......A1
Big Gun II...................................A4
Clear Day....................................C1
Cococon......................................E1
CTC-50..K1
Det-Chlor....................................A1
Heavy-D Steam-All....................A4
Hy-22..C2
Hydene 202.................................A4
In-Solv..K1
Lo-Foam......................................A4
Low Suds Laundry Compound.B1
LS Descale...................................A3
LS Descale-A...............................A3
LS Stripper.................................A2
LS-Gel Strip................................C3
LSC Delimer...............................A3
Pine Oil Disinfectant.................C1
Power Pine..................................C1
Real Kleen..............................C1,B1
Shur-Kleen..................................C1
Shur-Kleen II..............................A4
Shur-Solv....................................K1
Star-Lite Conc. Multi-Purpose
   Detergent..............................A4,A8
Steam-All....................................A4
Strike-Bowl Cleaner...................C2
TAS..A1
20% Coco.....................................E1

Lone Star Paper Co., Inc.
Concentrate.................................A4
Hand Soap Ready to Use............E1
Heavy Duty Cleaner Degreaser.A4
Kleer-Vu Glass Cleaner..............C1

Lone Star Supply Company
Dis-O-Pine..................................C1
Lemon-Brite Disinfectant...........C2
LS-100 Porcelain & Bowl
   Cleaner.....................................C2
LS-33...A1
Ready-To-Use Hand Soap...........E1
Star-Brite....................................C2
Star-Glo.......................................C2
Super D-C Degreaser Cleaner...A1

Long Chemical Co.
Mark 111.....................................A1
Quick-Strip Heavy Duty Floor
   Cl & Strip...............................A1

Long Chemical, Incorporated
#51 A. P. Cleaner.......................A1
Biodegradable Condenser
   Treatment #6...........................G7
Concrete Floor Cleaner..............A1
Long Chemical Formula WT
   66...G6
Long Gone Sulfuric Acid Drain
   Opener.....................................L1
No Foam......................................A1
No. 59 Heavy Duty Cleaner......A4,E1,A8
Product #170 Electrical Safety
   Solvent................................K1,K2
Product #171 Degreasing
   Solvent................................K1,K2
Product #172 Printers Helper...K1,K2
Spray Clean 81............................A1

Long's Preferred Products
Big Dog Concrete Cleaner.........A4
Break Clean.................................A4
Chlorinated Egg Cleaner...........Q1
Deluxe Machine Dishwashing
   Compound...............................A1
Dynamo Detergent......................A1
Extra Heavy Duty Concrete
   Cleaner.....................................A4
Golden Gator Concrete Cleaner C1
Golden Magic Cleaner................A4
Grime-Scat..................................A1
Liquid Hand Soap 15.................E1
Liquid Hand Soap 20.................E1
Liquid Hand Soap 40.................E1
Liquid Steam Cleaner................A2
Multi Blue...................................A4
Pink Cleaner...............................A4
Pink Satin...................................E1
Poultry House Cleaner Formula
   II..A1
Sewer Solvent.............................L1
Steam Cleaning Compound........A2
Wax Scat......................................A1
X-150 Chlorinated Machine
   Dishwash.................................A1
360 Milkstone Remover..............A3

Lorenz Chemical Company
Hy-Klor C...............................D2,Q4
Pyrenone Insect Spray #6...........F1
R-K Rat and Mouse Bait............F3
20% Lindane Emulsion
   Concentrate..............................F2
25% DDVP Emulsion
   Concentrate..............................F2
65% Chlordane Emulsion
   Concentrate..............................F2

Lori Chemical
Drain Cleaner Enzyme Type......L2
Duocide.......................................D1
LC-215 Dizer...............................C1
Power-Plus...................................A1

Los Angeles Chemical Co.
Lacco 8015 Cleaner.....................A1

Lou Fox Ind. Specialties Co.
Dri-Sil Silicone Spray.................H1

Louis Rich, Inc.
Rich's Chlorinated Equipment
   Cleanser...................................A1
Rich's Multi-Purpose
   Chlorinated Clnr.....................A1

Rich's Smoke House Cleanser.... A2

Louisville Chemical Co., Inc.
L-C Bowl Cleaner C2
LC 201 .... A1

Louisville Chemical Company
L C Solvent Cleaner &
  Degreaser K1
LC-601 Multi-Purpose Cleaner .. D1

Lovinger Company
Rescue Prilled Drain Opener L1

Lowe's Incorporated
Dri-Spot.. J1
Kwik-Dry J1
Nabsoil #1.................................... J1
Safety Absorbent J1

Lub-O-Line Industrial Oil Co.
Luboco F.M. Grease AA H1

Lube Systems, Inc.
AA Food Machinery-Number
  8002 .... H1

LubeCon Systems, Inc.
Series AA Grease H1
Series AA Oil.............................. H1

Lubri-Chem Products, Inc.
Lubri-Chem Safety Solvent K1
LC-328 Sfgrd No.2 EP FMG
  Anti-Wear................................. H1
Silicone Spray.............................. H1

Lubrical, Inc.
Laminall BB-222 Bearing
  Lubricant H2
Laminall Gear Oil SAE 140....... H2
Laminall Gear Oil SAE 90......... H2
111 USP Food Machinery
  Lubricant H1
112 White Food Machinery
  Lubricant H2
44-A Colson Caster Lube H2

Lubricating Specialties Co.
Gld Mdl All Wthr Gr Oil
  162P80/90/140 .... H2
Gld Mdl Ice Mach Oil Wax
  Free 4025 Hvy H2
Gld Mdl Ice Mach Oil Wax
  Free 4025 Lt........................... H2
Gld Mdl Ice Mach Oil Wax
  Free 4025 Med H2
Gld Mdl Moly Bentone Grs
  A.P. Grd 1............................... H2
Gld Mdl Moly Bentone Grs
  A.P. Grd 2............................... H2
Gld Mdl Moly Bentone Grs
  A.P. Grd 3............................... H2
Gld Mdl Prm/HD Mtr Oil
  SAE20/30/40.......................... H2
Gld Mdl White Litho Grs.
  Grade 1.................................... H2
Gld Mdl White Litho Grs.
  Grade 2.................................... H2
Gld Mdl White Litho Grs.
  Grade 3.................................... H2
Gold Medal All Prp. Lith
  Grease Grd 1........................... H2
Gold Medal All Prp. Lith
  Grease Grd 2........................... H2
Gold Medal All Prp. Lith
  Grease Grd 3........................... H2
Gold Medal Compressor Oil
  2932.. H2

Gold Medal OCT 500
  Compressor Oil H2
Gold Medal SR-1 Gear Oil H2
Gold Medal Wheel Bearing
  Grease S H2
Gold Medal 100.......................... H1

Lubrication Engineers,Inc.
Monolec Air Compresser Oil H2
Monolec Air Tool Lubricant
  Light...................................... H2
Nutrigent A1
1107 Dexron 11 D...................... H2
1155 Ford Type F
  Auto.Transmission Fluid......... H2
1225 Bearing & Chassis Lub.
  (Heavy)................................... H2
1225 Bearing & Chassis Lub.
  (Light)..................................... H2
1225 Bearing & Chassis Lub.
  (Medium)................................. H2
1250 Almasol High
  Temperature Lubricant H2
1275 Almaplex H2
175 White Machine Lubricant.... H1
2059 Monolex............................. H2
300 Monolec Industrial
  Lubricant H2
3751 Almagard Vari-Purpose
  Lubricant H2
3752 Almagard Vari-Purpose
  Lubricant H2
401 Almasol Pure Mineral Gear
  Lubricant H2
4024 (AA) Food Machinery
  Lubricant H1
4025 AA Food Machnry
  Lubricant H1
4701 Monolec Industrial
  Lubricant H2
600 Almasol Trans-Worm Gear
  Lubricant H2
601 (Almasol) (SAE 90)............ H2
602 (Almasol) (SAE 140).......... H2
606 (Almasol) (SAE 80)............ H2
607 (Almasol) (SAE 90)............ H2
608 (Almasol) (SAE 140).......... H2
609 (Almasol) (SAE 250).......... H2
6105A Monolec Hydraulic Oil .. H2
6110A Monolec Hydraulic Oil .. H2
6120 (SAE 20) Monolec
  Hydraulic Oil H2
6120A Monolec Hydraulic Oil... H2
6130A Monolec Hydraulic Oil... H2
6155 Monolec Fire-Resist. Hyd.
  Fluid....................................... H2
6202 Monolec Air Compressor
  Oil H2
6203 (SAE 30) Monolec Air
  Compresser Oil H2
6204 Monolec Air Compressor
  Oil H2
6222 Monolec Air Compressor
  Oil H2
6303 Monolec Air Tool
  Lubricant H2
6305 Monolec Air Tool
  Lubricant H2
6401 Monolec Turbine Oil......... H2
6402 Monoice Turbine Oil......... H2
6403 Monolec Turbine Oil......... H2
6404 Monolec Turbine Oil......... H2
6405 Monolec Turbine Oil......... H2
6455 Monolec Fire Resistant
  Hydra. Fluid........................... H2
6721 Monolec Refrigeration Oil. H2
6722 Monolec Refrigeration Oil. H2
6723 Monolec Refrigeration Oil. H2
701 Monolec Marine Gear Oil .. H2
9102 Monospray H2

9901 Extemp Lubricant.............. H2
9952 Synolec Rot. Air Comprs
  & Hydrl Oil H2
9963 Monolec Syntemp
  Lubricant................................. H2
9974 Synolec Reciprocating Air
  Comp. Oil H2

Ludale, Inc.
Atomic Liquid Steam Cleaner.... A2
Big "Q" A4

Lundmark Wax Company
All Surface Cleaner A1
H D B.................................... A4,A8
L.D.C. C1
Lundmark's Sanitizer D2
M.R.C. A1
Muscle Man.................................. A4
Suds-Less..................................... A4

Lundstrom Industrial Supply Company
Grease-Break A1

Luseaux Laboratories, Inc.
Delite A1
Floor Clean C1
Floor Clean No. 2 .... A4
Sur-Nuf....................................... A1

LuBar Company
Boiler & Steam Line Treatment
  FG G6
Boiler Compound FG G6
Bowl Cleaner C2
Chlortrol..................................... F2
Concentrated Dish Washing
  Detergent................................. A1
Defoamer...................................... A1
Defoamer Concentrate A1
Degreaser Concentrate K1
Detergent Concentrate A1
Digesto L2
Drain Pipe Cleaner & Opener ... L1
Foaming Ox Type Liquid.......... A3
Fuel Additive.............................. P1
Gen. Purpose Non-Sudsing
  Cleaner A1
General Purpose Sudsing
  Cleaner A1
Glycolene..................................... C2
Heavy Duty Floor Cleaner A4
Hi-Co A4
Ice Melt C1
Industrial Deodorant C1
Interior, Exterior Cleaner A1
Jell Degreaser Concentrate K1
Krystal D1
L-100... A1
L-101..................................... A4,A8
Liquid Conveyor Lubricant H2
Liquid Hand Soap FG E1
Lucide Mint C1
Non Flammable Safety Solvent . K1
Odor Control Lemon C1
Odor Control Lilac C1
Odor Control Mint C1
Oven Cleaner A8
Pine Lucide C1
Pine Oil Disinfectant C1
Porcelain Cleaner Concentrate .. A3
Pyrokill....................................... F1
Red Prilled Drain Opener L1
Safety Solvent K2
Scale Remover A3
Sewer Opener L1
Special Compound A3
Special Industrial Deodorant..... C1
Special Soap Powder B1

Stainless Steel............................ A7
Suracid 11 .... D2
Syntho Suds B1
Tops.. A1
Water Soluble Rat&Mouse
 Killer.................................... F3

M

M & B Products
Mop-N'-Bucket Blue Stuff.......... A1
Mop-N'-Bucket Complete
 Laundry Detergent.................. B1
Mop-N'-Bucket Denver Mint
 Disinfectant C2
Mop-N'-Bucket Glub
 Concntrtd. Cleaner A4
Mop-N'-Bucket Lemon Combat
 Disinf. C2
Mop-N'-Bucket Lotion Hand
 Soap E1
Mop-N'-Bucket Luxury Hand
 Soap E1
Mop-N'-Bucket Pine Oil
 Disinfectant C1
Mop-N'-Bucket Porcelain &
 Bowl Clnr............................. C2
Mop-N'-Bucket Product #250
 Disinf Cl............................... C2
Mop-N'-Bucket Product #512
 Dshwshng Det A1
Mop-N'-Bucket Pure-Powr A4
Mop-N'-Bucket Sewercide
 Sewer Solvent........................ C1
Mop-N'-Bucket Spectrum
 Cleaner-Disinf. C2
Mop-N'-Bucket Winner Clnr.
 Glass Clnr............................. C1

M & M Distributing Company
Detro A4

M & M Manufacturing Co., Inc.
M & M Car Wash Soap.............. A1

M & M Supply Company
Proclean................................... A4

M & N Chemicals Inc.
Foam-Away A1
Glisten C2

M & S Chemicals, Inc.
Big "G" A4
Chemtronic............................... A4
Descaler & Delimer A3
Descaler & Delimer................... C2
Discrete Safe Sure Cleaner......... A3
Iodo-175 Concentrated Iodine.... D2
No. 500 Tripe Cleaner N3

M A Chemical Laboratories,
 Incorporated
Degrease-R................................ A1
Degrease-R Sanitizer.................. D1
MA-104.................................... A1
Re-Move................................... A1

M B M and Supply Company
All-Clean A4
Aquasolve................................. A4
Liquid Hand Soap E1
Tuff-Strip................................. A4

M D Laboratories Inc.
Di-Gestit.................................. L1

M F A Oil Company
Food Machinery Grease............. H1
Penn Guard............................... H2
Rat and Mouse Killer................. F3
Sho-Me H2

M M Industries, Inc.
No. 350 HiFo-Cling Cleaner A1
No. 350 Tri-Chem Cleaner A1

M Oil International
M-105-FM H1

M.A.B. Paints, Inc.
Power Wipe A1
Wipe.. A1

Macco Adhesives/ SCM Corporation
Macco Liquid Seal Tub & Tile
 Caulk.................................... P1

Machemco, Inc.
Adjunct 100.............................. G6
Adjunct 101.............................. G6
Adjunct 200.............................. G6
ASC A1
Best-I A2
Blue Klean................................ A4
Bottle Wash.............................. A2
Chloro-Clean H.F...................... A4
Chloro-Suds A1
Clean Up 1-1 .... A2
Conquer A1
Conquer + Pine C1
Descale S.................................. P1
Descale 118 .... A3
Dishwash-No Foam A1
Flush Kleen.............................. L1
Foamore 11 .... A1
Foamy...................................... A4
Fresh-N-Clean........................... A1,B1
G.P.C.-CL................................ A2
G.P.C.-11................................. A1
Great-I..................................... A4
Grill-O-Brite............................ A8
GPC-III A1
GPC-1...................................... A1
Jet Spray A1
L.F. Spray Wash A1
Lemon Grove 10 .... C1
Macfactant BX........................... A4
Macfactant C.P.C. A3
Macfactant D.C. A1
Macfactant FLC......................... A1
Macfactant FPC A3
Macfactant S.F. A1
Macfactant S.P.W....................... A4
Macfactant S.S.C. A4
Macfactant W.W. A1
Macfactant 21 .... A2
Phos Clean LCTT....................... A1
Phostreet 1000L.......................... G6
Phostreet 807 .... G6
Pine Oil Cleaner C1
Poly-Treet 500 .... G6
Poly-Treet 5000 .... G6
Polytreet 180............................. G7
Polytreet 3000............................ G6
Polytreet 800............................. G7
Pressur-Clean A1
Quip Clean A2
Quip Clean 11 .... A2
Rapiclean A4
So-Clean A1
Sol-va-Sol A1
Solva-Suds................................ A1
Spank-N-Clean........................... A4,A8
Sparkle Brite A1
Sparkle Clean 303...................... A1

Sparkle Clean 404 .... A1
Sparkle Clean 606 .... A2
Sparkle Clean 707 .... A1
Sparkle Clean 909 .... A2
Spin-Off A2
Steamzit Off A4
Super Clean 1001 .... A4
Tough III A1
Tough IV A2
Tough IX A2
Tough V A2
Tough VI A2
Tough VII A2
Tough VIII A1
Tough-I A1
Tough-X A1
True Grit A6

Mackay Industries, Inc.
Multex A4

Mackwin Company, The
Ratorex F3
Ro-Dent F3

Macmillan Ring-Free Oil Co., Inc.
Supergard HiPerfor No.151
   Multi-Purp Grs H2

Macon Industrial Chemicals Corp.
All-Purpose Heavy Duty
   Cleaner A4

Madison Chemical Co., Inc.
Acid Clean A A3
Acid No. 3 .... A3
Acid No. 3 + ... A3
Acid No. 3 A A3
Additive A1
Aero-Foam A2
Aerofoam CL A1
Aerofoam D A1
Aerofoam Special A1
Alkaline Deruster A2
Antifoam B A1
Beads A1
C.I.P. A3
C.I.P. 181 .... A1,Q1
Can Wash T-142 .... A1
Chlor-Clean D2
Chlorinated CIP Cleaner A2
Chlorinated Egg Cleaner Q1
Chloro-Floor A1
Compound A-97 .... A1
Compound AD-8 .... A2
Compound AF-A A1
Compound AF-B A1
Compound AF-C A1
Compound AF-H A1
Compound AT-4 .... A4
Compound BC-14 .... A1
Compound CL-2 .... A1
Compound DC-14 .... A1
Compound DC-144 .... A1
Compound HS-30 .... A1
Compound HT 50 .... A2
Compound IDC-30 .... D2
Compound IDC-32 .... D2
Compound K-260 .... A4,H2
Compound L-14 .... A1
Compound L-144 .... A1
Compound L-144-NP A1
Compound MSP-10 .... A1
Compound NBR-2 .... A1
Compound PS-1-NP A1
Compound R-98X A1
Compound 14-XC A1
Compound 14-XP A1
Compound 14X A1

Compound 175-D A1
Compound 175-D-NP A1
Compound 175-F A1
Compound 175-F-NP A1
Compound 175-FJ A1
Compound 175-FJ-NP A1
Compound 175-FK A1
Compound 175-FK-NP A1
Compound 175-FW A1
Compound 175-FW-NP A1
Compound 175-X A1
Compound 175FWF A1
Compound 50-M A1
Compound 50ME A1
Conveyor Lube H2
Crystal Base E1
Dairy Cleaner No. 14 .... A1
Dairy Cleaner No. 14-NP A1
Dart 150 .... A3
Dart 151 .... A3
Dart 172 .... A3
Dart 254 .... H2
Dart 260 .... A1 .
Dart 286 .... A1,Q1
Dart 351 .... A1
Dart 701 .... A1
Dart 751 .... A1
Dart 800 R A2
Dart 822 .... A1
Dart 824 .... A2
Dart 825 .... A2
Dart 835 .... A2
Dart 875 .... A1
Decharacterized Sodium Nitrite ... G5
Egg Cleaner DF Q1
Fast Clean A1
Galvanite A3
General Purpose Heavy Duty A2
General Purpose Heavy Duty
   NP A2
General Purpose N A1
General Purpose N-NP A1
General Purpose 5 .... A1
Germicidal Hand Soap E1
Help A1
High AC A1
High AC-NP A1
Hog Scald N2
Hog Scald Powder N2
Hog Scald SJ N2
HT-25 .... A2
In Place A A2
In Place A-NP A2
Jel Terge A1
Lanosoft C2
Liquid AC A1
Liquid Bottle Wash A2
Liquid Caustic Soda A2
Liquid Hand Soap E1
Madisan 16 .... D1
Madisan 2080 .... D2
Madisan 32 .... D1
Madisan 90 .... D1
Madison Concentrate A4
Marvella A4
Marvella M A1
Marvella P A4
Oil Dip H1
Poultry Scald N1
Powdered Hand Cleaner C2
Powdered Hand Cleaner B C2
Power Scrub Concentrate A4
Residual Insecticide F2
Sewer Cleaner L1
Shroud Special A1,B1
Smoke House B A2
Smoke House B NP A2
Smoke House SS A2
Smoke House SS NP A2
Smoke House 2 .... A2

Smoke House 2C A2
Smoke House 3 .... A2
Smoke House 66 .... A2
Smoke House 99X A2
Spray Insecticide F1
Stainless Cleaner No. 3 .... A2
Steam Cleaner SN A2
Stripper A2
Stripper 07 .... A2
Stripper 20-N A2
Super Floor Cleaner 5 .... A2
Tech Oil H1
Triox Supreme A1
Tripe Bleach N3
Tripe Cleaner N3
Tripe Cleaner CD N3
Tripe Cleaner E N3
Vegetable Cleaner A1
Wagon Wash A1
1-2-3 .... A1
1-2-3-50 .... A1
10% Hyamine Sanitizing Agent .. D1

Madison Sales Company
Acid Cleaner A3
Floor Cleaner A4
General Purpose Cleaner A1
Iodine Sanitizer D2
Liquid Egg Wash Q1
PIP-Clor Egg Wash Q1

Madison-Bionics
pHydrate G6
Acrofon A1
Actinate G5
Actionize H1
Aerogen C1
Aerotape H2
Al-Right A3
Antidine D2,Q6
Antural L1
Atronal A2
Attract C1
Barz P1
Baxate A2
Bichlornate Q1
Bio-Care L1
Bio-Dyne D1
Bowl Bar C2
Carbotrol A1
Chem Rod L1
Chem 2 0 .... A6
Chem-Buds A6
Chem-Ox A3
Chem-Steam A1
Chemplex A4
Chemtrust K1
Chemtrust (Aerosol) K2
Chemtrust Alternative Aerosol .. K1
Chlorganic D2,Q4
Clophane C1
Cocinex Q1
Cold Steam K1
Cold-Steam K1
Combinoid F1
Command H2
Compatex(Aerosol) H1
Compatex(Bulk) H1
Contronate C1
Conversol A1
Correctol A1
Cyclophene C1
Darvex A1
Davenex A8
Davinate N4
De-Min A3
Declomate A2
Delate N1
Delate LF N1
Deo-Drex C1

Polarisin Bulk.............................. H2
Prephase A1
Prim E1
Process #11................................. G7
Process #12................................. G6
Procid A3
Proloid.. H2
Prophylex H2
Puritize D1
Quatrol...................................... D2,Q3
Radene C2
Reclamate.................................... A3
Recover A2
Respond A1
Sanacide Cleaner Concentrate ... A3
Sanatrox...................................... A3
Scoop.. C1
See Glass Cleaner C1
Sentry Mint C1
Shieldex C1
Shieldex (Alternative Aerosol)... C1
Shine-Again................................. A7
Silfax Alternative Aerosol H1
Silifax... H1
Silifax (Bulk).............................. H1
Solar Steam A1
Soothe... E1
Spectrasol.................................... A4
Steam Roller A1
Steel-One A7
Steel-One Alternative Aerosol ... A7
Sterilather................................... E1
Submers-able H2
Sundance C1
Surprocin.................................... A4
Symbex C1
Systeam....................................... A1
SFS Triangle................................ C2
Tarp C1
Techmate..................................... K2
Technate (Aerosol)...................... K1
Technate (Bulk).......................... K1
Thermatic.................................... L1
Thermelt..................................... C1
Thermolyte.................................. A2
Thrax L1
Touch-Up Alternative Aerosol .. C1
Tower Trol G7
Towerful...................................... G7
Toxcide....................................... F2
Toxcide(Aerosol)......................... F2
Tribite... N3
Trival N3
Tylon H2
Up-Tite P1
Vivax C1
Wite-Way.................................... H2

Madison, J.R. Maintenance Supplies

JRM Brite Blue Toilet Bowl
  Cleaner..................................... C2
JRM Brite White Conc. Toilet
  Bowl Clnr................................. C2

Mador Company, Inc., The

Bugum F1
C & D Cleaner............................. A3
Germicidal Cleaner C2
Greasy H2
Green Kleen C1
Green Kleen Toilet Bowl
  Cleaner..................................... C2
HD Drain Opener L1
Klensit! Cream Cleanser.............. A6
Liquid Hand Soap E1
Lotionized Hand Soap E1
Mador Concentrate Formula
  321... A4
Mador-Kleen Heavy Duty ... A1
Mador-Kleen Heavy Duty A1

Madorcide F1
Madorect D1
No Fume Drain Opener............... L1
Toilet Bowl Cleaner.................... C2
Ultra Clean................................. A1
Vandal Mark Remover C1
Waterless Hand Cleaner.............. E4
666 Odorless Disnf. San. Slmcd
  Algcd D2

Magee Chemical Company

Adios... A4
All Clean..................................... A4
Buster.. A4,A8
Condition A4
Life Guard Non-Acid Porcelain
  & Bowl Cln C2,A8
Mr. Power................................... A4
MC-27 NP................................... D2
No Suds....................................... A1
Strip Clean A1
Suspend A4
Take Over A4
Tear Drops................................... A4
103135 Cleaner............................ A4

Magic Chemical Co.

Magic Chem's Shur-Solv A4,A8

Magic Chemical Company

Magic Foam A1

Magicolor Paint Company

F-181-01 Neutral Cleaner E1
F-181-02 All Purpose Cleaner A4

Magnamark Corporation

Magnaclean-MP........................... A1

Magnolia Chemical Co., Inc.

Hari-Kari Type II......................... F1

Magnolia Products Co.

All Purpose Cleaner..................... A4
China Clean................................. A1
Liquid Hand Soap E1
Lotionized Hand Soap E1
Presurg E1
Sani Pak...................................... D1
Super Concentrate....................... A1

Magnum Research Corp.

MAG-44PF A1
Neutri-Kleen A1

Magnus Chemical Division, Economics Laboratory, Inc.

All Purpose Steam Cleaner A1
Car Wash #1................................ A1
Colloids No. 3.............................. G6
D-Scale-R-100.............................. A3
Hand Cleaner C2
Hot Strip #2 .... A2
IBTC-6L....................................... A2
IBTC-6P....................................... A2
IBTC-9L....................................... A2
IBTC-9P....................................... A2
Kling Grease H2
Kling Oil Regular........................ H2
Magkleen No. 2 .... K2
Magnatrol No. 2 .... G6
Magnatrol 100.............................. G7
Magnatrol 50................................ G7
Magnox Liquid G6
Magnu-Clenz 620......................... A2
Magnu-Clenz 630......................... A2
Magnu-Clenz 640......................... A2
Magnu-Clenz 650......................... A2
Magnu-Spray 205......................... A2

Magnufos G6
Neutra Wash A1
NZL.. A1
Pine Oil Cleaner C1
Plus Strip-1............................ A1
Spray All Liquid A2
Spray-All Powder A2
Waterless Hand Cleaner........... E4
1005-B A4
114.. A1
1156.. A1
1204.. A1
143-X3.................................... A1
147-X A1
155-X A1
215-D A2
23-X A2
55-P C1
61-XX A2
645-DX A2
800-X A3
92-XX A1
921-X3.................................... A2
94-XX A1

Main/Tex, Inc.
GPC-24.................................... A4
MT-120 .... A4
MT-235.................................... A4

Mainpro, Inc.
Kleen-Quik E4
Micro-Zyme L2

Maintenance Chemical Suppliers, Inc.
MCS Pressure Wash #1 .... A1
MCS Super 30 .... D1
214-H.D. Degreaser-
Concentrate............................. A4,A8

Maintenance Engineering Co.
Slip-Not J1

Maintenance Engineering Corp.
Boiler Water Dispersant 132 ...... G6
Mecolene G6
Monofilm Liquid G7
MECO Microbiocide C-11 .... G7
MECO Nix-Ox G6
MECO Sodium
Hexametaphosphate................. G6
MECO Treat H-Amine............... G6
MECO Treat NA-1..................... G6
MECO Treat 125....................... G6
MECO Treat 1406...................... G7
MECO Treat 249....................... G6
MECO Treat 498....................... G6

Maintenance Materials Co.
Macolite Boiler Water
Treatment #3016 .... G6
MBT #251 .... G6

Maintenance Mates
Toxene.................................... F2
2 M Disinfectant Cleaner........... A4
2 M Multi-Purpose Space &
Contact Spray F1

Maintenance Products Inc.
Kwik Foam Cleaner A1
MP-5000 .... A4

Maintenance Research Laboratory
Deck Detergent C1
Jet Pressure Wash Detergent
(M.R. 337)............................... A1
M.R. 200.................................. A4
M.R. 44 Delimer....................... A3

M.R. 489.................................. A4
M.R. 544 Cream Hand Soap....... E1
M.R. 747.................................. A4,A8
M.R. 88................................... G6
Ol Smokey A2
Task A2
Viking Tru-Quat....................... D2
Viking 1000 Conveyor Lube H2
Viking 3000 Conveyor Lube H2
Viking 4000 Conveyor Lube H2

Maintenance Supply Co., Inc.
Glo Clean C1
Grime Getter A1
Improved Grime Getter............. A4
Improved Nelcophene D1
Liquid Ded Eye........................ F1
Nelco Pine Odor Disinf.Deod.
Clnsr. C1
New Look C2
Northwoods Pine Odor
Disinf.Deod. Clnsr...... .... C1
Pine Odor Disinf Deod.
Cleanser C1
Pine Oil Disinf. Phenol Coef. 5.. C1
Pine Oil Disinf. Phenol Coef. 7.. C1
Premium Floor Cleaner A4
Sentinel C1
Smooth E1
Speed Clean A1
Supreme Bol-Soi....................... C2
Ward-O-Phene C2

Maintenance Supply Company A Division of Bunn Capitol Co.
Masco Break Thru...................... A1
Mascocide HD-25...................... D1

Maintenance Supply, Inc.
Enhance.................................. A4
Hydrosolve Cleaner A4
Super Stripper-Degreaser A4

Maintenance Warehouse
Drain Line Opener.................... L1

Maintex, Inc.
A.P.C...................................... A4
DDC Disinfectant-Cleaner
Sanitizer D1
Liquid Hand Soap E1
Mainkleen A4
Mainstrip A4
Spectrum Neutral Free Rinsing
Cleaner A4
Super Jetclean.......................... A4
Therm-O-Solv A1
Typhoon.................................. A4
7-11 Industrial Heavy Duty
Cleaner A4,C1

Major Supply Company
Foam-Kleen.............................. A1
40 Below A5

Major Supply, Inc.
40% Liquid Hand Soap E1

Malco Products, Inc.
#665 Concrete & Garage Floor
Clnr C1
Bio All Purpose Household
Cleaner.................................... A4
Cleaning Compound P.C. 00435 A1
CC 9502 Alkaline Chlorinated
Cleaner A1
Dairi-Brite Deluxe Auto.
Dshwshng Comp. A1
Deluxe Laundry Compound....... A1,B1

Do-It-All A1
Egg Washing Compound........... Q1
Floor Shampoo A4
H/A Concentrate A1
Laundry Compound A1,B1
Liquid Car Wash A1
Lotion Hand Cleaner E1
Malco Deluxe Automatic
Dshwshng Compound............. A1
Malco Disinfectant Cleaner
DC-100 .... D1
Malco Ice Melt C1
MC 9501 Manual Chlorinated
Cleaner.................................... A1
New Hand Dishwashing
Powder A1
Pressure Car Wash Compound .. A1
Sunburst Multi-Purpose Clnr. C1
SC-1070 Steam Cleaner............. A1
SC-1071 Steam Cleaner.............. A1
SC-1072 Steam Cleaner.............. A1
SC-1073 Steam Cleaner.............. A2
SC-1075 Steam Cleaner.............. A1
SC-1076 Steam Cleaner.............. A1
SC-1077 Steam Cleaner.............. A1
SC-1078 Steam Cleaner.............. A1
TC 9058 Alkaline Chlorinated
Cleaner A2

Mallet and Company, Inc.
Exa-Lube................................. H1
K-Lube.................................... H1
Prima-Lube H1
Trough Grease.......................... H1
Ultra-Lube............................... H1

Malone Chemical Co.
All Purpose Cleaner 41.............. A4
Ammoniated Wax Stripper 52.... C1
Anti Microbial Hand Soap 88 .... E1
Cleaner Degreaser 91................ A4
Malone No. 76 Cleaner Disinf
Deod Fngcd D2
Malone 710 Antiseptic Hand
Clnr Lotion E1
No.77 Conc Det, Sntzr, Fngcd,
Dsnft, Deo............................... D2
Protective Coating Cleaner........ A1

Malsbary Manufacturing Co.
All Purpose Cleaning
Compound A2
All Purpose Liq. Cleaning
Chemical A1
Heavy-Duty Cleaning
Compound A2
Heavy-Duty Liq. Cleaning
Chemical A1

Malter International Corp.
A-Plex Concentrate................... A4
Adios...................................... F2
Advantage A4
Airborne Space Spray F1
Ammene C1
Banish F2
Barrier Protective Skin Cleaner. E4
Big B....................................... G6
Blast....................................... A3
Blu-Tex................................... C2
Blue Power E4
Bug-Cide................................. F2
Carate P1
Chem-Terg A3
Chlorothion F2
Clear...................................... D1
Con-Cease P1
Control A2
CL-771 Cleaner & Dry
Lubricant K2

Wid-Out.................................... F2
Zing Extra-Strength Oven
    Cleaner.................................. A8
Zzap H2

Manufacturers Supply Corporation
Bake Pan Cleaner A1
Deluxe Concrete Cleaner A4
Food Plant Cleaner A1
Neutra Clean A1

Mapco Products
Lacko Metal Polish A3
Laundry Bleach B1
Liquid Concentrate A1
Liquid Concentrate 25 .... A1
Liquid Concentrate 38 .... A1
Mapco Conquest A1
Mapco Egg Wash Q1
Mapco Insecticide Spray No. 1.. F1
Mapco Laundry Detergent........... B1
Mapco Liquid Soap..................... E1
Mapco Mapclene-S....................... A1
Mapco Mapclene-SX..................... A1
Mapco Pine Oil Disinfectant C1
Mapeo Pynocide........................... C1
Mapco PSM Smokehouse
    Cleaner.................................. A2
Mapco Sanol................................ D1
Mapeo Sewer Line Cleaner........ L1
Mapco Super Sanol D1
New Phenex................................ E1
Trolley Magic A2

Mapps & Salk Inc.
Ultra Kleen All........................... A1
Ultra Kleen Hand Soap E1

**Mar-Chem, Incorporated/ Chemical &
Supply Company**
Gem Cutter A1
No Secret C1
Total Clean A1

Mar-Len Supply Inc.
ML-41 .... A1

Mar-Lo Chemical Company
MLB 270.................................... G6
MLB 330.................................... G6
MLB 360.................................... G6
MLB 370.................................... G6
MLB 380.................................... G6
MLB 410.................................... G6
MLB-350.................................... G6
MLS 180.................................... P1
MLS 250.................................... G6
MLT 200.................................... G7
MLT 222.................................... G7
MLT 400.................................... G7
MLW-90..................................... G2

Mar-Mo Chemical Co.
Pot-Lick..................................... A1
Zoom-It...................................... A4

Marathon Morco Company
Sontex 100NF H1
Sontex 130NF H1
Sontex 150NF H1
Sontex 19 USP........................... H1
Sontex 21 USP........................... H1
Sontex 33 USP........................... H1
Sontex 35 USP........................... H1
Sontex 475................................. H1
Sontex 55NF H1
Sontex 70-MP H1
Sontex 70NF H1
Sontex 7030............................... H1

Sontex 75-T............................... H1
Sontex 75NF H1
Sontex 80-MP H1
Sontex 85-T............................... H1
Sontex 85NF H1
Sontex 8530............................... H1
Sontex 95-T............................... H1
Sontex 9530............................... H1

Marathon, Inc.
Formula No. 89-D...................... G6
Micro-Bicide No. 42................... G7
Micro-Bicide #41 .... G7

Marco Equipment Company
#L-1163 Non Butyl Degreaser .. A1
Non-Butyl Degreaser A1

Marco Supply Company
New Improved Super Power A1

Marimik Industrial Supply
Big Blue.................................... A1
Big Blue Heavy Duty Cleaner/
    Degreaser.............................. A4
Dual-27 .... D1
Marimik Sting A4

**Marion Brush Mfg./ & Janitorial
Supply**
Foam-Kleen................................ A1
Marion Green Giant H.D.
    Cleaner.................................. A4

Mark Chemical Company
All Purpose Actshine.................. C1
Beadletts Mark 100-B................. A1
Bullox Mark 100........................ A1
Chem Plex Mark 100-X A1
Chemmark Magic Snake............ L1
Midas Auto Con F1

Mark Chemical Company
Automatic Dishwash.................. A1
Bio Clean 11 .... D1
Bio Clean 2................................ D1
Bio Clean 20.............................. D2
Bio-Clean 1................................ D1
Bio-Quat 20 .... D2
Breakthrough.............................. A1
Concrete Cleaner........................ A4
Enviro Insecticide F1
Glow... A1
Hand Dishwashing Compound .. A1
Heavy Duty Warewash............... A2
Lime & Scale Remover.............. A3
Lotionized Hand Cleaner
    Concentrate E1
LSC#3.. A1
Mark Bowl Cleaner.................... C2
Mark Special Degreaser............. A4,A8
Mark Super Degreaser............... A4,A8
Mark Super Degreaser
    Concentrate A8
Mark Ten D2,E3
Mark 25 .... D2
Mark 250 .... D2
Mark 5 .... D2
Mild Bowl Cleaner..................... C2
New Shine Stainless Steel Clnr
    & Polish................................ A7
Pot & Pan Wash A1
Remove...................................... A1
Remove Concentrate................... A4
Scale-Away A3
Surge Laundry Detergent B2
Win A1

Mark Manufacturing Corporation
DR-50.. A2

FS-404....................................... A1
FSA-606 .... N1
LHS-888 .... N2
Mar-Chlor................................... Q1
Mark-O-Fect............................... D1
Phenol Mark 10-X...................... D1
PEM-51 .... A2
PHS 444 .... N2
R-11 .... A1
R-22 .... A2
TC-66.. N3

Mark-L Corporation
Bowl Cleaner & Disinfectant C2
Mark-L Residual Insecticide F2
Mark-L Solv A1
Mark-L-Teen............................... D1,Q6
Super Strength Drain Opener &
    Maintainer L1

Markay Laboratories
Adjunct TS G6
Aquagard 334............................. G6
Aquagard 334-5.......................... G6
Aquagard 338............................. G7
Aquagard 338-5.......................... G7
Boil Out G6
Boiler Water Treatment BCS-2.. G6
Boiler Water Treatment BL-1C. G6
Catalyzed Sulfrite Organic G6
Easy Acid Formula A3
Formula BDF G7
Formula BSL-1........................... G6
Formula BSL-2........................... G6
Formula BSL-3........................... G6
Formula BSL-9........................... G6
Formula SRL-25......................... G7
Formula 101L............................. G6
Formula 102L............................. G6
Formula 103L............................. G6
Formula 105L............................. G6
Formula 106S............................. G6
Formula 201S............................. G6
Formula 202S............................. G6
Formula 205L............................. G6
Formula 521CS.......................... G5
One Shot G6
Polymer EMG G6
Polymer 104HL.......................... G6
Polymer 104L............................. G6
Polymer 520L............................. G6
Polymer 522-L............................ G7

Marko Chemical, Inc.
Marko Alum-Glo......................... A3
Marko CDD................................ D1
Marko Hand Kleen 40 .... E1
Marko KO Insect Spray F1
Marko Liquid Drain Kleener L1
Marko PCC-26............................ A4
Marko SC-100............................. A4
Marko 1000 Plus A4
Marko 1002 .... A2
Marko 1185 .... F1
Markocide.................................. D1
Markofresh................................. C2
Markosol.................................... C2
Pineteen C1

Marks Supply Co., Inc.
Axiom A4
Bio-Logical Lemon Disinfectant
    7... C1,A4
Bio-Logical Mint Disinfectant 7 C1
Bio-Logical Pine Disinfectant 6. C1
Du Mor C2
Fahrenheit A4
Neutral Cleaner A1
Tartan A4
Teepole A1

McCain Boiler & Engineering Co.
Compound #SA-100 .... G7
Compound #SP-100 .... G7
Compound #SPL-100 .... G7
Compound #1 .... G6
Compound #2 .... G6
Compound #3 .... G6
Compound #4 .... G6
Compound #5A G6

McCall's, T, Frank
Dynasol A1
Pink Lotion Dish Detergent A1

McConnon And Company
Cleanser & Water Softener A1,G7,G2
Mouse Rid F3
Rat-O-Mice F3
Zyme L2

McGlaughlin Oil Co., The
Petro-Gel H1

McGrayel Company
McGard #DL-1847-Boiler
    Water Conditioner G6
McGard #DL-1850-Boiler
    Water Conditioner G6
McGard #DL-1853-Boiler
    Water Conditioner G6
McGard #DP-1935-Boiler
    Water Conditioner G6
McGard #DP1925-Boiler
    Water Conditioner G6
McGard #DP1945-Boiler
    Water Conditioner G6
McVol #100-20AC-Steam Line
    & Cndnst Cond G6

McKesson Chemical Company
McKessol 8530 .... H1

McLaughlin Gormley King Co.
D-Trans Fogger And Contact
    Spray -2147 .... F1
Esbiol Ind & House.Space &
    Cont Spy 2201 .... F1
Formula 7243 .... F1
Multicide Concentrate 2120 ...... F2
Multicide Fogger And Contact
    Spray 2178 .... F2
MGK Soluble Powder w/
    Dipterex F2
Pyrocide Concentrate 7254 .... F2
Pyrocide Fogging Concentrate
    5628 .... F1
Pyrocide Fogging Concentrate
    7167 .... F1
Pyrocide Fogging Concentrate
    7192 .... F1
Pyrocide Fogging Concentrate
    7206 .... F1
Pyrocide Fogging Concentrate
    7211 .... F1
Pyrocide Fogging Concentrate
    7219 .... F2
Pyrocide Fogging Concentrate
    7257 .... F2
Pyrocide Fogging Formula
    7207 .... F1
Pyrocide Fogging Formula
    7216 .... F1
Synergized Pyrethrin
    Indust.Spray F-6630 .... F1

McManus Chemical Company
McManus H. D. Safe-Fireprf Cl
    Wax Rmvr A4

McManus Neutral Liquid
    Cleaner A4

McVicker, W. B., Company
Plainsol M A1

Meat Industry Suppliers, Inc.
Chemtrol D1

Meat Packers & Butcher Supply Co.
Long Life Prime White Oil H1

Mechanics Choice
Super Greasegone NB A1

Medallion Chemical Corp.
Bio-Kleen A4
De-Greez 21 .... K1
De-Vour "L" L2
De-Vour "S" L2
Digest "SP" L2
Digest "TD" L2
Super De-Zolv-19 .... A4

Megargel Bros.
Big Mac A1

Meinhardt Products Division
Ice-Ban (with color) C1
Ice-Ban(Colorless) C1

Mellocraft Company, The
#364 Flush L1
AE H 010 BWT G6
AE M 16 RLT Neutralizing
    Amine G7
AE 1201 DE-OX Liquid G6
AE 24 BWT G6
AE-H-20 BWT G6
AE-44-2 Cooling Tower
    Treatment G7
AE-44-4 Cooling Tower
    Treatment G7
F-27 Mellosheen A1
Lac-1300 .... A4
Lac-1302 .... A4,A8
Mello-Shac Q1
Metal-Brite A7
MC-200 Industrial Cleaner and
    Degreaser A4,A8
No. 1030 M C B A3
No. 1032 Mello-Dyne D2
No. 1038 Agri-Dyne D2
No. 1050 Sanitizer D1
No. 1093 Dissolv-It Prilled
    Drain Opener L1
No. 111 Hand Dishwashing
    Compnd A1
No. 115 Lan-O-Suds A1
No. 1307 N.B. Degreaser A1
No. 1310 Auto-Lube H2
No. 1311 Mello-Lube A1,H2
No. 1404 Mello-Power A2
No. 1405 Mello-Circ A1
No. 1412 Mello-Hac A1
No. 1413 Mello-Shae A1
No. 1420 Mello-Pan A1
No. 153 Pre-Soak De-Ox Liq
    Tarnish Remvr A1
No. 163 W.S.D. A4
No. 1651A LPR H2
No. 1656 A Never-Sweat P1
No. 1657 A New-Life P1
No. 1659 A Liq TEF Tape Pipe
    Thd Seal H2
No. 174 Lift-Off. A4
No. 175 Mellosheen
    Concentrate A1
No. 176 Mello-Cide D1

No. 178 Scum Remover A4
No. 192 Econo-Suds A1
No. 201 40% Pearl Liquid E1
No. 214 Creme-Lotion Liquid
    Hand Soap E1
No. 223 Mello-Kleen C2
No. 303 Pure Kleen A1
No. 335 Heavy Duty Mello-Sol . C2
No. 343 Foam Off A4
No. 346 Spra-Zit C1
No. 349 Chef's Choice A8
No. 435 Rinz-Ex Q1
No. 443 Thermo-Kleen A3
No. 444 Mild Mello-Sol C2
No. 484 Spra-Zit Surface
    Cleaner C1
No. 533 Lemon Scented
    Disinfectant C1
No. 612 Astro-Dis D1
No. 651 Glass Shield C1
No. 663 Institutional Metal-
    Brite A7
No. 675 Beta-Phene C2
No. 790 Liqua-Scale A3
No. 888 Mello-Solv C2
No.368 Sulf. Acid For.
    Dissolvo Drn Opnr L1
Odor Control L2
Pac-1400 .... A1
Pac-1402 .... A2
Pac-1450 .... C1

Menco Chemical Division
Mono-Molecular Water Treat
    No. 171 .... G6
Water Treatment No. 1 .... G6
Water Treatment No. 2 .... G6
Water Treatment No. 201H G6
Water Treatment No. 3 .... G6
Water Treatment No. 50 .... G6
Water Treatment No. 501H G6

Mendez & Company, Inc.
Action Stripper Cleaner A4
Menaco Antiseptic Hand-
    Cleaner Lotion E1
Menaco Hand Liquid Soap E1
Menaco Super Deterge A1
Menaco Wax Stripper
    Degreaser Cleaner A1
Pine Oil Disinfectant C1
Tuf-Strip Wax Stripper A4

Mentholatum Co., The
Baracaide E4

Mercury Chemical Co., Inc.
Fogging Concentrate 3610 .... F1
Fogging Formula #7216 .... F1
Merchem All Weather Bait
    Cakes F3
Merchem Bait Block. Rodnt w/
    Mol-Pea.But. F3
Merchem Bait Blockette Rodnt.
    w/Apple F3
Merchem Bait Blockette Rodnt.
    w/Fish F3
Merchem Bait Blockette Rodnt.
    w/Meat F3
Merchem Rat and Mouse Bait ... F3
Merchem Special Fogging
    Concentrate 511 .... F1
Merchem Syn-Tox F1
Mercomist Aerosol Insect
    Killer F1
Mercovap F2
Mercovap 5 .... F2
Mercury Fogging Concentrate
    #7167 .... F1

Mercury Fogging Concentrate
#7192 .... F1
Mercury Fogging Concentrate
#7206 .... F1
Multi-Purpose Fogging
Insecticide-100 .... F1

Meridian Petroleum Company
Dri-Quick J1
Dri-White J1

Merit Chemical Company, Inc.
Caustic-Jell Merit 675 .... A1
Concrete Kleen Merit 20RP C1
Concrete-Kleen Merit 20XP C1
Hi-Foam 1 .... A1
Kloro Kleen Merit Pink 2XW ... A1
Kloro-Kleen Merit 21XW .. A1
Kloro-Kleen Merit White 2XW . A1
Kloro-Kleen NR A1
Kloro-Kleen White 220 .... A1
Kloro-Kleen White 227 .... A1
Liquid Hand Soap 15 .... E1
Liquid Hand Soap 20 .... E1
Liquid Steam Kleen Merit
HD701 .... A2
Low Foam Merit 711 .... C1
Low Suds Merit 21XX B1
Merit Liquid Skin Cleaner E1
Merit 487X A3
Merit 487XC A3
Merit-Kleen Merit 11DX A2
Merit-Kleen Merit 12RX A2
Merit-Kleen Merit 14RX A2
Merit-Kleen Merit 2R A1
Merit-Kleen Merit 22BX A1
Merit-Kleen Merit 28BX A1
Merit-Kleen Merit 3 BDX A2
Merit-Kleen Merit 3DCX A2
Merit-Kleen Merit 3DX A2
Merit-Kleen Merit 325 .... A2
Merit-Kleen Merit 41 .... A3
Merit-Kleen 10RX A2
Merit-Kleen 2BX A1
Merit-Kleen 500 .... A1
Merit-Lube 1 .... H2
Vari-Kleen Merit 7BDX A4
Vari-Kleen Merit 7XX A1
Vari-Kleen Merit 705 .... A1

Merit Chemical, Inc.
AL-10 .... A1
AL-32 .... A1
BTC A1
Chloricide-12 .... D2
Chlorinated Kleen Suds A1
Circl-Acid A3
Circl-Alkali A2
Cirel-Phos G2
Egg Shine Q1
General Sheen A4
Hi-Shine Alkali A2
Iodibac D2
Kleen-Brite A2
PH-5 .... A3
Special Circl-Alkali A2
Suds-Rite A1
Super Chloro A1
Super-Kleen A1

Merit Corporation, The
"Merit" No-Mess Oven and
Grill Cleaner A8

Merit Maintenance Supply, Inc.
Hand 40% Pure Coconut Oil
Hand Soap E1

Merit Manufacturing Co., Inc.
Cold Water Immersion Cleaner . A1

Descaler A3
Equipment Cleaner A2
Foam Glo A1
Fryer Cleanser A2
Immersion Cleaner A1
Merit Hand Cleaner C2
Merit Liqui-Kleen A1
No. 11 Line Cleaner A2
No. 61 Hvy Dty Cleaning
Compound C1
P Chlorinated Hand Cleaner A6
Poultry Processor's Special
Cleaner A1
Smokehouse Cleaner A4
Sulfamic Acid A3
WC Chlorinated Equipment
Cleaner A1

Merit Paper Corp.
Merit Par Excellance Hand
Dish. Cmpd A1

Merit Sanitary Supply Co.
Hot Shot A4

Mesa Sanitary Supply Company
Mesa Strip-It C1

Metalene Chemical Company, The
Anticor No. 604 .... G6
Boiler Feedwater Treatment
No. 1111 .... G6
Boiler Feedwater Treatment
No. 306 .... G6
Boiler Feedwater Treatment
No. 401 .... G6
Boiler Water Treatment No.
1125 .... G6
BWT No. 1108 .... G6
BWT No. 1166 .... G6
BWT No. 550 .... G6
BWT 4-075 .... G7
BWT 4-165 .... G6
BWT 4-171 .... G6
BWT 4-174 .... G6
BWT 4-181 .... G6
Liq. Boiler Water Treatment
No.1115 .... G6
Organic Adjunct CD No. 1039 .. G6
Organic Adjunct CD No. 1040 .. G6
Organic Adjunct No. 1041 .... G6
Rust Foil C No. 2083 .... G6
Rust Foil No. 2093 .... G6
Rust Foil 5-1351 .... G6
Rust Foil 5-1361 .... G6

Metra Chem Corp.
NBC 77 .... A4,A8

Metro Chemical Supply, Inc.
General Purpose Cleaner A4
Heavy Duty Industrial Cleaner .. A1
Lemon Disinfectant C2
Lotion Hand Soap Ready For
Use E1
Mint Disinfectant C2
Pine Odor Disinfectant C1

Metro Products Company
Concrete Cleaner X-1 .... A1
Concrete Cleaner X-6 .... A1
Feen-O-Pine C1
Geolin Antiseptic Lotionized
Liquid Soap E1
HC-45 Disinfectant Cleaner A4
Insectrol No. 100 .... F1
Insectrol No. 75 .... F1
Liquid Drain Opener L1
Liquid Hand Soap E1

Liquid Smokehouse Cleaner
Formula I1 .... A1
Litnin Sol C1
M-O Concentrate A4
Metro Formula M-32 .... A1
Metro Pine-Sol C1
Metro-Sol C2
Metrodyne D2
Metrolite Digester L1
Metzolite A1
Mint Disinfectant C1
No. 100 Packing House General
Cleaner A1
No. 200 Smoke House Cleaner .. A2
No. 300 Acid Vat Cleaner &
Descaler A3
No. 400 Heavy Duty Det &
Neut & Vat Clnr A1
No. 500 Tripe Cleaner N3
No. 600 Hog Scald N2
Ozonene C1
S-D-C A1
S-F Cleaner A1
Santrol No. 25 .... D1
Stain-Off C2
Steam Cleaner KS A2
Steam Cleaner MSP A2
Super Five A4
Tempo Chlorinated Bleach
Cleanser A6

Metropolitan Refining Co., Inc.
C S Cleaner .... A1'
Clean-Away 300 .... A2
Corrodicide Dry G1
Corrodicide FP G1
Corrodicide Liquid G3
Corrodine 710 .... G6
Corrodine 940 .... G6
Drainkleen L1
Duboth A G6
Duboth A-38 .... G6
Duboth B G6
Duboth CA G6
Duboth CA-96 .... G6
Duboth H-6 .... G6
Duboth JA G6
Duboth JS G6
Duboth LP G6
Duboth OX G6
Duboth P G6
Duboth P-AC80 .... G6
Duboth S-5 .... G6
Duboth 89 .... G7
Kno Klog L1
Metro #340 .... G6
Metro #630 Water Supply
Treatment G1
Metro Anti-Foam U-125 .... G6
Metro Clean-Away Formula
200 .... G7
Metro Clean-Away Formula
400 .... A4
Metro Clean-Away Formula
500 .... C1
Metro Cleaner 129 .... A1
Metro Cleaner 671 .... P1
Metro Degreaser AP K1
Metro Descaler A3
Metro General Cleanser A1
Metro Solvent EM K1
Metro Steam Cleaner A1
Metrofloc 515 .... G6
Mejrosperse 269 .... G7
Seagull L1
Sterokleen A1
Vaporene C G7
Vaporene C-2 .... G7
Vaporene D-680 .... G7

Vaporene DT-216........................ G7
Vaporene FP.............................. G7
Vaporene K-5 .... G7
Vaporene SA-112 .... G7
Vaporene 216............................. G7
Vaporene 72............................... G7
Vaporene 89............................... G7

Meyer-Blanke Company See: Em-Bee Chemicals, Inc.

Meyer, H. B. & Son, Inc.

"IT" E4
Action A4
Chlorinated Super Clean............ A1
Clenzene A1
Cream Lotion Hand Soap........... E1
Dare A4
Dek-Ade................................... A4
Double XX A1
Great Southern Hnd
  Dishwash.Powd. A1
H.B. Meyer Hand Clnr. &
  Sanitizer................................ E2
Heavy Duty Floor Cleaner A1
Heavy Duty Steam Cleaning
  Comp. A2
Kleenswell................................ A4
Launderette B1
Lemon Odor Disinfectant C2
Lotion Clenzene A1
Low Suds Launderette B1
Metaclean K1
Meyer's Foaming Cleanser A6
Meyer's Pressure Spray Wash.... A1
Pepsene.................................... A1
Pineolene C1
Sani-Sope................................. E1
Sedoc C1
Sterylclene............................... A4
Sudzee..................................... A1
Super Clenzene A1

Mihar Inc.

Aqua Solve............................... A1
Boil Out G6
Chem Ox A3
Chlor Aid C1
Complete A4
Con-Pow A4
Crystal C1
Electri-Sol K2
Glitter...................................... A3
Jel Clean.................................. K1
Likwi Date............................... A3
Metro Clear.............................. L1
Mi Steam Powder..................... A1
Mist & Clean............................ C1
Mizyme 40................................ L2
Non Flam K1
Refuse C1
Safe A1
Spray & Wipe A8
Stripper.................................... A1
Super DE K1

Michigan Sales Co.

Pete's Stuf Cleaner A4,A8

Michigan Salt Company

Crystal Flow Brine Blox............ P1
Crystal Flow Gran. Water
  Softener Salt........................... P1
Crystal Flow Louisiana Rock
  Salt P1
Crystal Flow Prls Wtr Soft Salt
  Peltzd F P1
Non-Hardening No.1 Evap
  Crystal Flow Sal.................... P1
Pearls Crystl Flow Wtr Soft
  Salt w/Iron............................ P1

Micro Products Corporation

Heavy Duty Degreaser XX-
  1810...................................... A1
Micro All-Purpose Cleaner A1
Micro Car Cleaner A1
Micro Marine Cleaner................ A1

Micro-Gen Equipment Corporation

ULD BP-100 Insecticide............. F1
ULD BP-300 Insecticide.............. F1
ULD V-500 5% Vapona
  Insecticide............................. F2

Mid American Chemical Corporation

Car & Truck Wash #7430........ A1
Floor Cleaner 7120................... A4
Rapid Cleen 2000...................... A3
Super Cleen 1500...................... A1
Super Cleen 1630...................... A1
ThoroClene E1

Mid South Supply Co.

Aerodet Heavy Duty Cleaner.... A4

Mid-America Chemical Co.

B-D-G #110 Belt Dressing......... P1
Dyna-Mite Sulfuric Acid Type
  Drain Clnr............................. L1
Electra #155 Electronic
  Solvent Dgrsr......................... K2
Germicidal Cleaner &
  Deodorant C1
Glass & Utility Cleaner............. C1
Insecticide (w/residual insect
  control) F2
Liquid Boiler and Feed Water
  Treatment G6
Quatra Rinse D2
Slash #195 Nut & Bolt
  Loosener Penetrant H2
Super Fogging Insect Killer....... F2
Super Steam Steam Cleaner A2
Super-Kleen Heavy Duty
  Cleaner................................. A4
Three Gly................................. C1

Mid-America Chemical, Inc.

Tru-Kleen Sodium
  Hypochlorite Solution............. B1

Mid-America Dairymen, Inc.

Mid-Am A.P.C.-23................... A3
Mid-Am Acid Cleaner A3
Mid-Am Acid Cleaner
  (Concentrate) A3
Mid-Am Acid Cleaner 100........ A3
Mid-Am Acid Glow................... A3
Mid-Am Acid Sanitizer.............. D2
Mid-Am Acid 220 .... A3
Mid-Am Acid 400 .... A3
Mid-Am Alkali Cleaner A2
Mid-Am Bon-Fome A1
Mid-Am Bon-Lift 5000 .... A1
Mid-Am Britank A1
Mid-Am C-10............................ A2
Mid-Am C-12............................ A2
Mid-Am Can Wash (L-160)....... A3
Mid-Am Chelaton No. 4............ A2
Mid-Am Chlorinated Cleaner.... A1
Mid-Am Chlorinated TSP.......... A1
Mid-Am Circulation Cleaner..... A2
Mid-Am Circulation Cleaner C-
  5.. A2
Mid-Am Crystal Chlor............... D1
Mid-Am Flex-O-Wash No. 2...... A1
Mid-Am Flex-O-Wash No. 2NP A1
Mid-Am General Cleaner........... A1
Mid-Am General Cleaner NP A1
Mid-Am General Cleaner 880 .... A1

Mid-Am LPC-279...................... A1
Mid-Am Neutral Cleaner(L-
  143)....................................... A1
Mid-Am NXZ Defoamer............ A1
Mid-Am Tile Floor Cleaner....... A4
Mid-Am 5100 XL Odorless........ A1

Mid-America Indust. Products Co., Inc.

Super-Kleen 11.......................... A4

Mid-American Research Chemical Corp.

Marc Fiberglass Cleaner............ C2
Marc Power Off ....: B2,C1
Marc 247 Glass Cleaner............. C1
Marc 248 .... C1
Marc 249 Gentle Scrub.............. A4,C2
Marc 50-D Bowl Cleaner C2
Marc 71-H.P. A1
Marc 72 Steam Clean................ A1
Marc-103 .... K1
Marc-108 .... A7
Marc-11 .... K1
Marc-136 .... P1
MARC Double Strength
  Descaler................................ A3
MARC Tile & Grout Cleaner.... A3
MARC 11-NF Safety Solvent..... K1
MARC 300 Spray and Wipe
  Cleaner................................. C1
MARC 305 Glass Cleaner C1
MARC 38-AAA Marcicide......... A4
MARC 53 Got-A-Mess.............. C1
MARC 54 Odor Counteractant
  Cherry.................................. C1
MARC 54 Odor Counteractant
  Citrus................................... C1
MARC 59.................................. L1
MARC 61 Odor Control C1
MARC 62 Deodorant Granules.. C1
MARC 68 Deodorizer Cherry.... C1
MARC 68 Deodorizer Citrus
  Lemon C1
MARC 69 Flocculent & Sludge
  Dewtr. Cmpd......................... L1
MARC 71 Concentrated High
  Pressure Clnr......................... A1
MARC 75 Liquid Paint
  Stripper................................. C3
MARC 76 Jelled Paint Stripper.. C3
MARC 92-GS Concentrated
  Gel Degreaser K1
MARC 92-GW Gel Degreaser .. A4
MARC 92-LC Conc. H.D. Eng.
  Dgrsr.&Mtr Cl........................ K1
MARC 94-LP Mighty-Solv
  Metal Parts Clnr..................... K1
MARC 95 Descaler.................... A3
MARC-101 TFE Fluorocarbon
  Lubricant H2
MARC-109 Glass Cleaner C1
MARC-116................................ H1
MARC-127................................ H2
MARC-135................................ H1
MARC-138................................ K1
MARC-139................................ F2
MARC-154 Invisible Glove E4
MARC-156 Pipe Thread Sealer.. H2
MARC-158 Oven & Hood
  Protector............................... H1
MARC-16.................................. C1
MARC-214................................ G6
MARC-218................................ C1
MARC-38.................................. D1
MARC-41.................................. A4
MARC-47.................................. C1
MARC-50.................................. C2
MARC-55.................................. L1

MARC-63 Marc-Xion L1
MARC-65 Homogenizer and
Solubilizer A4,L1
MARC-66 .... C1
MARC-88 Jelled Descaler A3
MARC-89 .... L2
MARC-97 .... C1

Mid-Coast Chemical & Janitorial Supply

Brute Degreaser And Wax
Stripper A1
Hand Soap Lemon Scented E1
Mid-Co Citracid C2
Power Cleaner All Purpose
Degreaser A1

Mid-Continent Chemical Co.

Alpha "B" Chlorinated A2
Alpha A Chlorinated A1
Alpha C Chlorinated A1
Alpha Pledge A1
Alpha Silva-Brite A1
Alpha Suds A1
Alpha Super A1
Alpha X-Cel A1
Concentrate All Purpose
Detergent A4
D.C.C. D1
Green Power A4
GW Klene A4
Hygienal A1
Laundry Detergent B1
Lime Go A4
Mid-Con B A3
Mid-Con D-3000 .... A1
Mid-Con Laundry Two B1
Mid-Con Ox-O-Brite B1
Mid-Con 220 .... A1
Mid-Con 2500 .... A2
Mid-Con 330 .... A2
Mid-Con 3990 .... A3
Mid-Con 500 .... A2
Soil-Gon A1
Sonic Solvent A1
Yellow Magic A1

Mid-Continent Laboratories, Inc

A-P-W-5 Food Plant Cleaner A4
A-P-W-5 Heavy Duty Cleaner ... A1
A-P-W-5 Inhibited Chemical
Clnr A3
Apple Blossom Air Santzr. &
Surf. Disinf C1
Aqua-Solv A1
Aqua-X D1
Car Kleen A1
Chem Ox A3
Cinch E4
Clor-Aid C1
Crystal C1
Deo-Clear L1
Dis-N-Kleen A4
E Z Clean A4
Express Sewer Solvent L1
Flavor-Sect Insect Spray F1
Fleet Kleen A1
Free-N-Clear A1
Germex C2
Hook And Trolley Cleaner A4
Jel Clean K1
Kwik-Fog Insect Spray F1
Lana San E1
M.C. Boiler Compound G6
M.C. Concentrate A1
Metro-Clear L1
Metro-Clear L1
Mid-Sol K2
Midzyme 20 .... L2

Midzyme 30 .... L2
Mintize C1
Mist N' Clean C1
MC-6 Water Treatment G6
MID-10 .... A1
Neutral Food Equipment
Lubricant H1
New X-Can C1
Non Flam K1
Odor-Gone Lemon C1
Odor-Gone Lilac C1
Odor-Gone Mint C1
Old Smokey A1
Pent-X H2
Pine Oil Disinfectant Coef. 5 ..... C1
Pine-All Disinfectant C1
Sani Klean E1
Sani-Lube Food Equipment
Lubricant H1
Sectrol F2
Solv-Aid C1
Sparkle A1
Super De K1
Super Solve A1
Thaw Away C1
Tri-D Disinfectant Deodorant ... C1
Whiz C1

Mid-Lands Chemical Co., Inc.

AC Foamer A1
AC 103 .... A3
AC 104 .... A3
CC 100 .... A2
CC-101 .... A2
CC-102 .... A2
EO 202 .... H1
EO-203 .... H1
Hulk A1
Inferno L1
LB 48 .... B1
LC-01 .... C1
LD 60 .... B1
LD-61 .... B1
LHS 40 .... E1
SS-98 .... K2
Total A1
TW 300 .... N3

Mid-South Chemical Co.

MBT #102 .... G6
MCT #501 .... G2
MHT #106 .... G6
MOS #144 .... G6
Steam Treat #200 .... G6
Steamtrol G6

Mid-South Supply

No. 185 Cleaner A1
Odor Control C1

Mid-State Chemical & Supply Corp.

Acid Brightener A3
Acid Cleaner 5525 .... A3
Chain Lube 8420 .... H2
Chlor-Clean 1612 .... A1
Chlorinated Hand Scrubbing
Cleaner A1
C1P Acid Circulation Cleaner
NP A3
CIP Circulation Cleaner A2
Floor Cleaner 7110 .... A4
Foam Additive 1650 .... A1
Foam Cleaner A2
Heavy Duty Alkaline Drain
Opener L1
Heavy Duty Bottle Wash A2
Heavy Duty Chlorinated
Cleaner A2
Heavy Duty Floor Cleaner
7120 .... A4

Heavy Duty Laundry
Detergent B1
Heavy Duty Smoke House
Cleaner A2
Heavy Duty Solvent
Degreasing Cleaner K1
Hook, Trolley, And Tree
Cleaner A2
Low-Alkaline General Purpose
Cleaner A1
Pan And Mold Cleaner A2
Pan And Mold Cleaner NS A2
Rapid Clean 2000 .... A3
Raven Brand CIP Acid
Circulation Clnr A3
Raven Brand Gen.Purp.Powdrd
Clnr A1
Raven Brand Non Phos. Hog
Scald Comp. N2
Raven Brand Phos-Free Det
Chain Lube8430 .... H2
Raven Brand Phos-Free Tripe
Clnr N3
Raven Brand Rinse Aid 1675 ..... G5
Safety Solvent 14000 .... K1
Safety Solvent 14010 .... K1
Super Clean 1500 .... A4
Super Kleen 1630 .... A1
Truck Wash 110 .... A1
Truck Wash 110 A A1

Mid-State Chemical Corp.

Hydro-Clean All Purpose
Water Sol. Dgrsr A1
Mid-State A1
Zoom A1

Mid-States Laboratories, Inc.

Concentrated Cleaning Crystals. C1
Multi-Solv Liquid Concentrate .. A4
MSL 1003 All-Purpose Silicone
Spray H1
MSL 101 Organic Sewer Line
Opener & Mutr L1
MSL 103 Drain Line Opener
and Maintainer L1
MSL 107 Enzyme Sewer &
Drain Line Mntr L2
MSL 115 Organ. Drain Line
Opnr & Maintnr L1
MSL 204 Floor Sealer Acrylic
13% P1
MSL 206 Multi Solv Wtr Base
Dgrs. Conc. A4
MSL 208 Super Conc. All
Purpose Clnr. A1
MSL 216 Polyurethane Sealer ... P1
MSL 226 Hvy Dty Water Base
Dgrs Conc A4
MSL 227 Super Neutral
Cleaner A1
MSL 407 Liquid Hand Soap E1
MSL 408 Descaler-Delimer
Concentrate A3
MSL 412 Non-Abrasive Liquid
Cleanser C1
MSL 423 Oven Cleaner A8
MSL 427 Cuts Rust Remover A3
MSL 503 Safety Solvent
Cleaner & Dgrsr K1
MSL 533 Timed-Rel Ice & Sno
Mltr Pelts C1
MSL 834 Lift Station Cleaner ... A2

Mid-States Sales Company

Super Foam Locker Plant
Degreaser A1
Super Strenght Soil Emulsifier ... A1

175

Mid-Tex Chemical Company

All Purpose Cleaner and De-
Greaser C1,A8

Mid-West Chemical Company

Grease Eater and Concrete
Cleaner A4
Heavy Duty Floor Cleaner A4
Nu-Glo A4

Midco Products Company

Autobrite A1
Cleaner And Disinfectant A4
CDS Cleaner-Degreaser
Solvent..................................... A4
Degreaser K1
Disinfectant Deodorant.............. C1
Dry Moly Lubricant H2
Flying Insect Killer.................... F2
Fogging Insecticide F2
Glass Cleaner C1
H.T.20....................................... A2
Kleen-Well A1
Mud-A-Way Concentrate L1
MDW 45 .... A1
Penetrating Oil.......................... H2
Silicone Mold Release............... H1
Silicone Spray............................ H1
Silicone Spray Food Grade........ H1
Stainless Steel & Formica Clnr.
& Polish A7
Steam-Off A.P. A1
Steam-Off H.D............................ A1
Waterless Hand Cleaner............ E4

Midco Products, Inc.

Midco "AA" Concentrate C1

Midland Chemical Corporation

Bee-Clene A4
Bee-Clene Plus............................ A4,A8
Floor Condtion C1
Foam Additive............................ A1
Hand Soap.................................. E1
M.C. Concentrate A1
PHC-70....................................... A1
PHC-75....................................... A4
PHX.. E1
Spra-Clene.................................. A4
Twenty Percent Hand Soap....... E1

Midland Industrial Products/ A Div. of Rochester Germicide Co.

BFW-3 .... G6
BFW-31 .... G6
BFW-32 .... G6
BFW-4 .... G6
CCI-13... G6
CL-13 .... G7
CS-24 .... G7
Midsperse CS-33 .... G7
Midsperse CS-37 .... G7
Midsperse CS-39 .... G6
ML-13.. G7
ML-19.. G7
ML-21.. G7
ML-9.. G7
OS-13 .... G6

Midland Laboratories, Inc./ Div. of Rochester Germicide Co.

Bay-O-Cide F2
Blitz.. L1
Cleen-Sheen D1
Di-O-Cide................................... F2
Di-O-Cide E-3............................. F2
F-102.. A1
F-118 Low Foam Cleaner A1

F-180.. A1
F-204.. A4,A8
F-25 Sanitizer............................ D2
Gas-O-Cide................................. F4
Germ-O-Solv A4
Gloz C1
HiVis 15 .... A3
HiVis 27 .... A1
HiVis 33 .... A1
Krysloh....................................... A1
Lime-Sol A3
Liquid Shiloh C2
Lohfoam...................................... A4
Losuds.. A1
MedMelt Freezer Cleaner A5
Mi-Clean..................................... A4
Mi-Solv....................................... A4
Mid Melt Acid Freezer Cleaner A5
Mid Melt II Freezer Cleaner...... A5
Mid-Prep A5
Midox Disinfectant D1,G4
Mill-O-Cide................................. F1
Mill-O-Cide "100" F1
Mill-O-Cide "25" F1
Mill-O-Cide "28" F1
Mill-O-Cide "500" F1
Mill-O-Cide B-9.......................... F1
Mill-O-Cide Super Strength F1
ML-10.. D1
ML-11.. A4
ML-11 Surgical Soap E1
Paragon E1
Sano-Septic Antiseptic Lotion
Soap E1
Shiloh... A6
Super Strip A4
Super-C....................................... C1
Sure.. A1
Syn-O Germ................................ A4
Syn-O-Germ 1 .... A4
Ulo-Cide V-500 Insecticide F2
ULo-Cide 10 .... F1
ULo-Cide 30 .... F1
ULO-Cide 25 .... F2

Midland Paper Co. Maintenance Supply Div.

Midpaco Power Kleen A4
Midpaco Power Plus................... A4
Midpaco Vanish Bowl Cleaner .. C2
Midpaco 101-Cleaner A4

Midland Research Laboratories, Inc.

Chem-I-Cal Chemet Liquid G1
Chem-I-Cal Chemet Powder...... G1
Chem-I-Cal G210 .... G6
Chem-I-Cal G210P...................... G6
Chem-I-Cal G211 .... G6
Chem-I-Cal G212 .... G6
Chem-I-Cal G213 .... G6
Chem-I-Cal G220 .... G6
Chem-I-Cal G221 .... G6
Chem-I-Cal G222 .... G6
Chem-I-Cal G223 .... G6
Chem-I-Cal G250 .... G6

Midwest Biochemical Corporation

Embiozyme N-B........................... L2
Embiozyme N-STA....................... L2
Embiozyme SE-A.......................... L2
Embiozyme SE-L4........................ L2
Embiozyme SEB-A........................ L2
Embiozyme SEB-2........................ L2
Embiozyme 5-E............................ L2
MBC-250 .... A4

Midwest Chemical Company

Graygone Commercial Grade.... B1
Tetra D Commercial Grade....... A1

Midwest Chemical Company

Concentrate Deodorant C1
Deo-Gran C1
Deodorizer C1
Disinfctnt-Sanitizer-Deodorizer-
Conc.. D1
Drain Cleaner L2
Drain Cleaner Enzyme Type L2
Enzymes for Sewage Systems... L2
General Purpose Liquid
Cleaner..................................... A1
Germicidal Cleaner D1
Heavy Dty Liq. Organic Drain
Cleaner..................................... L1
Heavy Dty Organic Drain
Cleaner..................................... L1
Ice Melting Compound............... C1
Laundry Detergent..................... B1
Liq. Solvent Non-Flammable
Clnr.. A4
Liquid Dishwashing
Concentrate A1
Liquid Hand Soap 20%.............. E1
Liquid Hand Soap 40%.............. E1
Liquid Ice Melt.......................... C1
Low Pressure Boiler
Compound G6
MS-11 Safety Solvent K1
New Improved Rat Bait F3
Organic Acid Detergent A3
Porcelain and Bowl Cleanser C2
Powdered Cleaning Compound . A1
Sewer Solvent............................ L1
Three-Gly................................... C1
Window Glass Cleaner C1
Window Glass Cleaner
Concentrate C1

Midwest Food Supply Company

F.C. Floor Cleaner C1
Midflash...................................... A1
No. 150 Super Cleaner A1
No. 200 All Purpose Cleaner A1
No. 2000 .... A1
No. 300 General Cleaner A1
No. 700 Chlor Cleaner A1
No. 925 Chelated Alkali A2

Midwest LTD

Do-It A4

Midwest Maintenance Supply, Inc.

Tux Lotion Soap E1

Midwest Oil Company

Ace-Lube AA-AW........................ H2
Ace-Lube BB-AW........................ H2
Ace-Lube Compressor Oil AA .. H2
Ace-Lube Compressor Oil Ext.
Light H2
Ace-Lube Compressor Oil
Heavy H2
Ace-Lube Compressor Oil Hvy
Med.. H2
Ace-Lube Compressor Oil
Light H2
Ace-Lube Compressor Oil
Medium H2
Ace-Lube CC-AW........................ H2
Ace-Lube Extra Heavy AW H2
Ace-Lube EE-AW........................ H2
Ace-Lube EP GL-#0.................... H2
Ace-Lube EP GL-#1.................... H2
Ace-Lube EP GL-#2.................... H2
Ace-Lube EP GL-#3.................... H2
Ace-Lube EP GL-#4.................... H2
Ace-Lube EP GL-#5.................... H2
Ace-Lube EP GL-#6.................... H2
Ace-Lube EP GL-#7.................... H2

Ace-Lube 62 White Mineral Oil H1
Ace-Lube 630-AA H2
Ace-Lube 701........................... H1
Ace-Lube 930AA-15 .... H1
Ace-Lube 930AA-2 .... H1
Ace-Lube 95 WA H2

Midwest Pressure Cleaner Co.

Acid Clean #40 .... A3
Acid Clean P-42 .... A3
AG #70................................. A1
Char Off 76 .... A2
Floor Soap M-44 .... A4
Liq. Gen. Purp. Clnr................... A1
SHC #74 .... A2
Thermosol A1
20 Degreaser A1

Midwest Supply Company

#200 Insecticide F1
Algaecide and Sanitizer
    Solution............................ D1
Aqua-Solv Cleaner Degreaser.... A4
Bafix Germicidal Cleaner C1
Bafix Germicidal Cleaner C1
Dri-Tef Spray Lubricant H2
E-Z Go Penetrant...................... H2
Grease-Solv............................ A4
Lemon Sewer Sweetener............ L1
Med-I-Co............................... A4
Mid Zone Disinfectant
    Deodorant C2
Midwest Power Solv................... A4
Midzyme Digestant 250 .... L2
No-Ox Inhibited Chemical Clnr
    & Deox. C1
Open Up Drain Pipe Cleaner L1
PY Spray Pyrethrum Type Fly
    Spray............................... F1
Q-Tabs Sanitizing Tablet D1
Select Cleansing Cream Hand
    Cleaner............................. E4
Sentry Mark Remover K1
Sewer Solvent L1
Sila-Lube H2
Smoothy Waterless Hand
    Cleaner............................. E4
Spray-A-Way C1
Staph-X................................ A4
T.N.T. Liquid Drain Pipe
    Opener L1
Warpath Roach n' Ant Killer F2
White Wizard Emulsion Bowl
    Cleaner............................. C2

Midwest Textiles, Inc.

Westex Bleach N3
Westex Heavy Duty Cleaner...... A2
Westex Hog Scald N2
Westex Shroud Wash A1
Westex Tripe Wash N3

Milchem Laboratories Inc.

M-4961................................. A1
M-4962................................. N4

Milfred Company, The

Ammoniated Oven Cleaner A8
Diocene C1
End-Con P1
Metal Brite A7
Milfuso Fog Generator Insect
    Spray............................... F1
Milfuso No. 200 Residual Insect
    Spray............................... F2
No. 25 Insect Spray................... F1
Tef-Tape Pipe Thread Sealer H2

Miller Chemical & Fertilizer Corp.

Fly Away Spray Away Ready
    to Use.............................. F1
Liquid "55" Malathion............... F2

Miller Chemical Company, Inc.

Mill Chem 275 .... A3
Mill Chem 300 .... C2
Mill Chem 310 .... C2
Mill Chem 320 .... C1
Mill Chem 404 .... F1
Mill Chem 404 Aerosol............... F1
Mill Chem 500 .... A1
Mill Chem 600 .... E1
Mill Chem 700 .... A1
Mill Chem 88 .... A1

Miller Scales & Food Machines, Inc.

Sani-Lube H1

Miller Supply Company

Ben's Best Cleaner-Degreaser A1

Miller-Aldridge Chemicals, Inc.

Aluminum Cleaner A1
Brite-Lume A1
Chlorinated C.I.P. Cleaner A1,Q1
Chlorinated General Cleaner
    No. A-2............................. A1
Chlorinated General Cleaner
    No. 102............................. A1
Chlorinated Heavy Duty
    Cleaner............................. A2
Chlorinated Liquid C.I.P.
    Cleaner............................. A2,Q1
Cir-Chlor A1
Conveyor Lube No. 800 .... H2
Easy Pad A1
Fryer Cleaner.......................... A8
General Cleaner No. A-1........... A1
General Cleaner No. A-3........... A1
General Cleaner No. 101 .... A1,Q1
General Cleaner No. 103 .... A1
Laundry Detergent No. 900....... B1
Liquid Acid Cleaner No. 11....... A3
Liquid Acid Cleaner No. 11NC. A3
Liquid Acid Cleaner No. 22....... A3
Liquid Acid Cleaner No. 22NC. A3
Liquid Acid Cleaner No. 33....... A3
Liquid Acid Cleaner No. 33NC. A3
Liquid Acid Cleaner No. 44....... A3
Liquid All-Purpose Foam
    Cleaner............................. A4
Liquid Foam Cleaner No. 50 ..... A1
Liquid Foam Cleaner No. 60 ..... A1
Liquid General Cleaner No. A-
    4................................... A1
Liquid General Cleaner No. 104 A1
Liquid Self Foaming Acid #70.. A3
Lo-Foam Circulation Cleaner ... A1
Mac Bottle Wash...................... A2
Mac Complete........................... A2
Mac Low Foam Floor Cleaner.. A4
MAC Chelated Heavy Duty
    Cleaner............................. A2
MAC Defoamer.......................... A1
MAC Econo-Foam...................... A1
MAC Foam.............................. A1
MAC General Cleaner X........... A1
MAC General Cleaner XL......... A1
MAC Heavy Duty Cleaner No.
    1................................... A2
MAC Heavy Duty Cleaner No.
    2................................... A2
MAC Powdered Acid................. A3
MAC PL Bottle Wash A2
MAC Rust Stripper A2
MAC Smokehouse Cleaner A2
MAC Super Ride H2

MAC Super-AM.......................... C1
MAC Super-Solv........................ A1
MAC Trolley Kleen.................... A2
Oven Cleaner A1
Plastic Case Wash A1
Self Foaming Acid #75 .... A3
Shell Kleen #2........................ Q1
Shell-Kleen #1........................ Q1
Spectacular MAC Floor
   Cleaner................................ A2
Trolley Oil H1

Miller-Norris Company, Inc.
Val-U-Line Clear Lemon
   Disinfect. C2
Val-U-Line Glass Cleaner........... C1
Val-U-Line Lime Cleaner........... A4
Val-U-Line Pink Lotn Hand
   Dish.Det. A1

Miller-Stephenson Chemical Co., Inc.
MS-180 Freon TF Degreaser K2

Milport Chemical Company
Chloro Crystals........................ A1
Coil Clean A3
Detergent Zero-Zero A1,B1
Dry Acid Cleaner..................... A3
General Purpose Cleaner A1
Heavy Duty Cleaner A2
Heavy Duty Liquid Acid
   Cleaner................................ A3
Inhibited Chemical Water
   Meter Cleaner A3
Liquid Acid Cleaner A3
Mil-Dri J1
No Specx Alkali A2
No Specx Alkali #63W G7
No Specx Alkali No.52.............. A2
No-Specx Alkali No. 42............. A2
No-Specx Alkali No. 43............. A2
No-Specx Alkali No. 46............. A2
No-Specx Alkali No. 47............. A2
No-Sprecx Alkali No. 45 .... A2
No-Sperx Alkali No. 55............. A2
Phospho-Clean A3
Smoke House Cleaner A2
25% Coconut Oil Hand Soap..... E1

Milsolv Companies, The
Blandol NF 85 .... H1
White Mineral Oil 70 Technical H1

Mingo Sales & Service Co.
Rust-Solve and Mositure Guard H2

Minnesota Central Supply Company
D-600 Degreaser....................... A4
S-20 All Purpose Cleaner A4
Sani-Surf 2................................ D1

Mione Manufacturing Company
Special C2

Miracle Corporation
Alkaline Powdered General
   Cleaner................................ A1
Aluminum Cleaner A3
Liquid Alkaline Cleaner............. A1

Miracle Suns, Inc.
Power-Mite A4

Miral Chemical Corp.
Super Solv #2............................ A4,A8

Miranol Chemical Company, Inc.
C2M-SF Conc............................ A1
Mirapon FBS A1

Mirataine CB............................ A1
Mirawet B................................. A1

Miraq Corporation
Gold-Chem Cleaner 22NP........ A1
Gold-Chem Cleaner 22XX........ A1
Gold-Chem Cleaner 33TT......... A1
Gold-Chem Cleaner 44TT......... A1
Miraq B1

Misan Chemical, Inc.
Ammoniated Wax Stripper....... C1
Break Loose A1
Concentrated
   H.D.StmClng&Press.Wash.Cmpd A1
Concentrated Liquid Acid
   Cleaner................................ A3
Foaming Degreasing Cleaner..... A1
Heavy Duty All Purp Cl Wax
   Strip & Dgrs......................... A4
Micro Jet L2
Multi-Purpose Cleaner A1
Thermex L1

Misco Chemical
Unique A4,A8

Misco International Chemicals, Inc.
Abrasive Action Cleaner C2
All-Purpose Cleaning Solvent... K1
Anti-Static Spray....................... P1
Auger...................................... L1
Chain And Cable Lubricant H2
Chemical Cleaner A3
Concrete Seal........................... P1
D.I.G. Deodorant Insect.
   Granules F2
Disinfectant Deodorant Spray... C1
Dyna-Mic A4
Electrical Equip. Protector
   (Aer.) P1
Electrical Equip. Protector
   (Bulk) P1
Eliminate C1
Emulsifier/Homogenizer............ C1
Finest A1
Floor Prep P1
Foam....................................... C1
Foam Suppressant L1
Garb-Ex.................................... C1
Glycol Disinfectant Deodorant
   Spray................................... C1
Grease Trap Cleaner................. C1
Heavy Duty Concrete Cleaner .. C3
Instant Action L1
Liquid Boiler Treatment G6
Liquid Hand Soap E1
M.I.C. Bowl & Bathroom
   Cleaner................................ C2
Mighty.................................... A4
Misco Little Green Granules C1
Misco Pots 'N Pans A1
Misco Waterless Hand Cleaner.. E4
Mud & Silt Remover................. L1
MLC Laundry Compound B1
Nature's Method....................... L2
Neutra-Clean A1
No Flash.................................. C1
Non-Residual Space Spray Bug
   Killer................................... F1
Oven & Grill Cleaner................ A8
Paint & Varnish Remover C3
Power A1
Profl. Machine Dishwashing
   Compound............................ A1
Removes C1
Residual Roach, Ant&Bug
   Killer................................... F2
Safety Solvent.......................... K1
Sewer Block Buster................... L1

Sewer Deodorant C1
Spray-Clean Oven&Grill
   Cleaner................................ A8
Stainless Steel Treatment
   (Aerosol).............................. A7
Stainless Steel Treatment (Bulk
   Liquid)................................. A7
Steam Cleaner Concéntrate A1
Super Paint & Varnish
   Remover C3
Super Safety Solvent(Aerosol)... K1
Super Safety-Solvent(Bulk) K2
Super Steam Cleaner
   Concentrate A1
Sustain Multi-Purpose Floor
   Cleaner................................ A4
Tar Remmover K1
Tire Marks Remover................. K1
Touch of Mint C1
Triple C K1
Two Faze K1
W.E. Degreaser K1
Water Tower Treatment............. G7
4-Way Bowl & Bathroom
   Cleaner................................ C2

Misco Products Corporation
All Purpose Cleaner A4
Cleaner-Degreaser A1
Concentrated Pot and Pan
   Cleaner................................ A1
Creme Lotion Hand Soap.......... E1
Hand Soap Coconut Oil Base..... E1
Heavy Duty Butyless Cleaner-
   Degreaser A1
Liquid Detergent A1
Liquid Steam Cleaning
   Compound............................ A1
Quat Germicidal Cleaner........... D1

Mission Chemical Company
A-1 .... F1
Ammonia Number 2.................. C1
Ammonia, Sudsy Household C1
Bactoll D1,Q3
Big M Egg Washing Compound Q1
Compound D-6......................... A1
Compound KF.......................... A4
Compound SC-25 .... A2
Compound SC-26 .... A2
Fast-X-Crystal Cleaner A4
Liquid Soap #1.......................... A1
Mical....................................... A4
Milmite A4
Misco Suds A4
Mission Cleanser A1
Mito A1
Pal-O-Clean C2
Pala Bleachit D2
Pala Killer F1
Pipe and Drain Cleaner L1
Roachcide................................ F1
Sanitary Toilet Fluid................. C2
Spray-93 Concentrate............... A4
Tri Clean A1

Mission Kleensweep Products, Inc.
Eezo.. L1
Kleenol 15 .... A4
Kleenol 30 .... A4
Purinse.................................... D1
Wonder Coconut Oil Liquid
   Soap E1
Wonder Concrete Floor
   Compound............................ C1
Wonder General Purpose
   Cleaner................................ A1
Wonder Hvy Dty Steam Mach.
   Cmpnd................................. C1

178

Mobiltemp 78 .... H2
Mobiltherm 603 .... P1
Mobilube C-140 .... H2
Mobilube HD 80/90 .... H2
Mobilube HD 85 W 140 .... H2
Mobilux EP 0 .... H2
Mobilux EP 023 .... H2
Mobilux EP 1 .... H2
Mobilux EP 2 .... H2
Mobilux No 1 .... H2
Mobilux No. 2 .... H2
Mulrex M H1,H3
Sovarex Grease No. 1 W H2
Sovasol #5 .... K1
Vactra Oil AA H2
Vactra Oil BB H2
Vactra Oil Extra Heavy H2
Vactra Oil Heavy H2
Vactra Oil Heavy Medium H2
Vactra Oil Light H2
Vacuum Pump Oil H2
Velocite #3 .... H2
Velocite #6 .... H2
Velocite Oil No. 10 .... H2
Whiterex 425 .... H1
600 W Cylinder Oil H2
600 W Super Cylinder Oil H2

Mobile Electric Company, The

Mobisolve A1

Mocho Industries, Incorporated

Mo-Co Flying Insect Spray F1
Mo-Co Industrial Roach & Mill
  Spry F1
Mo-Co Roach Spray with
  Diazinon F2
Mo-Kleen Ammoniated Soap C1
Mo-Kleen Dish Detergent A1
Mo-Kleen Heavy Duty
  Detergent A1
Mo-Kleen Liquid Hand Soap E1
Mo-Strip Floor Stripper&Wax
  Remvr A4
Mo-Strip Flr Strppr & Wax
  Remvr Am. C1

Modern Chemical, Incorporated

Blue Gold A4

Modern Research Corporation

A B Aluminum Cleaner A3
Alert P1
All-Bright A7
Astro A1
B-All B1
Bay-Kill F2
Belt-D P1
Big-C A1
Big-Mac A1
Big-Name B1
Brave A3
Break Thru L1
Bright-View C1
Bullit L1
Clean-N-Glo A1
Coil-X A3
Digest L2
Diplomat K1
Easy-Stripp C3
End Zip L2
F.P. Lube H1
Fast Action L1
Foam-Stick A1
Formula MR 777 .... K1
Frigid Lube H2
Heat A2
HP Control G6
Kleen N' Soft E1
Liquid Sander C3

LP Control G6
Med-Chem D1
Modereen C2
Modereen XL C2
Modern-Ox A3
Moderns 99' C1
Mr-Strength A1
MR 101 .... H1
MR 610 .... A4
MR 611 .... A1
MR 613 .... A1
MR 625 .... A4
MR 628 .... C1
MR 777 .... K2
MR-678 .... K1
Nice D1
O-GO G6
Odor-End C1
Open-Up L1
Oven Klean A1
Ox-Out G6
Phenol Germicidal Cleaner D1
Pinen C1
Power A3
Protect E4
Rat Kill F3
Rich E1
RCT-Away A3
Sanicon D1
Scrub-A-Matic A4
Smell Away C1
Sta-Bright C1
Strength L1
Strip-All C3
Super All A1
Super CC A4
Super Kill F1
Supreme F2
SSR C1
Tiger A1
True-Sect F1
Zodiac Vaporizing Insecticide ... F1

Moellering Supply Co., Inc.

Releze L1

Mogul Corporation, The

Mogul A-421 .... G7
Mogul A-423 .... G7
Mogul A-492 .... G7
Mogul A-495 .... G7
Mogul AG-414 .... G7
Mogul AG-415 .... G7
Mogul AG-416 .... G7
Mogul AG-420 .... G7
Mogul AG-452 .... G7
Mogul AG-460 .... G7
Mogul AG-461 .... G7
Mogul AG-471 .... G7
Mogul C-632 .... A3
Mogul C-633 .... A3
Mogul C-640 .... P1
Mogul C-641 .... A1
Mogul CL-630 .... A3
Mogul CL-631 .... A3
Mogul CL-646 .... G7
Mogul CL-647 .... G7
Mogul CL-648 .... G7
Mogul CL-649 .... G7
Mogul CL-652 .... G7
Mogul CL-653 .... G6
Mogul CL-654 .... G7
Mogul CO-941 .... G1
Mogul CO-983 .... G1
Mogul CO-984 .... G1
Mogul CO-985 .... G1
Mogul CT-601 .... G7
Mogul CT-602 .... G6
Mogul CT-655 .... G7
Mogul CT-656 .... G7

Mogul CT-657 .... G7
Mogul E-5202 .... G6
Mogul E-5224 .... G6
Mogul E-5225 .... G6
Mogul E-5241 .... G6
Mogul E-5303 .... G6
Mogul EG-5200 .... G6
Mogul EG-5201 .... G6
Mogul EG-5205C G6
Mogul EG-5207 .... G6
Mogul EG-5209 .... G6
Mogul EG-5215 .... G6
Mogul EG-5216C G6
Mogul EG-5218C G6
Mogul EG-5220C G6
Mogul EG-5223 .... G6
Mogul EG-5235C G6
Mogul EG-5237 .... G6
Mogul EG-5239 .... G6
Mogul EG-5240 .... G6
Mogul EG-5245 .... G6
Mogul EG-5246 .... G6
Mogul EG-5247 .... G6
Mogul EG-5248 .... G6
Mogul EG-5250 .... G6
Mogul EG-5255 .... G6
Mogul EG-5256 .... G6
Mogul EG-5320 .... G6
Mogul EG-5340 .... G6
Mogul EG-5341 .... G6
Mogul EG-5343 .... G6
Mogul EG-5360 .... G6
Mogul EG-5371 .... G6
Mogul H-201 .... G6
Mogul H-234 .... G6
Mogul H-237 .... G6
Mogul H-256 .... G6
Mogul H-282 .... G6
Mogul H-284 .... G6
Mogul HP-220 .... G6
Mogul HP-257 .... G6
Mogul HP-263 .... G6
Mogul HP-270 .... G6
Mogul HP-271 .... G6
Mogul HP-271 NR G7
Mogul HP-272 .... G6
Mogul HP-273 .... G6
Mogul HP-273A G6
Mogul HP-274 .... G6
Mogul HP-278 .... G6
Mogul HP-280 .... G6
Mogul HP-281 .... G6
Mogul HP-283 .... G6
Mogul HP-287 .... G6
Mogul IS-669 .... L2
Mogul O-303 .... G6
Mogul PC 1437 .... G6
Mogul PC 5239 .... G6
Mogul PC-1291 .... G7
Mogul PC-5200 .... G6
Mogul PC-5201 .... G6
Mogul PC-5202 .... G6
Mogul PC-5209 .... G6
Mogul PC-5223 .... G6
Mogul PC-5224 .... G6
Mogul PC-5225 .... G6
Mogul PC-5240 .... G6
Mogul PC-5241 .... G6
Mogul PC-5245 .... G6
Mogul PC-5248 .... G6
Mogul PC-5250 .... G6
Mogul PC-5256 .... G6
Mogul PC-5303 .... G6
Mogul PC-5320 .... G6
Mogul PC-5340 .... G6
Mogul PC-5341 .... G6
Mogul PC-5343 .... G6
Mogul PC-5360 .... G6
Mogul PC-5371 .... G6
Mogul PC1110 .... G1

Mogul PC1111 .... G6
Mogul PC1220 .... G7
Mogul PC1221 .... G7
Mogul PC1230 .... G7
Mogul PC1231 .... G7
Mogul PC1262 .... G7
Mogul PC1281 .... G6
Mogul PC1290 .... G7
Mogul PC1292 .... G7
Mogul PC1311 .... G7
Mogul PC1320 .... G7
Mogul PC1340 .... G7
Mogul PC1350 .... G7
Mogul PC1404 .... G6
Mogul PC1405 .... G6
Mogul PC1412 .... G6
Mogul PC1413 .... G6
Mogul PC1417 .... G6
Mogul PC1418 .... G6
Mogul PC1421 .... G6
Mogul PC1425 .... G6
Mogul PC1640 .... G6
Mogul PC1642 .... G6
Mogul PC1643 .... G6
Mogul PC1652 .... G6
Mogul PC1901A G1
Mogul PC1911C G1
Mogul PC1921N G1
Mogul PC1950 .... G1
Mogul PC1952 .... G1
Mogul PC2010 .... P1
Mogul PC2011 .... A1
Mogul PC5205C G6
Mogul PC5207 .... G6
Mogul PC5215 .... G6
Mogul PC5216C G6
Mogul PC5218C G6
Mogul PC5220C G6
Mogul PC5235C G6
Mogul PC5237 .... G6
Mogul PC5246 .... G6
Mogul PC5247 .... G6
Mogul PC5255 .... G6
Mogul Sanitizer D2
Mogul SL-320 .... G6
Mogul SL-340 .... G6
Mogul SL-341 .... G6
Mogul SL-343 .... G6
Mogul SL-360 .... G6
Mogul SL-362 .... G7
Mogul SL-371 .... G6
Mogul W-132 .... G2
Mogul W-138 .... G1
Mogul W-139 .... G1
Mogul W-180 .... G7
Mogul W-195 .... G7
Mogul W-198 .... G7
Mogul WS-110 .... G6,G2
Mogul WS-119 .... G3
Mogul WS-123 .... P1
Mogul WS-142 .... G7
Mogul WS-178 .... G7
Mogul WS-181 .... G7
Mogul WS-182 .... G7
Mogul WS-192 .... G7
Mogul WS-193 .... G7
Mogul WS-194 .... G7
Mogul 6110 .... G6,G1
Mogul 6130 .... G6,G1
Mogul 6139 .... G2
Mogul 6142 .... G7
Mogul 6161 .... G7
Mogul 6180 .... G7
Mogul 6182 .... G7
Mogul 6192 .... G7
Mogul 6198 .... G7
Mogul 6200 .... G6
Mogul 6220 .... G6
Mogul 6222 .... G6
Mogul 6230 .... G6

Mogul 6231 .... G6
Mogul 6232 .... G6
Mogul 6237 .... G6
Mogul 6240 .... G6
Mogul 6242 .... G6
Mogul 6257 .... G6
Mogul 6260 .... G6
Mogul 6261 .... G6
Mogul 6262 .... G6
Mogul 6263 .... G6
Mogul 6272 .... G6
Mogul 6273A G6
Mogul 6275 .... G6
Mogul 6276 .... G6
Mogul 6280 .... G6
Mogul 6281 .... G6
Mogul 6282 .... G6
Mogul 6283 .... G6
Mogul 6284 .... G6
Mogul 6287 .... G6
Mogul 6300 .... G6
Mogul 6303 .... G6
Mogul 6310 .... G6
Mogul 6340 .... G6
Mogul 6341 .... G6
Mogul 6360 .... G6
Mogul 6362 .... G7
Mogul 6371 .... G6
Mogul 6601 .... G7
Mogul 6630 .... A3
Mogul 6632 .... A3
Mogul 6633 .... A3
Mogul 6641 .... A1
Mogul 6646 .... G7
Mogul 6647 .... G7
Mogul 6649 .... G7
Mogul 6652 .... G7
Mogul 6653 .... G6
Mogul 6655 .... G7
Mogul 6657 .... G7
Mogul 6903C G1
Mogul 6913A G1
Mogul 7110 .... G6,G1
Mogul 7130 .... G6,G1
Mogul 7139 .... G1
Mogul 7142 .... G7
Mogul 7168 .... P1
Mogul 7180 .... G7
Mogul 7192 .... G7
Mogul 7198 .... G7
Mogul 7200 .... G6
Mogul 7222 .... G6
Mogul 7230 .... G6
Mogul 7232 .... G6
Mogul 7237 .... G6
Mogul 7256 .... G6
Mogul 7257 .... G6
Mogul 7260 .... G6
Mogul 7261 .... G6
Mogul 7263 .... G6
Mogul 7272 .... G6
Mogul 7273A G6
Mogul 7275 .... G6
Mogul 7278 .... G6
Mogul 7280 .... G6
Mogul 7281 .... G6
Mogul 7283 .... G6
Mogul 7300 .... G6
Mogul 7303 .... G6
Mogul 7303 .... G6
Mogul 7360 .... G6
Mogul 7360 .... G6
Mogul 7362 .... G7
Mogul 7371 .... G6
Mogul 7414 .... G7
Mogul 7416 .... G7
Mogul 7420 .... G7
Mogul 7423 .... G7
Mogul 7460 .... G7

Mogul 7471 .... G7
Mogul 7601 .... G7
Mogul 7630 .... A3
Mogul 7633 .... A3
Mogul 7640 .... P1
Mogul 7641 .... A1
Mogul 7646 .... G7
Mogul 7647 .... G7
Mogul 7648 .... G7
Mogul 7649 .... G7
Mogul 7652 .... G7
Mogul 7653 .... G6
Mogul 7654 .... G7
Mogul 7903C G1
Mogul 7913A G1
Mogul 9001A G1
Mogul 9003C G1
Mogul 9013A G1
Mogul 9025N G1
Resin Cleaner P1

Mogul Maintenance Systems Div. Of The Mogul Corporation

Mogul Break-Foam A1
Mogul H.D.C. A4,A8
Mogul P.W.D. A4,A8

Molectron Company

Food Lube H1
Food Machinery Grease H1
Food Machinery Grease EP H1
Silicone Lube-Food Plant
  Grade H1

Mollen Chemical Company

Bowl Cleaner C2
Deodorized Toilet Fluid C2
Drain Solvent L1
Liquid Hand Soap E1
Moladine Odorless Disinfectant . D1
Molopine Phenol Coef. 3 .... C1
Molopine Phenol Coef. 5 .... C1
Molopine Phenol Coef. 5.5 .... C1
Super Clean Concentrate A4
Tri-Mol A1

Mom Chemical Company, Inc.

"Ezy" P1
"Wow" K1
Action Seal P1
Attack F2
Ban 22 Insecticide F2
Bang A4,A8
Bang w/a C1,A8
Bi-Chlor C1
Blast C1
Bleach-Eez D1
Blow Out G6
Blow Out 'A' A3
Borated Hand Soap C2
Bowl-Kleen C2
Bright Window Spray C1
Carb-O-Sol K1
Carbo-Tronic Super A8
Carbonic-Tronic A8
Cherish A1
Chloro C1
Chloro RF-2 .... C1
Commercial Ammonia C1
Concrete Cleaner A4
Control Tabs G7
Culture L2
D-Foam A1
D-Stroy P1
Deco Grip Belt Dressing P1
Degrease A4
Deluxe Chlorinated A1
Deodorizing Brightener A1,B1
Drain Out Liquid L1

Monterey Sanitary Supply, Inc.
"Power Foam" A4
Sparkle-Wash A1

Montgomery Chemical Co.
Swish Aerosol Electrokleen K1

Montrose Chemical Specialties Co.
Bio-Fecto C1
Bio-PNB 20 .... C1
Bio-San 10 No Rinse
  Disinfectant Santzr. D2
CS-752 .... D1
Monocem 1021 .... A4
Monocem 1812 .... A4,A8
Monocem 217 .... A4
Monocem 545 .... A3

Moore Research, Inc.
Bowl Cleaner C2
Germa-Klenz A4
Moore Kill Insect Killer F1
Moore Kill Insecticide F1
Moore No. 18 .... A1
Moore No. 36 .... A4
Moore Residual Roach 'N Ant
  Killer F2
Moore Stainless Steel Cleaner A7
Morcept Hospital Disinfectant
  Deodorant C2
No-Mor-Marx K1
Power Kleen C1
Super Strength Insecticide F1
Triad Germicidal Cleaner
  Deodorant C1

Moorman Manufacturing Co.
Fly Bait F2
Fly Spray F2

Morco Chemical Company
Amigo A1
Big Blue Kreme Kleanser A6
Big Dog C1
Coil-Ite A4,P1
Combat C2
Granular Sewer and Drain
  Solvent L1
Industrial Aerosol Insecticide F1
Iodine-175 .... D1
L.P.S. Liquid Pot Soap A1
L-Tox Spray F1
Liminox F2
Odo-Rout C1
Pan Right A1
Pine-Fresh C1
Reflect B1
Sani-Lube H1
Splash A1
Super Glo A1
Ten & Out L1
Ultra Glo B1

Morgan-Gallacher, Inc.
Acid #40 .... A3
Acidisol D1
All Brite A1
ALK-1 .... A2
B-Brite #2 .... A2
Bio-Clor D1
Bio-Quat D2
Biodisan D2
Blue Foamer A1
Brite A6
C W C 2 .... A1
Candidate Egg Wash Q1
Canners AP A1
Chem Brite A2
Cir Lac A3

Cir-Q-Sol A1
Clor-VL. A2
Concentrated All Purpose
  Cleaner A1
CSD A2
Duo-Clor A1
Duo-Clor-Pink A1
Egg Cleaner Sanitizer D1
Floor Cleaner A4
FM-1000 Pan Washer A1
Gleam A1
GSD A2
Heavy Duty MSR A3
Hi-Glow A2
HM-6 .... A1,G1
Improve A1
Kount Down D1
M & G Foam Breaker Q5
Nu Brite A2
Organic Remover STU A2
Pan Safe A1
Powdered MSR-Inhibited A3
PDQ All Purpose Clnr
  Concentrate A1
Resque A4
Shur-San D1
Suds-All A1
Super Duty A2
Super Gleam A1
Tri-Clor A1
TSL-45 .... A2

Morrell Chemical Company/ Division of John Morrell Co.
All Purpose Liquid Detergent ... A1
All Purpose Liquid Detergent
  Conc Cln A1
Aluminum Pan Cleaner A6
Anti-Stick Silicone Spray H2
Beef Shroud Detergent B1
Car and Truck Wash
  Compound A1
Chain Saver No. 120 .... H2
Chain Saver No. 60 .... H2
Chlorinated Scouring Powder A6
Circuit A1
Clean Hand FF E2
Cold Motor Cleaner K1
Commercial Laundry Soap B1
Complete Foam A2
Dish Wash Cleaner A1
Drain Cleaner L1
Floor Cleaner A4
Freezer-Locker Cleaner A5
General Purpose Chlorinated A1
General Purpose Cleaner A1
H.D. Sanitizer D1
Heavy Duty Laundry
  Detergent B1
Heavy Duty Stripper A2
Hog Scald C.S. N2
Hydrated Lime N2,N3
Kettle Wash 12.25 .... A1
Laundry Additive B1
Liquid Alkaline Cleaner #10 ..... A2
Liquid Smokehouse Cleaner A8
M-11 Liquid Hand Soap E1
M-6 Bowl Cleaner C2
Mild Suds Detergent A1
Miton Chlorinated Detergent A1
Mor Acid 61 .... A3
Mor Alkaline 50 .... A2
Mor-Chlor A1
Mor-O-Dine D2
Mor-San D2,E3,Q3
Moracid A3
Moracid 30 .... A3
Morrell Alkaline Cleaner No.
  10 .... A2
Morrell Alkaline Cleaner No. 5 . A2

Morrell Baygon 311 .... F2
Morrell D.S. Liquid D1
Morrell Egg Wash...................... Q1
Morrell Foamer A1
Morrell Liquid Alkaline
　Cleaner No. 5............................ A2
Morrell Liquid Auto Wash........ A1
Morrell Liquid Circuit A1
Morrell Liquid Hog Scald N2
Morrell Malathion Concentrate . F2
Morrell Multi Purpose Liquid
　Cleaner.................................... A1
Morrell Pyrenone B F1
Morrell Residual #1................... F2
Morrell Residual "D"
　Insecticide............................... F2
Morrell Scald Plus..................... N2
Morrell Semi-Concentrate #2 .... F1
Morrell Space Spra #2 .... F1
Morrell Space Spra #3 .... F1
Morrell Special Scald Aid N3
Morrell Tanners' Special Form.
　Insect..................................... F2
Morrell 57% Malathion
　Concentrate F2
Neutralon-1002.......................... N3
Pro Strip.................................... A2
S.F. Acid Brightener A3
Scal-Curtale................................ G5
Smoke House Cleaner A2
Soda Ash N2,N3
Soft Soak A1
Special Additive A1
Special Contact Spra.................. F1
Special Mill Spray..................... F1
Special Tripon N3
SC-15 Sewer Cleaner L1
SC-15 Sewer Cleaner L1
Tripon.. N3
Trolley Wash and Rust
　Remover A2
Water Deposit Remover............. A3
Whiton 101................................ N3

Morris Litman

Mor-Eez Detergent-Cleaner-
　Disinf. D1
Mor-Eez Lotionized Pink
　Detergent................................ A1
Mor-Eez Mint Disinfectant C2
Mor-Eez No. 195....................... H2
Mor-Eez No. 200........................ E1
Mor-Eez Pine Oil Disinfectant... D1
Twist-Eez A1
X-Cel-Z Concentrate.................. A1

Morris Paper & Chemical Co.

Aci-Det...................................... A4
Big-Bull A1
Blue Butyl Cleaner A4
Bowl Kleen C2
Foxy Blue Concentrate.............. A4
Golden Gator C1
Golden Pine C1
H-D 321 Concentrate.................. A4
Lift A4,A8
Red Devil Concrete Cleaner...... A1
Superior Concrete Cleaner A2
Total Break-Up.......................... A1

Morrison Products, Inc.

Auto-Kut A4
Bomb-Bard A1

Morse, F.J., & Company, Inc.

Scoot.. A4

Morton Pharmaceuticals, Inc.

Silicone Sanitary Lubricant
　Spray...................................... H1

Spra-Lub Sanitary Lubricant
　Spray...................................... H1

Morton Salt Company

Morton Pellets P1
Purex Salt P1
White Crystal Northern Rock
　Salt P1
White Crystal Solar Salt............ P1
White Crystal Southern Rock
　Salt P1

Motomco, Incorporated

Aquadyne EC A1
Pival Concentrate....................... F3
Water Soluble Pivalyn F3
2% Pival Concentrate................. F3

Moulder-Oldham Company

No. 40 Pearl Liquid Toilet Soap E1
Pearl Liquid Toilet Soap E1
Wallop Institutional Foaming
　Degreaser A1

Moultrie Distributors

Caustic Clean #10...................... A1
Rich n' Thick Hand Cleaner E1
Super Clean 500......................... A1

Mozel Chemical Products Co.

CIP-MID Shroud Cleaner A1
Galvoline A3
Mozel No. 351 .... A1
Mozel No. 472 .... A3
Mozel No. 52 .... A1

Mr. Aerosol

No. 25 Insect Spray.................... F1
Sani-Lube Food Equipment
　Lubricant................................ H1

Mr. Janitor

Blue Power Concentrate.............. A4
Dynamo Foam............................ A1
Grime Scat A4

Mt. Hood Chemical Corporation

Al-Kleen.................................... A1
Boomerang A1
Bulldozer A1
C-80.. A1
Cascade Granular Bleach B2
Cascade Granular Bleach B1
Challenge................................... B1
Chlor-A-Sanitizer B1
Crown Sodium Hypochlorite D2,N4,G1
D.G.C.. A2
Dandi-Suds................................ A1
Deo Solv B1
Detergo-Suds............................. B1
Diamond.................................... A4
Dri-Walk C1
Dynamo..................................... A1
Floor Matic A4
Foam Kleen A3
Foamz-It.................................... A1
Formula 100 .... A1
Formula 21................................ A1
Formula 26................................ A2
Formula 39................................ A2
Formula 43................................ A1
Formula 510.............................. A1
Formula 540.............................. K1
Formula 560.............................. A1
Formula 561.............................. A1
Formula 563 Steam Cleaner A1
Formula 699.............................. L1
Formula 711.............................. A1
Formula 750.............................. A2

Formula 790............................... A2
Gone.. A1
Gorilla....................................... A2
Hi Jet A1
Hoodi-Kleen.............................. A1
Ice Away.................................... C1
Ido-Sept Disinfectant D2
IOF Sanitizer D2
Klenz-All................................... B1
Kreme 'N Kleen A6
Lift A4
Liquid Sani-Fluff....................... B2
Liquid Soft & Fresh B2
Mineral D.................................. B2
Mira-Tex................................... B2
Mira-Tex................................... B1
No Spot A1
Now... D2
OMO.. A1
Power Scrub A1
Quadrex E1
Ramrod A1
Re-Nu.. A1
Rinse-Well A2
S-44... B1
S-77... B1
S-88... B1
S-99... B1
Sani Det Sanitizer...................... D1
Sani Fluff.................................. B2
Sani Guard A4
Sani Wash Sanitizer................... D1
Sani-Cleaner Sanitizer............... D2
Scalex.. A3
Shell Game................................ Q1
Shell White Q1
Silver Brite B1
Soft & Fresh.............................. B2
Star.. A1
Sterling A1
Strike A2
Strip A1
Strip Eaze C1
Super Alk A2
Super Re Nu.............................. B1
Super Re Ny.............................. B1
Super Shine Brite A1
Super Trio A1
Super Tripe Kleen N3
Super White B1
Super White B2
Surf Sanitizer D1
Surgatol E1
Surgatol Ready to Use............... E1
Total B1
Tri-Tex C2
Trio A1
Tripe Kleen N3
Triumph..................................... B1
Triumph..................................... B2
Trix B1
Washing Powder A1
Wonder...................................... A3
Zippo A3

Mullen, E. C. Company

777 Food Machinery Grease...... H1

Multi-Chem Industries, Inc.

B-102 Phase Two Bowl Cleaner C2
BCA-3535 Shower Program
　Step 1..................................... C1
BTC 400 .... P1
D.D. 107.................................... C2
Deep Kleen C2
DLF-1275 Shower Program
　Step 2..................................... A4
F-E-L Food Equipment
　Lubricant................................ H1
F-E-L Food Equipment
　Lubricant................................ H1

Zeolite Steam & Return Line
  TMT 1000-4L G7
Zeolite Steam &Return Line
  TMT #1000-2-L G6
Zeolite Steam &Return Line
  TMT #1000-3-L G6
Zeolite Trpl"S"Plus Non-
  Poltng 1300-13L....................... G7
Zeolite-BFWT-1200-2P.............. G6

National Chemical Labs. of PA., Inc.

Acti-Pine Disinfectant C1
All-Off....... .... E1
Astro-Chem Degreaser Cleaner A4,A8
Bacti-Chem.................................. D1
Bacti-Cream E1
Big Punch Oven & Grille
  Cleaner................................... A8
Blast A1
Bullseye A1
C-22 Stainless Steel Cleaner
  Polish A7
Chem-Eez.................................. A1
Conkleen 204............................. A4
Deo Pine Oil Deodorant............. C1
Dermacide.................................. E2
E C Conc. Dish Wash A1
Economy Butyl Cleaner
  Degreaser A4
Flicker N1
Formula NBC All Purpose
  Degreaser Clnr........................ A4
Fresh Air C1
GC12 Protective Shield Cleaner A7
Hurrah A1
Ionox Disinfectant Cleaner D2
Kleer-Brite C1
Lazer Wax & Floor Finish
  Stripper A4
Liquid Hand Soap E1
Micro-Chem................................ D2
Mighty Giant Cleaner A1
Mirage A1
Nac Lube................................... H2
Nac Mint Disinfectant................ C2
Nac Pine Odor Disinfectant C1
Nac-X Disinfectant-Cleaner A4
Naca Pine Oil Disinfectant C1
Nactrol...................................... C1
Neutral Cleaner A4
Nu Hide A7
NAC-Safe Super Safety Solvent K1
NL Nuclear Cleaner Degreaser . A4
Pine Scrub Soap C1
Pink Lotion Hand Soap.............. E1
Pink Panther All-Purpose Conc.
  Cl w/Ion................................. A4
Pooff Foam Degreaser and
  Cleaner................................... A1
Pow... D1
Power-Eez Detergent A4
Quarry Tile Cleaner A4
Rainbow B1
Razo Cleaner.............................. A4
S.D. Ammoniated Wax Stripper A4
Sani-Derm E3
Scum Remover and Renovator.. A1
Septisurge E1
Septisurge Concentrate............... E1
Steam & Pressure Cleaner A1
Strike Oven & Grill Cleaner A8
Super Cherry Deodorant............. L1
Super Lube D H2
Super Nac Concentrate............... A1
Syn 40 Cleaner.......................... A4
Synglide Conveyor Lubricant.... H2
Synglide H.F. Conveyor
  Lubricant................................ H2
SS 5 Safety Solvent................... K1
Wash-Brite................................. B1

Wild Cherry Deodorant C1
Yukon Heavy Duty Solvent
  Cleaner................................... C1
10X Concentrate......................... C1
800 Conveyor Lube.................... H2

National Chemical Labs, Inc.

BW-202...................................... G6
BW-203...................................... G6
BW-214...................................... G6
BW-216...................................... G6
BW-218...................................... G6
BW-219...................................... G6
BW-224...................................... G6
BW-231...................................... G6
Formula No. Special #43-T G6
Formula No. Special #43-59 ...... G6
Formula No. Special #55-71 ...... G6
Formula No. Special NL-25....... G6
Formula No. 2-52-T Boiler
  Water Treat............................. G6
Formula No. 52-50A-T Boiler
  Water Treat............................. G6
RL-301...................................... G6
RL-303...................................... G6
Special Formula.......................... G6
Triangle Brand Anti-Foam......... N1

National Chemical Products

Blue Solvent Cleaner A1

National Chemicals, Inc.

Blu-Magik.................................. A3
D-B-C Drainboard Cleaner A1
LFD Low Foaming Detergent.. A1
Q.A... D2
T-D-C Detergent......................... A3

National Chemsearch/ Division of NCH Corporation

A-Qulex 288 .... P1
A-Qulex 312 .... P1
Actamine C1
Aerolex H2
Aerosol Concentrate F1
Al-Chek..................................... G7
Aquafog F1
ABF-42...................................... G7
B-Lube....................................... P1
Bactrol (Aerosol)........................ C2
Bactrol Concentrate C2
Bio-Pure A4
Bio-Quad C1
Blue Wizard Cream Cleaner....... C2
Boil-Tane G7
Brace Concentrated Jel
  Degreaser K1
Brex Concentrate Chemical
  Cleaner................................... A3
Camelia C2
Cease... H2
Chem-A-Lube ML-770 .... H2
Chem-A-Lube WL-660 Multi-
  Purp. Wht Lub........................ H2
Chem-Aqua 100.......................... G6
Chem-Aqua 1444 .... G7
Chem-Aqua 1500 .... G7
Chem-Aqua 200.......................... G6
Chem-Aqua 250.......................... G6
Chem-Aqua 300.......................... G7
Chem-Aqua 4000 .... G7
Chem-Aqua 8000 .... G7
Chem-Aqua 900.......................... G6
Chem-Fog................................... F1
Chemalube F-880........................ H1
Chemene-X.................................. D2
Chemlyte.................................... A3
Chemstrip................................... K3
Chemstrip (Aerosol).................... C3

Chemzyme I................................ L2
Cleanamatic................................ A4
Clix Degreaser K1
Coil-Trate P1
Concentrate A1
Control 77 .... C1
Convoy....................................... H2
De Luxe Liquid Hand Soap........ E1
Decade....................................... A1
Deox A3
Di-Ron....................................... F2
Di-Sect...................................... F2
Dichloron Insecticide................. F2
Dimon....................................... H2
Dishine...................................... A1
Duo-Power................................. A1
Dura-Gard................................. P1
Enerlex Multi-Grade Hydraulic
  Oil H2
Enforce...................................... A4
Etch Klenz A3
Everbrite.................................... A4
Eze-Way..................................... H2
Fad A1
Flair C1
Flair (Aerosol)........................... A1
Foam-Ad A1
FL-100 Packing House Cleaner . A2
Gearco Extrm Presr Gear Oil
  SAE140.................................. H2
Gearco Extrm Presr. Gear Oil
  SAE90.................................... H2
Gearco Extrm Press. Gear Oil
  SAE 80W-90........................... H2
Gearco Extrm Press. Gear Oil
  SAE 85W-140.......................... H2
Germene A4
Get Out B1
Gex H2
Glisten C1
Glo-SS A7
Gly-Con..................................... A1
Great Giant A1
Heaven-Soft B2
Hi-Brite..................................... A1
Hydrocide.................................. A4
I-So-Sect.................................... F2
Killzol Brand F1
Kilzone...................................... F2
Lexite.. K2
Limex.. A1
Liquid Chem-Zyme..................... L2
Lube-Plus................................... H2
Lube-San.................................... H1
Lube-Shield................................ H2
Lubrease.................................... H2
Lustra A3
Maldane..................................... F2
Melts... C1
Method...................................... A1
Micro-Treat Parts A&B............... L2
Neldracide................................. F2
New Can San Concentrate C1
Nu-Concept................................ C2
NC-123 (Aerosol) K2
NC-123 (Bulk)........................... K2
ND-150....................................... A1
ND-20... A2
ND-25 Compound A3
ND-66... L1
NDK... A2
Open Road C1
Ov-Care A8
P-O-W Wasp Spray..................... F2
Patrol F1
Pirate K1
Power Plus Germ. Food Plant
  Cleaner................................... A4
Power Plus Heavy Duty
  Cleaner................................... A1

188

Odo-Rout.. C1
Trak Lube H2

National Detergents, Inc.
Deodorant Nuggets...................... C1
Fast Oven & Grill Cleaner A8
Gentle Power................................ A1
Glisten C2
Great... A4
Heavy Duty Steam Cleaner A1
Lemon Hand Soap E1
LF Suds... B1
Quardizyme.................................. L2
S.A.I.D. II..................................... E1

National Environmental Chemical Corp.
Blue Boy Liquid Cleaner
  Concentrate.............................. C2
Blue Boy Solid.............................. C2
Chem-Mite A1
Citrusol A4
Di-Date... F2
Di-Date Residual Insect Spray... F2
Exothermic.................................... L1
Spotless... A4
V.I.P. 1000 .... F1
V.I.P. 600 .... F1
VIP No. 500 Insect Spray F1

National Institutional Food Distributor Associates, Inc.
Ammoniated All Purpose
  Cleaner...................................... A4
Bowl Cleaner C2
Concentrated Dry Bleach........... B1
General Purpose Cleaner............ A1
Grill-Oven & Fat Cleaner A1
Heavy Duty General Purpose
  Cleaner...................................... A1
Heavy Duty Machine Scrubber
  Concentrate.............................. A1
Institutional Liq Drain Pipe
  Opn &Maintr............................ L1
Liquid Hand Dish Wash.............. A1
Machine Dish Wash Compound A1
Supreme Low Suds Laundry
  Detergent.................................. B1
335 Institutional Deep Fat
  Fryer Cleaner............................ A8
945 Alkaline Degreaser & Oven
  Clnr Conc.................................. A8
950 Extra Heavy Duty
  Degreaser A4,A8

National Janitorial Supply
The Eliminator A1

National Janitors Supply Co.
Nasco Ordorless Cleaner
  Disinfectant D2
Nasco Quat 80 .... D2

National Laboratories
Aerozone Mist C1
Amphyl.. A4
Amphyl Spray C2
Barrage All Purpose Cleaner A1
Barrage Concentrate Bowl
  Cleanse...................................... C2
Brisk.. A1,Q1
Con-O-Syl Disinf.-Detergent..... A4
Conpact A1
Duel Crystal Fragrance C2
Duel Golden Fragrance............... C2
Fasolv K1
Fasolv A1
Galley Oven-Grill Cleaner A8

Institutional Lysol Brand Disnf
  Spray.. C1
Lime Shine A3
Liquid Power................................ A1
Lysol.. C2
LF-10... D1
N-L Concentrate A4,E1
N-L Creme Cleanser A6
N-L Degreaser101 .... K1
N-L No Phosphate Cleaner........ A4
Neutrodor..................................... C2
New Roccal Mist Air Sanitizer
  Module...................................... C1
New Tergiquat Germicidal
  Cleaner...................................... D1
Osyl... A4
Premeasured Tergisyl
  Detergent Disnf. C1
Quatsyl 256................................... D1
Roccal Brand Sanitizing Agent
  10% D1,Q3
Roccal II 10% D2,Q3,E3
Roccal Mist C1
Scene... C1
Scene Six C1
Shimmer A7
Shimmer (Aerosol) A7
Shimmer Multi-Purpose Cleaner
  Polish A7
Shop Industrial Waterless Hand
  Cleaner...................................... E4
Soaperfect..................................... A4
Soaperior E1
Tergiquat D1
Tergisyl A4
Titan Bulk Liq. Fogging
  Insecticide................................ F1
Titan Concentrate Residual
  Insecticide................................ F2
Titan Dual Action....................... F2
Titan Fly Bait (Blue).................. F2
Titan Insect Killer F1
Titan Mist Insecticide Module ... F1
Titan Supreme F2
Vani-Blocs.................................... C2
Vani-Sol Bowl Cleanse............... C2
Vani-Sol Bulk Disinfectant
  Washroom Clr.......................... C2
Vani-Sol Disinfectant
  Washroom Cleaner.................. C2
Vani-Sol Per Diem C2
Vani-Sol Washroom Drain
  Opener...................................... L1

National Milling & Chemical Co.
"447"... A1
AL2 Cleaner A1
B-263 Star Chain Lube H2
Bake Pan Cleaner A/S................ A1
Borax Base Hand Soap C2
C-25 Hand Soap........................... C2
Cling Det....................................... A6
Coco Tallow 40% Liquid Soap. H2
Compound 458.............................. A1
CLD A1
Deep Fat Fry Cleaner................. A2
Delimer... A3
Deluxe D.W................................... A1
FDA Hand Soap A1
General Purpose Cleaner........... A1
Gentle Giant Powdered Hand
  Soap C2
H. D. Deluxe A1
H.D. Chlorinated Cleaner........... A1
Hand Suds A1
Hi-Suds A6
K-2... A1
Mechano-Det A4
Namico Abrasive Detergent...... A6
Namico CT Flakes B1

189

Namico Green Bar Soap............. H2
Namico Power-Chlor................. A1
Namico Sour "B"...................... B1
Namico SPTX A2
Namico 444 .... A4
Namifluff............................... A1
Namilube............................... A1,H2
Namilube 1058 .... H2
Namisuds............................... A1
Namisyn Flosuds A1
Pot and Pan Cleaner A1
Powdered Grill Cleaner............. A2
PR-261-P A1
Rinse Aid............................... A1
Smoke House Cleaner Heavy
  Duty.................................. A2
Smokehouse Cleaner A2
Special "B" Cleaner A1
Special C Compound A1
Super H.D. Chlorinated
  Cleaner.............................. A2
Supreme Steam Cleaner
  Granules............................ A1
Supreme Steam Cleaner Liquid . A1
Sure Suds............................... A1,B1
15% Cocoanut Oil Soap E1
40% Cocoanut Oil Soap E1
462-G B1

National Products Co.

Coach House Non-Butyl
  Cleaner.............................. A1

National Purity Soap & Chemical Co.

#65 Prime Oil Soap H2
Ace Bar Lube H2
Arrow A3
Aseptic-Clean......................... A4
Desolve.................................. A1
Emulso Degreaser A1
Fabri-Soft Green B2
Garfloor................................. C1
Glass Shine............................. C1
Glo Chlor B Dishwash A1
Kwi Kleen Hvy Dty Acid
  Detergent............................ A3
No. 20 All Coco Liquid Hand
  Soap E1
No. 40 Coco Castile Liquid
  Soap E1
No. 60 Detergent Cleaner......... A1
NPS Fabri-Soft Powser PN
  #180.................................. B2
NPS Glownite Reg. Dishwash... A1
NPS Pipeline And Bulk Tank
  Cleaner #711....................... A1
NPS Proto-Clean...................... B1
NPS Proto-Wash B1
NPS Super Glownite Dishwash. A1
NPS 100 Floor Cleaner C1
NPS 20 Liq. Surgical Soap........ E1
NPS 40 Liq. Surgical Soap........ E1
NPS 8A-15 Lube PN #625 .... H2
P.N.#213 Wall Cleaner-Heavy
  Duty.................................. A1
Plastic Gloss Cleaner A4
Purity Glass Wash A1
Rex Lo Suds B1
Rex-Heavy Duty B1
Rex-One-Shot........................... B1
Royal Oil Soap #50 .... H2
Satin Suds.............................. B1
Spik Cone. D5 Det. Floor
  Cleaner.............................. A4
Super Kwik Shine A1
Super Whip P.N. #106.............. A1
Syntha No. D2......................... A4
Trio C2
3-V Liquid Detergent A1

National Research & Chemical Co.

Lemon Liquid Soap E1
M & F Pot and Pan Cleaner A1
Rustex L-7.............................. A3
Tyfo Butchers 500.................... A1
Tyfo Chek.............................. A1
Tyfo Defoamer Q5
Tyfo Egg Destainer Q2
Tyfo Sewer Compound L1
Tyfokleen #300 .... A2
Tyfokleen CIP A1
Tyfoscald............................... N1
Tyfowash "A" Q1
Tyfowash "B" Q1
Ultrasonic Egg Cleaner............. Q1

National Sales And Distributing Co.

G P "66" A1

National Sanitary Supply Co.

Action-D A4
Big Break............................... A1
Blitz Residual Roach'N'Ant
  Killer................................. F2
Blitz 1932.............................. F2
Cloudy Ammonia C1
Crystal Glass Cleaner............... C1
D-Fen.................................... C2
D-Fen 30-40 .... A4
Electro Solve K2
Foam Up A1
Four-In-One C2
Household Bleach..................... B1
Industrial Extra Strength
  Bleach B1
Lemon D-Fen 1763 .... C2
Liqui-Terge............................. C1
Liquid Lemon Hand Soap E1
Lubri-Can 3055 .... H2
Majestic No. 11 Dish-Clean....... A1
Measure Up............................ A1
NBC 2107 Conc. Non-Butyl
  Deo. Clnr/Dgrsr.................... A1
NSS Cement Cleaner A4
NSS Concentrate Cleaner.......... A4
Pathfinder.............................. A4
Power Stripper A4
Power Wash 2623 Truck And
  Equipment Cl....................... A1
Satin 3250.............................. A7
Scum Clean A4
Slick Silicone Spray H2
Sud-N-Kleen A1
Teflex.................................... H2
Teflon Thread Sealer 3070........ H2
Vanguard 2660........................ K1
Wind-O-Clean......................... C1
Xtra Solve K2

National Sanitary Supply, Inc.

National Solv
  Ex.Hvy.Dty.Clnr.Degreaser... A1
Oven Brite Oven & Griddle
  Cleaner.............................. A8

National Sanitary Supply, Inc.

Foaming Action....................... A1
N Pink Suds A4,A8
NBS 180 Cleaner A4

National Specialty Products Co., Inc.

Odo-Chlor C1

National Wax Company

Flexoil 100.............................. H1

Nationwide Chemical Co., Inc.

All-Purpose Cleaner Blue........... A4
All-Purpose Cleaner Clear.......... A4

All-Purpose Cleaner Opal........... A4
Aluminum & Stainless Steel Cl.
  & Bright............................. A3
Block Buster Cleaner, Strip &
  Dgrsr Blue.......................... C1
Block Buster Cleaner, Strip &
  Dgrsr Opal.......................... C1
Block Buster Cleaner,Strip &
  Dgrsr Clear......................... C1
Compactor & Kitchen Aqueous
  Spray #4005........................ F1
Concrete Remover & Masonry
  Cleaner.............................. A3
Deodorant Granules.................. C1
Dishwashing Liquid (Hand)
  #509.................................. A1
Drain Flo,............................... L1
Flo-Go 308 .... C1
Germacidal Cleaner.................. C1
Heavy Duty All Purpose
  Cleaner.............................. A1
Heavy Duty Dgrs Tar &
  Asphalt Remover.................. K1
Heavy Duty Emulsifiable
  Degreaser K1
Heavy Duty Steam Cleaning
  Compound A2
Hvy Dty Wtr Sol Indust Cl&
  Dgrsr Purple....................... A1
Hvy Dty Wtr Soluble Indust
  Cl& Dgrsr Clr...................... A1
Hvy Dty Wtr Soluble Indust
  Cl&Dgrsr Red...................... A1
Industrial Aqueous Spray
  #4004................................. F1
Industrial Spray #4000.............. F1
Liquid Air Freshener #1038 ...... C1
Liquid Hand Soap A1
Mill Spray#4001...................... F1
Pine Odor Disinfectant C1
Pine Oil Disinfectant................ C1
Powdered Heavy Duty All
  Purpose Cleaner................... A1
Quat D1
Refuse Cleaner C1
Safety Solvent......................... K1
Sewer-Treatment L1
Simmons Silicone Lubricant
  #1073................................. H2
Soap Base Lubricant H2
Special Stripper #1 .... A1
Special Stripper #2 .... A1
Steam Cleaner Pressure Wash
  Blue A1
Steam Cleaner Pressure Wash
  Clear A1
Super Food Plant Fogging
  Insect. #4003....................... F1
Super Hvy Dty Wtr Sol Indust
  Cl & Dgrsr A1
Super Industrial Cleaner &
  Dgrsr. Clear........................ A4
Super Industrial Cleaner
  &Degreaser Blue................... A4
Super Industrial Cleaner
  &Degreaser Pink................... A4
Super Power Flo 1031 .... A4
Synthetic Cleaner A4
Vaporizer Concentrate #4002.... F1
X-Tra Heavy Duty All Purpose
  Clnr Blue............................ A4
X-Tra Heavy Duty All Purpose
  Clnr Clear........................... A4
X-Tra Heavy Duty All Purpose
  Clnr Pink............................ A4

Nationwide Chemical Corp.

Enzo-Gest............................... L2
Enzo-Gest Drain & Trap
  Cleaner.............................. L2

191

Liquid Steam Cleaning
Compound A1
LB 105 .... G6
LB 106 .... G6
LB 107 .... G6
LB 150 .... G6
LB 151-C G6
LB 152-C G6
LB 201 .... G6
LB 210 .... G6
LB 212 .... G6
LC 400 .... G6
LC 505 .... G6
Malathion Fly Bait F2
Mint Disinfectant C2
Nu-Air C1
Nugloss Beads A1
Odorless Wildcat Cleaner A4
Oven & Griddle Cleaner A1
Oxsolv A3
Pep A1
Pink Magic Concentrate A1
Pot & Pan Cleaner A1
Powdered General Purpose
Cleaner A1
Pyrethroid Insecticide F1
Quickwash B1
Rins-O-Dine D1
Rust Remover A2
Rust Stripper P1
S A Cleaner A2
S.K. Concentrate A1
Safety Solvent K1
Safeway Bleach B1
Safti -Sol C2
Sani-Rinse D1
Sanitex ..../................... A1
Sewer Solvent L1
Shackle Cleaner A3
Smokehouse Cleaner A2
Special Foamer A1
Spray Mist Deodorant C1
Stack Odor Control C1
Super Diazinon Residual Spray . F2
Super Foamer A1
Tetralite A1
Thermal Fog Spray F1
Tile & Grout Cleaner C2
Touch "N" Go A1
Tripe Cleaner N3
Tropico Antiseptic Liquid Hand
Soap E1
Tropico 40 .... E1
Tropico 40% Liquid Hand
Soap E1
Tuf-Stuf E1
Velvet Suds A1
30% Malathion In Oil F2
300 Acid Cleaner A3
710 Technical, Non-Butyl
Cleaner/Dgrsr A1
710-B Technical, Non-Butyl
Cleaner/Dgrsr A1
710-G Technical, Non Butyl
Cleaner/Dgrsr A1

Newbridge Chemical Corp.
Multi-Clean XP-101AA A1

Newport Chemical Company
Newport 77 .... C1

Niagara Blower Company
Niagara No-Frost Liquid KV P1
Niagara No-Frost Liquid LV..... P1

Niagara Lubricant Co. Inc.
White Oil-AA H1

Niagara National Corporation
Accusol................................. A4
Niagara NT-50........................ A1
Niagara TK-10 Concentrate A4
Niagara 101 Product 1000 .... A1
Niagara 330 Trail-Clean
Concentrate A1
Niagara 446 WY-59 Conc.
Dgrs&Stm Cl.Cmpd A1
2700 OD D1

Nichols Chemicals Inc.
CC-53.................................... A1
Dual-27 .... D1
Mint Disinfectant C1
NC-50 .... A4
Orange Deodorant................... C1

Nick Muller Co.
"Powerhouse" Steam Cleaner A1
Liquid Pressure Wash M276L..... A1
M-36 Cleaner Degreaser A1
Pressure Wash Detergent M-
176 .... A1

Nino's, Incorporated
Challenger, Degreaser&Floor
Clnr..................................A4,A8

Nobel, Incorporated
APD A1

Noboil Company
Noboil Bleach D2

Nodor Chemical Company
Nodor FP A1

Nokomis International, Inc.
Nokomis 3 .... A1
Nokomis 4 .... A1

Nonfluid Oil Corporation
#150 NFO White Lubricating
Oil H1
Air Lube AA H2
Chem-Plex FM #1 .... H1
F-925.................................... H2
F-925/EP............................... H2
Grade LS-683.......................... H1
KF #132................................. H1
KF-133.................................. H1
Non-Fluid Oil A-#5................. H1
Non-Fluid Oil Grade #1315....... H1
Non-Fluid Oil H-#12............... H2
Nonfluid No. G-60 .... H2
S-5....................................... H2

Nor-Chem
Nor-Solv A4

Nor-Scott Manufacturing Co.
Sud-Zy.................................. A1

Norchem Corporation
Dinaclean A1

Norco Manufacturing And Distributing Co.
Pizza Inn Ammoniated All
Purpose Cleaner A4
Pizza Inn Inst Liq Drain Pipe
Opnr & Mnt L1
Pizza Inn Liquid Hand Dish
Wash A1
Pizza Inn Machine Dish Wash
Compound A1

Norman Chemical Company
Boiler Water Treatment No.
610 P G6
Boiler Water Treatment No.
610X................................. G6
Boiler Water Treatment No.
610XP G6
Filt R Kleen P1
Form.#830 Domstc Hot Wtr
Treat G6
Formula #10........................... G6
Formula #16........................... G6
Formula #200.......................... G6
Formula #202.......................... G6
Formula #241.......................... G6
Formula #242.......................... G6
Formula #251.......................... G6
Formula #30........................... G6
Formula #300.......................... G6
Formula #40........................... G1
Formula #400.......................... G6
Formula #50........................... G6
Formula #51........................... G6
Formula #52........................... G6
Formula #53........................... G6
Formula #54........................... G6
Formula #60........................... G6
Formula #61........................... G6
Formula #610.......................... G6
Formula #611-W G6
Formula #62........................... G6
Formula #63........................... G6
Formula #655.......................... G6
Formula #66........................... G6
Formula #811.......................... G6
Formula #815.......................... G6
Formula #816.......................... G6
Formula #841.......................... G6
Formula #841 C Liquid Gel G6
Formula #853.......................... G6
Formula #90........................... G7
Formula #900-P G7
Formula #91........................... G6
Formula #910.......................... G7
Formula #920.......................... G7
Formula #936.......................... G6
Formula #940.......................... G1
Formula #950.......................... G7
Formula #970 Threshold
Treatment G2
Liquesol D P1
Lyme-Kleen A3
No-Cor A Steamline Treatment. G6
No-Cor B Steamline Treatment. G7
No-Cor D Steamline Treatment. G6
Nor-Kleen Liquid C1
NV 600 .... G6
610L..................................... G6
815L..................................... G6
974-D A4

Norman Manufacturing Co.
Boiler Compound BWT-1 .... G6
Boiler Water Treatment No. 7
BWT 7 .... G6
H-100 Oxygen Inhibitor BWT-
100.................................... G6
H-300 Treatment BWT-300........ G6
No. 11 Boiler Treatment BWT-
11..................................... G6
No. 77 Sludge Conditioner
BWT-77.............................. G6
Oxygen Catalyst BWT-107......... G6
Poly-Ox BWT-125.................... G6
Saf-T-Solvent K1
Sol-3 Norman......................... A4

Norman, Fox & Co.
Norfox CB Lube 12 .... H2

192

195

Oil Mop Inc.
Sheenex A1

Oil Specialties & Refining Co., Inc.
AD-BAC-22 .... D1
AD-BAC-4227 .... D1
Concentrated Pressure Washer
  Cleaner A1
CoCo Liquid Hand Soap E1
Freezer Kleen A5
Mint Disinfectant C1
OSR Heavy Duty Cleaner
  Degreaser A4
OSR Powdered Dish Wash
  Detergent A1
Ping A1
Power Foam Degreaser A1
Saflex Concentrate Synthetic
  Floor Clnr. A4
Sanflex Germicidal Cleanser A4
Strip XX A A1
Stripp XX A4

Oil-Dri Corp. of America
Oil-Dri J1

Oil-Sorb Industries
Oil-Sorb Oil & Grease
  Absorbent J1

Oilchem Company
All Purpose Dairy Cleaner A1
All Purpose Laboratory
  Cleaner A1
Bake Pan Cleaner A1
Biocide C2
Blen A4
Blue Gold A1
Blue Pressure Spray Wash A1
Bora-Clean E1
Chain & Cable H2
Che-een A4
Chef-ette A4
Chlorinated In Place Dairy
  Cleaner A1
Chlorinated Mach. Dishwshng
  Cmpnd A1
Cleanup A3
Dakiree C2
Delite E1
Deluxe Concrete Cleaner A4
Dermaterg 100A A1
DC-20 Pressure Spray Wash A1
DC-40 Automatic Spray Wash .. A1
Flora-Mint C2
Foaming Scouring Cleanser A6
Food Plant Cleaner A1
Germisyl A4
Glass Cleaner C1
GP-2000 .... A1
Hand Dish Washing Cmpnd
  No. 1 .... A1
Handco E1
Handco 20% E1
Heavy Duty Steam Cleaning
  Cmpnd A2
Hydroxade A1
Hydroxade Concentrate A1
HPH Pink Automatic Spray
  Wash A1
Lemon Disinfectant Coef. 6 .... C2
Low Suds Laundry Compound . B1
Lubon H2
Machine Dish Compound No. 1 A1
Medium Duty Steam Cleaning
  Cmpnd A1
Medium Suds Laundry
  Compound B1
Neutrovapor Steam Cleaning
  Cmpnd A1

No. 2 Food Machinery Grease .. H1
Non-Suds Detergent A1
Oilchem A-5 Bob Kat Concrete
  Cleaner A4
Oilchem-5006 Extra Hvy Dty
  Stm Clng Pdr A1
Oilchem-5012 Premium Fleet
  Wash Powder A1
P.C.D.-10 .... A4
Packing House Cleaner A1
Pine-Dis-5 .... C1
Pine-Feen C1
Pink Pearl E4
Pinky A4
Purge A4
Royal Concentrated Detergent .. A1,B1
Safety Hard Water Scale
  Remover A3
Safety Solvent K1
Sewer Pipe Cleaner L1
Shamrock 880 .... A4
Special Concrete Cleaner
  (Improved) A4
Steri-Fog D1
Steri-Pine C1
Super Rig Wash A4
Super Sapon A4
Super Sapon Concentrate A4
Super Strip A4
Synall A4
Syncon A4
SR-158-SZ A1
SSC Metal Polish A7
Technical APC A4
Technical D.W. Base A4
Touch-Up All Purpose Cleaner . C1
Triclenal A4
Truck and Car Wash No. 1 .... A1
Truck and Car Wash No. 2 .... A1
Truck and Car Wash-Brown A1
TFE Release H2
U.N.C.L.E. Insecticide F1
Victory Cleaner A4
W.C.-170 .... C1
W.C.-85 .... C1
Washing Powder A1
904 Concentrate A4

Ojserkis Paper & Janitor Supply
Rocket 8 .... A1

Oklahoma Paper Co.
Opaco Cleaner Disinfectant D1
Opaco Mint Disinfectant C2
Opaco Pure Pine Oil
  Disinfectant C1

Okun Company, Incorporated
Liquid Industrial Cleaner
  Concentrate A1
Rust Bust A3

Old Dominion Paper Company
Odpaco Liq. Drain Conditioner
  & Cl. L1

Old Hickory Food and Chemicals, Inc.
Bowl Safe Bowl & Porcelain
  Cleaner C2
Foam Away C1
Pounce Hospital Disinfectant
  Deodorant C2

Olde Worlde Products, Inc.
Germicidal Concentrate D1
Heavy Duty Concentrate A4,A8
Laundry Compound A1,B1
Laundry Compound Phosphate
  Free A1,B1

Natural Concentrate A1

Oldham Chemicals Co.
Diazinon 4E Insecticide F2
Oldham's Rat and Mouse Kill... F3
P-40 Aerosol Insect Killer F1

Oldham Chemicals Co., Inc.
Oldham's Rodent Cake F3

Olin Corporation
AD-Dri Bleach B1
Caustic Soda A2
Disodium Phosphate G6,N2,G5
HTC Granular Dry Chlorine .. D2,Q4,N4,G4
HTH Tablets D1,Q3,N4,G4
Improved Teox Compound A1,B1
Lo-Bax Special D1,Q3
Monosodium Phosphate G6
MCC Flake A2
Polyphos A1,G6,N2,G5
Scav-Ox G6
Sodium Acid Pyrophosphate N2
Sodium Hexametaphosphate..... A1,G6,N2,G5
Sodium Tripolyphosphate .. A1,G6,N2,G5
Tetrasodium Pyrophosphate A1,G6,N2
Trisodium Phosphate A1,G6,N2,N3
Trisodium Phosphate
  Chlorinated D1,Q1,Q4
Trisodium Phosphate Chlorntd.
  Pink D1
58% Dense Soda Ash A1,G6
58% Light Soda Ash A1,G6

Olin Water Services
Clean-Out A1
Olin Scale-Out A3
Olin T1013 .... G6
Olin T1015 .... G6
Olin T2201 .... G7
Olin U1019 .... G6
Olin 1000 .... G6
Olin 1001 .... G6
Olin 1002 .... G6
Olin 1003 .... G6
Olin 1005 .... G6
Olin 1006 .... G6
Olin 1008 .... G6
Olin 1010 .... G6
Olin 1011 .... G6
Olin 1017 .... G6,G5
Olin 1020 .... G6
Olin 1021 .... G6
Olin 1025 .... G6
Olin 1026 .... G6
Olin 1028 .... G6
Olin 1501 .... G6
Olin 1502 .... G6
Olin 1503 .... G6
Olin 1504 .... G6
Olin 1505 .... G6
Olin 1507 .... G6
Olin 1511 .... G6
Olin 1514 .... G6
Olin 1516 .... G6
Olin 1517 .... G6
Olin 1519 .... G6
Olin 1520 .... G6
Olin 1522 .... G6
Olin 1524 .... G6
Olin 1600 .... G6
Olin 1601 .... G6
Olin 1602 .... G6
Olin 1603 .... G6
Olin 1604 .... G6
Olin 1605 .... G6
Olin 1606 .... G6
Olin 1609 .... G6
Olin 1610 .... G6
Olin 1750 .... G6

198

Rustam P1
San-D Iodine Hand Soap E2
Sanadine D1
Scan A1
Scrubber Concrete Floor
    Cleaner A4
Scrubsol A4
Select A2
Silicare Silicone Spray Lube H1
Slur A1
Soil Mist A1
Soilex-50 .... A2
Soilsolv A2
Sudsy Plus A1,H2
Surge A3
Tame B2
Trak-2 .... A2
Tri-Tak A1
Viro-X Disinfectant E2,D2,E3,Q6
199 Alkali Heavy Duty Cleaner A2

Ozark Chemical Company, Inc.
410 Boiler Water Treatment G7
410-C G6
450-S Boiler Water Treatment ... G6

P

P & M Service
PM #455 .... G6

P B I-Gordon Corporation
Diazinon Roach Spray F2
Fly Bait F2
Industrial Emulsifiable Conc.
  Stable(1) F1
Institutional Area Spray F1
Institutional Mist Spray F1
Malathion (5lbs/gal) F2
New Last Meal F3
Pyrenone Multi-Purp.
    Knockout Spray 826 .... F1
Pyrenone Multi-Purpose
    Knockout Spray F1
Quick Draw Fly Bait F2
Quick Seep L2
Roach Rid Brand F2

P D F, Incorporated
New Sure Clean A1

P J Supply Co.
Stainless Steel Cleaner & Polish. A3

P M Chemical Company
Astro A1
Cleaner-Stripper A1
Degreaser A4,A8
Liquid Hand Soap E1
Liquid Organic Cleaner A4,C1
Reelite A4
Sofwite B1
Sofwite 10 .... B1
Steam Cleaner #51 .... A1

P N & P M Corp.
Re-Fresh C1

P P G Industries
Caustic Soda Coarse Flake A2,G6,L1,N2,
         N3
Caustic Soda Fine Flake A2,G6,L1,N2,
         N3
Caustic Soda Liquid 50% A2,G6,L1,N2,
         N3
Caustic Soda Medium Flake A2,G6,L1,N2,
         N3

Caustic Soda Small Flake A2,G6,L1,N2,
         N3
Granular Soda Ash A1,G6,N3
Hydrogen Peroxide 35% SP
    Grade N3
Light Soda Ash A1,G6,N3
Pels 180 .... A2,G6,L1,N2,
         N3
Pittabs D1,Q4,G4
Pittchlor D1,Q4,G4

P Q Corporation
Alkadet B1
Alkasohp B1
Beachrite B1
Besco B1
Brawn A1,B1
Brawn "PF" A1,B1
Enerdet B1
Fluorium B1
G A1
GD B1
Hi-Built Soap B1
Hycon B1
Hycon "PF" B1
ISP Detergent Booster B1
ISP Fabric Brightener B1
ISP Softener SS B2
Lydet B1
Lydet 100 .... A1
Metso Anhydrous A1,N1,G6,N4,
         N2,N3
Metso Anhydrous 60 .... A1,N1,G6,N4,
         N2,N3
Metso Beads 1048 .... A1,N1,G6,N4,
         N2,N3
Metso Beads 2048 .... A1,N1,G6,N4,
         N2,N3
Metso Granular A1,Q1,N1,G6,
         N4,N2,N3
Metso Pentabead 10 .... A1,N1,G6,N4,
         N2,N3
Metso Pentabead 20 .... A1,N1,G6,N4,
         N2,N3
Metso 20 .... A1,N1,G6,N4,
         N2,N3
Metso 200 .... A2,N1,N4,N2,
         N3
Metso 55 .... A1,Q1
Metso 66 .... C1
Metso 66-A C1
Metso 99 .... A1,N1,N4,N2,
         N3
N A1,G6,G5
O A1,G6,G5
Prime Sohp B1
Qudet B1
Sayfbrite B1

P. H. Chemical Co.
Activated Alkaline Powdered
    Drain Opener L1
Blast A4
Flo-Eze C1
Kleenzall A4
Kleenzeall Heavy Duty All
    Purpose Cl. A1
Liquid Hand Soap E1
Odor Control C1
Quat D1
Safety Kleen Safety Solvent K1
Sewer Solvent L1
Stalume Aluminum & Stainless
    Steel Cl. A3
Steamy A4

P.B. & S. Chemical Co.
Sno-Glo Bleach D2

P.L.C. Corporation

1008 Scouring Powder A6
201 Heavy Duty Cleaner A4
203 General Purpose Liquid
  Cleaner A4

P.S.F. Organics, Inc.

Environmental-Aid C1

Pace Ltd.

Action C1
Activated Alkaline Powdrd
  Drain Opener L1
Aerosol Disinfectant C1
All-Purpose Non Flammable
  Safety Solvent K1
Belt Dressing P1
Bio-Kleen A1
Bowl Cleaner C2
Chain & Conveyor Belt
  Lubricant H2
Cotes-It H1
D.O.A. Super Fogging Insect
  Killer F2
De Greez C1
DIgestant 250 .... L2
Drain-Away L1
E-Z Out H2
E-Z Steam A1
Foamer A1
Foamy A1
Frigi-Kleen A5
General Purpose Liquid
  Cleaner A1
Gladiator F1
Graphite Lubricant H2
Grease-Off C1
Grease-X K1
H D Cleaner A1
H.D. Power A4
Heavy-Duty Floor Cleaner C1
Hook and Trolley A2
Husky Jr. E4
I-O-Dine D2,E3
Kleen C1
Laundry Detergent B1
Lemon Sewer Sweetener L1
Liqui-Liner G3
Liquid Dishwash Concentrate ... A1
Liquid Hand Soap E1
Maxi-Glow A1,A4
Mity Lube AA H1
Mura-Clean A3
MP Chlor D2
MP 101 .... A1
MP 22 .... A3
MP 33 .... A2
MP 44 .... A3
Pace 5 .... H2
Paint and Varnish Remover C3
Palm Balm E1
Pine Oil D C1
Porcelain and Bowl Cleaner C2
Powerhouse A4
S.S. Shine A7
Seize A4
Seize-N D1
Silicone Lube H1
Solv-It C1
Speedy Spray A1
Spray Fog F1
Squeaky Kleen A1
Stericide D1
Super Power A4
Thrust A4
Tri-Kleen K1
View II A7
Virucide D1
Waterless Hand Cleaner E4

Wintermint C1
4D A4
4DN D1

Pace Products, Inc.

Hylo-Zyme L2
Kandu 110 .... A4

Pacer Lubricants, Inc.

Pacer Com-Cyl 130 .... H2
Pacer Goltex Indust. Gear Oil
  AGMA 6 EP H2
PAcer Hi-Q FlUid 500 .... H2
Pacer Mineral Gear Oil AGMA
  6 (SAE90) H2
PAcer Non-Coke HD 65 .... H2
Pacer Power V 315 .... H2
Pacer Puritan Oil 170 .... H2
Pacer Rock Drill Oil 600 .... H2
Pacer Syndrol 46 .... H2
Pacer Syntherm 700 .... H2
Pacer Thermal T 315 .... H2
Pacer-Therm 500 .... H2
Synfilm 4X H2

Pacific Chemical

Adhere CD A1
Adhere LK A1
Adhere N-72 .... A1
Alkalact N-25 .... A4
All Soil A4
Alldet-76 .... A1
Balance Detergent A1
Bar Cor N-128 .... G6
Bar-Cor CWS-4 .... G6
Bar-Cor-OWS G3
Bar-Tox L-1 .... G6
Bar-Tox L-44 .... G6
Bold A1
Brawn A2
Brew Clean MF-56 .... A2
Calsol A3
Chemprocide D1
Conlube N-10 .... H2
CONTROL M-61 .... A1
Convey N-100 .... H2
Cope N-97 .... A1
Creet N-33 .... C1
Degreaser-S K1
E-Chlor C1
Esteem MF-38 .... A1
Expel L-49 .... A1
Fist A1
Force A2
Friction H2
Genteel E1
Handsome C2
Hercules A1
Hi-Press A2
Io-Cide D2,E3
Liquid Scraper A4
Liquid Tru-Wash A1
Master A A1
Master BXX A2
Master CA D1
Master CAX D1
Master M-H-W A1
Microcide CA-39 .... G7
Microcide L-36 .... G7
Pace A-D A3
Pace C-16 .... D1
Pace MF-63 .... A1
Pace MF-81 .... A3
Pace MF-82 .... A1
Pace MF-88 .... G6
Pace N-127 .... A2
Pace N-51 .... A1
Pace N-53 .... A1
Pace N-63 .... A1
Pace S-L A1

Pace SR A2
Pace T-10-X A1
Pace T-20X D1,Q3
Pace T-30X A1
Pace T-60XX A1
Pace T-70-CL A1,Q1
Residex MF-34 .... A1
Ridall A4
Scale Pro L-38 .... G7
Scale Sol WT-29 .... A3
Scale-Cycle D-30 .... G7
Scalex D-12 .... G6
Scalex D-13 .... G6
Scalex D-17 .... G6
Scalex D-400 .... G6
Scalex DC-46 .... G6
Scalex DCH-47 .... G6
Scalex DO-45 .... G6
Scalex DP-42 .... G6
Scalex DPLA-48 .... G6
Scalex DS-40 .... G6
Scalex DS-41 .... G6
Scalex EB-5 .... G6
Scalex L-20 .... G6
Scalex L-22 .... G6
Scalex L-23 .... G6
Scalex L-56 .... G6
Scalex L-57 .... G6
Scalex LCP-53 .... G6
Scalex LOA-43 .... G6
Scalex LOA-52 .... G6
Scalex LP-51 .... G6
Scalex LP-54 .... G6
Scraper A4
Scraper-P C1
Solub MF-50 .... A1
Solvit A4
Strike MF-79 .... A3
Top Hand E4
Tru Wash A1
Vacate A1

Pacific Chemical Company

GP-100 Cleaning Concentrate A1

Pacific Chemical Company

Formula 22 .... A4
Formula 22-M A1

Pacific Chemists, Incorporated

Sequoia Dura-Chlor Q1,Q4
Sequoia Egg Cleaner "B" Q1
Sequoia Egg Cleaner "C" Q1
Sequoia Egg Cleaner "E" Q1
Sequoia MSR A3
Sequoia NF 9 .... Q5

Pacific Coast Chemical Corp.

Formula MC-3 .... A1

Pacific Janitor Supply

Non Butyl Degreaser A1
Pacific Alkaline Cleaner A1

Packard Chemical & Equipment Co.
Division Of Pacco, Inc.

Eliminate A1
Formula 66 .... C1

Packard Chemical Co.

SS-50 Solvent Degreaser K1

Packard Chemicals/ Division of
Packard Industries, Inc.

Aggregate MR A3
Belt-Tex Belt Dressing P1
Boil-Pak G6
D.P.W. C1
Grease Light'ning C1

Hands-Up Waterless Hand
 Cleaner...................................... E4
Image Glass Cleaner C1
Invader A3
KDS Non-Fuming Cleaner &
 Descaler.................................... A3
Lectra Safe................................. K1
Lectro-Pur.................................. K2
Lubra-Cone Silicone Spray H1
Metal-Glo A7
Oxy-Pak...................................... G6
Pak-Kill Insect Spray F1
Pink Submarine.......................... A1
Pipe-Pak P1
Quat-Pak Cleaner Disinfectant
 Deo. Fngcd A4
Rat-Pak Rat Bait........................ F3
Scour... A4
Shield Kleen Anti-Fog............... C1
Steam & Return Line
 Treatment G6
Stron Arm Oven Cleaner.......... A8
Strong-Arm Industrial Strength. A8
T.S.R. A4,A8
Tef-Lube..................................... H2
Therma-Pak................................ L1
Thunderbolt................................ L1
Toughman (Bulk) A4
Toughman All Purpose Cleaner
 (Aerosol)................................... C1

Packer-Scott Company

Blue Lightning............................ A4
Break-Thru.................................. A1
Compound 1908.......................... A1
De-Grease C1
Down-N-Out............................... L1
Formula 20-10............................ A4
K-28 Automat Detergent............ A4
Magic Mist-Plus......................... A4
Municipal Sewer Cleaner L1
Pascocide #64 Odorless............. D1
Pascophene................................. A4
Pascot Blast-Off...................... A4,A8
Pascot Kwik-Solv...................... K1
Pascot Stop-N-Go Bowl Clnr.
 Disinfectant C2
Pascot Super-Solv K1
Pascot Syn-Det...................... A4,E1,A8
Prep.:.. A3
Reward A4
Reward Neutral Fragrance........ D1
Steamex SK-747......................... A1
Wash Off Remover C3
40 Below A5

Packers Development Corp.

Padeo Packers Grease................ H1

Packers Engineering & Equip. Co., Inc.

Electro-Honer Oil H2
Pak-Co No. 20 .... E1
Pak-Co No. 40A E1
Warco #44 .... N3
Warco #44A N3
Warco #55 .... N3
Warco General Purpose
 Cleaner..................................... N3
Warco No. 99-LC....................... N3

Pagoda Industries, Incorporated

Alsolv .ᵢ....................................... A1
Eazolv (Concentrate)................. A4
Exsolv,... A2
Kezolv (Concentrate)................. A4
Pago-10 (Concentrate) A1

Pak West Paper & Chemical Corporation

Non Butyl Cleaner A1

Palm Beach Chemical Co.

Lemon Lite A4
New Royal Heavy Duty All
 Purpose Clnr. A4

Palmer Company, Incorporated

#1005 Water Soluble Degrsng
 Solv... C1
#1010 Liquid Steam Cleaner..... A1
#1015 Tough Job Cleaner.......... A4
Aero Clean Germicidal Cleaner
 & Deod..................................... C1
Brilliant No. 13 .... C2
D Liquid Soap............................ E1
Easy Wipe Multi-Use Cleaner
 Degreaser C1
Everything Goes A3
Guardian Disinfectant
 Detergent................................. A4
Lime and Scale Off.................... A3
Palm-A-Tize............................... D1
Palm-X.. D1
Sparkle No. 15 Disinfecting
 Porcln Clnr C2
Super Foam Clean...................... A1

Palmer House Chemical Corp.

General Purpose Detergent
 (GPD)....................................... A1
Low Foam General Purpose
 Det.. A1

Palmetto Chemical & Supply Co., Inc.

Ammoniated Oven Cleaner A8
Aquasect F1
Clean-Rite 200L......................... G6
Cocoanut Oil Hand Soap
 Concentrate.............................. E1
Drip-Dri P1
Egg Wash.................................... Q1
F-11 Concentrate A4
Fly Tox F1
Fry-Kleen A1
He & She Concentrated Liquid
 Skin Clnr E1
Heavy Duty Floor Cleaner A4
Hi-P Contact Cleaner................ K2
Hog Scald................................... N2
Ideal... A4
Killo Insect Spray F2
Kleen All..................................... A2
Lanolin Waterless Hand
 Cleaner..................................... E4
Lectra-Clean K1
LB #1.. C2
Metal Brite A3
Milk Stone Remover A3
Mint-O-Phene............................. C1
Mr. Brite Porcelain Cleaner C2
Now A1
Pal Kleen 10............................... K1
Palco 76...................................... A1
Perma Kill.................................. F2
Phene-O-Pine C1
Preen.. A1
Pride A1
Red Hot...................................... C1
Release.. A4
Safety Sheen A4,A8
Safety Solvent............................ K1
Sani-Lube H1
Seal-Ease H2
Sewer Sol L1
Shackle Kleen Liquid................. A1
Shackle Kleen Powder............... A2

Skin Shield Cream...................... E4
Stainless Steel and Metal Polish. A7
Sur-Grip..................................... P1
Thermo Kill F1
Trailer Brite A3
Treatrite 100L............................ G6
Tri-Tex A1
Triple Power............................... D1
TFE Fluorocarbon Lubricant H2

Pan American Chemical

Liquid Toilet Soap E1
Pacco Aerodet A1
Pacco No. 25 Insect Spray........ F1
Pacco Pure-Lube......................... H1
Pacco Silicone Spray H1
Pine Odor Disinfectant C1

Pan American Chemical Co.

Blue Magic.................................. A4

Panoramic Industrial Chemicals Co., Inc.

Antiseptic Liquid Soap E1
Blockbuster................................. C1
Cyclo Therm Degreaser Form.
 999.. K2
Cyclo-Jell................................... K1
Cyclozyme A1
Hand Soap.................................. E1
Lustre-Lite Formula #300
 Cleaner..................................... A4
Lustre-Lite Formula #3220
 Cleaner..................................... A4

Panther Chemical Company, Inc.

Action... A2
Brighter B................................... A3
D-22 .... A1
Econ-O-Blitz A1
HD-250.. A2
HP Stripper................................ A2
Liquid Hot Soap......................... A2
LD-3 .... E1
LHT-100...................................... A2
N-ergy 3310 .... A4
Panco AS-33 .... A2
Panco Filter Coat....................... P1
Panco Mean Green..................... A4
Panco Pure.................................. H1
Panco RR-50-NF......................... A2
Panco Trolley Cleaner............... A2
Panco WD 506 .... A4
Panco 618................................... G7
Panco 619................................... G7
Panco 642................................... G6
Panco 691................................... G7
Pangold...................................... A1
Panther Brightener S................. A3
Panther C-450 .... A1
Panther CD-64............................ A4
Panther Hot Soap...................... A2
Panther HD-25 .... A1
Panther HLD.............................. B1
Panther Old Yeller A1
Panther SW-440.......................... A2
Panther SW-440 CL.................... A2
Panther WE-160 .... A1
Panther WE-525 .... A1
Panther WED-729....................... A1
Panther WED-137....................... L1
Panther 111 .... A1
Panther 388 .... A3
Panther 399 .... A3
Panther 600 .... G6
Panther 610 .... G2
Panther 611 .... G7
Panther 612 .... G6
Paw Power.................................. F1

Patele Chemicals

Alltreat-300 .... G7
Boilerite 100 .... G6
Command A1
CLT-100 .... G6
CWT-500 .... G7
Descaler #1 .... A3
Super Reprosolv A4
Tridet A4

Paterson Card & Paper Co.

Prelco Concentrated Cleaner A4

Patrick Cudahy Incorporated

7601 Standard Detergent A1
7602 Chlorinated Detergent A1
7606 Smokehouse Detergent A2

Patterson Chemical Co., Inc.

Fogging Concentrate F1
Mill & Food Plant Spray F1
Pyrenone 20 New F1

Patterson Company, Inc.

Adjunct 'A' G6
HT G6

Paul Koss Supply Company

C.P.C. A7
Colonel Slick A7
Crystal Clear Lqiuid Glass
   Cleaner C1
Down & Out L1
Dynamic Industrial Cleaner A4
Foam Off A1
Glo-Power Truck And
   Equipment Cleaner A1
Green Stuff C1
Heavy Duty Industrial Cleaner .. A4
K-Kreme A4
Kleen-O-Fect A4
Laundramagic Alkaline Builder . B1
Laundramagic Blood & Protein
   Remover B1
Laundramagic Chlor Brite B1
Laundramagic Chlorine
   Neutralizer B1
Laundramagic Detergent
   Booster B1
Laundramagic Laundry
   Detergent B1
Laundramagic Softener-Sour B2
Laundramagic Solvent
   Degreaser B2
Laundramagic Wash-All B1
Magic Film & Scum Remover ... A4,C2
Moon Shine A4
New Palm Liquid Hand Soap E1
P.O.G. Remover B2
Purple Stuff D1
Red Stuff C1
Ring-Gone Cleaner-Descaler C2
Shower-Wash Cleaner-Descaler . A3
Yellow Stuff A4
Zonk Degreaser Cleaner-
   Deodorant A1

Paulsen & Roles Laboratories

Super-Strip Fireproof Cleaner
   Degreaser A4,A8

Payette Industrial Laboratories

Attain A1
Brace A2
Clean-Up A4
Commit A1
Compel A3
Compliment A1
Dignify A3

Engage A1
Foam-Up A1
Habit A1
Intense A2
Onset A3
San-Foam A1

Peat Belting & Supply Company

Belt Dressing P1

Peck's Products Company

#200 Centramatic Lube H2
#450 Dishwashing Compound ... A1
Aseptex E1
Aseptrol E1
Blu-Fome A4
Britex A2
BNC A1
BNC-A A1
BNC-B A1
BNC-M P1
BNC-P A6
Ca-Mag Solve Code #492 .... A3
Cleanser AC A1
Compound 702 .... A1
Deo Clean B A1
Duo Lube H2
Foamtrol A1
Formula 113 .... A1
Formula 118 .... A1
Glycerole 17% E1
Glycerole 20 .... E1
Glycerole 20% Liquid Soap E1
Glycerole 40% Concentrtd Liq.
   Soap E1
Grease Off A1
Heavy Duty Pexide A4
K-3 .... A4
L.G.C. Lemon Odored
   Germicidal Cleaner A4,C1
Laboratory Glass Cleanser A1
Lemon Odored Pepco DC C1
Liquid "Squeek" A1
Liquid Pexchlor D1
Liqulube Conveyor Lubricant .. H2
M.G.C. Mint Odored
   Germicidal Cleaner C1
Master Muscle A4,A8
N.B.C. A1
No. 201 Jell-Lube H2
No. 213 Jell-Lube H2
No. 216 Jell-Lube H2
No. 431 Bar-Lube H2
No. 710 Chlorinated Dshwshng
   Cmpnd. A1
No. 77 Detergent Liq. Cleaner .. A1
No. 80 Liquid Cleaner A1
P.G.C. 13 Pine Odored
   Germicidal Cleaner C1
P.G.C. 6 Pine Odored
   Germicidal Cleaner C1
Peck-Kleen A4
Peckleer A2
Pepco Add A1,N1
Pepco AP C1
Pepco Can Cleaner 717 .... A1
Pepco Concentrated General
   Purpose Clnr. A1
Pepco D-C Plus D1
Pepco DC D1
Pepco Foam A1
Pepco N C #463 .... A1
Pepco Pan Acid 719 .... A3
Pepco Pan Alkali 718 .... A2
Pepco Pexene 10 .... D2
Pepco QK A4,A8
Pepco Shine P1
Pepco Spray 143 .... A1
Pepco SQK A4,A8
Pepco Winter Lube H2

Pepco 107 .... A4
Pepco 152 .... A1,Q1
Pepco 295 .... E1
Pepco 315 Hand Soap E1
Pepco 408 .... A1
Pepco 410 .... A1
Pepco 415 .... C2
Pepco 418 .... C2
Pepco 460 .... A1
Pepco 477 Egg Wash Q1
Pepco 485 .... A2
Pepco 490 Acid Detergent A3
Pepco 493 .... A2
Pepco 553 .... A4
Pepco 707 .... A1
Pepco 714 .... A2
Pepco 716 .... A2
Pepco 720 .... A3
Pepco 724 .... A1
Pepco 725 .... A1
Pepco 732 .... A3
Pepco 914 .... A1
Pepco-Lube H2
Pepcochlor A2
Pepcocide D2,Q3
Pepcodex A4
Pepcodine E2
Pepcol C-110 .... A1,H2
Pex #1 .... A4
Pex #26 .... E1
Pex #4 .... A4
Pex #8 .... A4
Pex General Cleaner 715 .... A1
Pex Granular A1
Pex Liquid Detergent A1
Pex-O-Dine D1
Pex-O-Matic Liq. Conveyor
   Lubricant H2
Pexan A1
Pexchlor D1
Pexell A1
Pexide Heavy Duty Chlorinated A1
Pexlar A2
Pexolv #489 Acid Detergent
   Sanitizer A3
Pexsyn 586 .... A4
Pexsyn 596 .... A4,A8
Premier A1
Puraphen D1
Puraphen Plus A4
Restaurall A1
Scaltrol A1
Skinny A1
Special 40 .... E1
Spray 66 .... A4
Sprazo A1
Super Britex A2
Super CC Code P-478 .... A1
Super Pexide A1
Super Pridex Code P-471 .... A1
Super 703 .... A4,A8
Syn-Lube H2
Syn-Lube Special H2
Terge-Off A4
Truly A3
White Lotion A4,A8
138 .... A1
700 Steam Cleaner A2
76% NA20 Caustic Soda A2

Peerless Chemical Company

#536 Kleenzall C1
#812-C Liquid Coconut Hand
   Soap E1
#812-S Liquid Hand Soap E1
Bulldozer C1
Peerless No. 333 Al-Pine All
   Purp. Clnr. A1
Spraykleen No. 901-1 .... A1

Peet Packing Company

Farmer Peet's Acid Cl.
Galvanized A3
Farmer Peet's Acid Cl. Stuls
Steel Only A3
Farmer Peet's General Purpose
Clnr. A1
Farmer Peet's Liq. Gen.
Purpose Clnr. A1
Farmer Peet's Roller Cleaner A2
Farmer Peet's Smokehouse
Cleaner A2

Pegler & Company

Peg Disinfectant Deodorant C2
Peg Germicidal Foam Cleaner ... C1
Peg Insecticide for House and
Garden F1
Peg Prilled Drain Opener L1
Peg Silicone Spray H2

Penetone Corporation

Brawn A1
Formula 611 .... A1
Formula 724 .... K1
Formula 990 .... A3
Formula 991 .... A3
Inhibisol (Spray) K1
KAR-M-1 .... A1
Lube 2017 .... H2
Lubrisil H1
Mystone E1
Pen-Strip A P1
Penesolve #1 Blue A1
Penesolve HTC A2
Penesolve 1009 .... A1
Penesolve 1022 .... A1
Penesolve 12 .... A1
Penesolve 202 .... A2
Penesolve 208 .... A2
Penesolve 220-A A2
Penesolve 229 .... A2
Penesolve 5 .... A2
Penesolve 814 .... A1
Penesolve 902 .... A1
Penetize A4
Penetone A4
Penetone DC C1
Penetone 128 .... A2
Penetone 2088 NF H2
Penetone 426 .... A4
Penetone 58 .... A4
Penetone 66 .... A1
Pengard H2
Pennesolve 206 .... A1
Pennesolve 209 .... A2,C3
Pl-998 .... A3
Power Cleaner A1
Power Cleaner 155 .... A4
Powertone A1
Professional Digester L1
PC-6 .... A4
QXL A4
Salubrite D1
Senior A4
Slix A1
Steam Kleen A2
Supar A4
TPC Solvent K1
VAluclean A1

Penick Corporation

SBP-1382 Liquid Spray 0.50% ... F1
SBP-1382 3% Multipurpose
Spray F1
SBP-1382/Bioallethrin
Insct.Aer.20% + .10% F1
24.3% SBP-1382-2 E.C. F1

Penick, S. B. & Company

SBP-1382/Bioallethrin
(.20% + .40%)A.P.S. F1

Peninsula Maintenance Supply

Super Degrease A1

Penn Chemical

#115 Organic Drain Line
Opener&Maintnr. L1
Polyurethene Sealer P1
Safety Solvent Cleaner and
Degreaser K1
Super Enzymes L2

Penn Tech Corporation

Erase A1
Extrol Drain Opener L1
Flash A3
Gemini C2
Hydroseal K2
Penn Fog F1
Penntrol C1
Streak A4,A8
Ultrasol A4
Wipe Out A3

Penn Valley Chemical Co., Inc.

NB-1 .... A1

Penner, L. C. Company

C.I.P. Cleaner A1
CGC A1
General Cleaner #3 .... A1
H.T.S.T. No. 5 .... A2
Milkstone Remover A3
Natol A1
Powdered MSR A3
Soaker Alkali No. 2 .... A2
Solvex Cleaner A4
Thirty-Three Alkali A2

Pennstate Chemical & Solvent Co.

Cleaning Compound A1
Penn State Concentrate A1

Pennsylvania Paper & Supply Co.

PV-712 .... C3
Sta-Kleen Liquid Hand Soap E1

Penntex, Inc.

Penntex Plus A1

Pennwalt Corporation

Accomplish D1
ACMR LF A3
BanKal A3
Besk A1
Brykleen G2
BK Belt Cleaner P1
BK Defoamer Q5
BK Egg Wash #2 .... Q1
BK Laundry Detergent B1
BK Liquid D2,Q4
BK LFI D2,E3,Q6
BK Powder D2,Q4
BK Retort Additive G5
BK Solvent-Degreaser K2,K3
BK Veg Wash N4
BK Vegpeel N4
BK 101 Chlorinated CIP
Cleaner A1
BK 202 Liquid Chlorinated CIP
Cleaner A1
BK 303 Chlorinated Manual
Cleaner A1
BK 404 Heavy Duty Alkali A2
BK 505 .... A1,H2

BK 909 Liquid Hand Soap E1
BK-BK A2,N4
BKD Stain P1
C.E.C. #4 .... A1
Caliber A1
Carvac A1
Clear Dry P1
Clor D1
Clorclean Liquid A1,Q1
Cloree B2
Cloreze B1
Clorital A1,Q1
Clorital Super A1,Q1
Color-Mix B2
Cosmic A1
Crystal Caustic Soda A2
CIP Chlor A1
CR 219 .... K1
De Min XL A1,Q1
Delklor Q1
Delscaler A A3
Dispatch K1
Dri-Clor B1
Dynahue B1
Echelon A1
Enzomatic A1
Exalt A3
Exception A2
Flake Caustic Soda A2
Foamore A1
Fresh E2
FCD 1 .... A4
G.A.C. #2 .... A3
G.A.C.(B) A3
Globest A1,Q1
Gon K1
GL Alkali Liquid A2
GL Alkali Liquid LF A2
GL Alkali Solid B A2
H.D. Chlor A1,Q1
H.D. Circulation A1
Hy Dry P1
Indal B1
Jato Strip B1
Jex K1
Kaycee A1
Kee-Phos A2
Keep A1
L.P. Bleach B2
Lewis Lye Drain Cleaner L1
Lexicon A1
Liqua Terg A1
MC-3 .... A1
MC-7 Chlorinated A1
Nu Dreme A1
Nu-Trox B1
NP-200 .... A1
NP-301 .... A1
Orthosil B1
Packleen A1
Pen Glo A1
Pendette F N1
Pengel A1
Pennbild B1
Penncarb K1
Pennchem Stericide Sanitizer ... Q3,D2,E3
Pennchem 14 .... A2
Pennchem 333 .... A1
Pennchem 4034 .... P1
Pennchem 81 .... A2
Pennchem 91 .... A2
Pennclean A3
Pennclor D2
Pennquest A2
Pennsan XXX D1,Q3
Pennswim Clor-Tabs D2,Q4
Pennswim Sentry D2,Q4
Pennwalt BPC A1
Pennwalt Hy-Con A1
Pennwalt Hydropel A1

Pest Control Supplies

PCS Blue-Green Fly Bait F2
PCS Pyrethrum Oil
   Concentrate 3-6-10 .... F1

Peter Eckrich and Sons, Inc.

Acid Cleaner A3
Eckrich #100 Cleaner A1
Eckrich #110 .... A1
Eckrich #20 Cleaner A2
Eckrich #30 Cleaner A1
Eckrich #40 Cleaner A1
Eckrich #50 Cleaner A3
Eckrich #60 Cleaner A3
Eckrich #70 Cleaner A1
Eckrich #90 Cleaner A2
Eckrich Acid Cleaner-NP-S1 ..... A3
Eckrich Cleaner #2-NP A1
Eckrich Cleaner #4-NP A1
Eckrich Cleaner #4K-NP A1
Eckrich Cleaner #5-NP A1
Eckrich Cleaner #9 NP A2
Eckrich Hook and Trolley   —
   Clnr-NP A2
Eckrich Liquid Cleaner #1 ...... A1
Eckrich Lubricant H2
Eckrich No. 88-NP A2
General Cleaner #2 .... A2
Heavy Duty Cleaner #9 .... A2
Mechanical Cleaner #4 .... A1
Mechanical Cleaner #4K A1
Mechanical Cleaner #5 .... A1
Smokehouse Cleaner #80 .... A2

Peterson/Puritan, Inc.

P/P Disinfectant Deodorant
   Spray "G" C1

Petrochemicals Company, Inc.

Petro AA N4
Petro BAF A1
Petro 11 .... N4
Petro 22 .... A4

Petrofax

Derma Pro E1
Garbosol C1
Groutex A4
GR-600 .... H1
Laundrite B1
Lectrofax K2
Oxidex A3
Petrocoil P1
Petrodyne A4,C1
Petrosol A4,A8
Powerfax 111 .... A1
Powerfax 444 .... A3
PLC-300 .... K1
Softsteam A1

Pfizer, Incorporated

Anti-Germ 55 .... D2

Pharmacal Research Laboratories

Clout A1
Quatricide A4
Trail C1

Phillips & Jacobs, Incorporated

Sulfamic Acid A3

Phillips Petroleum Company

Automatic Transmission Fluid
   (F) H2
Automatic Transmission Fluid
   Dexron II H2
Baltic Grade 1000 .... H2
Baltic Grade 1500 .... H2
Baltic Grade 2150 .... H2

Baltic Grade 3150 .... H2
Baltic Grade 465 .... H2
Baltic Grade 700 .... H2
Baltic Grade 75 .... H2
Baltic Oil Grade 105 .... H2
Baltic Oil Grade 150 .... H2
Baltic Oil Grade 215 .... H2
Baltic Oil Grade 315 .... H2
Baltic Oil Grade 500 .... H2
Baltic Oil Grade 60 .... H2
Condor Grade 1000 .... H2
Condor Grade 105 .... H2
Condor Grade 1500 .... H2
Condor Grade 75 .... H2
Condor Lubrcat. Oil Grade 315
   SAE 20 20W H2
Condor Lubricating Oil Grade
   150 SAE 10W H2
Condor Lubricating Oil Grade
   215 .... H2
Condor Lubricating Oil Grade
   465 SAE 30 .... H2
Condor Lubricating Oil Grade
   700 SAE 40 .... H2
Corona W Lubricating Oil SAE
   30 .... H2
Corona W Lubricating Oil SAE
   40 .... H2
Heavy Duty Motor Oil SAE
   10W H2
Heavy Duty Motor Oil SAE
   20-20W H2
Heavy Duty Motor Oil SAE 30 H2
Heavy Duty Motor Oil SAE 40 H2
Heavy Duty Motor Oil SAE 50 H2
Hector Oil 1000S H2
Hector Oil 2000 .... H2
Hector Oil 2000S H2
Hector Oil 3000 .... H2
Hector Oil 3000S H2
HDG Motor Oil Grade 10W H2
HDG Motor Oil Grade 20-20W H2
HDG Motor Oil Grade 30 .... H2
HDG Motor Oil Grade 40 .... H2
HDS + 1 Motor Oil SAE 10W H2
HDS + 1 Motor Oil SAE
   2020W H2
HDS + 1 Motor Oil SAE 30 .... H2
HDS + 1 Motor Oil SAE 40 .... H2
HDS Motor Oil SAE 10W H2
HDS Motor Oil SAE 20-20-W .. H2
HDS Motor Oil SAE 30 .... H2
HDS Motor Oil SAE 40 .... H2
HDS Motor Oil SAE 50 .... H2
HG Fluid H2
Magnus -A- Oil Grade 215 .... H2.
Magnus Oil Grade 1000 .... H2
Magnus Oil Grade 105 .... H2
Magnus Oil Grade 150 .... H2
Magnus Oil Grade 1500 .... H2
Magnus Oil Grade 215 .... H2
Magnus Oil Grade 315 .... H2
Magnus Oil Grade 465 .... H2
Magnus Oil Grade 700 .... H2
Magnus-A-Oil Grade 150 .... H2
Magnus-A-Oil Grade 315 .... H2
Magnus-A-Oil Grade 465 .... H2
MM Motor Oil SAE 10W H2
MM Motor Oil SAE 20 .... H2
MM Motor Oil SAE 20-20W H2
MM Motor Oil SAE 30 .... H2
MM Motor Oil SAE 40 .... H2
MM Motor Oil SAE 50 .... H2
Philkool Soluble Oil H2
Philube A NLGI Grade No. 2 ... H2
Philube All Mineral Gear Oil
   SAE 85W-90 .... H2
Philube All Minrl Gr. Oil SAE
   140 .... H2
Philube All Minrl Gr. Oil SAE
   250 .... H2

Philube All Minrl Gr. Oil SAE
   80 .... H2
Philube All Minrl Gr. Oil SAE
   90 .... H2
Philube All Purpose Gear Oil
   SAE 140 EP H2
Philube All Purpose Gear Oil
   SAE 80W EP H2
Philube All Purpose Gear Oil
   SAE85W-90EP H2
Philube All Purpose Gr. Oil
   SAE 250 .... H2
Philube All Purpose Gr. Oil
   SAE 90 .... H2
Philube ASM H2
Philube B-2 Grease H2
Philube C-3 .... H2
Philube C-4 .... H2
Philube CB-0 Grease H2
Philube CB-00 Grease H2
Philube CB-1 Grease H2
Philube CB-2 Grease H2
Philube IB&RB NLGI Grade
   No. 2 .... H2
Philube L-0 Multi-Purpose H2
Philube L-1 Multi-Purpose H2
Philube L-2 Multi-Purpose H2
Philube M-2 Grease H2
Philube PF H1
Philube S.M.P. Gear Oil SAE
   80W-90 .... H2
Philube S.M.P. Gear Oil SAE
   85W-140 .... H2
Philube SMP Gear Oil SAE
   140 .... H2
Philube SMP Gear Oil SAE 80 . H2
Philube SMP Gear Oil SAE 90 . H2
Sixty Six Motor Oil SAE 10W .. H2
Sixty Six Motor Oil SAE 20-
   20W H2
Sixty Six Motor Oil SAE 30 ..... H2
Sixty Six Motor Oil SAE 40 ..... H2
Sixty Six Motor Oil SAE 50 ..... H2
Sixty Six Special Mtr Oil SAE
   20W-40 .... H2
Sixty Six Special Mtr Oil
   10W20W30 .... H2
Soltrol 170 .... H1
Super HD II Low Ash Motor
   Oil SAE 10W H2
Super HD II Low Ash Motor
   Oil SAE 15W-40 .... H2
Super HD II Low Ash Motor
   Oil SAE 20-20W H2
Super HD II Low Ash Motor
   Oil SAE 30 .... H2
Super HD II Low Ash Motor
   Oil SAE 40 .... H2
Super HD Motor Oil SAE 10W H2
Super HD Motor Oil SAE 20-
   20W H2
Super HD Motor Oil SAE 30 .... H2
Super HD Motor Oil SAE 40 .... H2
Super HD Motor Oil SAE 50 .... H2
Thetis Oil Grade 315 .... H2
Thetis Oil Grade 465 .... H2
Thetis Oil Grade 700 .... H2
Trop-Artic All Sea. Mtr Oil
   SAE 10W 30 .... H2
Trop-Artic All Sea.Mtr Oil
   SAE 20W40 .... H2
Trop-Artic All Seasn Mtr Oil
   10W20W30 .... H2
Trop-Artic All Season Mtr Oil
   SAE 10W-40 .... H2
Trop-Artic All Season Mtr Oil
   SAE 5W-30 .... H2
Trop-Artic Motor Oil SAE
   10W H2

211

Foaming Cleanser A6
Glacier A5
Go L1
GW-61 .... L1
Hand Shield E4
Hantex Cleansing Cream E4
Hook & Trolley Cleaner A2
HP-673 .... G6
L.A.B. Liquid Alive Bacteria L2
Lemon Scented Sterene
  Disinfectant Deod C2
Light Duty Cleaner A1
Liquid Hand Soap 15% E1
Liquid Hand Soap 40% E1
Liquid Multi-Purpose Cleaner A1
Lube-2 .... H2
Lustre A7
Mint-O-Green C1
Multi 2000 .... H2
Nice N'Easy C1
Non-Caustic Spray Cleaner A1
Open Gear Lubricant H2
Paint & Varnish Stripper C3
Penetral H2
Penetral-Industrial Penetrant H2
Power Foam C1
Power-Gel C1
Power-Plus A1
Precision Concrete Clnr.-Hvy-
  Dty. Fl. Cl C1
PL Stop Odor Control
  Chemical Granules C1
PL-Concentrate A1
PL-Food Machinery Lubricant .. H1
PL-Fresh (Mint) C1
PL-Laundry Detergent B1
PL-Mud-Flo L1
PL-Power Clean A1
PL-Tef-Tape H2
PL-Teflex H1
PL-100 Degreasing Solvent K1
PL-109 .... K1
PL-200 .... A4
PL-34-P A4
PL-345 .... A3
PL-350 .... A3
Real-Ease H2
Respond A4
Rhino L1
S.T. F1
Sana Clean Special Formula C2
Sani-Rinse 11 .... D2
Scentamint Disinfectant C1
Sewage Enzymes L2
Smash F2
Smoke House Cleaner A2
Spray And Wipe C1
Steam Clene A2
Sterene C2
Sterene II Air Sanitizer C1
Super L.I.M. C1
Super Sanaclean C2
Super Tuff Concrete Cleaner A4
SM-38 .... L1
T-K-O F1
The Better Way E1
Toro K1
Ultimate F2
Vat Cleaner A1
Victory Coefi-5 .... C2
View C1
1830 Cleaner A2

Premax Laboratories

Drops of Magic D1
Plumb Magic L2

Premier Chemical Corporation

BW Special A2
Germicidal Liquid Hand Soap ... E1

Liquid Hand Soap E1
Liquiklor A1
M-S-R-C A3
Montex 3A A1
Monti AC A1
Monti AC 505 .... A1
Monti AC 505 W/Color A1
Monti Klor Cleaner #3 .... A1
Monti Tone Liquid Lubricant
  Conc H2
Monti Tone 114 .... A2
Monti Tone 135 .... A4
Monti Tone 228 .... A2
Monti Tone 232 .... A2
Monti-Base Conc A1,H2
Monti-Base Conveyor Base
  Lubricant H2
Premchem 116 .... A1
Premchem 120 .... A1
Premchem 123 .... Q1
Premchem 131 .... A1
Premchem 133 .... A2
Premchem 134 .... A2
Premchem 136 .... A4
Premchem 146 .... A1
Premchem 205 .... A1
Premchem 216 .... A4
Premchem 217 .... A4
Premchem 225 .... A1
Premchem 227 .... A1
Premchem 502 .... A2
Premchem 516 .... A1,B1
Premchem 545 .... A3
Premchem 546 .... A3
Premchem 846 .... A2
Premfome 1021 .... A1
Premfome 1022 .... A1
Premfome 1722 .... A1
Premkleen 1522 .... A1
Premkleen 1612 .... A4
Premkleen 1712 .... A4
Premklor 1305 .... A1
Premklor 1321 .... Q1
Premklor 1331 .... A2
Premklor 1502 .... A1
Premklor 501 .... A1
Premloob 117 .... H2
Premloob 71 .... H2
Premloob 75 .... H2
Premscale 1545 .... A3
Premscale 1645 .... A3
Premsoke 1131 .... A1
Premsoke 1641 .... A2
Premspra 1611 .... A1
Premsteem 1005 .... A1
Premsteem 1631 .... A4
Premstrip 1648 .... A2
Premstrip 1649 .... A2
Sanpan A1

Premier Chemicals, Inc.

Premier Solvent Degreaser K1
Spritz's Shine A4
Strippette A1

Prentiss Drug & Chemical Co., Inc.

Co-Rax Pelleted Bait F3
Prentox Blue Powder F2
Prentox D.D.V.P. Five F2
Prentox Diazinon 4E
  Insecticide F2
Prentox Diazinon 4S Insecticide F2
Prentox Diazinon 6E
  Insecticide F2
Prentox Dursban 2E Insecticide F2
Prentox Dursban 4-E
  Insecticide F2
Prentox DDVP-Ronnel
  Emulsifiable Conc F2
Prentox DPBM Concentrate
  #4425 .... F2

Prentox Emulsifiable Spray
  Conc. #101 .... F1
Prentox Fogging Concentrate
  #1 .... F1
Prentox Fogging Concentrate
  #2 .... F1
Prentox Insect Spray "A" F1
Prentox Insect Spray "B" F1
Prentox Insect Spray "C" F1
Prentox Insect Spray "D" F1
Prentox Liquid Household
  Insect Spray #1 .... F2
Prentox Mill Spray Concentrate
  #562 .... F1
Prentox Pyronyl Oil
  Concentrate #225 .... F1
Prentox Pyronyl Oil
  Concentrate #3610 .... F1
Prentox Pyronyl Oil
  Concentrate #525 .... F1
Prentox Pyronyl Oil Concntrt.
  OR-3610-A F1
Prentox Pyronyl Pltry
  House&Barn Fly Spy F1
Prentox Pyronyl 1.2-2.4-4.0
  Emulsf. Conc F1
Prentox PB Concentrate #125 ... F1
Prentox PB Concentrate #20 ..... F1
Prentox Residual Concentrate
  DV-One F1
Prentox Residual Concentrate
  DV-Two F2
Prentox Residual Insect Spray
  #2 .... F2
Prentox Residual Spray
  Contains Baygon F2
Prentox Ronnel 24E Insecticide. F2
Prentox Spray Concentrate No.
  1575 .... F1
Prentox Spray Concentrate No.
  625 .... F1
Prentox Synpy-Diazinon
  Emlsfbl. Conc F2
Prentox Vapon 2 .... F2
Prentox Vapon 20% Emulsifi.
  Concentrate F2
Prentox Profi Dual Synpy
  Chlrpyrifos Eml Conc F2

Presco Food Products, Inc.

Caldina N2
Grease Remover A2
Super-Cleanser A2
Two-In-One Cleanser A1

Press Chemical & Pharmaceutical Labs.

La Vecze Concentrated
  Sanitizing Clnsr D1

Pressure-Lube, Inc.

Jax Battery Saver and Cleaner ... P1
Jax Chain & Cable Lubricant H2
Jax Dry-Slide H1
Jax Lift Truck And Sliding
  Tandem Lub H2
Jax Non-Toxic Lube H1
Jax Penetrating Oil H2
Jax Spray Lube for Open Gears H2
Jax Spray-Kote P1

Prestige Chemical & Supply Co., I

APD All Purpose Detergent B1

Presto-X-Company

Our Own Insect Spray F2
Our Own Rat And Mouse Bait .. F3
Pyrenone Insect Spray No. 6 ..... F1

Prime Chemicals

AT 306-L A3
CW-107-P A1
D-100 .... A1
HTS-203P A2
Liqui-Base with-107 .... A1

Prime Laboratories, Inc.

"Spotless" A1
C-Thru C1
Derma-Suds E1
Glisten A4
Grime Stopper A4
Irish Mist C1
Multi-Klean A1
Power-House A4
Rust Off C1
Sila-Guard G3
Speedy Spray A1
Spray Klean K1
Steam-Klean A1

Prime Solutions, Inc.

Aqua-Vent Ventilator Cleaner-
  Dgrsr Conc. A4,A8
Low Foam Cleaner/Degreaser .. A1
Prime "Equipment Spray" H2
Prime "Stick-Less" Snow
  Release Spray H2
Prime Acid Tile and Wall
  Cleaner A4
Prime Air Sanitizer Mint C1
Prime Air Sanitizer Spice C1
Prime All Purpose Cleaner A4
Prime Anti-Crawl Residual
  Insecticide F2
Prime Anti-Seize Lubricant H2
Prime Bath And Tile Cleaner C2
Prime Chewing Gum & Candle
  Wax Remover C1
Prime Citrus Air Sanitizer C1
Prime Cleaner-Degreaser A4,A8
Prime Cleaner-Degreaser
  Concentrate A4,A8
Prime Disinfectant &
  Deodorant C1
Prime Disinfectant Foam
  Cleaner C1
Prime Drain Cleaner (Enzyme
  Type) L2
Prime Elktra "S" Contact
  Cleaner K2
Prime Foaming Metal Polish A7
Prime FPD A1
Prime Glass Cleaner C1
Prime Grill & Oven Cleaner A1
Prime Ice Melt C1
Prime Liquid Pot &
  Dishwashing Detergent A1
Prime Long Shot 11 Wasp &
  Hornet Killer F2
Prime Low Temp Auto Scrub ... A5
Prime Multi-Purpose Insecticide F1
Prime Oven And Grill Cleaner .. A8
Prime Penetrating Lube &
  Demoisturizer H2
Prime PW 3 .... A4,A8
Prime Safety Solvent K1
Prime Silicone Lubricant Spray . H1
Prime Spot Remover K1
Prime Stainless Steel Cleaner A7
Prime Super All Purpose
  Cleaner C1
Prime Tile and Grout Cleaner ... A4
Prime Time Air Sanitizer Mint .. C1
Prime Vandalism Mark
  Remover K1
Prime Waterless Hand Cleaner .. E4

Primrose Oil Company

#216 Primrose Plus White
  Mineral Oil H1
#314 Primrose Superior Cannry
  Grs H1
#610 P.D.S. Concentrate A1
#620 Quik-Solv A1
No. 314 Primrose Plus Cannery
  Grease H1
Quik A1

Princeton Chemical Co. Inc.

Floor Wiz A1
Kitchen-Wiz A4,A8
Oven Wiz A8

Prinova Co. Inc.

Ladrin CO-DAC-PCM B1
Ladrin Ecolo Blend B1
Ladrin Ecolo-Break Plus B1
Ladrin Ecolo-Det B1
Ladrin Ecolo-Det S B1
Ladrin H-Chlor B Bleach B1
Ladrin Super Sil B1
Ladrin X P-12 .... B1
R.T.U. Starch B2

Pritchard Sales Company

A-150 Superior Concentrate All
  Purp. Cln A1
A-151 Pink Concentrate A1
A-153 Atomic Action A1
A-161 Sof-T-Suds Hand
  Dishwashing Cmpd A1
A-162 Beaded Hand
  Dishwashing Compound A1
A-225 Blue Sea Beads A1
A-225 Pink Sky Beads A1
A-225 White Foam Beads A1
A-230 Super Chlorinated Mach
  Dshwsh Det. A1
AA-106 Chain Lub & Cleaner ... H1
ABC A-290 Po Pac Ho Cleaner A1
ABC A-310 Labor Saver
  Chloro Cleaner A1
B-165 Liquid Machine
  Dishwash Detergent A2
B-194 White Odorless Steam
  Cleaning Cmpd A2
B-195 Steam Cleaner "Heavy
  Duty" A2
B-196 Medium Duty Steam
  Cleaner A2
B-197 Extra Heavy Duty Steam
  Clng Cmpd A2
B-250 Restaurant King DFF
  Cleaner A2
D-156 Odorless Concrete
  Cleaner A4
G-100 Cocoanut Oil Liquid
  Hand Soap E1
G-105 Econo Cocoanut Oil Liq.
  Hand Soap E1
General Disinfectant C1
Germ-I-San D2
Heavy Duty Sewer Solvent &
  Drain Opener L1
Insecticide No. 111 .... F1
L-210 Hog Scald Compound N2
M-157 Heavy Duty Magic
  Concrete Cleaner C1
Malathion-50 Spray F2
Pine Odor Disinfectant C1
Pine Oil Disinfectant C1
Space Spray No. 3 .... F1
Super Germicide D1

Private Label Chemicals, Inc.

APC Super Concentrate A4

DO-104.. C1
1A-200... A3
LDDC-89.. A1
LDSO-50... L1
LSDO-88 Liq Sulfc Swr&Drain
  Opnr Sulf Sn L1
OC-90 Oven Cleaner..................... A8
PC-33... A1
PC-33 E.B. A4
Silicone Spray.............................. H1
SE-75 Safety Solvent Degreaser K1
SECC-33... Q1
SF-400.. A1
SF-500.. A1
SFDC-16 .... D1
TBC-77 .... C2
TCE 500 .... K2

Products Chemical Company

Cling & Clean ....,............... C2
Green Genie Bowl Clnr. w/
  Mint Fragrance......................... C1
Pro Chem 200 Germicidal
  Cleaner...................................... A4

Professional Chemical Company, Inc.

Professional Spray Concentrate . F1

Professional Chemists, Inc.

Realclean Spray Concentrate C1

Professional Cleaning Services/ & Supply Corporation

Pro-Chem Formula 800 .... A1

Professional Maintenance Center, Inc.

PMC Flying Insect Killer........... F1
PMC Glass & Utility Cleaner C1
PMC 910 Disinfectant Bowl
  Cleaner...................................... C2
Silicone Lubricant Spray H2

Professional Maintenance Supply

Big Dutch...................................... A1
Big Dutch II A1

Professional System Supply

Blaz Off A1
Sensuous Lotion Hand Soap E1

Proform Products Corporation

Pollution Control Agent A4

Progress Chemical Co., Inc.

Blend 17.. Q1
Blend 17 WA Egg Detergent..... Q1
Blend 5 Egg Detergent............... Q1
DE 1 Egg Detergent.................... Q1
DE 2 Antifoam Formula Q5
DE 3 Detergent Booster............. Q1
DE 4 Egg Destainer Q2
DE 5 Chlorine Sanitizer
  Powder Q4
Foam-Away Q5
Klor Plant Cleaner A1
Linear Alkane Sulfonate A1
Odor Quest.................................... C1
One Step Detergent..................... A1
Power Terge Egg Detergent Q1
Power Terge Formula Q1
Power Terge Plus Formula........ Q1
Pro-San 10..................................... D2
Pro-White Egg Detergent Q1
PCC Alk Plant Cleaner A1
PCC Chlorine Sanitizer D1,Q4
PCC CIP Cleaner......................... A2
PCC CIP Cleaner 300................. A2
PCC Detergent Booster............. A1,Q1
PCC Egg Destainer...................... Q2

PCC Feather Penetrant 50 .... N1
PCC Foamadd A4
PCC Klor 150 Plant Cleaner...... A1
PCC Liq. Milkstone Remover
  Conctrt...................................... A3
PCC Liquid Hand Soap.............. E1
PCC One Step Disinfecting
  Detergent.................................. D1
PCC Plant Cleaner P-196 .... A1
PCC Plant Cleaner P296............ A1
PCC Plant Cleaner 100.............. A1
PCC Plant Cleaner 18................ A1
PCC Plant Cleaner 20................ A1
PCC Plant Cleaner 200.............. A1
PCC Plant Cleaner 30................ A1
PCC Plant Cleaner 8.................. A1
PCC Plant Cleaner 80................ A1
PCC Plant Cleaner 90................ A2
PCC Plant Cleaner 99................ A2
PCC Shackle Bright.................... A2
PCC Shackle Cleaner................. A2
PCC Shackle Cleaner CC........... A2
Spray Clean.................................. A4

Progressive Chemical Corp.

Pro-Solve 7-11 .... K1
RH-63 .... A1

Progressive Enterprizes

"21" Concentrate........................ A1
Aci-Det.. A3
All-Kleen Heavy Duty A1
Bowl Kleen C2
Extra Hi Concentrate................. A1
Grease Trap Formula.................. L2
Industrial & Municipal Waste
  Formula L2

Progressive Hydrology

Boiler Purge................................. A3
Hy-Blend Liquid 201.................. G6
Hy-Blend SS G6
Hy-Blend 20................................. G6
Hy-Blend 25................................. G6
Hy-Blend 30................................. G6
Hy-Poly 400................................. G6
Hy-Poly 521................................. G6
Rezonator P1
Sea-Quest...................................... G6
Sea-Quest Pro-44 .... G2
Steam-Aid..................................... G6
Steam-Eze G6
Steamline Treatment G6
Tri-Blend 20................................. G7
Tri-Blend 30................................. G7

Progressive Products

Aqua-Plus...................................... G2
Aqua-Plus Pro-44 .... G1
Blend-Dex Dyno Stix................. G6
Blend-Dex Pro-Cat SS Oxygen
  Control...................................... G6
Blend-Dex Pro-15 SLT.............. G6
Blend-Dex 2 .... G6
Blend-Dex 3 .... G6
Boiler Power Liquid G6
Boiler Power 2............................. G7
Boiler Power 3............................. G7
Boiler Power 350......................... G7
Boiler Purge................................. G7
Dyno-Stix G7
Eco Blend...................................... G7
Eco-Blend FG G6
Oxygen Control G6
Poly Blend 400 .... G6
Poly Blend 521 .... G6
Tower Purge................................. G7
Tower Treatment G7

215

Promerca

Destroyer L2

Property Chemical Products

Acrofoam C2
Alumasol A3
Autumn Gold.............................. C1
Awair C1
Chem-Mate.................................. A4
Chem-Ram L1
Control 21 .... H2
Desist F1
Di-Phonal C1
Direct.. C2
Direct Non-Perfumed C2
Disonate....................................... L1
Disoterge...................................... A4
Dyapen F2
Electroloid (Aerosol) K2
Electroloid (Bulk)...................... K1
Evax.. F1
Extar C1
Flexite.. H1
Gar-Buds C1
Hard Surface Cleaner No. 205 ... A4
Methagran L1
Mint Condition C1
Mirrex.. C1
Pathatrol D2
Power-Melt C1
Release.. H2
Rinseless Cleaner No. 200 .... A1
Ronite A1
Sila-Lube (Aerosol) H1
Sila-Lube (Bulk) H1
Steel-Tone A7
Steelene A7
Sure Bowl Cleaner C2
Thermax A8
Thrift C1
Thrusty... H2

Prosclean, Incorporated

Acid P-22 .... A3
Acid P-33 .... A3
Acid-M-44 .... A3
Foam-O A1
Gal.-122 .... A1
General L A1
General Purpose 10 X
  Chlorinated............................... A1
H-Duty-L100 .... A2
Implace L-P A2
L. Hog Scald N2
Machine A-Spray A1
Machine Spray A2
Machine Spray P A2
P. Hog Scald................................ N2
Ruster-P-99.................................. A2
Safe General A1
Safe General 101......................... A1
Sewer Cleaner 9 .... L1
Shroud Wash A B1
Smoker-P-88................................. A2
Smoker-P-881............................... A2
Tree Wash.................................... A2
Tripe-W-66................................... N3

Protex Janitorial Products

Super Protex A1

ProClean Inc.

All-Purpose Cleaner.................... A1
Degreaser-Cleaner....................... A1
Industrial Degreaser-Cleaner...... A1
Pot Licker A1
Stripper... C1

ProLine Chemical & Supply Div.
George Kech & Company

D.U.D.. C2

Prudential Chemical Corp.

Aluminum & Stainless Steel
  Cln. & Brtnr.............................. A3
Arctic Clean................................. A5
Clean It-All Purpose Cleaner A1
Clean It-All Purpose Cleaner
  #1... A1
Clean-It... A1
Concentrated Inhibited
  Chemical Cleaner A3
Deep Fry Pot & Pan Cleaner..... A4
Drain-Flo...................................... L1
Emulsifiable Cleaner K1
Emulsifiable Degreaser K1
Extra Strenght Hvy Duty Wtr
  Soluble Cln................................ A1
Flo-Go No. 308 .... C1
Foam Kleen A1
General Purpose Liquid
  Cleaner....................................... A1
Germicidal Cleaner C1
Hand Soap.................................... E1
Heavy Duty Organic Drain
  Cleaner....................................... L1
Ice Melting Compound............... C1
JT-650... A4
Liquid Alive Bacteria.................. L2
Liquid Hand Soap 40%
  Concentrate E1
Liquid Solvent Non-Flammable
  Cleaner....................................... A4
Neutral Cleaner A1
Organic Acid Detergent A3
Pine Odor Disinfectant C1
Prudential Agricultural Enzyme L2
Prudential Butyl Power Cleaner A4
Prudential Concrete Cleaner A4
Prudential Controlled Suds B1
Prudential Deodorant Granules . C1
Prudential Germicidal Cleaner... D1
Prudential Lemon Hand Soap.... E1
Prudential Odor Control
  Granules C1
Quat Disinfectant........................ D1
Safety Solvent K1
Sewer Solvent.............................. L1
Super-Kleen A4
Synthetic Cleaner A4
Three-Gly C1

Prudential Chemical Corp.

Handyman..................................... A1
Success.. A1

Public Health Equipment & Supply
Co.

Oil Concentrate #3610............... F1

Public Health Systems, Inc.

Kleen Up...................................... A1

Pulmore

Slipmor H1

Puratex Company, The

Stainless Steel Cleaner A1

Purdue Frederick Company, The

Betadine Skin Cleanser Foam E2
Betadine Surgical Scrub/Skin
  Clnsr... E2

Purdy Products, Inc.

Trophy Concentrated Toilet
  Bowl Clnr................................... C2

Trophy Porcelain Grout Tile
  Bowl Clnr................................... C2
Trophy Renovator A4

Purdy, N.B. Products Company

Chlorinated Par A1
Par French-Fryer & Filter
  Cleaner....................................... A1
Regular Dairy Par....................... A1
Thoro-Cleen A1
X-Ton A3

Pure Water Products, Inc.

Formula BL-154A G6
Formula BL-160A G6

Pure-Chem Products Company

Bottle Sheen................................. A2
Clean Away A1
E-Z Clean..................................... A1
Pur Chlor A1
Pur Lac... A3
Pure Cir Q................................... A1
PC Chain Lube 41....................... H2
PC Conveyor Lube...................... H2
PC No. 1 Lube H2
PC-VL... A2
PC-50.. A2
PC-51.. A1
PC-52.. A1
PC-53.. A1
PC-54.. A1
PC-55.. A3
PC-6.. A1

Purex Corporation

All Fabric-All Color Dry
  Bleach B1
All Purpose Concentrated
  Cleaner....................................... A4
All Purpose Low Foam Cleaner A4
All Purpose Spray Cleaner........ A4
Ammoniated Wax And Finish
  Remover C1
Blue Barrel Soap H2,B1
Bowl Cleaner C2
Brillo Bowl Cleaner C2
Brillo Foaming Cleaner-
  Degreaser A1
Brillo Pine Creme Deodorant
  Cleaner....................................... C1
Brillo Pinosan Disinfectant
  Cleaner....................................... D1
Brillo Vanquish........................... C2
Cold Room Cleaner A5
Cold Solvent Degreaser.............. K1
Dutch All Purp
  Liq.Ammoniated Clnr.............. C1
Extra Strength Wax & Floor
  Finish Remvr A4
Fleecy White Liquid Bleach D2,N4
Franklin Blu-Lite........................ C2
Franklin Clearinse Foaming
  Clnr Dgrsr................................. A1
Franklin Disappear C2
Franklin Dutch C1
Franklin F-37 Multipurpose
  Acid Cleaner............................. A3
Franklin Foam-Xit...................... A4
Franklin Formula 707 .... A1
Franklin Formula 900 .... A4
Franklin Friendly Air C2
Franklin F9890 Acid Cleaner.... A3
Franklin G-O-E-S........................ A1
Franklin Heavy Duty Cleaner
  Degreaser A4
Franklin Heel Mark Remover ... A4
Franklin Liquid Detergent A4
Franklin Out-Strip...................... A4
Franklin Phenomysan................. A4

Ten-Four F1
Thermo-Jet C1
Thrust A1
Top Secret...................... A4
Trounce D2
Tuffer A7
Twinkle A1
Ultimate C1
Valiant C1
Vortex............................ A1
Zodiac............................ A1
Zot L1
4-D 25............................ D1

Purity Chemical Company Div/Ion Exchange Products, Inc.

Dry Acid No. 88 .... A3
P.B.T. No. 100 .... G6
Product No. 175 .... G7
Puramine No. 60............... G6
Purity Product No. 102 ..... G6
Purity Product No. 103 ..... G6
Purity Product No. 104 ..... G1
Zero-Klenz...................... P1

Puro Chemical Company

Blue Tint Toilet Bowl Cleaner .. C2
Creme Cleanser................. C2
Drain-A-Go-Go L1
Drain-E-Z........................ L1
Odorless Concrete Cleaner
P554 .... A4
Perfumed Deodorant Blocks C2
Powdered Drain Solvent L1
Sana-Bole Drain Pipe
Cl.&Maintnr L1
Sana-Bole Emulsion Toilet
Bowl Cl. C2
Toilet Bowl Deodorant........ C2
Wall Block Deodorants C2

Puromation Industries, Inc.

KO-200 Liq Steam & Pressure
Clng Compud................. A1
Mus'L-X Clnr.-Stripper-Dgrsr.
Liq Compnd................. A4
Myriad 100 Do-All Cleaner A4
Puro 10-B Boiler Water
Treatment G7
Puro 20 B Boiler Water
Treatment G6
Puro 30 B Boiler Water
Treatment G6
Puro 40 B Boiler Water
Treatment G6
Puro 60 B Closed System Hot
Water Treat................... G6
Puro 80-B Adjunct G6
Puro 90-B G6
Puro-Scrub A1
Puro-Washdown A1

Q

Q-22, Inc.

Detergent-Disinfectant
Deodorizer-Fngcd D2

Qadrex, Incorporated

Q-Ten-2 .... G2
Q-429 Fast Action Stripper/
Cleaner........................... A4

Quaker Chemical Co. Inc.

"Wipe Out" A4,A8
Chem-Terge A4,A8

Quaker City Paper And Chemical Co.

All Purpose Cleaner A4,C2
Degreaser Concentrate A1
Industrial Degreaser
Concentrate A1
Sink/Hand Concentrate A1

Qualcon, Inc.

Qual-Chem Sanitizer D2

Qualichem Incorporated

Biodor C1
Bora-Clean E1
Chef-Ette A4
Conductive Cleaner
Disinfectant D1
Conductive Detergent Sanitizer. D1
HD-100 .... A4
Lemon Fresh Odor.................. C1
Lotion Hand Soap.................. E1
Mint Fresh Odor C1
Neutral Cleaner A4
Pine Scent Disinfectant C1
Pine-Dis-5........................... C1
Quali-Blue A4
Supurple Concentrate.............. A4
U.N.C.L.E............................ F1
20% Liquid Hand Soap E1
40% Liquid Hand Soap E1
40% Surgical Liquid Hand
Soap E1

Qualified Laboratories, Inc.

Quali Hospital Cleaner A4
Quali Septic Degreasing
Cleaner.............................. A1
Quali-Sil Silicone Spray H1
Safety-Solvent K1
Skin Shield Cream.................. E4
Stainless Steel and Metal Polish. A7
Super-Solv............................ K2
Touch-Up All Purpose Cleaner. C1
TFE Fluorocarbon Lubricant H2

Quality Chem Products

#391 .... Q1
A.C.30 #B1001 .... A3
A-1011 Pan Wash A1
A-1013 Foamer A2
A-1112-Endlich A2
A-1112A-Endlich Additive........ A1
A-1196 Tray Wash A1
A-1199 Soak A1
A-1231 Ultra Clean A2
A-1231 XP............................ A2
A-1231A Ultra Clean Additive .. A1
Activate A1
Anti-Foam Q5
AAA#100 .... Q1
AC-60 .... A3
AC-90 .... A3
B-1219-Tanken Schein............. A1
Blitz L #B1017 .... A1
Blitz S #B1019...................... C1
Bottle Clean #A1021 .... A2
Chlorospray A1
Clean A.C. #A1041X............... A1
Clean Deluxe #A1043 .... A1
Con Clean #A1061.................. C1
Con Clean HD #A1065............ C1
Conveyor Lube....................... H2
Descale L A3
Descale P A3
Descale 25 .... A3
Dyna Clean HD #A1103 .... A2
Exceed #B1113...................... A1
Exceed LF A1
Exceed Plus #B1115 .... A4
Foamer A A2

Jell All A1
Lube Deluxe H2
Lube Deluxe #B1137............... H2
Lube Special #B1139............... H2
Lube Special LF..................... H2
Machine #1 #A1205............... A2
Saluta #B1187....................... A1
Sewerex L1
Spray Chlor D #A1209............ A2
Spray Chlor S #A1207............ A2
Spray L #1215....................... H2
Super Six LF H2
Ultra-Chlor........................... A1

Quality Chemical Company

Speed Sol............................. A4
Speed Sol Extra..................... A4

Quality Chemical Corp.

Heavy Duty Industrial Cleaner
& Degrease A1
Kleen C1
Quat Kleen D1

Quality Labs.

Heavy Duty Stripper Cleaner &
Degreaser C1
Q.L. Emulsifier A1

Quality Research Labs.

All-Kleen A1
Ceram-Glo............................ C1
Coil-X P1
De-Scale C1
Degree-Sor A4,A8
Drane X................................ L1
Drane-Free L1
Met-L-Brite A7
Oven-Grill A1

Quality Supply Company

Dissolve A4
Ream L1

Quasar Corporation

Moon Shine C2
Solube................................. K2
800..................................... K1
880..................................... K1

Quipu Corporation

Q-Mist Air Sanitizer(Hospital
Type).............................. C1
Q-Mist Bug Killer.................. F1
Q-Mist Insecticide Indust. Type. F2

R

R + R Square Company

Hand Soap-R 247................... E1
Nu Clean-R Degress-R 149....... A4

R & D Products Corporation

Big G A4
Citrosene 10 .... C1
L-Tox Spray F1
R&D Liquid Steam Concentrate A1
R&D Truck Soap A1

R & H Enterprise Co., Inc.

AG Detergent........................ A1
AGP.................................... A1
Bowlex................................. C2
Bowlite................................ C2
Clear Concentrate.................. A1
Clear-Out............................. L1

De-Scale A3
Dissolve A4,A8
Foam Add A1
Fresh B1
Grease Cut A1
Grease Release..................... A1
GP-401 .... A1
Hy-Foam A1
HTP A2
Laundry Brite B1
Liquid Hand Soap A1,E1
LCMR A3
LCMR Concentrate A3
LCMR Concentrate Low Foam A3
LCMR Low Foam A3
LHT A1
Medi-Clean E1
Q-Ten................................... D2
Remove A1
RH-55 .... G6
RHS-90 .... G6
Sani-Clean 11 .... D1
Sani-Clean 20 .... D2
Sani-Ten D2,E3
Shiny Brite A3
Soak-Lean A1
Special C A1
Spra-Away C1
Strip Away............................ A2
Strip Kleen A2
Super AGP A1
Super Chlor A1
Super Chlor Plus A1
Super Heat Drain Pipe Opener.. L1
Wash Brite B1

R & K Chemical Co.
Nu-Power A1

R & P Chemicals
Shi-nee Industrial Strength All
Purp Cl A4

R & R Manufacturing Co.
Hi-Trac-Lube H2

R & R Sales Company
R & R Brand Solvent Cleaner
& De-greaser K1

R B Chemical Company
All Right A4,A8
Hand & Hand E1
Hand & Hand Antiseptic Lotion
Hand Soap.......................... E1
Hot Shot............................... B1
R-B Concentrate A4
R-B 212................................ A1
Rapid A1
Shine A1
Vigor E1

R M C Products Co., Inc.
RMC Super Bar F3

R. & D. Enterprises
F-103 Foaming Cln. A1
F-105 Blue Foamer............... A1
F-109 Boil Out A2
F-111 C.I.P. Acid Heavy Duty.. A3
F-112 All Purpose Cleaner A1
F-113 Kettle Klean................ A2
F-115 Pan Wash.................... A1
F-116 Power A1
F-118 Protein Stone Remover.. A3
F-119 Hook & Trolley A2
F-132 Clorinated Cleaner...... A2
F-136 Super Power A1

R. & E. Martin
Marsolv................................ A4
Remkleen Can-Do K1
Remkleen WS K1

R. & S. Enterprizes
RS 519 .... A1

R. C. P., Incorporated
Aqua-Zyme L2

R. L. R. Industries
Gentle A4
Natural................................. A4
Perform................................ A1
Qualem................................. C1
Steem Kleen A2

R.C. Research & Development Corp.
"High Concentrate" Sil Lub &
Mold Release...................... H1

R-Square Chemical & Coating, Inc.
B.C.-10................................. A4
Cherry Odor-Mask C1
Chlor-Clean.......................... C1
Coconut Oil Hand Cleaner.... E1
Mor-Chlor D2
R-Quat-15 .... C2
R-Quat-25 .... D1
RA-101 .... A1
RA-102 General Purpose
Powder A1
RA-103 .... A1
RA-122 .... A1
RA-123 .... A1
RA-125 Manual Cleaner A1
RA-150 Powder A1
RA-43 General Cleaner A1
RB-104 .... A2
RB-105 .... A1
RB-106 .... A1
RB-124 Cleaner Booster A2
RB-160 Powder A1
RB-70/30 Shackle Cleaner A2
RBB-115 Lubricant H2
RBB-116 .... H2
RBB-118 Lubricant A1,H2
RC-107................................. A3
RC-126 Acid Cleaner A3
RCC-100 Laundry Detergent.. B1
RD-113 .... A4
RD-50M Floor Cleaner A4
RFB-121 Foam Booster A1
RFB-701 Foam Booster A1
RH-112 .... Q1
RH-114 Shell Egg Wash
Detergent........................... Q1
RL-210 Feather Penetrant N1
RL-220 Feather-Penetrant N1
RL-230 Scald Vat Media N1
RP-170 Detergent A1
RP-175 All Purpose Cleaner ... A1
RP-180 Cleaner A1
RP-185 General Cleaner A1
RP-190 Neutral Detergent...... A1
RP-192 Neutral Cleaner......... A1
RP-195 General Cleaner A1
RP-215 Hi-Foam Power A1
RP-245 General Cleaner A1
RP-405 HF Acid A3
RP-405C Concentrated Acid
Cleaner.............................. A3
RP-565 H.D. Cleaner A1
RP-701 Chelate A1
RP-703 Chelate A1
RQ-109................................. E1
RY-200 Tripe Cleaner............ N3
Scat A4

Sof-Glo Hand Cleaner
Sof-Glo Hand Cleaner E1
Sof-Tex Hand Cleaner............ E1
Softsuds Hand Cleaner........... E1
Super-C A1
Tray Wash............................ A1
White Oil 35 USP.................. H1
70/30 Extra A1

R/B Industrial Supply
R/B 42 Germicidal Cleaner D1

Raban Supply Company
Concentrated Cleaner A1
R.S.C. Liquid Soap E1

Rae, J. H. Oil Company, Inc.
FPM Grease.......................... H1

Raechem Products, Inc.
Breakdown A1
Clean-Sept............................ A1

Ralston Purina Company
All Purpose Cleaner A4
All Purpose Cleaner F.G......... A1
C & S Powder D1
Chlorine Sanitizer-11............. D2
Chlorine Sanitizer-0-40-11 ..... D2
Cygon 2E F2
Disinfectant Q3
Egg Cleaner Q1
Liquid Anti Foam.................. Q5
Morlex Corrosion Inhibitor
A(Morpholine) G6
Powdered Hand Cleaner C2
Purina Acid Cleaner.............. A3
Purina Acid Cleaner #2.......... A3
Purina Acid Cleaner F.G. A3
Purina Acid Cleaner HD......... A3
Purina Acid Cleaner LP.......... A3
Purina Acid Cleaner(Powdered) A3
Purina Alkaline Cleaner......... A1
Purina Ammoniated General
Purpose Clnr. A4,C1
Purina Bowl Clean C2
Purina Bowl Clean S.............. C2
Purina Caustic Cleaner 116 .... A1
Purina Caustic Cleaner 25P A2
Purina Caustic Cleaner 299 .... A2
Purina Caustic Cleaner 60P..... A2
Purina Caustic Cleaner-294 .... A2
Purina Chlordane 2-E
Insecticide.......................... F2
Purina Chlorinated Cleaner D1
Purina Chlorine Egg Rinse...... Q4
Purina Chlorine Sanitizer 0-10.. D1
Purina Chlorine Sanitizer 50 .. D1,Q4
Purina Cleaner A1
Purina Conveyor Lubricant A1,H2
Purina CC-244L.................... A2
Purina CC-25 .... A2
Purina CC-294 .... A2
Purina CC-60 .... A2
Purina CHC-1 .... A1
Purina CW-1 .... A1
Purina D-P Cleaner A1
Purina Descaler A3
Purina Disinfectant Concentrate
4X D2,Q3
Purina Drain Pipe Opener L1
Purina Drain Pipe Opener S L1
Purina DW-1 .... A1
Purina DW-2 .... A1
Purina Egg Cleaner-1............. Q1
Purina Fast Clean A4
Purina Floor Cleaner A4
Purina Fly Patrol (Fly Bait)...... F2
Purina Food Industry
Insecticide Conc. F1

219

Purina Fresh Aire Odor Neutrl.
Strip C1
Purina Fryer Cleaner A2
Purina FM-100...........................A1
Purina General Cleaner
(Powdered)................................A1
Purina General Purpose Cleaner A1
Purina Glass Clean C1
Purina Hand Cleaner..................E1
Purina HPC................................A1
Purina HPC-Hi Foam A1
Purina I-O-Concentrate..............D2
Purina I-O-Dairy D2
Purina Insecticide Mist F1
Purina Insecticide Mist Special..F1
Purina Iodine Concentrate 3.5....D2
Purina KS-27..............................A1
Purina KS-30.............................A2
Purina KS-30 Part B A4
Purina L.E.C. A1,Q1
Purina Liquid Detergent.............A1
Purina Liquid Egg Cleaner Q1
Purina Malathion Spray F2
Purina Mouse-Kill F3
Purina Odor Control C1
Purina P.E.C. A1
Purina Powdered Detergent/
Sanitizer D1
Purina Pura-Kleen A1
Purina Quaternary Disinfectant .D2,Q3
Purina Rat Control Pellets F3
Purina Residual Insect Spray F2
Purina Retort & Cooling Water
Cond. G5
Purina Rinse Aid A1,G5
Purina Roach and Ant Killer F2
Purina Roach Spray F2
Purina RP 300...........................A2
Purina RP 302...........................A2
Purina RP 303........................A1,Q1
Purina RP 304...........................A2
Purina RP 401...........................A1
Purina RP 402............................D1
Purina RP 403............................A1
Purina Sanitizer (Chlorinated)...D2
Purina Sanitizer 0-40.................D2
Purina Solu-Clean..................A1,Q1
Purina Space Mist Insecticide ...F1
Purina Spray and Dip F2
Purina Spray Disinfectant..........C2
Purina Spray Fresh Bouquet C2
Purina Spray Fresh Floral..........C2
Purina Spray Fresh Lemon C2
Purina Stainless Steel Polish/
Lub. H1
Purina Starch/Fat Cleaner A1
Purina Tool Spray.......................H1
Purina 400 LD Concentrate A1
Quebracho Tannin......................G6
Rat Kill.......................................F3
Sodium Carbonate G6
Sodium Phosphate G6
Strip Clean A4

Ram Enterprise, Inc.
Lem-O-Nal A4
Rage..F1
Ram Foam-Kutter A1
Release..A1

Ram Industries
Fogging Spray Concentrate F1

Ramrod Chemical Company
Ram-Cide II D2

Ran/Scott Enterprises
Freez-O-Clean.............................A5

Ranco Chemical
AC 30 ....A3
Blitz..C1
Blitz L...A4
Car...A1
Clean Deluxe A1
Endlich...A2
Exceed Plus..................................A1
Exceel A1
Foamer...A2
Machine L....................................A1
Mild...A1
Saluta A1
Spray Klor....................................A2
Tray Wash....................................A1

Randall's United Petroleum Company, Inc.
"C#362" Grease H2
Randall's White Grease #2 ....H2

Randik Paper Company
All Purpose Cleaner....................A1
All' N One D1
Pineol 5..C1

Rapid Chemical & Color Company
Conveyor-Ease H2
Cos All Purpose Steam Clng
Comp A1
Cos Concrete Floor Cleaner A4
Cos Heavy Duty Steam Clng
Comp A4
Liquid Spec.................................A2
Rapid Floor Clean II A4
Rapid Spec..................................A2

Rath Packing Company, The
"G" Cleaner A2
All Purpose Farm Detergent......A1
AC Cleaner A1
General Purpose "C" Cleaner....A2
General Purpose Cleaner A1
GP "S" Cleaner A1
GP "W" Cleaner A1
Mesh Glove Cleaner A1
Rail Car Cleaner.........................A1
Revolving Smokehouse Cleaner A1
S & S Cleaner A1
Scouring Powder.........................A6
Smokehouse 500 Cleaner A2
Stainless Steel Cleaner A3,A6
Tankhouse Cleaner......................A2
Tree Cleaner A1
Trolley Cleaner............................A1

Ravan Products, Incorporated
All Purpose Cleaner (Liquid).....A1
All Purpose Cleaner (Powder)....A1
B.C.-10...A4
Bright Glo Stainless Steel
Cleaner......................................A2
Bright Line..................................A1
Bright Line Special A2
Centralized Cleaner-Detergent....A1
Cherry Odor-Mask C1
Chlor-Clean..................................C1
Coconut Oil Hand Cleaner E1
Iodine Disinfectant D1
Mor-Chlor D2
Odorless Concrete Cleaner
Special.......................................A4
PLS Super C................................A1
R-Quat-15 ....C2
R-Quat-25 ....D1
Ravan Bleach...............................D2
Ravan Odor Control C1
RA-101 ....A1
RA-102 ....A1

RA-103 ....A1
RA-122 ....A1
RA-123 Neutral Cleaner A1
RA-125 ....A1
RA-150 Powder A1
RA-205 General Cleaner A1
RA-43 General Cleaner A1
RB-104...A2
RB-105...A2
RB-106...A1
RB-124 Cleaner Booster A2
RB-160 Powder A1
RB-70/30 Shackle Cleaner A2
RBB-115 Lubricant H2
RBB-116 ....H2
RBB-118 Lubricant A1,H2
RC-107...A3
RC-126 Acid Cleaner A3
RCC-100 Laundry Detergent.....B1
RD-113 Concrete Floor Cleaner A4
RD-50M Floor Cleaner A4
RFB-121 ....A1
RFB-701 Foam Booster A1
RH-112 ....Q1
RH-114 Shell Egg Wash
Detergent..................................Q1
RL-210 Feather Penetrant N1
RL-220 Feather-Penetrant N1
RL-230 Scald Vat Media.............N1
RP No.106 Stripper Detergnt A2
RP-All Purpose Cleaner A1
RP-Odorless Concrete Cleaner ..A4
RP-170 Detergent A1
RP-175 All Purpose Cleaner A1
RP-180 Cleaner A1
RP-185 General Cleaner A1
RP-190 Neutral Detergent A1
RP-192 Neutral Cleaner A1
RP-195 General Cleaner A1
RP-215 Hi-Foam Powder A1
RP-220..C1
RP-221..C1
RP-245 General Cleaner A1
RP-302...Q1
RP-305 All Purpose Low Foam A1
RP-339 ....A1
RP-339 Special............................A1
RP-405 Acid Cleaner A3
RP-405 HF Acid A3
RP-405C Concentrated Acid
Cleaner......................................A3
RP-505 Chlorinated Heavy
Duty Cleaner............................A2
RP-515 Chelated Heavy Duty
Cleaner......................................A2
RP-525 Smoke House Cleaner ..A2
RP-543 ....A2
RP-545M A2
RP-565 H.D. Cleaner A2
RP-575 Heavy Duty Foam
Cleaner......................................A1
RP-701 Chelate............................A1
RP-703 Chelate............................A1
RP-715 Jell Additive A1
RQ-109...E1
RY-200 Tripe Cleaner N3
Scat A4
Sof-Glo Hand Cleaner E1
Sof-Tex Hand Cleaner E1
Softsuds Hand Cleaner................E1
Special AA Fly Spray F1
Super Diazinon Spray F2
Super Fog Spray F1
Super-C.......................................A1
Tray Wash...................................A1
White Oil 35 USP.......................H1
70/30 Extra A1

Rawleigh, W. T. Company, The
Superior Bowl Cleanser..............C2

D'Oxid A3
De-Lime A3
De-Odo-Gest C1
Dot A3
E-Z Kleen Concentrate A4
Economic B1
Facet C2
Flash A2
Flush-A-Way L1
Foma-Cid A3
Food Equipment Lubricant H1
FPD Odorless Disinfectant
  Cleaner D2
Gar-Ban C1
Germitol Concentrate D1
Glass Kleen Super Concentrate . C1
Grease Trap Cleaner L1
Green Suds A1
Halo-Gen A4,B1,C2
Herculex C1
Hi Power B1
In & Out Insect Killer F1
Industrial Oven Cleaner A8
Ionogen A3
Kleen-Flush L1
Kleerall A1
Lavar C2
Luma Shield P1
M-C Gel A4
Mark-Oft C1
Mate C1
Mecanix E1
Mint Power C1
Minty C1
New Shine Metal Polish
  (Aerosol) A7
New-Shine A7
O-So-Brite C2
Off-En A8
Ovenate A8
P.M.C. C2
Petro-Lub H2
Pine Odor Disinfectant C1
Pine 75 .... C1
Pondate L1
Purge L1
Pyretocide F1
Refco A3
Reo-Clean 100 .... C1
Sani-Tize D1
Servate A4,A8
Shur-T-Solv K1
Silicon-Lub H1
Six Shooter High Concentrate .. F2
Six Shooter Insect Killer F2
Skal-Zo A3
Soak Hot Dip Cleaner A2
Softaskin E1
Solv-X-17 .... K1
Spritz A8
Star A1
Super Bleach D1,B1
Super Bol Brite C2
Super Carpet Concentrate P1
Swipe A6
Synthetic Auto Wash A1
SOF B2
Uni-Gest L2
Universal 512 .... A1
Vale C1
Versatol A4
Washex B1

Regent Chemical Corporation
C-28 .... A4
CS-343 .... K1
CS-927 Steam Cleaner A1

Reichold Chemicals
Newport Nairos P1

Penbro P1

Reily Chemical Company
"Sani-Clear" Liq. Chlor. Mach.
  Dishwash A1
Alkaline Liquid Drain & Sewer
  Line Opnr L1
Biodor C1
Bora-Clean E1
Cemex Concrete Remover A3
Chef-een A4
Chef-ette A4
Chlorinated Mach Dish Wash
  Comp. A1
Concrete Cleaner A1
Creme Antistatic Cleaner A1
Delite Extra 20% E1
Delite Extra 40% E1
Delite 20% E1
Desoiler and Descaler A3
Desoiler and Descaler #200 ...... A3
Disinfectant No. 918 .... A4
E-Z Greasy A1
Flora-Mint C1
Food & Dairy Cleaner All
  Purpose A1
General Purpose Cleaner A1
Germisyl D1
GP-2000 Cleaner A1
Handco 20% E1
Handco 40% E1
Hi-Clean A1
Hydroxade Concentrate A1
Hydroxade Regular A1
HD-100 .... A4
HPF 904 Concentrate A4
Ion 407 .... K1
Lemon Beaute-Cair E1
Lemon Disinfectant No. 10 .... C1
Lemon Disinfectant No. 13 .... C1
Lemon Disinfectant No. 20 .... C1
Lemon Lite Cream Lemon
  Scented Disinf. A4
Lo Foam 904 Heavy Duty
  Cleaner Conc. A4
Lo Foam 904 Heavy Duty
  Clnr. Super Conc. A4
Lotionized Hand Soap
  Unperfumed E1
Low Suds Laundry Detergent ... B1
Lubon H2
Magic Muscle C1
Manogel 11 .... E4
Mint Disinfectant C1
Oven Cleaner A8
Phosphate Free Hand Dish
  Wash. Comp. A1
Phosphate Free Mech. Borax
  Hand Clnr. E4
Pine-Dis-5 .... C1
Pine-Feen C1
Pinky A4
Power Pack A4
Presurg-Plus E1
Purge A4
Redress C1
RCS-69 .... K1
RCS-69 Safety Solvent K1
Safety Hard Water Scale
  Remover A3
Shamrock 880 .... A4
Special Pink Super Sapon
  Concentrate A1
Steam-X-Steam Cleaner A2
Steri-Fog D1
Steri-Pine C1
Super Rig Wash A4
Super Sapon A4
Super Sapon Concentrate A4
Super Strip A4

Synall A4
Syncon A4
SR-158-SZ A1
Technical A.P.C. A4
Technical D.W. Base A4
Total 100 .... D1
Total 200 .... D2
Triclenal A4
U.N.C.L.E. F1
Utensils Cleaner A1
Victory Cleaner A4
W.C.-170 .... C1
W.C.-85 Window Cleaner C1
Workhorse A4
40% Delite E1
904 Concentrate A4
904 Regular A4

Reitech Corporation
CD-30 Cleaner Degreaser A4
PB-10 Liquid Cleaner A1

Release Coatings, Incorporated
M-12 Release Cote H1
Mr. Release Food Release
  Cmpd No. 1701 .... P1
Mr. Release Food Release
  Cmpd No. 1704 .... P1

Reliance Brooks, Inc.
Alkaline Deruster Formula No.
  736 .... A2
Anti-Foam Formula No.
  M12073-1 .... G6
Aqua Turge No. 744
  Disinfectant D2
Aqua-Turge Formula No. F-
  6179-1 .... A1
Aqua-Turge Formula No. 740 ...... A3
Aqua-Turge Formula No. 741 ...... A3
Aqua-Turge Formula No. 747 ...... A4
Aqua-Turge Formula No. 750 ...... A1
Aqua-Turge Formula No. 752 ...... A1
Aqua-Turge Formula No. 754-
  M A1
Aqua-Turge Formula No. 757 ...... A2
Aqua-Turge Formula No. 758 ...... A1
Aqua-Turge No. 756 .... A1
Aqua-Turge 745-NP D1
AFS A4
Baygon Insect Residual Spray
  No. 922 .... F2
Bio-Mix Formula No. B L2
Brooks Protectamine Corsn
  Inhib For.No.6 .... G6
Chain Lube Formula No. 20 ...... H2
Chain Lube Formula No. 22 ...... H2
Chain Lube Formula No. 23 ...... H2
Chain Lube Formula No. 24 ...... H2
Chain Lube No. Z-5195-1 .... H2
Cold Solvent #778 .... K1
Cold Solvent No. 788 .... K1
Defeathering Agent N1
Drain Clean Formula No. 324 .... L1
Dual Synergist Pyreth.
  Insect.For.No.920 .... F1
Dual Synergist Pyreth.
  Insect.For.No.923 .... F1
Foam Additive A4
Formula No. PM M12073-1 ...... G6
Formula No. PM 105 .... G7
Formula No. PM 110 .... G6
Formula No. PM 113 .... G6
Formula No. PM 413 .... G2
Formula No. PM 453 .... G6
Formula No. PM 454 .... G6
Formula No. PM 460 .... G1
Formula No. PM 462 .... G7
Formula No. PM 464 .... G7
Formula No. PM 465 .... G7

Klintol K1
Manogel E1
Manogel II E4
Nonal Phenolic Type Cleaner
  Disinfectant A4
Procsteam G6
Promano 62
RDI 50 .... A4
Sanhae Lemon C1
Sanbac Mint C1
Secelec Safety Solvent K1
Swimmex A3
Treaboil G7
Watelec H2

Research Industries Corporation
Tam General Purpose Cleaner ... A1

Research Products Company
Detia Gas-EX-B F5
Max Kill Malathion 57-WE F2
Max Kill PPB 5 .... F1

Research Products Company
Acid Detergent A3
Ammoniated Wax Stripper C1
Aqua-Kill Insecticide F2
Astrodet Super Concentrate A4
Automatic A1
B.W.T. 101 .... G6
B.W.T.-100 .... G6
Bio-Syn C1
Blue Whiz C2
Bolex C2
C.L.T.-100 .... G6
C-10 Algaecide G7
Citra-Syn C1
Clear Lemon 10 Disinfectant C2
Clear Mint-10 .... C1
Clear-Pyne Disinfectant
  Deodorant C1
Command A1
CoCo 40% Hand Soap E1
CWT-300 .... G7
CWT-500 .... G7
D-70 Concrete Cleaner A3
Descaler #1 .... A3
Descaler #2 .... A3
Dish-A-Clean A3
Dual Synergist Space Spray
  Insctcd F1
Easy Rinse Filter Spray P1
Enviro 40 .... D1
Flash Klenze LC-12 .... A3
Foam-It A4,A8
Fog Kill Oil Base Insecticide F1
Frosty Pearl E1
Glass Shine C1
Husky A4
HDC-303 .... A1
Klink A1
Liquidate L1
LDS-20 .... A1
LF-100 .... A1
LF-310 Low Foam Concentrate A1
Mediclean E1
Neo-San D1
Neutra Clean A1
NAB C1
OCS C1
Pine Cleaner C1
Pine Oil Disinfectant C1
Porcelain & Stainless Cleaner ... A3
Pro-D-Foam Q5
Quat-22 .... D1
Repco-Dyne C2
Repco-Tox Space Spray Insctcd
  Conc F1
Repco-Tox Space Spray
  Insecticide F1

Repro-Chlor C1
Reprosolv A1
Residual Insecticide F2
Secure C2
Sewer Solvent C1
Sludge Floc L1
Space Spray F2
Steam O Clean A1
Super Creme C2
Super Mint C1
Super Pine Odor Disinfectant C1
Super Reprosolve A4
Take-Off A1
Tridet A4
20% Cocoall Hand Soap E1
406 Concrete Cleaner A4
408 Sewer Pipe Cleaner L1
501 Heavy Duty Steam Cleaner A1
511 Extra Heavy Duty Steam
  Cleaner A2
603 Chlor. Mach. Dish Washing
  Compound A1
604 Deluxe Mach. Dish
  Washing Compound A1
622 Laundry Powder Low Suds B1
622 Low Suds Laundry
  Compound B1

Residex Corporation
Banner A4
Blitz L1
Bulls Eye D1
Charge L2
Countdown L2
Direct HIT A4
Dursban Roach Concentrate F2
Fly Fog F1
Fogdex F2
Household Insect Spray F2
Magnify L1
Principal A4
Prolin F3
Resicide Emulsion Concentrate . F1
Resicide Fogging Compound F1
Residex Brand Diazinon 4E
  Insecticide F2
Residex Brand Diazinon 4S
  Insecticide F2
Residex Foam Clean A1
Residex Malathion 5lb. Emulsi.
  Concntrt. F2
Residox A4
Resivap Roach Concentrate F2
Resrattus F3
Resrattus Di-Pax F3
Resrattus En-Pax F3
Resrattus-Plus F3
Rodent-Blox F3
Super Resicide Concentrate F2
Target L2
Triple Threat A4
Tumblebug F2

Reslabs, Incorporated
SX-3 .... D1
SX-5 .... D1

Rex Chemical Corporation
Ammoniated Oven Cleaner A8
Borax Rex C2
Concrete Cleaner C1
Contact Cleaner K2
Denaturant Green Dye M2
Devil A4
Diazinon 4E Insecticide F2
Dursban 2E Insecticide F2
Duz All Detergent A1
Hot Stuff Cleaner A2,A8
Hot Vat Cleaner A2
Iodi-Clean D2

Liquid Hand Soap E1
Metal Brite Cleaner & Polish A7
Penetr-In L1
Rex Bowl Cleaner C2
Rex Light Mineral Oil H1
Rex Mineral Oil H1
Rex Premium Machine
  Dishwash Det. Chlor. A1
Rex Rat Glue F3
Rex 99-K A4
Rose Magic A1
S O S A1,B1
Sanilite A3
Silicone Spray H1
Steel Brite D1
Stop Insecticide F1
Super 00 .... A1

Reyns Chemicals, Inc.
R-408 Reco Conquest A4
Reco #3 .... A4
Reco Bio Syn A4
Reco Blue Flash A1
Reco Steam Kleen A1
Reco Tiger D1

Rhiel Supply Co.
Break Down A1
Lotion Skin Cleaner E4

Rhimi Company, Incorported
All Purpose Cleaner A1
Antiseptic Hand Soap E2
Heavy Duty Cleaner A2
Lotion Hand Soap E1

Rhodes Chemical Company
Bug-O-Spray F2
Fly Bait F2
Prime Professional Insecticide ... F1
Pyrethrins ULV Concentrate F1
Rat & Mouse Killer F3

Rhodia, Incorporated
Alamask 200E C1
Alamask 200EY C1
Alamask 308A C1
Alamask 308AX C1
Alamask 410A C1
Alamask 429A C1
Alamask 429AX C1
Alamask 500A C1
Alamask 500AX C1
Alamask 510A C1
Alamask 510AX C1
Alamask 518B C1
Alamask 518BY C1
Alamask 520A C1
Alamask 520AX C1
Alamask 520AY C1
Alamask 520B C1
Alamask 535A C1
Alamask 535AX C1
Alamask 903B C1
Alamask 903BX C1
Alamask 904A C1
Alamask 904AX C1
Alamask 906A C1
Alamask 906BY C1
Rhodorsil 3B Silicone
  Construction Seala P1

Rice Chemical Co., Inc.
Algaetret-300 .... G7
Big Red G6
CWT-333 .... G7
CWT-69 .... G7
Hard Mak-1 .... G6
Liqu-Phos G2

River City Chemical
RC 100 Non Butyl Degreaser A1
Super Kleen A1

Riverside Chemical Company, Inc.
ARSECO-22 .... C1

Ro-Vic Inc.
Rovic Drainout............................. L1

Road Runner Chemicals, Inc.
Tri-Jel L2

Robeco Chemicals, Incorporated
Calcium Hypochlorite Granular D1

Roberson Pools
Odorless Disinfectant D1

Robert A. Campbell & Associates
Grea-Di A1

Robert Langer Company, Inc.
Aquafroth Antibacterial Skin
   Cleanser E2
Aquafroth Antiseptic Liquid
   Soap E1
DSD-10 .... D2
DSD-25 .... D1
Formula 70 .... A1
HD-Detergent.............................. A4

Robert Orr & Co.
Kleen-All Heavy Duty
   Maintenance Cleaner A1
Kleer-Glo...................................... A1
Klor-A-Matic................................ A1
Silver Brite................................... A1
Solv-All A1
Sparkle Glo A1
Ultra Beads.................................. A1

Roberts Harvester, Inc.
Crown Internal Detergent A1

Roberts Laboratories
'Rodex' Pelleted Bait................... F3
Roberts Cygon 2-E Systemic
   Insecticide................................ F2
Roberts Pyrenone Dairy&Food
   Plant Spray............................... F1
Rodex Pelleted Bait With
   Coated Warfarin F3
Super Fly Bate.............................. F2
Synergized Pyrethins F1

Roberts, G. B. Maintenance Supply Inc.
Suds A1

Robfogel Mill-Andrews Corp.
Power-Foam A1
Power-Plus A1
Power-Sol..................................... A1

Robins, G. S. & Company
Robins Alkali No. 5..................... A2

Robinson Chemical Co., Inc.
RC-203... C2
RC-209... A4,H2
RC-210... H2
RC-211... A1
RC-213... A2
RC-215... A1
RC-219... A1
RC-221... A1
RC-222... A1

RC-223.. A1
RC-224.. A1
RC-225.. A1
RC-227.. A1
RC-228.. A1
RC-229.. A1
RC-230.. A1
RC-231.. A1
RC-232.. A2
RC-234.. A1
RC-238.. A3
RC-243.. A3,Q2
RC-246.. A3,Q2
RC-251.. E1
RC-252.. A1,H2
RC-253.. A1,H2
RC-254.. E1
RC-255.. H2
RC-256.. H2
RC-257.. A1,H2
RC-258.. H2
RC-259.. H2
RC-260.. H2
RC-261.. H2
RC-262.. H2
RC-263.. H2
RC-264.. H2
RC-280.. A1
RC-501.. A1
RC-504.. A1
RC-506.. A3
RC-507.. A1
RC-508.. A1
RC-511.. A1
RC-512.. A2
RC-514.. L1
RC-516.. A4
RC-518.. A2
RC-519.. A2
RC-520.. A2
RC-521.. A1
RC-522.. A1
RC-523.. A1,Q1
RC-524.. A1
RC-525.. A1
RC-527.. A1
RC-536.. A2
RC-538.. A2
RC-539.. A1
RC-540.. A1
RC-543.. A1
RC-544.. A1
RC-545.. A2
RC-546.. A1
RC-547.. A1
RC-548.. A1
RC-549.. A1
RC-550.. A1
RC-551.. A1
RC-552.. A2,Q1
RC-553.. A1
RC-554.. A1
RC-555.. A2
RC-556.. A2
RC-557.. A2
RC-558.. A2
RC-558A.. A2
RC-561.. A3
RC-562.. A1,Q1
RC-564.. A1
RC-565.. A1
RC-566.. A1
RC-567.. A1
RC-568.. A2
RC-570.. A2
RC-572.. A1
RC-576.. A1
RC-580.. A2
RC-581 Concentrate All
   Purpose Cleaner....................... A1

RC-582 Poultry Scald N1
RC-585 Pine Oil Disinfectant C1
RC-586 Pine Odor Disinfectant . C1
RC-587 Fortified Pine Odor
   Disinfectant............................. C1
RC-588 Mint Disinfectant........... C2
RC-591 Detergent Cleaner-
   Disinfectant............................. D1
RC-592 Detergent Cleaner-
   Disinfectant............................. D2
RC-594 Disinfectant Cleaner...... D2
RC573.. A1
RC574.. A2
RC575.. A1
RC576.. A1
RC577.. A1
Sodium Hypochlorite Solution... D2

Roby's Janitor Supply Corp.
Roby's 7-11 .... A1

Rochester Germicide Company
Aqua-Creme Lotion Skin
   Cleaner..................................... E1
Extermo Mill Spray Insecticide . F1
Extermo Vaporant Concntrt
   Insect....................................... F1
F-347 Solvent Degreaser K2
Formula 260................................. A4
Glo San Bowl Cleaner C2
K-100 Concentrate A4
K-99 Low Foam Cleaner............ A4
Karon Concentrate All Purpose
   Cleaner..................................... A4
Lanolated Mexo Powdered
   Hand Cleaner C2
LaBelle Liquid Hand Soap........ E1
Liquid Mexo Lotion Hand
   Cleaner..................................... C2
Now L1
Olive Leaf Liquid Hand Soap..... E1
Palmetto Emollient Hand Soap.. E1
Rocadyne Iodine Disinfectant.... D2
RG Aluma Solv............................ A1
RG Electro-Spray K2
RG Pena Solv A1
RG Powa-Spray A1
Sanor Chem D C2
Sanor Chem G C2
Sanor Chem K C2
Sanor Chem Line L C2
Sanor Chem M C2
Sanor Chem WC C2
Sanor Delux C2
Sanor X Fluid C2
Soil-Solv A4

Rochester Sanitary Products Co.
Aer-O-San A4
Hi-Clear Foamy Glass Cleaner .. C1
Hy-Kil.. F1
Klean-O-Lite D.T. Heavy Duty
   Cleaner..................................... C1
Kleen-O-San Foaming Cleanser. A6
Lo-San E1
Medicide...................................... C2
Ro-Sil H2

Rocket Dispensers & Fluids, Inc.
Rocket Knok Concentrate C1
Rocket Mist Fluid #1-C.............. C1

Rockford Chemical Company Inc.
Formula 210................................. A1

Rockfort Industries/ Division of Corporation
All-Star... E1
Bio-Rock Germicidal Cleaner A4

226

Bolt-Save H2
Brill-X A3
BC-10 .... C2
Clincher Concentrated
  Deodorizing Clnr C1
Diamond Rock P1
Dividend H2
Dual-Kil F2
Duty K2
Econo-Treat G7
Fighton F2
Flo-Free L2
Fortify E4
Gentle Giant C2
Get-Im Wasp Spray F2
Heat-Tect H2
Heavyweight C1
Horizon H2
Insecti-Mist F1
Insta-Fog F1
IS-1000 .... A1
Lift-Off P1
Lucent C1
Mandate Extrm Press Gear Oil
  SAE 85W-140 .... H2
Moly-Mist H2
My-O-Cide F2
MAL-101 .... L1
Oil Foil A4
Oven-Fresh A8
Poly-Glide H1
R.K.-2000 .... G7
R.K.-300 .... G6
R.K.-750 .... G7
Rebel A1
Reef C1
Rock-New K3
Rocklyte A3
San Fort Aerosol Lubricant H1
San-Aire C1
San-Aire (Aerosol) C1
Scrub-Free A8
Sect-Out F1
Seize-Eze H2
Shelter Aerosol Industrial
  Solvent K2
Skin-Tone C2
Spec-5 .... C1
Splitz L1
Thaw-Out C1
Tribune A1
Variety A1
Wet-Seal (aerosol) K2
Wet-Seal (bulk) K2
Xtra-Steam A1

Rockland Chemical Co., Inc.

Penn-Mist F1
Penn-O-Pine C1

Rockland Corportion, The

Coil Medic A3
Edge B1
Formula 519 .... A1
Ozark Mountain Magic A1
Random Wash A1
Razzle Dazzle A1
Rockland RD 10 .... A4
Super Duper A1
Super XYZ A1
White Wall Detailer A4

Rocol Limited

White Chain & Drive Spray H1
White Food Grease H2

Roger Popp, Incorporated

Blue-Skil A1
Germi-Skil Detergent Clnr.-
  Disinfectant D1

Klout Industrial Degreaser-
  Cleaner A4,A8
Popp Liquid Hand Soap E1
Popps Bleach Cleanser A6
111 Cleaner A4

Rogue Valley Distributors

Rogues No One A1

Rohm & Haas Company

Hyamine FHP-80%
  Concentrate P1
Paraplex G-62 .... P1
Triton X-114 .... A1

Roisman Chemical & Drug Sundries, Inc.

Durox Bleach D2,B1

Roman Chemical Corp.

Roman Bakery Floor Cleaner A4
Roman Rack Wash A1
Romansheen A1

Roman Cleanser Company

Leto's Roman Bleach D2,B1
Roman Cleanser Bleach C2,Q4,B1

Rome Building Maintenance & Sanitary Supply Co.

Floor Cleaner A4
Stripper Floor Cleaner A2

Rominda Chemicals, Inc.

All-Purp A1
Florr-Grripp J1
Formula #23-Y C1
Formula #4 .... A1
Formula #9 .... A1
Quick-Clean A2
Rominda Formula #70 .... A4
RAL-#105 .... A1
RAL-#107 .... A2
RAL-#108 .... A2
RAL-#109 .... A4
RAL-#110 .... A1
RAL-#112 .... A2

Romro Corporation

"Blast" Multi-Purp. Steam
  Cleaning Conc. A1
"Max" Liquid Steam Cleaning
  Concentrate A1
"Monster" Ultra Hot Sewer
  Solvent L1
Acid-X No. 1 .... A3
Acidex Powder Cleaner A3
All-Foam A1
Alpha 5000 .... A8
Big "C" Concentrate A4
Boil Clean Liquid Strip A1
Chain Lube #1 .... H2
Chlor Kleen A1
CF-84 No. 1 .... A2
C1P 300 .... A2
DS-444 No. 1 .... L1
DS-444 Sewer Chemical L1
Foam Plus No. 1 .... A1
Foam Press A1
Foam-Plus A1
Foam-Press No. 1 .... A1,N1
Freezer Cleaner A5
Freezer-Kleen No. 1 .... A5
GC-54 General Cleaner A1
GC-54 No. 1 .... A1
Heavy Duty Formula Liquid
  Stm Clng Conc A1
HD Scald No. 1 .... N2,N3
HD-52 Heavy Duty Cleaner A2

HPS No. 1 .... A1
HPS-2 Liq Cleaner A1
Inhibited Chemical Cl. &
  Oxidation Rmvr. A3
LITIC Liquid Alkali A2
MCQ No. 1 .... D2
MW-622 Mold Wash A2
MW-622 No. 1 .... A1
No. 26 BAFS H2
No. 38 BAFS H2
O-Chlor No. 1 .... A1
O-Chlor No. 2 .... A1
O-Dine No. 1 .... D2,E3
O-Foam A1
Organic Acid Detergent A3
Pro-Tap Acidulated Cleaner A3
Pro-Tap No. 1 .... A3
Red Hot Sewer Solvent L1
Rom-Foam A1
Romro #270 Liquid Alkali A1
Romro AMC-9 .... A3
Romro APC-2 .... A1
Romro CC-34 .... A4
Romro GP-280 .... A1
Romro H.D. Alkalene A2
Romro Lubricant TL H1
Romro LD-515 .... A1
Romro MP-550 .... A1
Romro No. 1 General
  Smokehouse Cleaner A2
Romro No. 1 Heavy Duty
  General Cleaner A2
Romro No. 2 Heavy Duty
  General Cleaner A2
Romro No. 23 Milkstone &
  Scale Rmvr Conc A3
Romro No. 5 Heavy Duty
  Smokehouse Clnr A2
Romro No. 8 Extra-Heavy
  Duty All Purp Cl A2
Romro R-10 .... A1
Romro RDX C1
Romro S-311 .... A2
Romro Superus A1
Romro SC-6 .... A1
Romro SGC-3 .... A1
Romro SHS-9 .... D2
Romro TC-4 .... A2
Romro 150 .... A1
Task Master E1
TC-480 Deruster A2
TC-480 No. 1 .... A2
TC-76 Alkaline Cleaner &
  Deruster A2
TC-76 No. 1 .... A2
TD-12 No. 1 .... A2
TD-12 Trolley Cleaner A2
TF1L No. 2 .... C1
TFIL-1 Liquid Cleaner A4
TRC-248 No. 1 .... N2,N3
Uni-Kleen A1

Ron Chemicals, Inc.

Cler-X 77 .... A1

Ronald Alan Industries

Foaming Cleaner A1

Roubar Lab. Inc.

Pots & Pans 114 .... A1

Ronell Industries

Attack Fly & Insect Spray F1
CLD A1
Guardian E1
Klean Air D1
Lift-Off A1
Mighty Comet A1
Odorol D1
Open-Sesame L1

Pot and Pan Cleaner A1
R-L Concentrate......................... A1
Rinse Aid.................................. A1
Sani-Lav Bowl Cleaner............... C2
Sentry Antiseptic....................... E1
Solu-Phene A4
Wash-Up................................... E1

Rose Chemical Products, Inc.
Chlor-Det D1
Lotionized Hand & Skin
  Cleaner E1
Rose Able A1
Rose Akidet A3
Rose Automate A4
Rose Brand 606-X Insecticide F1
Rose Challenge Hvy Dty
  Degreaser A4,A8
Rose Concentrated Ice
  Remover C1
Rose D-Matic............................. A1
Rose Dazzle A1
Rose Decide............................... D2
Rose Drain Cleaner................... L1
Rose Fling A1
Rose Fórmula II......................... A1
Rose Formula IV A1
Rose Formula VIII A1
Rose FPC No. 5 .... A4
Rose FPC No. 7 .... D1
Rose Hi-Chlor............................ A1
Rose IGR Iodine Germicide D1
Rose Kleer Kleen A4
Rose Kreme Kleener................... A4
Rose Liqua Concentrate E1
Rose Liqua 7B E1
Rose Luma-Brite A1
Rose LPWD................................ A1
Rose M-6 Cleaner...................... A1
Rose Many A4
Rose New Formula Laundri-
  Matic................................... B1
Rose P.C. Cleaner C1
Rose Power Pak Detergent......... A4
Rose Power Wash Detergent A1
Rose Pre-Dip............................. A1
Rose Re-Nu................................ A3
Rose Sparkleen A1
Rose Steam-O-Clean A1
Rose Swat Insecticide F1
Rose Task-Master A1
Rose Warebrite A1
Rose Waterless Hand Cleaner.... E4
Rose Wetso A1
Rose 202 Insecticide.................. F2
Rose 404 Insecticide.................. F1
Rose 606 Insecticide.................. F1
Rose 808 Insecticide.................. F1
Rosecide A4
Rosephene A4
Rosephene 128 .... A4
Rosept...................................... E1

Rose Exterminator Company
Rat and Mouse Bait with
  Fumarin F3
Rexco Multi-Purpose Kil-Fog/
  II.. F1

Ross Industries, Incorporated
Ross 601 Oil H1
601 White Mineral Oil
  (Aerosol)............................... H1

Rotanium Products Co. Div. of Premier Indust. Corp.
Food Machinery Grease E.P. H1
Rotanium Food Grade Silicone . H1
Rotanium Pure Lube.................. H1

Roth & Deng Supply Co.
R & D...................................... A1

Rothlan Corporation
Marla Spray Lubricant H2

Rovac Industries, Inc.
Out-of-Sight C1

Rowley Brothers, Inc.
White Food Machinery Grease
  AA H1

Roy Wilson Manufacturing Co.
All Purpose Liquid 4-1 Cleaner . A1
Amber No. 2 Oil Soap A1,H2
Forstar Heavy Duty Cleaner A1
Green Glo 20% Liquid Hand
  Soap E1
Green Glo 40% Concentrate
  Liq. Hand Soap.................... E1
John D. Jr. C2
John D. Jr. Powdered Hand
  Soap C2
Nu-Blend A4,B1
Red Glo Cleaner No. 2............... A1
Ristokrat Heavy Duty Cleaner .. A4,A8

Royack, M. Company
AM-100-M................................. G6
Formula 611 .... G6
HDL Alkaline Cleaner................ A2
Ox-T G6
R-1403..................................... G6
209.. G6
413.. G6
425.. G6
601 B....................................... G6
611.. G6
613.. G6
700.. G6
715.. G6
717.. G6
721.. G6
722.. G6
805.. G6
807 AF..................................... G6
900 FOT K1

Royal Chemical Company
Detergent Concentrate............... A1
Liquid Hand Soap FG E1
RC-027 Safety Solvent.............. K1
RC-245 Boiler Compound G6
RC-250 Boiler&Steam Line
  Comp. G6

Royal Chemical Company, Inc.
CS-44 .... L1
Dish Wash................................ A1
Fast Kleen A1
Golden Concrete Cleaner A4
Heavy Duty Cleaner A1
Liquid Steam Cleaner A2
Premium Fly & Roach Spray..... F1
Royal Flush............................... L1
Royal Sure Kleen A1
Stain Remover A3
Surgical Liquid Hand Soap E1
Toxene..................................... F1
Wiz.. A4

Royal Chemical Corporation
Crete Clean C1
L-Tox Spray F1
Lavo-Brite Porcelain Cleaner..... C2
Light N'Sudsy Powdered
  Washing Compound A4

Many Uses Germicidal Cleaner
  & Surf. Deo........................... C1
No. 600 Hog Scald N2
Surfa Sept Hospital Disinfectant
  Deo. C2

Royal Chemical Corporation
Jet Wash.................................. A1
Ro-Pol Cleaner A1
Royal Jet A1
Royal Lube H2
Royal 46 .... A1
Royal 55-LS A1
Royal 8 .... A1

Royal Enterprises
Roy-O-Cide D1
Roy-O-Con................................ A4
Roy-O-Mint C2
Roy-O-Pine C1
Roy-O-Roach F2
Roy-O-Scale A3
Roy-O-Tar C1
Royal Dish A1
Royal Dish P Odorless A1
Royal Flush............................... C2
Royal Fog-O-Sect F1
Royal Hickory Klenz.................. A2
Royal Ide.................................. D1
Royal M A1
Royal Nu-Mold A3
Royal Nu-Troll A2
Royal Odor Out......................... C1
Royal Orange............................ C1
Royal Powerhouse A4
Royal Scald............................... N2
Royal Scald Plus N2
Royal Sect 2-0 .... F2
Royal Sect 2W F2
Royal Sect-W............................ F2
Royal Slay Zem F1
Royal Soak A1
Royal Steam A1
Royal Steam Zip........................ A1
Royal Super Jet L1
Royal Total Klenz...................... A1

Royal Oil Company
E.P. Industrial Bearing &
  Chassis Lube......................... H2
Food-Machinery Grease A.W.... H1
Green Fury A1
Royal Command......................... A1
Straight Mineral Trans. Lub.
  SAE-140............................... H2
Straight Mineral Trans. Lub.
  SAE-80................................. H2
Straight Mineral Trans. Lub.
  SAE-90................................. H2
Torque Trans Fluid.................... H2
Universal Gear Lubricant SAE-
  140...................................... H2
Universal Gear Lubricant SAE-
  80.. H2
Universal Gear Lubricant SAE-
  90.. H2
646 Supreme Gear Lubricant
  SAE 75 .... H2
646 Supreme Gear Lubricant-
  SAE-140............................... H2
646 Supreme Gear Lubricant-
  SAE-90................................. H2

Royal Supply Company, Inc.
RS-100 Blue Cleaner&Stripper... A1
RS-44 Concentrate Cleaner A4
RS-88 Steam & Pressure Liq.
  Concntrt. A1

229

230

Odo Rout C1
Orange Deodorant........................ C1
Pine Oil Disinfectant (Phenol
  Coef. 5)...................................... C1
Power Concentrate...................... C2
Power Plus................................... C2
Quik Clenz C1
R.I.S.-15...................................... F2
Red Hot.. L1
Red Ram Drain Opener.............. L1
Remuv-Ox A3
Saphene 11 .... E1
Smite.. F1
Steam Cleaner Liquid A2
Steamrite 202 .... G6
Steamrite-200 .... G6
Super Solv.................................... K1
SM Industrial Deodorant............ C1
Tank Saver................................... G7
Thrift Detergent A4
Tough Touch Concentrated
  Liquid Skin Cl E1
Tower All NC Concentrate G7
Trak-Lube H2
Vegetable Oil Base Soap 40-
  45%.. H2
Vita-Pine (Phenol Coefficient 5) C1
W.T.C. Algaecide and Algal
  Slimicide G7
Whirlpool Drain Opener............. L1
Whomp.. L1
6/49 Concentrate......................... A4

Sanaco Chemical Manufacturers, Inc.

Sanaco-TM-10 Acid Cleaner
  Concentrate A3
Sanaco-100 All Purp. Clnr.
  Conc... A1
Sanaco-100 High Foam All
  Purpose Cl Conc...................... A1
Sanaco-300 Multi-Purpose Cl.
  Super Conc............................... A4
Sanico Little Tiger...................... A1

Sanchem, Inc.

No-Ox-ID "A Special" H2
No-Ox-ID 490.............................. H1

Sanco Products Company

Bulldog Cleaner........................... C1
Dynamic Drain Opener &
  Maintainer L1
Industrial General Purpose
  Insect Spray F1
Paraffin Oil.................................. H2
Regal Concrete Cleaner.............. A4
S-25G... A4
Sanco Cherry Bowl Cleaner C2
Sanco Chloro-Terg...................... A1
Sanco Emulsion Bowl Cleaner... C2
Sanco N-85 Cleaning
  Concentrate A1
Sanco P-40 Fogging Spray......... F1
Sanco Rainbow Steam Cleaner.. A1
Sanco Rainbow Steam Cleaner
  Liq.. A1
Sanco Regal Hand Soap E1
Sanco Rins-O-Dine...................... D2
Sanco San-O-Sol.......................... A4
Sanco Sewer Solvent L1
Sanco Smoke House Cleaner A1
Sanco Trolley Cleaner A2

Sanco Products Company, Inc.

Mark Remover............................. C1

Sanco Supplies, Incorporated

C-260 All Purpose Cleaner......... A4
Concrete Floor Cleaner C1

Dry Chlorinated Bleach SP-210. B1
General Purpose Synthetic
  Cleaner C-131 .... A1
Heavy Duty Bowl Cleaner C-
  280... C2
San-A-Quat C-339 .... D2
Sanco C-285 Non-Acid Bowl
  Cleaner..................................... A6
Sanco Heavy Duty Cleaner
  Degreaser A4
SP-505 Hand Cleaner E4
SP-540 Hand Cleaner E4
SP-560 Hand Cleaner E1
SP-570 Hand Cleaner E4

Sand Mountain Exterminating Co.

Marshall Rat Killer F3

Sanders, L. E. Co., Inc.

Bi-O-Act...................................... L2
Od-R-Act...................................... C1

Sandoz-Wander, Incorporated

Sandopan DTC............................. A1

Sanfax Chemicals

Act A3
Aero-Fect..................................... C2
Alkaflow-HD A2
Bargasan A1
Blue Treet G7
Brute A2
Bull's-Eye F1
Cac-137.. A3
Candoo A1
Combo A1
Corrodisolve A3
Counteract................................... C1
D-Scal.. A3
Denude-XL N3,P1
Derma-Fax E1
Drop-M NR F1
Fail-Safe E2
Faxain-HM A1
Flite-Lub H2
Floor-Stock A4
Foamax A1
Fry-Zo A2
Gelusan Concentrate A1
Gypsy-DR.................................... A2
GS 6... D1
Insto Scrub A4
Klor-Scrub A1
Kontrol G7
L-Toro A1
Laundra-Brite.............................. B1
Laundra-Fax................................ B1
Laundra-Magic B1
Laundra-Soft B2
Laundrasan B1
Liquid Skrub A1
Lypidax L2
Marauder F2
Med-829....................................... A3
No-Add A1
Odor-Ender.................................. C1
Poli-Fax....................................... A1
Poly-Floc..................................... L1
Polysan BT.................................. G6
Retalsan A1
Saf-Skrub A1
Safe-Brite A3
San-Brite...................................... C2
Sana-Fome................................... A1
Sana-Klenz A2
Sanacid N A3
Sanamist C C1
Sanamist V C1
Sanasol-L..................................... L1
Sanfax Aluminal A3

Sanfax AA Smooth-Lub H1
Sanfax Caution Acid A3
Sanfax Chloroboost D2
Sanfax Eradisan D2,E3
Sanfax Faxiod D2,E3
Sanfax Ferret D2,E3
Sanfax Malasan R F2
Sanfax Mirro-Shack.................... A2
Sanfax Nutramine 150................ G6
Sanfax Nutro............................... B2
Sanfax Organisan-XL.................. D2
Sanfax Softec B2
Sanfax 3-D Acid A3
Sanzyme L2
Scatter....................................... N1,L1,N2
Shackle-Magic............................ A2
Smooth-Lub................................ H2
Solusan....................................... G1
Solvo.. A1
Sonic Smokehouse Cleaner A2
Sparkl-Brite B1
Stormy-XL.................................. A2
Super-Glo A3
Sur-White N3
Tar-Free A2
Toro... A2
Tower G7
Tripe-White................................ N3
Trol-Eze H1
Trolsan-DR A2
Ultra-Fome................................ A1
Visca-Foam A1
Wed-L C1
Wed-O C1
Wed-V C1
X-Pel HD L1
Yield N3
1135... P1
222... L1
522... A2

Sanford Chemical Co., Inc.
Loft #447 Detergent.................. A1

Sani Kleen Chemicals, Inc.
#747 .... A4
Klean Aid.................................. A1

Sani Toil Products Inc.
De-Grease All............................ P1

Sani-Chem, Incorporated
S.C. 739 Germicidal Cleaner...... A4
SC-600 .... L1

Sani-Clean Distributors, Inc.
Ultra-Clean................................ A1

Sani-Clean Products, Inc.
Improved Formula C25 Indust.
Liq. Deo. C1

Sani-Fresh
Sani-Fresh Antiseptic Hand
Soap E2
Sani-Fresh Body Shampoo E4
Sani-Fresh Canela Liq.
Deodorant C1
Sani-Fresh Cherry Blossom Liq.
Deodorant C1
Sani-Fresh French Bouquet Liq.
Deodorant C1
Sani-Fresh Hand Lotion E4
Sani-Fresh Heavy Duty Soap E1
Sani-Fresh Lemon Twist Liq.
Deodorant C1
Sani-Fresh Mint Ice Liq.
Deodorant C1

Sani-Kem Chemical Corporation
H.T.S.T. Pasturizer.................... A2
Heavy Duty Spotless A1
Machine Cleaner....................... A1
Sani-Cide.................................. D1
Sani-Glo.................................... A3
Sani-Kleen A1
Sani-Klor A1
Sani-Solv A4
Scale Control G2,G5
Solar Lube................................. H2
Supreme Lube............................ H2

Sani-Kem East, Inc.
Gleam-Team All Purpose
Cleaner.................................. A4,C2
Gleam-Team Degreaser
Concentrate........................... A1
Gleam-Team Industrial
Degreaser Conc. A1
Gleam-Team Sink/Hand
Concentrate A1
Sani-Gleam All Purpose
Cleaner.................................. A4,C2
Sani-Gleam Degreaser
Concentrate........................... A1
Sani-Gleam Industrial
Degreaser Conc. A1
Sani-Gleam Sink/Hand
Concentrate A1
Sani-Glo All Purpose Cleaner.... A4,C2
Sani-Glo Degreaser
Concentrate A1
Sani-Glo Industrial Degreaser
Conc. A1
Sani-Glo Sink/Hand
Concentrate A1

Sanico
Air Mist Odorless (Bulk) C2
Airmist Odorless (Aerosol)......... C2
Aquanism Liquid Live Micro-
Organisms.............................. L2
AC-40 .... A3
AP-54.. C1
BC-98.. G6
Car-Sheen A1
Count Down F1
CL-25.. H2
CR-27.. A1
D.P.O.. L1
Dri-Nu P1
E.S.P. H2
Easy-Going C1
Formula #600 .... C2
Hylite.. A4
Lano-San (Cream) E4
Lano-San Aerosol...................... E4
LH-88 .... E1
Mr. Sanico................................ A1
N-4.. D1
OC-24 .... C1
Pan-Glo A1
PS-22... C3
RCT-9.. K1
RCT-9 (Aerosol)......................... K1
RM-11....................................... A3
RP-49.. A3
Skin-Tone E4
Sparkle Concentrated Glass
Cleaner.................................. C1
Sparkle Concntrd Glass
Cl.(Aerosol)............................ C1
Sparkle Ready to Use C1
Super Charge............................ L1
SC-7 Steam Cleaner Compound A2
SS-21... L1
Top Notch................................. A1
TC-3.. C2

UG-33.. A4
UG-35.. A4

Sanifas Products
Accept....................................... C2
Cherry Bowl Cleaner.................. C2
Discrete..................................... A3
Honey-Creme............................ E1
Liquid Hand Soap 15................ E1
Namifluff Beads A1
Sewer Solvent............................ L1
100 Plus Drain Pipe Opener..... L1

Saniplex Laboratories, Inc.
A-186 Chlorinated Krystals....... A1
A-225 White Foam Beads........... A1
Foam-Up A4
L-210 Hog Scald Compound...... N2
Space Spray No. 3 N-212........... F1

Sanitary Chemical Supply, Inc.
Brite Wash #105........................ B1
Brite Wash #2........................... B1
Hi-Test Brite-Wash #300........... B1
Hi-Test Brite-Wash #4.............. B1
Hi-Test Grease Solvent.............. B1
Hi-Test Super Drychlor Bleach. B1
Hi-Test Super Lime Solvent B1

Sanitary Products Corp.
Force A4,A8

Sanitary Specialties Company
Hurry-Klean.............................. A1
Sani-S-S Concentrate A4
Sani-Solv K1
Sani-Spray A4
Sani-Super Strip........................ A4

Sanitary Supply Co.
NB 131...................................... A1
NB-31.. A1

Sanitary Supply Co.
Do-It All C2

Sanitary Supply Co. of Tucson, Inc.
Double SS Brand Down & Out
Drain Opener L1
Double SS Brand Lemon
Disinfectant C2
Double SS Brand Lime-Off De-
Liming Cmpd.......................... A3
Double SS Brand Liqui-
DishHnd-Dshwsh Det............. A1
Double SS Brand Oven And
Griddle Cleaner A8
Double SS Brand Palm Soap..... E1
Double SS Brand Pine Oil
Disinfectant C1
Double SS Brand Ready-To-
Use Hand Soap E1
Double SS Brand Sani-Pine
Pine Odor Disn C1
Double SS Brand Sani-Solv
H.D. Cln-Dgrsr....................... A1
Double SS Brand
40%PureCoco.Oil Hnd Soap.. E1
Foam-It..................................... A1

Sanitary Supply Specialties
Samson...................................... A4
Tornado..................................... A1

Sanitary Systems & Supply
Advance Multi-Purpose
Germicidal Cleaner D1

Sanitation Associates, Inc.
Little Tiger........................A4

Sanitek Products, Inc.
Chain Lubricant 5286-1A H2
Saf-Dee.............................C1
Sani-Kem...........................A4
Sanicide...........................Q3
Saniphene..........................E1
Saniphene Antiseptic Hand
  Soap(Lot.)........................E1
Saniphene Surgical Soap............E1
Sanitek Sanitize...................A4
Spraytek...........................A1
Tuff-Kleen.........................C1

Saniturg Chemical Co.
Insta-Kleen........................A4

Sanivan Laboratories, Inc.
Drain and Sewer Cleaner DC-1.L1

Sanolite Chemical Corporation
Ammoniated Oven Cleaner A8
Bake Pan Cleaner HH..........:....A1
Boiler Water Treatment.............G6
Bulls Eye..........................A1
BWT-PDR Conc.......................G6
Chlorinated Jester.................A1
Chloro.............................C1
Crystal............................A1
Dual Chlor.........................D1
EDC Boiler Treatment G6
Fantasia...........................A1
Flash..............................A4
Foaming Disinfectant Cleaner... C1.
Formula No. 605 ....D1
Freeze-Kleen.......................A5
Grease Kicker......................A8
Heavy Duty Floor Cleaner A4
Hy Pressure Cleaner A2
Iodex..............................D2
Jester NF..........................A1
Jester OF..........................A2
Jester Star........................A1
LC-1151............................B1
Magic..............................A1
Magic M............................A1
Magic ML...........................A1
Magic Pink Lotion A4
Mint O Green C1
New GR Germicidal Rinse D2
Ovenew A8
Oxygen Scav PDR G6
Oxygen Scavenger L G6
Peak LF............................A3
Peak NF............................A3
Pot Guard A1
Pot Luck A1
Purge..............................A1
Ram Rod NF C1
Return Line Treatment A...........G6
Reward.............................A1
Rinse-It...........................A1
Roach & Bug Killer F2
Sanisol............................A1
Sano-Kleen.........................B1
Sano-Shine.........................A7
Sanobac............................E1
Sanolac #10 ....A2
Sanolac #5 ....A1
Sanolac #60 ....A1
Sanolac #7-0.......................A1
Sanolac #8 ....A1
Sanolac BW A2
Sanolac C1P No. 17.................A2
Sanolac Foam All...................A1
Sanolac Formula 510................A4
Sanolac FP-100.....................A2

Sanolac FP-101 ....A1
Sanolac FP-103.....................A1
Sanolac FP-104.....................A2
Sanolac FP-105.....................A3
Sanolac FP-106.....................A2
Sanolac FP-107.....................A2
Sanolac FP-108.....................A2
Sanolac FP-109.....................A1
Sanolac FP-700.....................A4
Sanolac Hy Pressure Cleaner A2
Sanolac HMC A1
Sanolac HS N2
Sanolac HT A2
Sanolac LC-SC......................A1
Sanolac No. 803....................A1
Sanolac PDQ A2
Sanolac Sewer Solvent L1
Sanolac TC N3
Sanolac 22BG A4
Sanolac 23.........................A2
Scale & Film Remover A3
Sentry.............................A2
Sentry Star A1
Shell Game.........................Q1
Shroud Cleaner B1
Silver Kleen A1
Slide Silicone Spray H1
Speedy.............................A4
Splash.............................A1
Splash With Ammonia................A1
Spraize............................A4
Sudsall............................A1
Super Sanolite.....................A1
Super Soap Compound #71 ....A1
Super Soap Compound #75 ....A1
Treat..............................A1
Triumph Bleach.....................B1
Ultimate...........................A1
Vaporizer No. 2 1/2 Insect
  Spray.............................F1
Vaporizer No. 375 Insect Spray F1
Varsity............................A1
Wham-O.............................L1

Santana Chemicals
De-Limer...........................A3
Orbit-200 Heavy Duty Cleaner
  & Deodorant A4
Santana Lagoon Enzymes L2
Unlock.............................L1

Santek Chemicals
Alkasan-10.........................A2
Alkasan-20.........................A4
Sanquest RD A4,A8

Saverite Water Treatment Co.
Formula A-210......................G6
Formula B-210......................G6
Formula 210-A Liquid G6
Formula 210-B Liquid...............G6
300................................G7

Savin Chemical & Industrial Supply
Corp.
Ammoniated Oven Cleaner A8
Cold Grille & Oven Cleaner A8
Dishwash Concentrate A1
Industrial Cleaner Degreaser...A4,A8
Insect Spray F2
Institutional Coconut Oil Hand
  Soap..............................E1
Kleen-Glo Concentrate..............A1
Stainless Steel and Metal Polish.A7
Steam-All..........................A1
Super Wax Strip Concentrate A4
White Ease Lithium Grease H2

Savogran Company
Dirtex.............................A1

Dirtex Cleaner Spray C1
TSP Cleaner........................A4

Savoie Supply Company, Inc.
Foaming Force A1

Savor Chemical Laboratories
All-Surface Cleaner................A1
Bowl Cleaner C2
Concentrated Floor Conditioner C1
Drain Opener And Maintainer... L1
Hand Soap Liquid E1
Hand Soap Lotion E1
Heavy Duty Steam Cleaner
  Liquid A2
Oven And Grill Cleaner K1
Pink Lotion A1
Power Cleaner A4
Stainless Steel Cleaner & Polish. A7
Surface Scum Remover A4
Wind-O-Shine C1

Scarry, E. J. & Company
C-13 Concentrtd Clnr &
  Disinfect. D1
Modern Disinfectant D1

Schaefer Chemical Products Co.
Boiler-Aid MP-210 Liquid G6
Boiler-Aid MP-300 Crystals G6
Boiler-Aid MP-320 Crystals G6
Boiler-Aid MP-340 Crystals G6
C W Crystals G2
Condens-Aid G6
Condens-Aid-A G6
Cool-Treat A-150 ....G7
Oxygen Remover - L G6
Oxygen Remover - P G6

Schaeffer Manufacturing Co.
Moly Ultra 800 Grease H2
R-100 Universal Gear Lubricant H2
Schaeffer's Moly Universal
  Gear Lube H2
Schaeffer's R & O Hydraulic
  Oil H2
White Food Mach. Grease
  #530AA H1

Schilling Sanitary Systems T/A
Griffith Company, Inc.
Ice Cold Surge.....................A5
Turbo 777..........................A1
Turbo-Foam.........................A1

Schloss & Kahn, Inc.
Deep Clean A2
Hi-Power...........................A1
L.S. Controlled Suds Saundry
  Deterg............................B1
S & K Deliming Solution A3
S & K Liquid Bactericide...........D2
S & K Quick-N-Ezy A8
S & K Sanni-Rinse..................D2
S & K Toilet Bowl Cleaner.........C2
S & K's T-Rific....................A1

Schmidt, C. Company, The
Slick Cleaner......................A1
Slick Hog Scald N2
Slick Poultry Scald N1

Schneid, I., Incorporated
A.I.O. 1500 Concentrate G6
Alltreat - 300.....................G7
Alltreat 700 ....G7
Alltreat 800 ....G6
Aquacide F1

233

B.A.C. Bio Active Cultures In
Liquid L2
Big "G" A1
Big Blast A1
Black Disinfectant(Phenol Coef.
2) C1
Black Disinfectant(Phenol Coef.
3) C1
Black Disinfectant(Phenol Coef.
5) C1
Black-Out Drain Opener L1
Blue Sewer & Drain Solvent L1
Boilerite 100 .... G6
Boilerite 150 .... G6
Bole Aire C2
Bowl Klean C2
Brandex A4
Chemtronic A1
Citrosene 10 .... C1
Coconut Oil Hand
Soap(Concentrtd) E1
Coil-Ite P1,A4
Combine A4
Concrete Cleaner C1
Dual-27 .... D1,Q1,Q3
Fast Klean A3
Foamzit A1
Fogging Spray Concentrate F1
FZ-3 .... A1
G B Breakdown L1
G. P. C. A4
Gent Concentrate A4
Gentle Scrub A6
Grill-O A8
GPC-24 .... A4
Hangs In There Bowl &
Porcelaine Cleaner C2
Heavy Duty Cleaner (Pine
Odor) C1
Hi-Count Bacteria Enzyme
Complex L2
Hydrolytic Enzyme Bacteria
Complex L2
JJC-16 .... A1
Kleer-Vue Glass Cleaner C1
L-Tox Concentrate F1
L-Tox Spray F1
Lectra Solv K2
Lemon Disinfectant C2
Lemon Fresh 20 .... A4
Lemon Tree Deodorant &
Cleaner C1
Lemonene C1
Liquid Conveyor Lube H2
Lo-Sudz A4
Magic Mike A1
Mighty Mike Concentrate A1
Mint Disinfectant C1
Mint-O-Dis C1
Miracle Cleaner A4
Mr. Brite Porcelain Cleaner C2
Nale-Um Insecticide F2
Neutral Cleaner (Sass Odor) C1
Neutrox 400 .... G7
Odo-Rout C1
Orange Deodorant C1
Para-Phyll Deod. Blox Plus
Chlorophyll C2
Para-Phyll Deod. Cakes Plus
Chlorophyll C2
Perfumed Deodorant Blox C2
Perfumed Deodorant Cakes C2
Perfumed-Wired Bowl Blox C2
Pine Oil Disinfectant, Phenol
Coef 5 .... C2
Power "Plus" C1
Power Concentrate C2
Quik Clenz C1
R.I.S. - 15 .... F2
Red Ram Drain Opener L1

Red-Hot L1
Remuv-Ox A3
Saphene II E1
Scrub a Dub A6
Smite F1
Steam Cleaner Liquid A2
Steamrite 200 .... G6
Steamrite 202 .... G6
Super Solv K1
SD-101 .... C1
SM Industrial Deodorant C1
Tank Saver G7
Thrift Detergent A4
Tough Touch Concentrated
Liquid Skin Cl E1
Tower All NC Concentrate G7
Trak-Lube H2
Vegetable Oil Base Soap 40-
45% H2
Vita-Pine C1
W.T.C. Algaecide and Algal.
Slimcide G7
Water Tower Conditioner G7
Whirlpool L1
Whomp L1
6/49 Concentrate A4

Schuster Industries LTD.
Promenade Controlled Suds
Automatic Det. A1

Scien-Tech Company, The
2-DO A4

Scientific Boiler Water Conditioning Co.
Boiler Water Treat. Form. No.
10 .... G6
Boiler Water Treat. Form. No.
104 .... G6
Boiler Water Treat. Form. No.
11 .... G6
Boiler Water Treat. Form. No.
15 .... G6
Boiler Water Treat. Form. No.
203 .... G6
Boiler Water Treat. Form. No.
301 .... G6
Boiler Water Treat. Form. No.
405 .... G6
Boiler Water Treat. Form. No.
405-E G6
LSC #1 .... A1

Scientific International Research, Inc.
Bio-Brite A1
Disinall D1
Disinall Cleaner D1
Glass-Glo C1
Grease-Off C1
Ice Off C1
Liquamelt C1
Neutralize C1
New S.I.R. Safety Solvent K1
New Solidzout L1
Probe L1
Probex L1
Rat-Begone F3
Rids-It C2
S.I.R. Aid L1
S.I.R. Concentrate A1
S.I.R. Delimer A3
S.I.R. Deo-Gran C1
S.I.R. Kill F1
Scent-A-Way C1
Sila-Slide H1
Solv All A4
Top C1
Vroom C2

Wiz-Kleen C2

Scientific Packaging Corporation
Pop & Clean Chlorine Clnr.
Sanitizer D1
Pop & Clean Heavy Duty
Cleaner & Degr. C1
Pop & Clean Liq. Neutral
Cleaner A4
Pop & Clean Machine
Dishwasher Detrgnt. A1
Pop & Clean Safety Cleanser A1

Scientific Supply Co., Inc.
ACW-15 .... A1
Best Of All B1
Boiler Water Treat. No. SC-1A. G6
Boiler Water Treat. No. SC-1B. G6
Boiler Water Treat. No. SC-1C. G6
Boiler Water Treat. No. SC-1D. G6
Boiler Water Treat. No. SC-1E. G6
Boiler Water Treat. No. SC-1G. G6
Boiler Water Treat. No. SC-1J. G6
Boiler Water Treat. No. SC-1K. G6
Boiler Water Treat. No. SC-4A. G6
Boiler Water Treat. No. SC-4B. G6
Cooling Tower Treat. No. SC-
3E G7
Cooling Tower Treat. No. SC-
3K G7
Formula 333 Meat Hook & Car
Cl A1
Gitz Spray Wash A4
Heavy Duty 60 .... C1
IC-99 .... A4
JA-DA A1
Liquid Hand Soap Concentrate
40% E1
Liquid Hand Soap 10 Percent E1
Liquid Hand Soap 20 Percent E1
Pine-O-Lene C1
Scien Ace B1
Scien Pac A.P. A1
Scien Pac Chlorinated A1
Scien Pac H.D. A2
Scien Pac Klor Klean A1
Scien Pac S.B. B1
Scien Pac S.C. A2
Scien Pac S.H. A1
Scien Pac S.W. A2
Scien Pac Sour B1
Scien-Foam A1
Scien-Thetic A1
Sewer Solvent L1
Sudsation A1
Super Gold All Purp. Cold
Water Det. A1
Super Gold Alphadyne D1
Super Gold Creme Soap E1
Super Gold Glass Cleaner C1
Super Gold Hospital Spray
Disinf. Deo. C2
Super Gold Hvy Dty Liquid
Drain Solvent L1
Super Gold Insect Killer F1
Super Gold Neutral Chemical
Clnr A1
SC-5A G1
SC-5B G1
SC-6B A3
SC-6C A3

Scoles Systems
PDQC A1

Scot Supply Company
Copasetic A1
Surgiphene 128 .... A4
Surgiphene 256 .... A4

Scotch Manufacturing Co., Inc.
Instant Plumber L1

Scotch Plaid Chemical Co.
Grease Gitter E4
Red Car Wash CW 17 .... A1

Scott Chemical Company
Pace Setter Foamade A1

Scott, I. Chemical Corp.
Blue Butyl Cleaner A1
Scottchem's Air Pure C1
Scottchem's AQDF...................... D2
Scottchem's Belt Dressing.......... P1
Scottchem's Digestant 250.. .... L2
Scottchem's Disinfectant #78...... D2
Scottchem's Drain-Away............. L1
Scottchem's ESS Safety Solvent K1
Scottchem's Fluoro Carbon........ H2
Scottchem's Freezer-Cleaner...... A5
Scottchem's Frigid Cut H2
Scottchem's Graphite Fast Dry
  Lube H2
Scottchem's Liquid Hand Soap.. E1
Scottchem's Mechanic's Friend . H2
Scottchem's Mint Deodorant C1
Scottchem's Non-Acid Lime &
  Scale Remvr............................. A1
Scottchem's Odor Ban Odor
  Contl w/Mask........................... C1
Scottchem's Oven & Grill
  Cleaner.................................... A8
Scottchem's Paint And Varnish
  Remover C3
Scottchem's Penetrating Oil H2
Scottchem's Pinky Lotion Skin
  Cleanser E1
Scottchem's Pipesafe Drain
  Pipe Opener L1
Scottchem's Polysiloxane
  Release Agent H1
Scottchem's Precision Parts
  Cleaner.................................... K2
Scottchem's Sewer Sweetener ... E1
Scottchem's Sila-Lube................ H1
Scottchem's Stainless Steel
  Cleaner.................................... A7
Scottchem's Super Power........... A4
Scottchem's White Lube
  Lithium Grease H2

Scott, J. Company
Scott Hot Tank Cleaner (100).... A2
Scott Safety Degreaser 325 .... A4
Scott 111 Wine Tank Cleaner A2

Scranton Chemical Company
Cocolene Liquid Soap................. E1

Sea-Air Chemical Corporation
Formula 1727 .... D1

Seagull Chemical Co., Inc.
Anti-Bacterial Hand Soap........... E2
Freezer Cleaner A1
Heavy Duty Degreaser Cleaner. A1
Spray-On Cleaner Degreaser A1

Sears, Roebuck and Company
SC 402 HD All Surface Cleaner A4
SC 403 Pure Vegetable Oil
  Soap.. A4
SC 404 Cleaner Degreaser......... A1
SC 405 Low Sudsing HD Conc. A4
SC 407 Industrial Concrete
  Clnr.. C1
SD 460 Lemon Disinfectant C2

SS 420 Pink Hand Dishwash
  Lotion..................................... A1
SS 436A Concentrated Glass
  Clnr.. C1
SS 437 Q Glass Cleaner (Ready
  To Use).................................... C1
104 Deodorant Blocks................ C1
204 Toilet Bowl Deodorant
  Blocks..................................... C1

Seatex Corporation
Optibrite B1

Security Chemicals
Anti-Crawl F1
Creature Killer............................ F2
Lift It-II..................................... A1
Lift-It.. A4

Security 4 Chemical Company
A.P.S. #LBK 002...................... A1
Aire-Care................................... C1
All You Need.............................. A1
Aqua Three L2
Aqua-Zyme L2
Artic Clean................................. A5
B J Heavy Duty Cleaner A4
Big Jim A1
Cav Auto Scrubber
  Concentrate A1
Check It..................................... A1
Clean Cream E1
Contac Prilled Drain Opener.... L1
Conveyor Lub H2
Dis-'N'Dat................................. C2
Dual S.C. Concentrate.............. A1
Dy-Sul Liquid Drain Pipe
  Opener L1
Formula #900 Heavy Duty
  Cleaner.................................... A4
Formula Creme Lotionized
  Liquid Soap E1
G.P.N. #LBK 001...................... A1
Glycolene M.S............................ C2
Gotcha Plus A1
Gotcha 60 Insect Killer F1
Grease Gun H2
Great Stuff................................. E4
Heavy Duty Cleaner A1
Heavy Harvey Concentrated A4
Higher Up A1
Hospital Disinfectant
  Deodorant C2
J C P S & L............................... A4
Jell Degreaser K1
Look Glass And Utility Cleaner C1
M 11 Concentrate Mncpl-Ind-
  Inst.Det.Conc. A1
M.P. Concentrate A1
Marvo 111.................................. C1
Medi-San E1
Mint Magic................................ C1
Moisture & Corrosion
  Preventer H2
My Favorite All Surface
  Cleaner.................................... C1
MP 72 More Power D2
Nox #4 All Purpose Detergent . A4
Pi-No-Germ C1
R.M. 40 P.................................. A3
R4CT... K1
S.C. Concentrate....................... A1
S-4 Teffy Dry Spray Lubricant . H2
Scale Clean Inhib. Chem. Clnr
  & Deoxidiz A3
Sewer Solvent............................. L1
Sno Ice C1
So Mild-150............................... E1
So Mild-400............................... E1

Spritz Multi-Use Cleaner
  Degreaser C1
Stainless Steel Cleaner A7
SC7 Hot Soak A2
S4 Foam Buster A4,A8
T/HAR GLD.............................. A4
The Fog AA Grade Fly Spray.. F1
Twin Power................................ A3
Vandal Mark C1
007 Concentrate......................... A1
4 Sure Concentrate.................... A1

Sefco Chemical Company
Batt F1

Selby Chemicals, Incorporated
Pyro-San Formula SDL-10 .... D2

Selig Chemical Industries
"All Pro".................................... E4
"Clean Up"................................. C1
A-M Plus C1
Ambush F2
Antiseptic Velvo Liquid Hand
  Soap.. E1
Aseptene 11................................ D1
AFP A1
B-G-T P1
Baccene C1
Bano.. L1
Bano 6AW C1
Bano 6AW SC80 Chemical
  Deodorant C1
Barrier Cream E4
Big "S"...................................... A1
Big Joe....................................... A1
Big R C1,B1
Brite Creme Cleaner C2
Concentrated Liquid Hand
  Soap.. E1
Conveyor Soap H2
Coral A1
Coralan E1
CP-43.. A4
CP-43 SC80............................... A8
D.I.G. Deodorant Insectex
  Granules F2
D-Foam ST A1,Q5
D-Lime...................................... A3
Destainer #100 .... Q2
Dox-33....................................... A3
Du-Kil F1
Du-Kil X-T................................ F1
Dukil SP.................................... F1
DDVP Fly Bait F2
DFC-I.. A2
DRB-74 .... F1
E.M.C.13 .... K2
Ex-Con P1
Ex-Foam 322.............................. Q5
Excello....................................... A2
Exit L1
EMC-13 Special......................... K1
Flo Vapor-ets............................. C2
Flo-Free..................................... L1
Formula 229............................... K1
Free-All..................................... L1
Free-Lon H2
Germicidal Dynamite................. D1
Glitter A1
Glycol Air Sanitizer................... C2
Glycolin Air Spray..................... C1
Gosh C1
Green Demon L1
Gresolv No. 6 .... C1
Grout And Tile Cleaner A4
GP-50.. A1
Helo Spray C1
Hi-Kleen No. 1001..................... A3
HDP Gear Lube......................... H2

Insect Killer F1
Kare Concentrate A1
Kare Steam Cleaner
   Concentrate A4
Kare SC80 Concentrate A4
Kare 24 .... A1
Kitchen Master A4,A8
Kleenall C2
Kleenapart P1
Kleenasolv K1
Kleenaspeed K1
Klenexo A1
Klenite A1
Kutzit Cleaner C1
Lectron-13 Contact Cleaner K2
Lem-O-Dis Disinfectant-
   Deodorant A4,C2
Liquid Dynamite A4
Liquid Ice Go C1
Liquidizer A1
Lumabrite A3
LD-86 .... A1
Mel-O.. E4
Mold And Mildew Stain
   Remover P1
Mr. Pest Control............................ F1
Muscle C1
No. 110 LD Neutro-Jel................ C1
No. 110 R Neutro-Jel.................. C1
No. 110C Neutro-Jel A4
No. 20KXG Troy A4
O.V.Cleaner A8
Oven Cleaner A8
Petrosorb J1
Pinetax C1
Pipe Patrol L2
Poliwash A1
Port Steam Cleaner A1
Port SC80 Low Foam
   Detergent A1
Port SC80 Steam Cleaner A1
PB-100 .... F1
PB-100 SP F1
PLB-20 .... F3
Quick-N-Easy................................ L1
Red Detergent A4
Rip-Off.. A8
Rust Rip A2
Safe-T-Lube H1
Safe-White Lube............................ H1
Scrubzol A4
Se-Fly-Go-33.................................. F1
Selco Concentrate A1
Selco Suds A1
Selco-Zyme L2
Selcol B D1
Selcol Disinfectant-Sanitizer
   Fngcd-Deo. D2
Selig's Bode Cote C1
Selig's E.M.C. 13 Solvent
   Degreaser K1
Selig's Foam 50............................ A1
Selig's Green Magic...................... E1
Selig's No. 110 C Neutro-Jel...... A4
Selig's Rust-Off.............................. A3
Seligsupreme All Purpose Clnr.. A1
Seligsupreme Electric Motor Cl
   & Saf Sol K1
Seligsupreme HD Clnr. &
   Stripper A4
Seligsupreme Insecticide Spray.. F1
Seligsupreme Steam Cleaning
   Comp. A2
Silicone Spray H1
Sio-Dex.. D2
Slick A6
Snow A1
Stainless Steel Cleaner & Polish. A7
Stat-Scrub...................................... E2
Sun Drops C1

Super Free All L1
Super Liquid Dynamite A4,A8
Super Sniper.................................. F2
Super Speed HS............................ A1
Super Speed LS............................ A1
Super 800.. A1
SC-90 .... A1
STX-100 .... D1
Tincture Green Soap.................... E1
Troy Brite #66 .... C1
TX-99.. A1
Up Tight Tef-Tape........................ H2
Vaporizer.. F1
Vaporizer Spray X-T.................... F1
Velvo E1
Vixol C2
Vorex.. C1
VPA Emulsible Concentrate F2
Waffle Iron and Grill Cleaner.... A8
Washkleen A1
Washkleen XXXX A1
White Grease H2
Whiz Wash-All Purpose
   Cleaner...................................... A1
Whiz Wash-Degreaser A4
W103 Coolant Concentrate H2
17-SX-79 .... A1
20-SX 77 .... C1
25-SX-78 Boiler Treatment.......... G6
44-SX-69 .... A4
5-SX-68.. A1

Sellers Chemical Corp.

Flo-Mo AJ 100 .... A1
Flo-Mo AJ 85 .... A1

Senter Chemical Corporation

Alka RP Strip A2
Alka-Etch.. A2
Aqua-Proof...................................... A1
AP-30.. A4
AP-70.. A4
Bio-Treat L2
Bottle Kleen C.............................. A2
Bottle-Kleen M.............................. A2
Coco-Concentrate.......................... E1
Coco-Sope...................................... E1
Concrete-Kleen.............................. C1
Dari-San A1
Deterge A1
Dia-Chlor A1,Q1
Diamond 77 Bowl-Help............. C2
Diamond-77 Drain-Aide L1
Dish Kleen A1
Dish Treat A1
Feathers Off N1
Flor-Maid.. A1
Formula 51 .... A2
Formula 52 .... A2
GPC-1 .... A1
GPC-2.. A1
GPC-73.. A1
GPC-75.. A1
HD Kleen.. A2
HD 80 .... A2
Kleen Krete.................................... A4
Kleen Krete HD............................ A2
Kleen-Al.. A1
Kleenal (Ammoniated).................. A4
Liqui-Lube H2
MD-Kleen A1
MD-45 .... A1
MH Kleen A1
Orange Bactrsl Germ Cl. Hosp.
   Form.. C1
Pine Odor Disinfectant
   Coefficient 5.............................. C1
Pine Oil Disinfectant
   Coefficient 5.............................. C1
Pink Satin Lotionized Hand
   Soap E1

Safety-Kleen 15.............................. K1
Sani Kleen A2
Sewer Free...................................... L1
Space Kill 3.................................... F1
Space Kill 4.................................... F1
Space Kill 6.................................... F1
Ston Terge A3
Surface-Kil F2
SC-70 .... A3
SK-95.. K1
S7046 .... A1
TR-28.. A1
Wax Kleen A4

Sentinel Chemical

DB-55 .... G6
LB 151-C.. G6
LB 206 .... G7
LB-150 .... G6
LB-300 .... G6
152 C.. G7

Sentinel Chemical Company

No. 9 Cooker and Cooler
   Compound.................................. G5

Sentinel Chemical Corp.

Series 1020...................................... G7
Series 1060...................................... G6
Series 421 .... G6
Series 422 .... G6
Series 605 .... G6
Series 610 .... G6
Series 850 .... G6

Sentinel Soap & Chemical Co., Inc.

Sentinel All Purpose Cleaner A1

Sentry Chemical Company

Aero Tape.. H2
Bactran E1
Begone F2
Boiler Feed Water Treatment G6
Boiler Water Rust Inhibitor G7
Cannibal.. L2
Chain Sentry H2
Chief-of-Staff.................................. A1
Chlorinated Egg Wash.................. Q1
Coil Conditioner P1
Coil Life P1
Contact K1
Control C1
De-Scaler A2
Defoamer.. Q5
Dermalan E4
Digest A1
Disolv L1
Drop Out A2
Dustat P1
Egg Britener Q1
Electric Sentry................................ K1
Electromagnetic.............................. K1
Exscale A3
Filter Cleaner P1
Flak.. F1
Formula #75.................................... A1
General A1
George Bernard Thaw.................... C1
Germ-I-San...................................... D2
Grease Gun A4
Gypsy Rose A4
GK #1.. A2
Halt.. C2
Handy.. E4
Jetaway.. A1
Kloro-Kleen All Purpose
   Cleaner...................................... A1
Lemon Guard A4
Machine Gun F1

Serv-Quat-N D1
Service Foam A1
Super Bowl-Brite Toilet Bowl
   Cleaner A2

ServiceMaster Industries Inc.
DeepClean P1
FloorStar DuoClene A4
G.S.R. A2
GlideRinse D2
Odor Neutralizer C1
OdorGo P1
SaniScrub D1
Sermac G.P.D. A1
Soft and Clean E1

Servisco, Inc.
Servisco Antiseptic Hand Soap .. E2

Setter Chemical
De-O-Sept D1
Setter Shell Egg Cleaning
   Compound Q1

Seven Hills Maintenance Supply, Inc.
Blue Butyl Cleaner A4,A8
Bowl Kleen C2
Break-Away A4,A8
Golden Gator C1
Grime Scat S/D-G A4

Sexauer, J. A., Inc.
Mule-Kick Enzyme L2

Sexton, John & Company
All Kleen Detergent A1
Chlorinated CIP Cleaner A1
Defenso A1
Dependable Disinfectant D1
Invinso A1
K-75 Alkaline Cleaner A2
Liquid Toilet Soap E1
Neutral Green Floor Cleanser .. A4
P.F.D Detergent A1
Sexton Heavy Duty Cleaner A2
Sexton Kleen D1
Sexton Stainless Steel Cleaner A3
Snowball A1
X-It Spray Cleaner and
   Degreaser A4

Seymour of Sycamore, Inc.
Heavy Duty Silicone Lubricant . H1

Seymour Chemical Associates, Inc.
Cunilate #2419(Wallwash) P1
Formula 645 Penetrant Sealer ... P1
Nylate-10 .... P1

Shack Sanitary Supplies
Ter-if-ic Golden Grime Buster ... A2

Shaklee Corporation
At-Ease A6
Basic H A1,E1
Basic I A1
Basic-G A4,C1

Shalco Chemical Corporation
Chain & Conveyor Belt Lub. H2
Concentrate Deodorant C1
Cyclone A1
Deo-Gran C1
Deodorizer C1
Drain Cleaner (Enzyme Type) ... L2
DC-66 Liq. Mach. Dshwshng
   Comp. A1
Enzymes for Sewage Systems ... L2
FC-11 Filter Conditioner P1

Gen. Purp. Low Sudsng Clnr. C1
General Purpose Liquid
   Cleaner A4
Germicidal Cleaner D1
Ice Melting Compound C1
Klean-O 99 .... D1
Liq. Steam Clng Comp. Hvy
   Dty Form. A1
Liquid Ice Melt C1
Low Pressure Boiler Comp. G6
MS 11 Safety Solvent K1
MU Hvy Dty Organic Drain
   Opener L1
Porcelain and Bowl Cleanser C2
Rat Bait F3
Sand & Silt Remover L1
Sewer Solvent L1
Shalco Porcelain Cleaner C2
Super Heavy Duty Industrial
   Clnr. A4
Three-Gly C1
Unblock L1
Window Glass Cleaner Conc. C1

Shanette Chemical Research
Qwik Power A4

Shannon Chemical Corporation
Shan-No-Corr G5

Share Corporation
Airborne Insecticide F1
Airborne Vaporizing Insecticide F1
Bacteria Cultures L2
Belt Dressing P1
Bowl and Porcelain Cleaner C2
Clean And Thaw A5
Commander A4
Concentrated Odor Control C1
Contact Cleaner K2
D-S-D Disinfectant D2
Degreaser K1
Degreaser & Motor Cleaner
   Conc. K1
Demoisturizer H2
Deodorant Blocks C1
Dual Action Odor Control C1
Electro-Spray Deodorant-
   Disinf. C1
Enzymes 300 .... L2
Evergreen Air Sanitizer C1
Foaming Cleaner C1
Food Grade Grease H1
Four Way Disinfectant D1
Gasket Cement P1
General Purpose Liquid
   Cleaner A4
Germicidal Cleaner D1
Germicidal Disinfectant
   Sanitizer & Deo. C1
Granular Deodorant C1
Graphite Lubricant Dri-Film H2
Grease Eliminator L2
Heavy Duty Industrial Cleaner .. A4
Heavy Duty Liquid Drain
   Opener L1
High-Glo A1
Indoor Mist or Fog Insecticide .. F1
Inhibited Chemical Water
   Meter Cleaner A3
King Size Deodorant Blocks C1
Lemon Air Sanitizer C1
Liquid Dishwashing
   Concentrate A1
Liquid Drain Opener L1
Liquid Hand Soap E1
Liquid Steam Cleaning
   Concentrate A1
Machine Dishwashing
   Compound A1

Machine Dishwashing
   Concentrate A1
Medicated Liquid Hand Soap E1
Non Acid Lime & Scale
   Remover A1
Non-Flammable Safety Solvent . K1
Odor Control and Degreaser C1
Open Gear Lube H2
Organic Acid Detergent A3
Organic Water Systems Cleaner G7
Penetrating Oil (Aerosol) H2
Pine Odor Disinfectant C1
Powdered Hand Soap C2
Primer P1
Rodent Bait F3
Room Deodorant Blocks C1
Safety Solvent K1
Silicone Lubricant H1
Snow and Ice Inhibitor C1
Soap and Scum Remover A4
Spray Disinfectant &
   Deodorizer C1
Spray Disinfectant And
   Deodorizer C1
Stainless Steel Cleaner A7
Surface Disinfectant And
   Deodorizer C1
Surface Insecticide F2
Teflon Fluorocarbon H2
Tri Power Deodorant Spray C1
Triple Duty Boiler Treatment G6
Triple X Sewer Compound L1
Water Soluble Non-Flammable
   Cleaner A1
Waterless Hand Cleaner E4
Window Glass Cleaner C1

Sharp Chemical Company
Sewer Cleaner #12 .... L1
Sharp Clean L 55 .... D1
SharpClean L-91 .... D1

Shawnmark Industries
M-100 .... A1

Shea Research Corporation
Dirt Fighter All Purpose
   Cleaner Conc. A4,E1
Dirt Fighter Concentrate A4,E1
Fat, Oil And Grease Remover ... A4,A8
Glass Cleaner C1
Ready To Use Liquid Hand
   Cleaner E1
3 X Concentrated Liquid Hand
   Cleaner E1

Sheila Shine, Inc.
Sheila Shine A7

Sheldon Supply Company
Shelco-Germa-Clean A4

Shell Chemical Company
No-Pest Strip Insecticide F2
Pest Strip Insecticide F2

Shell Oil Company
Aeroshell Grease 14 .... H2
Alvania Grease 1 E.P. H2
Alvania 1 .... H2
Alvania 2 .... H2
Clavus Oil 100 .... H2
Clavus Oil 68 .... H2
Darina AX H2
Darina 1 .... H2
Darina 2 .... H2
Delima 69 .... H2
Delima 71 .... H2
Diala Oil AX H2

Donax T-6 Dexron B-10111 ...... H2
Donax T-6 Dexron B-10475 ...... H2
Donax T-6 Dexron B-11317 ...... H2
Irus Fluid 905 .... H2
Omala Oil 220 .... H2
Omala Oil 680 .... H2
Rotella 20/20W H2
Rotella 30 .... H2
Rotella 40 .... H2
Shell Hydrogen Peroxide 35%... N3
Surface Disinfectant and
   Deodorizer C1
Tellus Oil 10 .... H2
Tellus Oil 22 .... H2
Tellus Oil 23 .... H2
Tellus Oil 25 .... H2
Tellus Oil 29 .... H2
Tellus Oil 33 .... H2
Tellus Oil 41 .... H2
Thermia Oil C H2
Turbo Oil 100 .... H2
Turbo Oil 150 .... H2
Turbo Oil 220 .... H2
Turbo Oil 32 .... H2
Turbo Oil 320 .... H2
Turbo Oil 46 .... H2
Turbo Oil 460 .... H2
Turbo Oil 68 .... H2
Valvata Oil J460 .... H2
Vitrea Oil 220 .... H2

Shelley Manufacturing Company
Shelleymatic Stainless Shine....... A7

Shelly-Andrews Company, The
Butyl Cleaner A4
Dawn A2
E&R Cleaner & Stripper A4
HD-206 .... A2
Industrial Lathrex E1
Lathrex E1
Lathrex Cocoanut Oil Hand
   Soap E1
Lotion Soft E1
Lyn Mar Cleaner A2
Shafe C2
Steel Brite A6
Synthetic Concentrate A1
SA-70 .... A1
SA-71 .... A1
SA-72 .... A1
T.J. Scrubmobile Cleaner A1

Shepard Bros.
Hi-Temp A3
Oven Cleaner A8
Shear Liquid C.I.P. (LS 17) A2
Sheen (LS-11) A3
Shine (LS-9) A3
Sho (DS-14) A4
Shotgun A1
Shur (DS-22) A1
Spout (LS-23) A1
Suds A1
Suppress A1
Swing (DS-9) A1

Sher-Gran Industries Ltd.
Wizzard A1

Sherry Enterprises, Inc.
Sherry Kleen A4

Sherry Pharmaceutical Company, Inc.
Break-Thru Liquid L1
Break-Thru Powder L1
Emulsifiable Degreaser K1
Germicidal Cleaner C1
Hand Soap E1

Hard Surface Cleaner.............. A4
Pine Odor Disinfectant C1
Pine Oil Disinfectant C1
Quatinary-Deodorizer Disinf.
   Clnr. D1
Safety Solvent K1
Super Industrial Cleaner &
   Degreaser A1
Synthetic Cleaner A4

Sherwood Chemicals Ltd.
General Purpose Insecticide....... F1

Shield Chemical Corp.
CDD Neutral Fragrance D1

Shield-Brite Corporation
Shield-Brite Defoamer N4

Shield, Incorporated
#40 General Purpose Cleaner A1
Mark 11 .... A1
RS-70 .... A2
Shield 88 .... A1
Shield-88 Liquid A1

Shields' Co.
Shield-O-Cide Fogging
   Concentrate 3000 .... F1
Shield-O-Cide Fogging
   Concentrte 101 .... F1
Shieldocide Mark 11 Fogging
   Formula F1
Shieldocide Point 50 Synrgzd
   Pyrthrn Spy F1

Shiloh Laboratories
Purple Power A1

Shiloh Products, Inc.
SB-2069 All Purpose Cleaner A1

Ship Shape Chemical Specialties
Blast Off A4
Blast Out A4

Shocket Chemical Corp.
D-Trans Fogger And Contact
   Spray 2147 .... F1
Formula 200 Silicone Release
   Spray H1
Formula 200B Silicone Release .. H1
Shocket's Fogging Concentrate
   #15 .... F1

Shoulder & Associates
Husky Oven Cleaner A8

Shrader Chemical Co., Inc.
#1-10 .... A1
#1010 .... Q1
#1014 .... Q1
#1014-1 .... Q1
#1014-2 .... Q1
#1110 .... Q1
#400 .... A1,Q1
#400-X A2,Q1
#422 Chlorinated Cleaner A1
#450 Chlorinated Cleaner A2
#578 Acid Cleaner for Metal A3
#61 .... Q1
B-123 .... A1
Big K A2
Car Shampoo #1 .... A1
Car Shampoo #4 .... A1
Car Wash A1
Concrete Cleaner #10 .... C1
Concrete Cleaner #15 .... A4
Concrete Cleaner #15-B A4

Concrete Cleaner #15-P C1
CC#8 Concrete Cleaner A1
De-Stainer Q2
Defoamer Q5
Defoamer B Q5
Defoamer XX A1
Descaler A3
DC Special A1
DM-45 .... A1
Foamer A1
Foamer Special A1
G-124 .... A1
G-125 .... A1
G-126 Additive A1
HC-102 .... E1
HC-103 Hand Lotion E1
HC-104 .... E1
HC-105 .... E1
HC-63 Plus A1
HT-2 .... A2
HT-4 .... A1
HT-5 .... A2
HT-5-X A2
JG-11-1 .... C1
JG-11-5 Solvent Cleaner K1
JG-11-77 Stainless Steel Clnr &
   Polish A7
Kleen-It A4
Kleen-It #202 .... A4
Kleen-It-N.S. A4
Kleen-It-Plus A4
MC-64 .... A1
OV-51 Oven Cleaner A1
Quik-Wet A1
Rug Shampoo A1
S-1-Klor Chlorinated Cleaner ... A1
S-2 Klor A1
S-2-Klor-A Chlorinated Cleaner A1
S-3 Klor A1
S-4 Klor A1
S-5 Klor D1
S-6 Klor D2
S-7-Klor Chlorinated Cleaner ... A1
S-8-Klor Chlorinated Cleaner ... A1
Shaaco C2
Shine-Kleen A1
Shur-Dri A1
Shur-Dri-Plus A1
Strat-O-Lube H2
Strat-O-Lube P H2
Strat-O-Lube Special H2
Sump Cleaner L1
SC-15 .... A2
SC-16 .... A1
SC-180 .... A1
SC-200 .... A4
SC-201 All Purpose Cleaner A1
SC-203 General Purpose
   Cleaner A1
SC-205 .... A1
SC-206 General Purpose
   Cleaner A1
SC-210 .... A1
SC-211 General Purpose
   Cleaner A4
SC-212 .... A1
SC-215 .... A1
SC-247 Spray Washer
   Detergent A1
SC-30 .... A1
SC-30-C Chlorinated Cleaner..... A1
SC-30-D Chlorinated Cleaner A1
SC-31 .... A2
SC-32 .... A2
SC-33 Packers Cleaner............ A1
SC-33-A Heavy Duty
   Chlorinated Cleaner A2
SC-33-C Heavy Duty
   Chlorinated Cleaner A2
SC-35 .... A4

SC-37 General Purpose Cleaner A1
SC-41 .... A2
SC-42 .... A2
SC-42 Plus A2
SC-42 Special A2
SC-42-X A2
SC-43 .... A2
SC-44 Cleaner A1
SC-45 .... A1
SC-46-A A1
SC-46-B A1
SC-47 .... A1
SC-47-X A1
SC-48 .... A1
SC-49 .... A2
SC-50 Additive A4
SC-500 .... A3
SC-500 Special A3
SC-500-F A3
SC-500-X Acid Cleaner A3
SC-502 .... A3
SC-503 Descaling Jelly A3
SC-504 Acid Cleaner A3
SC-505 .... A3
SC-510 .... A4
SC-515 Acid Cleaner A3
SC-52 Additive A1
SC-520 .... A3
SC-55 Bottle Wash Additive ... A1
SC-55-Plus A3
SC-555 .... A3
SC-567-A A3
SC-575 .... A3
SC-575 Special A3
SC-590 .... L1
SC-60 .... A1
SC-63 All Purpose Cleaner A1
SC-65 Freezer Cleaner A4,A5
SC-699 .... F1
SC-7 .... A1
SC-703 .... A4
SC-710 .... G7
SC-711 .... A1
SC-725 .... D1
SC-730 .... D1
SC-740 .... D1
SC-745 .... D1
SC-75 General Purpose Cleaner A1
SC-750 .... A3,D1
SC-76 .... A1
SC-760 .... D1
SC-762 .... D2
SC-78 Alkaline Cleaner A1
SC-9 .... A1
SC-91 .... A1
SC-99 .... A1
X-70 .... A2
X-80 .... A1
X-85 .... A1

Shur-Chem Industries
Coil Cleaner and Conditioner A3
Concrete Floor & Driveway
  Cleaner.............................. A4
Hot Soak Tank Cleaning
  Powder A2
Pressure Spray Washing
  Detergent........................... A1
Steam Cleaning Liq. Ext.
  Heavy Duty A1
Steam Cleaning Liq. Multi-
  Purpose A1
Steam Cleaning Powder A1
Steam Cleaning Powder Heavy
  Duty A2
Steam Cleaning Powder
  Medium Duty....................... A1

Shur-Gloss Manufacturing Co.
Liquid Car Wash Concentrate ... A1

Powder Car Wash Detergent A1
Spra-N-Rinz A1

Silco Oil Co.
75-T Technical White Mineral
  Oil H1

Simmons Industries/ Simmons Engineering Company
Speci-Lube 2 .... H1

Simon Brothers, Incorporated
Sim-Tex All Purp. Oil&Grease
  Absorbent J1
Sim-Tex Institutional Foaming
  Clnsr. C2
Sim-Tex No 77 Liquid Cleaner .. A4

Simple Products, Inc.
Simple Green A4,B2,A8

Simplex Products
Food Grade Degreaser General
  Purp. Clnr........................... A1
Plex-Cide D2
Simplex 200 .... A4,A8

Sinclair Refining Company
Cadet Oil A H2
Cadet Oil B H2
Cadet Oil C H2
Cadet Oil D H2
Cadet Oil M H2
Caldron Grease No. 1 EP.......... H2
Caldron Grease No. 2 EP.......... H2
Castor Machine Oil Extra
  Heavy H2
Castor Machine Oil Heavy........ H2
Castor Machine Oil Light......... H2
Castor Machine Oil Medium H2
Commander Oil A H2
Commander Oil B H2
Commander Oil C H2
Commander Oil D H2
Commander Oil E H2
Commander Oil F H2
Commander Oil H H2
Duro AW Oil No. 16 .... H2
Duro AW Oil No. 21 .... H2
Duro AW Oil No. 31 .... H2
Duro Cylinder Oil H2
Duro Oil No. 100 .... H2
Duro Oil No. 150 .... H2
Duro Oil No. 160 .... H2
Duro Oil No. 200 .... H2
Duro Oil No. 250 .... H2
Duro Oil No. 295 .... H2
Duro Oil No. 300 .... H2
Duro Oil No. 400 .... H2
Duro Oil No. 55 .... H2
Duro Oil No. 600 .... H2
Duro Oil No. 900 .... H2
Extra Duty Gear Lubricant
  SAE 140 .... H2
Extra Duty Gear Lubricant
  SAE 80 .... H2
Extra Duty Gear Lubricant
  SAE 90 .... H2
Gascon Oil A H2
Gascon Oil B H2
Gascon Oil C H2
Gascon Oil D H2
Gascon Oil F H2
Gascon Oil H H2
Ice Machine Oil Heavy............. H2
Ice Machine Oil Medium H2
Ice Machine Oil Medium
  Heavy H2
Light Mineral Valve Oil H2

Litholine Multipurp Grease
  NLG1#2 .... H2
Mineral Gear Oil SAE 140........ H2
Mineral Gear Oil SAE 80.......... H2
Mineral Gear Oil SAE 90.......... H2
Modoc Cylinder Light.............. H2
No. 29 Valve Oil H2
No. 87 Valve Oil H2
Rocket Oil B H2
Rocket Oil C H2
Rocket Oil Z H2
Rocket Oil 60......................... H2
Rubilene Oil-Extra Heavy H2
Rubilene Oil-Extra Light........... H2
Rubilene Oil-Heavy.................. H2
Rubilene Oil-Light H2
Rubilene Oil-Light Medium H2
Rubilene Oil-Medium............... H2
Rust-O-Lene 10....................... H2
Rust-O-Lene 30....................... H2
Sinco Prime 200...................... H1
Sinco Prime 70 N.F. H1
Sinco Prime 90 N.F. H1
Sinco Tech 10 .... H1
Sinco Tech 6 .... H1
Sinco Tech 8 .... H1
Superheat Valve Oil................. H2
Transformer Oil HD H2
Valve Oil Dark H2
Valve Oil Light H2

Sioux
Liquid A Steam & Pressure
  Cleaning Cmpd. A1

Six M Company
Lemon Disinfectant No. 13 .... C1

Skasol Corporation
Brite Bowl Conc. Toilet Bowl
  Clnr. C2
Concentrated Toilet Bowl
  Cleaner............................... C2
Dyno-Drain............................ L1
Jet Wite Toilet Bowl Cleaner C2
Krome Kleen Porcelain Cleaner C1
Nu-Coil................................. P1
Pink Glo Bowl&Porcelain
  Cleaner............................... C2
Porcelain Tile & Enamel
  Cleaner............................... C2
Thix Porcelain Cleaner C2

Skasol, Incorporated
Biocide No. 2 .... G7
Biocide No. 8 .... G7
Boiler Treat. Formula No. 257 .. G6
Boiler Treat. Formula No. 2578. G6
Boiler Treat. Formula No. 275 .. G6
Boiler Treat. Formula No. 275-
  S G6
Boiler Treat. Formula No. 3000. G6
Boiler Treat. Formula No. 54 .... G6
Boiler Treat. Formula No. 5759. G6
Boiler Treat. Formula No. 7558. G6
Boiler Treat. Formula No. 92 .... G6
Boiler Treatment Formula No.
  224 .... G6
Boiler Treatment Formula No.
  850 .... G6
Boiler Treatment Formula No.
  850 R G6
Condensate Cor. Contrl Form
  #CCC7 .... G6
Corrosion Inhibitor Formula
  No. 1881 .... G7
Drain Pipe Solvent L1
Quick Scale Solvent A3
Tower Treat. Formula No. 100 . G7

240

Tower Treat. Formula No.
   100B G7
Tower Treat. Formula No.
   100L G7

Skat Products, Incorporated
Skaticide F1

Skip, Incorporated
Skip Concrete Cleaner 1418 .... A4
Skip 1300 .... A3
Skip 1386 .... A1
Skip 1399 .... Q1
Skip 1414 .... A1
Skip 1426 .... A2
Skip 1434 .... A1
Skip 1458 .... A2
Skip 1675 .... A1
Skip 1995 .... A3
Skip 1997 .... A1

Sky Enterprises
Blast A1

Sky Manufacturing, Inc.
Acid Detergent (Inhibited) A3
Heavy Duty Concentrate A1
Low Foam Scrub Soap A4
Twenty Eight Concentrate A4,A8

Sla Chem Corporation
Power Fog N.T. Insect Spray F1
S.C. 105 .... K1
Silispray N.T. H1

Slifer Manufacturing Co., Inc,
Hotsy Coil Conditioner A3
Hotsy Fleet Clean A1
Hotsy Germicide D1
Hotsy Pow-R-Det A2
Hotsy Ripper A1
Hotsy Super A1
Hotsy Triumph A1

Slip-Not Corporation
Sure-Step J1

Smart & Final Iris Company
Coconut Oil Liquid Soap E1

Smart Chemical Company
FP210 Heavy Duty Alkaline
   Cleaner A2
FP211 Heavy Duty Alkaline
   Descaler A2
FP212 Heavy Duty Alkaline
   Cleaner A2
FP215 Heavy Duty Chlor.
   Alkaline Cln. A2
FP220 General Purpose Cleaner A1
FP222 General Purpose Cleaner A1
FP225 Chlorinated C.I.P.
   Cleaner A1
FP226 Smart Chlor A1
FP227 Chlorinated CIP.Cleaner A1
FP228 Chlorinated General
   Purpose Cln. A1
FP231 Milkstone Remover-LF .. A3
FP235 Powdered Acid Cleaner . A3
Heavy Duty Chlor. Alkaline
   Cln. FP 217 .... A2
Smart FP 230R Milkstone
   Remover A3

Smith Supply Co., Inc.
Bleach D1,B1

Smith Supply Company
Sani-Lube Food Equipment
   Lubricant H1
Sewer Solvent L1

Smith, E. W. Chemical Company
Ems Acid Salt Cake Remover ... A3
Ems Algaecide #322 .... G7
Ems Algaecide Pellets "S" G7
Ems All Purpose Can Cleaner .. A1
Ems Boiler Water Treatment
   #605 .... G6
Ems Boiler Water Treatment
   #628 .... G6
Ems Boiler Water Treatment
   #632 .... G6
Ems Boiler Water Treatment
   #633 .... G6
Ems Boiler Water Treatment
   #634 .... G6
Ems Boiler Water Treatment
   #635 .... G6
Ems Boiler Water Treatment
   #636 .... G6
Ems Boiler Water Treatment
   #643 .... G6
Ems Boiler Water Treatment
   #651 .... G6
Ems Boiler Water Treatment
   #676 .... G6
Ems Boiler Water Treatment Z. G6
Ems Cool. Twr&Condensr
   Treat.#304 .... G7
Ems Cool. Twr&Condensr
   Treat.#315 .... G7
Ems Dart A1
Ems Descaling Neutralizer X A1
Ems Dock & Floor Cleaner
   #102 .... A2
Ems General Purpose Cleaner .. A1
Ems Genie P1
Ems Hook and Roller Cleaner .. A2
Ems Laundry Compound 20-S .. A1
Ems Liquid Emsope E1
Ems LCO A1
Ems Meat Packers Cleaner
   #250 .... A1
Ems Meat Packers Super
   Cleaner A1
Ems Milkstone Remover #837-
   A A3
Ems Oxygen Scavenger G6
Ems Powdered Scale Cutter
   #829 .... A3
Ems Powdered Scale Cutter
   SRC A3
Ems Pyrenone Fly Spray F1
Ems Quel Insect Spray F1
Ems Return Line Treat. #647 .... G6
Ems Return Line Treat. #647
   (Conc) G6
Ems Return Line Treat. #647
   (Spel) G6
Ems Rust Off #825/Scale
   Cutter#825 Sp A3
Ems Scale Clean A1
Ems Scale Cutter #821-E A3
Ems Smokehouse Cleaner A2
Ems Smokehouse Cleaner 257A A2
Ems Ster-O-Fect C2
Ems Supplmtl Boil Wtr
   Treat.#612-MP G6
Ems Supplmtl Boil Wtr
   Treat.#613 .... G6
Ems Supplmtl Boil Wtr
   Treat.#614 .... G6
Ems Supplmtl Boil Wtr
   Treat.#615 .... G6
Ems Target A4

Ems Tripe Cleaner 287 .... N3
Ems Triumph A1
Ems Water Control Crystals G7
Ems Water Control Treat No.
   314 Conc. G1
Ems Water Control Treat No.
   324 Conc. G1
Ems Wafer Control Treat. No.
   314 .... G1
Ems Water Control Treat. No.
   323 .... G1
Emsbactocide D1
Emsbreak "S" A1
Emscoat 169 .... H2
Emscoat 169A H2
Emsdrain L1
Emsfast Concrete Floor Cleaner A4
Emsflow D1
Emsguard D1
Emsmighty A1
Emsnap A1
Emsprite A1
Emspronto A1
Emsteam (Super) A2
Emsteam Heavy Duty Cleaner .. A2
Emsteam Heavy Duty Liq.
   Cleaner A1
Emstrol "H" D1
Emstrol "R" D1
EMS Cleaning Solvent #16 ...... K1

Snee, W. G. Company, Inc.
Power-Guard RD-10 .... D2

Snider's Atlantic Service Inc.
Dermagard E1
GT 21 Grease Off A1
JB 31 Germicidal Cleaner A4
Snider's Easy Quick Germicidal
   Cleaner A4

So-White Chemical Company
So-White Bleach and
   Disinfectant D2,B1

Soakup Company, The
Soakup Oil,Grease&Water
   Absorbt J1

Sofield Supply Company
Norlo Non-Butyl Super Power
   Cleaner A1

Soil Chemicals Corporation
Methyl Bromide 100 .... F4

Solar Chemicals, Inc.
S-1492 Shroud Detergent B1
S-3000 Lotion Hand Soap E1
Solar Hook And Trolley
   Cleaner A2
Solar Shroud Bleach B1
Solar Tripe Cleaner N3

Solar Chemicals, Inc.
Solar-Might A1

Solar Industrial Products, Inc.
Solargrease H2

Solarine Company, The
C-99 Concentrated Cleaner A1
Castile Liquid Hand Soap 15% .. E1
Castile Liquid Hand Soap 40% .. E1
Deodorant Disinfectant Cleaner D1
Equipment Cleaner EC-130 .... A4
Heavy Duty Stripper A4
Jel Additive A1
Low Foam Cleaner C1

Neutral Cleaner A1
Pure Coconut Oil Soap E1
Solar-Glo A1
Solaraction A1
Solarbrite A3
Solarchlor A1
Solarclean A2
Solarterg A1
Ultimate Concentrated Cleaner .. A1

Soloco, Inc.
Powersol A4,A8

Soltek Division/ Tech Manufacturing Co.
Emulsol #282 .... A4
Min-T-Cide C1
Tech Charge L1
178 Hy-Ox A3
276 Tech Zymes L2
278 Mint-Off A3

Solvent Manufacturing Co., Inc.
Boiler Solvent #3 .... G6
Boiler Solvent AMO-44 .... G6
Boiler Solvent BW-60 .... G6
Boiler Solvent O-1 .... G6
Boiler Solvent PK-7 .... G6
State F-1 Special Boiler Solvent G6

Solventol Chemical Products, Inc
#2058 .... A3
#2059 .... A3
#2060 .... A3
#2068 Ham Mold Cleaner A3
#2409 Powdered Acid
  Detergent A3
#2776 Liquid Acid Cleaner A3
#301 Pot and Pan Wash A1
#3015 .... A1
#3050 .... A1
#3618 Animal Cage Wash A1
#3654 Heavy Duty Cleaner A2
#3944 .... B1
#3955 .... A2
#3956 .... A2
#3997 .... A1
#601 All Purpose Cleaner C1
#8088 General Purpose Cleaner A1
#8388 Auto Laundry
  Compound A1
#8502 .... A1
#8513 .... A1
#8545 Ham Mold Cleaner A1
#8565 All Purpose Liq.
  Detergent A1
#8589 .... A1
#8592 Smoke House Cleaner A2
#8595 Smoke House Cleaner A1
#8602 All Purpose Cleaner A4
#888 Bake Pan Cleaner A2
#889 Drain Cleaner L1
#979 .... A1
Bottle Wash-R A2
Can Wash A1
Chlorinated Egg Wash Q1
Chlorinated Egg Wash #966 ...... Q1
CDC A1
Dairy Cleaner A1
Doughnut Machine Cleaner A1
Egg Wash Q1
Egg Wash #965 .... Q1
Fome Solv A2
H.T.S.T. Circulating Cleaner A2
H.T.S.T. Type II A2
Heavy Duty Liq. Steam
  Cleaner A1
Hi Suds A1
Hog Scald N2

Liquid Chain Lubricant H2
Liquid Hand Soap 15 .... E1
Low Suds Concentrate B1
Mechanical Dish Wash MDW
  201LF A1
Medium Duty Liquid Steam
  Cleaner A1
Milkstone Remover "Ol"-Non
  Foam A3
Milkstone Remover "Ol"-
  Regular A3
Powdered Acid Detergent A3
Smoke House Cleaner A2
Solo San 120 .... D1
Solo San 304 .... D1
Solo San 310 .... D1
Solo San 500 .... D1
Solo San 510 .... D1
Solvene PT-20 .... A1
Solventol #701 .... D1
Solventol Powdered Bleach B1
Solventol 8507 .... A4
Solventol 8508 .... A4
Solventol 8518 .... A4
Solventol 8560 .... C1
Sparkleen HD A1
Sparkleen S A1
Sparkleen 20 .... A1
Steam Cleaner #590 .... A2
15% Organic Laundry Bleach ... B1
8505 Cleaner Disinfectant C1

Solvex Chemical Company
Solvex A4

Solvit Chemical Co., Inc.
Clean-All A1
Industrial Aerosol Insecticide ... F1
Rat & Mouse Killer F3
Rinse Away A1
S.C.C. A4
Solvit All-Purpose Cleaner C1
Space Spray Insecticide F1
Super Clorinated, Clean-All A1

Solvox Mfg. Company
Solterge 70 .... A1
Solvox 22 .... A1

Sorb-All Company
Saco Detergent 30366 .... A1

Sorensen Building Service, Inc.
Attack A4
Capture A4

South Coast Distributing Company
De-Grease It A1

South Coast Products, Inc.
Blue Might A1
Food Grade Cleaner S-53 .... A1
Pack P1
Scalon P1
Thread Sealant No. 10 .... H2
Thread Sealant No. 20 .... H2
Thread Sealant No. 30 .... H2
Una-Seal with Teflon H2
White Food Grade Grease FM
  431 .... H2

Southeast Sanitary Supply Co.
Ox Power Clean A4

Southeastern Chemical Company
Isis Room Deodorant (Aerosol). C1

Southeastern Chemical Corporation
Secco Antifoam #2 .... P1

Secco Antifoam #4 .... P1
Secco Antifoam #6 Special P1

Southeastern Cleaning Products
SCP-1030 .... A1
SCP-1040 .... A1

Southeastern Laboratories, Inc
A-3 Cooling Water Treatment ... G7
A-4 Cooling Water Treatment ... G7
A-5 Corrosion Inhibitor G7
A-6 Cooling Water Treatment ... G7
B-081532 Boiler Water
  Treatment G6
B-081532A Boiler Water
  Treatment G6
B-081532AB Boiler Water
  Treatment G6
B-1 Boiler Water Treatment G6
B-10 Boiler Water Treatment G6
B-11 Boiler Feedwater
  Supplement G7
B-11 F Boiler Feedwater
  Supplement G7
B-191 Boiler Water Treatment ... G7
B-2 Boiler Water Treatment G6
B-3 Boiler Water Treatment G6
B-4 Boiler Water Treatment G6
B-7 Boiler Water Treatment G6
B-8L Boiler Water Treatment G6
B-9 Boiler Water Treatment G6
B-900 Boiler Scale Treatment ... G6
B-901 Boiler Water Treatment ... G7
D-1 Steam Line Treatment G6
D-2 Steam Line Treatment G6
D-3 Steam Line Treatment G7
D-4 Steam Line Treatment G7
D-5 Steam Line Treatment G6
D-6 Steam Line Treatment G6
D-7 Steam Line Treatment G7
F-1 Boiler Scale Solvent A3
F-2 Cooling Tower Scale
  Solvent A3
F-3 Cooling Tower Scale
  Solvent A3
G-1 Neutralizer A2
G-4 Resin Cleaner P1
H-2 Algaecide G7
H-3 Algaecide G7
K-2 Threshold Treatment G2

Southeastern Sanitary Supply Co.
A.P. Cleaner A1
Hurricane Cleaner Liquid A1
Hurricane Cleaner Powdered A1
Instant Stripper Cleaner C1
Rebel A1
Release Meat Room Cleaner A1
Sewer Solvent L1
Tiger Flash-135 .... A1

Southern Agricultural Chemicals, Inc.
Big "G" A4
Black Disinfectant Coef. 2 .... C1
Black Disinfectant Coef. 3 .... C1
Black Disinfectant Coef. 5 .... C1
Dual Disinfectant-Cleaner D1
Lemon Tree C1
Mint-O-Dis C2
Pine Oil Disinfectant Coef. 5 ..... C1
Power Concentrate C2
Vita Pine Coef. 5 .... C1

Southern Chemical Company
Sano-Cide F1
Sano-Cide 64 .... A4
Sano-Phene 256 .... A4
SomoLube H2

Southern Chemical Products Co.

Abrasive Detergent A6
Antiseptic Hand Soap 15 .... E1
Antiseptic Hand Soap 20 .... E1
Bacteriasol Germicidal Cleaner.. A4
Big Dog Concrete Cleaner A4
Big Jim Lemon Foam
  Germcdl.Cl.&Disinf............. C2
Blue Blazes Sewer Solvent L1
Blue SP 1105 .... A1
Brightline Special A2
Cherry Bowl Cleaner C2
Chlorinated Egg Cleaner Q1
Clean-It..................................... A1
Cooler Coil Cleaner P1
Deep Chlor D2
Delux Machine Dishwashing
  Compnd A1
Discrete A3
Dynamo Detergent..................... A1
E-Z Strip C3
Egg Wash No. 1000 .... Q1
Electrosol Safety Solvent K1
Emulsol Bowl Cleaner C2
Extra Hvy Dty Concrete
  Cleaner.................................. A4
Foam-Aid A1
Golden Gator Concrete Cleaner C1
Golden Magic Cleaner A4
Grime Scat A1
Hospital Disinfectant................. C1
Industrial Cleaner A4
Iodo-175................................. D2,E3
Laundry Detergent..................... B1
Lemon Bathroom Cleaner &
  Disinfectant C2
Lemon Disinfectant No. 6......... C1
Liq. Smoke House Clnr. Form.
  II... A1
Liquid Chlorinated CIP Cleaner A1
Liquid Hand Soap 15 .... E1
Liquid Hand Soap 20 .... E1
Liquid Hand Soap 40 .... E1
Liquid Steam Cleaner A2
Low Phos Alkaline Detergent ... B1
Multi-Blue A1,E1
No. 100 .... A1
No. 100 Plus.............................. L1
No. 200 Smoke House Cleaner .. A2
No. 300 .... A3
No. 400 .... A1
No. 500 Tripe Cleaner N3
No. 600 Hog Scald N2
No. 80 Space Spray F1
No. 90 Space Spray F2
Orange Bacteriasol Germicidal
  Cl Form 2.............................. D1
Orange Bacteriasol Germicidal
  Clnr. C1
Oven Cleaner A8
Pene-Lube Penetrating
  Lubricant H2
Pine Odor Disinfectant C1
Pine Odor Disinfectant Coef. 5
  Formula 2................................ C1
Pink Cleaner........................ A4,E1
Pink Rainbow Liquid Dishwash A1
Pink Satin E1
Pot and Pan Cleaner A3
Poultry House Cleaner............... A1
Poultry House Clnr. Form. II.... A1
Premium Fly & Roach Spray..... F1
Pressure Truck Wash Powdered A1
Pressure Wash............................ A1
Prometheus Alkaline Cleaner..... A2
Pyrethroid Fly & Roach Spray.. F1
Pyrethroid 351 .... F2
Quaternary Ammon. Germ.
  Conc..................................... D1
Quick Scrub A4

Residual Roach Killer................ F2
Residual Roach Spray................ F2
Safety Solvent........................... K1
Scald Vat Cleaner A1
Scrub Wash................................ A1
Special Chain Lubricant H2
Stainless Steel Cleaner C1
Steam Cleaning Compound A2
Super Enzymes for Sewerage
  Systems................................. L2
Super Heavy Duty Cleaner........ A1
Sweeping Compound (Red) Oil
  Base....................................... J1
Wax Scat A1
X-100 Chlorinated Machine
  Dishwash A1
X-150 Chlorinated Machine
  Dshwsh.................................. A1
Y-150 Chlorinated Machine
  Dishwash A1
360 Milkstone Remover A3

Southern Maid Chemical Co.

Bora-Clean E1
Chef-Een Liquid Detergent........ A4
Cocoanut Oil Soap E1
Concrete Cleaner........................ A1
Delite 20%................................. E1
Desoiler And Descaler................ A3
Germisyl..................................... D1
Lemon Disinfectant No. 13 .... C1
Low Foam No. 904 Super
  Concentrate A4
Lubon Chain Lubricant H2
Pine-Feen Pine Scent
  Disinfectant C1
Presurg-Plus............................... E1
Special Pink Super Sapon
  Concentrate A4
Super Strip A4
Total 200 .... D2
Victory Cleaner.......................... A4

Southern Mill Creek Products of OH, Inc.

Clemco Pyrocide Fogging
  Concentrate 5628 .... F1

Southern Mill Creek Products Co., Inc.

Bor/Act...................................... F2
Bug Boy F1
Clemco Food Plant Spray F2
D/V 217 Insecticide F2
Diazinon 4E F2
Diazinon 4S................................ F2
Dursban 1E Insecticide.............. F2
Dursban 2E Insecticide.............. F2
General Purpose Household
  Spray #10............................... F1
Malathion 57% Premium Grade F2
Para-Blox................................... F3
Pyrenone General Purpose
  Spray....................................... F1
Residual Roach Spray................ F2
SMCP Pyrethrins ULV
  Fogging Concentrate............... F1
SMCP SBP-1382-2.E.C............. F1
Trap Stik F3
Warehouse Fog Insecticide........ F1
Xtraban Roach Concentrate....... F2

Southern Oklahoma Janitor Supply Co.

Sohoma Cleaner Disinfectant D1

Southern Products Co., Inc.

Bowl Cleaner C2

Concrete Cleaner H.D.
  Odorless................................. A4
Foam Klean A1
So-Pro Chlorinated Machine
  Dish Cmpd A1
So-Pro Concrete Cleaner........... A1
So-Pro General Purpose
  Cleaner.................................. A1
So-Pro Germisyl Coef. 10 .... D1
So-Pro Lemon Disinfectant No.
  10... C1
So-Pro No. 35 .... E1
So-Pro Phos Free Mech. Borax
  Hand Clnr. E4
So-Pro Pink Lotioned Liquid
  Detergent............................... A4
So-Pro Purple X Heavy Duty
  Cleaner.................................. A4
So-Pro Super Fog Mist
  Insecticide............................. F1
SP-Concentrate Neutral Cleaner A4

Southern Products Company

Air O Phene C2
Dairy San D1
Kill Fly F2
Klene-O A1
Mill Spray F1
Ortho Sol................................... C1
So-Pro-Co "100"........................ A1
So-Pro-Co Air-O-Pine............... C1
So-Pro-Co Pine-O-Phene........... C1
Spar-Germ D1
Super Spar-Germ....................... D1
Velva Lathe E1

Southern Sanitary Co., Inc.

All Purpose Cleaner And
  Maintainer A4
Fast Clean A1
San Quest A4
Southern Sanitary's Neutral
  Cleaner.............................. A4,A8
Special Cleaner A1
3 In 1 Cleaner-Disinfectant-
  Deodorant A4
3 In 1 Neutral Fragrance........... A4

Southern Saw Service, Inc.

ASA Jetspray............................. H1

Southern Specialties

Tom Terrific Concentrate A1

Southern States Chemicals, Inc.

S.S. No. 36 Xtra-Clean.............. A4

Southern-Aire Chemical Co., Inc.

Gleem Clean Stainless Steel
  Clnr & Plsh A7
Long Needle Pine Oil
  Disinfectant C2
Mighty Joe A4
Mor-EZ...................................... A1
Mor-O-Mint............................... C1
Mor-Plus Concentrate A4
Moroscolor F2
Morticide Concentrate F1

Southland Corporation, The

A.C. #9 .... A3
Alpac A1
AC#1.. A3
AC#10.. A3
AC#16.. A3
AC#2.. A3
AC#39.. A3
AC#4.. A3
Bactidine.................................... D1

Boiler Water Treatment 11-71.... G6
Boiler Water Treatment 17-11.... G6
BW#1 .... A2
BW#8 .... A2
C.G.C. A1
Cent-O-Mint F2
Chlor-All #30 .... A1
Chlorinated All Purpose Gen.
  Cleaner........................... A1
Cip-San A3
Circle-8 .... A1
Con-Cleen.......................... A4
CIP Cleaner A2
CPL #17 .... A1
CW-15............................... A1,N4
Defoamer #160..................... A2,Q5
Defoamer #34...................... A1,Q5
Di-Kill F2
Enzyme-400 .... L2
Foam-Ad A1
Fog-It F1
GB-120.............................. A1
GB-755.............................. A1
H.D.S. A2
H.D.S.-W A2
Hy-Q 10 .... D1,Q3
HD-11 Egg Washing
  Compound......................... A2
HD-15............................... A2
HD-17............................... A2
HD-24............................... A2
HD-33............................... A2
HD-34............................... A2
HD-35............................... A2
HD-36............................... A2
HDC-50............................. A2
I.B. 25 .... A4
I.B. 31 .... A1
I.B. 36 .... A1
I.B. 40 .... A1
Iodex............................... E2
Johnny.............................. C1
Kase Kleen A2
Klor Kleen A1
Laundry Detergent................. B1
Loc.................................. D2
LD 7-11 .... A2
Multi-San A3
MAL-30.............................. F2
MHC-10.............................. A1
Neutralizer #311................... B2
Neutralizer Sour A3,B1
NA-Complex A2
NR-4................................ F1
Old Tiger........................... A2
Phosbrite........................... A3
Phosflake A2
Polychlor........................... D1
PC-10 Egg Washing Compound A2,Q1,Q2
PC-20 Egg Washing Compound A2,Q1,Q2
PC-40E A1
PC-60............................... A1
PC-70............................... A1
PC-85 Egg Washing Compound A2,Q1,Q2
Residacide F2
Sani Chlor D2,Q3
Sanidine D1
Silky............................... E1
Sparkle A.C. A3
Thermo Kill F1
TC-200 Tripe Cleaner............... N3

Southland Paper & Supply Co.,Inc.
Sanitare............................ D1
Strip-Rite.......................... A4
T.N.T. A4

Southwest Commercial Chemical Corp.
SWC-101A E1
SWC-33 .... C1

SWC-35 .... C1
SWC-37 .... A4
SWC-39 Virucide D1
SWC-44A A3
SWC-47 .... K1
SWC-62 .... G6
SWC-74 .... C1
SWC-75 .... C1
SWC-76 .... C2
SWC-99 .... L1

Southwest Distributing Company
Non Butyl Degreaser A1

Southwest Petro-Chem, Inc./ A Division of Witco
Aluminum Cmpx Fd Mach
  Gres-EP Code B6672 .... H2
Aluminum Cmpx Fd Mach
  Gres-EP Code C6112 .... H2
Aluminum Cmpx Fd Mach
  Gres-EP Code 15212 .... H2
Aluminum Cmpx Fd Mach
  Gres-EP Code 2387................. H2
Aluminum Cmpx Fd Mach
  Gres-EP Code 3358................. H2

Southwest Sanitary Company, The
Crystal C1
Lemon Odor Disinfectant............ C2
Liquid Detergent................... A1
Liquid Hand Soap Ready For
  Use E1
Mint Germicide-Deodorant C2
Mug The Bug Liquid
  Insecticide...................... F1
Mug The Bug Residual
  Insecticide...................... F2
New Life A4
Old Limey A4
Pine-Odor Disinfectant C1
Pink Lotion Hand Soap E1
Porcelain Power C2
Potent Power A4
Pure Pine Oil Disinfectant......... C1
Silicone Spray (Aerosol)........... H1
Spray 88 Heavy Duty Clnr. &
  Degreaser A1
Super Bowl C2
Titanic............................ C1
Window Shine Glass Cleaner........ C1
40% Concentrated Liq. Hand
  Soap E1

Southwest Saw Corporation
S-S-S-Spray-Oil..................... H1

Southwestern Boiler Service
Caustic Soda G6
T-11 S G6
T-18-S G6
T-19................................ G6
T-75................................ G6
918................................ G6

Southwestern Chemicals Co.
New Scalex G6

Southwestern Petroleum Corp.
All-Purpose Cleaner................ A4
Concrete Etcher A3
Emulsifiable Degreaser C1
Germicidal Detergent A4
High Heat Grease #104 Heavy . H2
High Heat Grease #104 Light ... H2
Hot Tank Degreaser A4
Lithium Grease #107 Heavy...... H2
Lithium Grease #107 Light H2
Steam Cleaner...................... A1

Swepco Barium Grease No.
  105, Heavy H2
Swepco Calcium Grease
  FM#109............................ H2
Swepco Calcium Grease
  FM#115............................ H1
Swepco Compressor Oil #707-1 H2
Swepco Compressor Oil #707-2 H2
Swepco Compressor Oil #707-3 H2
Swepco Compressor Oil #707-4 H2
Swepco Concentrated Cleaner... A1,E1
Swepco Disinfectant Detergent . D1
Swepco Food Machinery
  Grease 115....................... H1
Swepco Gear Lube #201 SAE
  140.............................. H2
Swepco Gear Lube #201 SAE
  250.............................. H2
Swepco Gear Lube #201 SAE
  80............................... H2
Swepco Gear Lube #201 SAE
  90............................... H2
Swepco Hydraulic Oil #704
  Heavy H2
Swepco Hydraulic Oil #704
  Light A H2
Swepco Hydraulic Oil #704
  Light B H2
Swepco Hydraulic Oil #704
  Light Med. H2
Swepco Industrial Oil #702 No.
  0................................ H2
Swepco Industrial Oil #702 No.
  00............................... H2
Swepco Industrial Oil #702 No.
  1................................ H2
Swepco Industrial Oil #702 No.
  1-A.............................. H2
Swepco Industrial Oil #702 No.
  2................................ H2
Swepco Industrial Oil #702 No.
  3................................ H2
Swepco Industrial Oil #702 No.
  4................................ H2
Swepco Industrial Oil #702 No.
  5................................ H2
Swepco Industrial Oil #702 No.
  6................................ H2
Swepco Industrial Oil #702 No.
  7................................ H2
Swepco 101 Moly Grease........... H2
Truck & Rig Wash.................. A4

Space Age Sales
56-C Floor & Wall Cleaner........ A4

Spale Products, Inc.
Miracle Cleaner A4,A8

Sparkel
Sparkel #1......................... A1
Sparkel Floor-Cleaner #3.......... C1

Spartan Chemical Co., Inc.
All Purpose Cleaner............... A4
BH-38.............................. A4
CR-2 Insecticide F2
DA-70.............................. C1
DFP-32............................. A1
F-1 Industrial Strength
  Insecticide...................... F1
F-6................................ F1
Germicidal Bowl Cleanse C2
Glass Cleaner C1
Goin' Home E1
Golden Glo Liquid................. A1
H2-D2.............................. A4
Lotionized Liquid Hand
  Cleaner.......................... E1
Lube-All........................... H2

M 95 .... A3
M.L.D. Bowl Cleanse C2
Metaquat Germicidal Cleaner D1
New Liquid Steam Cleaner A2
Oven & Grill Cleaner A8
Pathmaker A1
Pressure Washer Cleaner A1
PD-64 .... A4
Sani-T-10 .... D2
Sparquat 256 .... A4
Spartan-Reinol E4
Stainless Steel Cleaner & Polish. A7
Sterigent A4
SC-200 Industrial Cleaner A1
SD-20 .... A4

Spartan Environmental Services, Inc.
Spartan Formula "Alk-Treat"
  (Liquid) G6
Spartan Formula "Boiler-Treat"
  (Liquid) G6
Spartan Formula "HCW"
  (Liquid) G7
Spartan Formula "Oxy-Treat-
  D" (Powder) G6
Spartan Formula "Phos-Treat-
  K" (Liquid) G6
Spartan Formula "Phos-Treat"
  (Liquid) G6
Spartan Formula "Sludge-
  Treat" (Liquid) G6
Spartan Formula "Spartamine"
  (Liquid) G6
Spartan Formula "Tower-
  Treat" (Liquid) G7

Spec-Chem Inc.
Special Formula #79-208
  Kitchen-Kleen A4
Special Formula #79-230
  Phosphatize A3
Special Formula #79-293 Spec-
  Foam A1

Specialized Industrial Products
DX 500 .... A4

Specialty Applied Chemicals
Descale A3
Lectrik K1
Liquify L1
Nothing A4
Spot-Shot C2

Spectral Chemical Co. Inc.
Disinfectant-Sanitizer-
  Deodorant D2
Lemon Scented Detergent-
  Disinfectant A4
Spectral Pine Disinfectant C1
Spectral Poultry Scald N1
Spectrasol A1
Spectraterge A A1

Spectro Chemical Specialties
Big "G" A1
Big Gee A4
Cocoanut Oil Hand Soap
  Concentrate E1
Gentle 7 .... C1
Hydrolytic Enzyme Bacteria
  Complex L2
HHH Freeze-Kleen A5
KP Kitchen Power A4
KP Kitchen Renovator A1
Mr. Brite Porcelain Cleaner C2
Remov-Ox A3
Steam Cleaner Liquid A2
Super 67 .... A1

Spectro-Chem Unlimited, Inc.
Aquacide F1
Big "G" A1
Black Disinfectant (Phenol
  Coef. 2) C1
Black Disinfectant (Phenol
  Coef. 3) C1
Black Disinfectant (Phenol
  Coef. 5) C1
Cocoanut Oil Hand Soap
  Concentrate E1
Dual-27 .... D1
Dura-Trol C1
Fogging Spray Concentrate F1
Gentle 7 .... C1
Hydrolytic Enzyme Bacteria
  Complex L2
HI-P Contact Cleaner K2
Kill Brand Insecticide F1
Kleen-Hands E4
KP Kitchen Renovator A1
L-Tox Spray F1
Mint Disinfectant C1
Mr. Brite Poreclain Cleaner C2
No Sweat P1
Oven-Brite Oven Cleaner A8
Remuv-Ox A3
Safety Solvent K1
Slide 'N Glide H1
Sta-Kleen H1
Steam Cleaner Liquid A2
Steel-Brite A7
Super 67 .... A1
Whirlpool L1

Spectrowax Corporation
#20 Liquid Hand Soap E1
Germ-Aside D2
Grime-Aside D1
Kontrol D1
Meat Room Degrsr Gen Purp
  Cleaner A1
Quat 99-L A4,C1
Spectrocide D2
Star Blue Label A1

Spencer Kellogg Division
Diamond Quality Castor Oil H1
Gold Bond Castor Oil H1

Spinelly's Chemical Products
Vaporizer Insect Spray F1

Spiro-Wallach Company, Inc.
No. 150 Power Cleaner &
  Degreaser A1

Spray "Cleen" International
Spray Cleen A4,A8

Spray-Dyne Corporation
Dyne-A-Mite L1

Sprayon Products, Inc.
No. 210 Food Grade Silicone H1
Penetrating Lubricant H2
PGR Paint & Gasket Remover .. C3
2002 T.F. Elect. Contact &
  Tape Head Cl K2
214 Paintable Mold Release
  Aerosol H1

Sprayway, Incorporated
C-60 Solvent Cleaner & De-
  Greaser K1
D-100 Disinfectant Spray C1
Glass Cleaner C1

Hi-Pressure 55 Industrial
  Aerosol Bomb F1
No. 940 Spray Lubricant H1
No. 946 Silicone Spray H1
Silicone Spray H1

Springfield Water Conditioning Co.,Inc.
Hy-Test Sodium Hypochlorite... D2,G4

Spruce Chemical Inc.
Dura 530 .... A4

Spruce Industries
"Blast" A1
Power X A4

Spur-Tex Products
1006 Spur-Tex Liq. Hand Dish
  Wash A1
1055 Spur-Tex Liq. Steam Clnr
  Conc. A1
1085 Spur-Tex Acid Drain Trap
  Opnr. L1
1090 Spur-Tex In. Liq. Dr. Pipe
  Op/Mnt. L1
330 Spur-Tex Machine DshWsh
  Comp A1
335-5 Institutional Deep Fat
  Fryer Clnr A8
341 Spur-Tex Spm Low Suds
  Laundry Det B1
345 Spur-Tex Concentrated
  Dry Bleach B1
350 Spur-Tex Delx Pwd Hand
  Dsh Wsh A1
366 Spur-Tex Gen. Purp.
  Pwdrd Clnr. A1
375-X40 Spur-Tex High Suds
  Hand Dsh Det A1
405 Spur-Tex Steam Clng.
  Compnd. A1
410 Spur-Tex H.D. Stm Clng
  Comp. A2
805 Spur-Tex Sanitizer Cleaner.. D1
810-1 Sanitizer Rinse D2
911 Spur-Tex Hvy Dty Mach.
  Scrubber Conc A1
915 Spur-Tex General Purp.
  Cleaner A1
920 Spur-Tex Hvy Dty Gen.
  Purp. Cl. : A1
925 Spur-Tex Ammntd All
  Purp. Cl. A4
940 Spur-Tex Bowl Cleaner C2
945 Spur-Tex Alk. Dgrs. &
  Oven Cln.Conc. A8
950 Spur-Tex Extra Heavy
  Duty Degreaser A4,A8
955 Spur-Tex Ready To Use
  Glass Spray C1

Squibb, E.R. & Sons, Inc.
IoQuat D1
Mineral Oil H1
Odorquell C1
Pitch Pack Tag M F3

St. Clair Custodial Supply Co.
S. C. All Purpose Cleaner A1
S. C. Chlorinated Cleaner A1
S. C. H.D. Caustic Cleaner A2
S. C. H.D. Cleaner C1
S. C. H.D. Liquid Cleaner A1
S. C. P. Cleaner A1
S. C. Steam Cleaner A1
Silicone Spray H1
Silicone Spray H1
Soil Off A4

St. George Sales & Service, Inc.

Dragon Brand Assault	A4
Dragon Brand Attack	A4
Dragon Brand Blue Flash	A4
Dragon Brand Blue Knight	A4
Dragon Brand Challenge	A4
Dragon Brand Conqueror	A4
Dragon Brand Freezer & Locker Cleaner	A5
Dragon Brand Gallant	A4
Dragon Brand High Foam "A"	A1
Dragon Brand Lo Foam	A4
Dragon Brand Neutral Cleaner	A1
Dragon Brand Warrior	A4

St. Lucie Chemical & Supply, Inc.

Dysolv	A1

Sta-Lube, Incorporated

Hand Cleaner Formula 11	E4
High Pressure FM Grease	H2
Industrial Pink Hand Cleaner	E4
Kleer FM Grease No. 1	H1
Multi-Purpose FM Grease No. 1	H2
Multi-Purpose FM Grease No. 2	H2
Sta-Lube Lotion Hand Cleaner	E4
Sta-Lube 10-42 Compressor Oil	H2
Superwhite FM Grease No. 0	H2
Superwhite FM Grease No. 1	H2
Superwhite FM Grease No. 2	H2
Superwhite FM-HT Grease No. 2	H1

Stabilization Chemicals

TLC-25	L2

Staci Chemical Inc.

Satellite	A1

Staco Industries

Food Grade Silicone Spray	H1
Prilled Drain Pipe Opener	L1

Staley, A.E. Manufacturing Co.

Sno-Bol	C2

Stan's Janitorial Supply

The Big One	A1

Stanbio Laboratory Inc.

Bio-Clean Laboratory Detergent And Deo.	A1
Bio-Clean-II	D1

Standard Chemical

Sani-Lube	H1

Standard Chemicals, Inc.

Boiler Water Treat. Control #GB-60	G6
Boiler Water Treat. Control #RL-300	G6
Control N2-SL	G7

Standard Disinfectant Company

Nu-Lite	A1

Standard Dry Wall Products

Acryl 60	P1

Standard Industrial and Automotive Equip., Inc.

M-P Multi Purp Steam & Pressure Cl Cmpd	A1
XHD Extra Hvy Dty Steam & Press Cl Cmpd	A1

Standard Maintenance Supply Co. Inc.

Disinfect Plus	D1
Double Power Plus	A1
Erase-All	A1
Insect Spray	F1
Power Plus	A4
Standard Silicone Spray	H1

Standard Oil Co. of California

Chevron Cylinder Oil 135PX	H2
Chevron Cylinder Oil 155PX	H2
Chevron Cylinder Oil 190PX	H2
Chevron CT Oil No. 70	H1
Chevron CT Oil No. 95	H1
Chevron Delo 100 Motor Oil SAE 10W	H2
Chevron Delo 100 Motor Oil SAE 20W-20	H2
Chevron Delo 100 Motor Oil SAE 30	H2
Chevron Delo 100 Motor Oil SAE 40	H2
Chevron Delo 100 Motor Oil SAE 50	H2
Chevron Delo 200 Motor Oil SAE 10W	H2
Chevron Delo 200 Motor Oil SAE 10W-30	H2
Chevron Delo 200 Motor Oil SAE 20W-20	H2
Chevron Delo 200 Motor Oil SAE 20W-40	H2
Chevron Delo 200 Motor Oil SAE 40	H2
Chevron Delo 200 Motor Oil SAE 50	H2
Chevron EP Hydraulic Oil MV	H2
Chevron EP Hydraulic Oil 32	H2
Chevron EP Hydraulic Oil 46	H2
Chevron EP Hydraulic Oil 68	H2
Chevron EP Industrial Oil 100X	H2
Chevron EP Industrial Oil 150X	H2
Chevron EP Industrial Oil 220X	H2
Chevron EP Industrial Oil 46X	H2
Chevron FM Grease 0	H1
Chevron FM Grease 000	H1
Chevron FM Grease 1	H1
Chevron FM Grease 2	H1
Chevron FM Lubricating Oil 19X	H1
Chevron FM Lubricating Oil 22X	H1
Chevron FM Lubricating Oil 80X	H1
Chevron FM Lubricating Oil 9X	H1
Chevron Grease BRB 2	H2
Chevron Industrial Grease Heavy	H2
Chevron Industrial Grease Light	H2
Chevron Industrial Grease Medium	H2
Chevron Inhibited Insulating Oil	H2
Chevron Insulating Oil	H2
Chevron Marine Oil 27X	H2
Chevron Marine Oil 31X	H2
Chevron Marine Oil 38X	H2
Chevron Multi-Motive Grease 0	H2
Chevron Multi-Motive Grease 1	H2
Chevron Multi-Motive Grease 2	H2
Chevron NL Gear Compound 100	H2
Chevron NL Gear Compound 1000	H2

Chevron NL Gear Compound 150	H2
Chevron NL Gear Compound 1500	H2
Chevron NL Gear Compound 220	H2
Chevron NL Gear Compound 2200	H2
Chevron NL Gear Compound 320	H2
Chevron NL Gear Compound 460	H2
Chevron NL Gear Compound 68	H2
Chevron NL Gear Compound 680	H2
Chevron OC Turbine Oil 100	H2
Chevron OC Turbine Oil 150	H2
Chevron OC Turbine Oil 220	H2
Chevron OC Turbine Oil 32	H2
Chevron OC Turbine Oil 46	H2
Chevron OC Turbine Oil 5	H2
Chevron OC Turbine Oil 68	H2
Chevron OC Turbine Oil 7	H2
Chevron Pinion Grease MS	H2
Chevron Refrigeration Oil 32	H2
Chevron Refrigeration Oil 68	H2
Chevron Rust Preventive	P1
Chevron RPM Delo 200 Mtr Oil Gr. 30	H2
Chevron Soluble Oil	H3
Chevron Special Motor Oil 10W	H2
Chevron Special Motor Oil 20W	H2
Chevron Special Motor Oil 30	H2
Chevron Special Motor Oil 40	H2
Chevron Special Motor Oil 50	H2
Chevron SRI Grease 2	H2
Chevron Universal Gear Lub. SAE 80W-90	H2
Chevron Universal Gear Lub. SAE 85W-140	H2
Chevron Universal Gear Lubricant 80	H2
Chevron Utility Grease 1	H2
Chevron Utility Grease 2	H2
Chevron Utility Grease 3	H2
Chevron Utility Grease 4	H2
Chevron White Grease 1	H1
Chevron White Grease 2	H1
Chevron White Grease 3	H1
Chevron White Oil No. 1 NF	H1
Chevron White Oil No. 11 USP	H1
Chevron White Oil No. 15 USP	H1
Chevron White Oil No. 23 USP	H1
Chevron White Oil No. 3 NF	H1
Chevron White Oil No. 5 NF	H1
Chevron White Oil No. 7 NF	H1
Chevron White Oil No. 9 USP	H1

Standard Sanitary Supplies, Inc.

Sanico Deluxe Liquid Hand Soap	E1

Standard T Chemical Company, Inc.

Stanlux Foaming Cleaner-Degreaser	A1
Stanlux Foaming Cleaner-Dgrsr 116-669	A4,A8
Stanlux 512 Disinfectant-Sanitizer	D2

Standardized Sanitation Systems, Inc.

Boro-Guard	C2
Hand-Guard	C2
Heavy Duty Lotion Skin Cleaner	E4
Lano-Guard	C2
Liquid Hand Soap	E1

Puricide C2
R.K. IX.............................. A1
Rust-O-Way B1
RTU................................... E1
Sequa-D A1
Smokehouse Cleaner A2
Solvit A4
Sta-Kleen........................... A1
Star Dry Bleach.................. B1
Star Lo-Foam B1
Star Lo-Foam FE................ B1
Star Plus A1
Star-PK 202....................... K1,K2
Star-PK.200........................ C1
Starcide No. 2 .... F1
Stardyne D2
Starlube............................. H1
Starmaid A4
Starsteel A7
Sterasol D1
Super Starcide.................... F1
SF-105............................... G7
SF-903............................... G6
SF-904............................... G7
SF-913............................... G6
Thrust A1
TR-Cleaner......................... N3

Star Chemical Company, Inc.
Acid Cleaner...................... A3
Chlorinated Cleaner D1
Foamer A1
General Purpose Cleaner........... A1
Heavy Duty Alkaline Cleaner.... A1
Heavy Duty Concentrate
    Liquid Cleaner A4
Hog Scald.......................... N2
Hook and Trolley Cleaner.......... A2
Liquid Hand Soap E1
Liquid Steam Cleaner A1
Malathion Fly Bait F2
Odorless Concrete Cleaner......... A4
Shackle Cleaner A2
Smokehouse Cleaner A2
Star Sterocide A4
Super Insecticide Concentrate.... F1
Thermal Fog Insecticide............ F1
Tripe Cleaner..................... N3

Star Industries, Inc.
Star-Clean GRA A2
Star-Clean GRB.................. A2

Star Laboratories, Inc.
Super Star Silicone Spray........... H1
Super Star White Grease............ H2

Star Maintenance Supply
Star Burst Cleaner Degreaser..... A1

Star Products Company
Star Brand Dishtergent A1

Star Sales Div., LWJ, LTD.
Tri-S................................. G2

Starco Chemical, Incorporated
Dyno.................................. A4
FC-44................................ A4
Gripon A1
Laundry Detergent J............. B1
Liquakleen.......................... A1
Liquid Hand Soap E1
No. 3 N.Y.......................... B1
Starkleen FL-34.................. A1
Starkleen No. S-100............. A2
Starkleen No. S-100 GH.............. A2
Starkleen No. S-85............... A1
Starkleen No. S-90............... A1

Starkleen S-15 .... A1
Starkleen S-20 .... A1
Starkleen S-80 .... A2
Starkleen S-83 CL A2
Starkleen 50........................ A3
Z-14 Starglo....................... A1

Stark, Wetzel & Company, Inc.
S & W Hook & Trolley Cleaner A2
S & W Phosphate Free Cleaner. A2
Stark & Wetzel Foam Cleaner ... A2

Starr Manufacturing & Chemical Company
All Purpose Cleaner.................. A1

Starr National Manufacturing Corp.
Action Market Cleaner A4
Action Safety Solvent A4
Glo Grittle n' Fryer Cleaner...... A2

Stat Enterprises/ A Div. of Pride Labs. Inc.
Advance K2
Alka-Con-13 .... G6
Aqua-Col 10........................ G6
Bacteria Complex 209 .... A4
Big "G" A1
Bowl-O 134 .... C2
D-Clog-QTS. 732 .... L1
Digest L2
Dual-27 .... D1
Jam................................... K1
Jaws L2
Lazy Boy............................ A1
Liquid Sharks Liq. Live Micro
    Organisms........................ L2
Mr. Everything..................... K1
Rid C1
Sani-Lube Food Equipment
    Lubricant...................... H1
Sno-Go C1
Space Shot 212 .... L1
Speed Kleen A4
Sponge 11........................... G6
Spra-Kleen 103 .... A4
Targo 707 .... A4
Tornado.............................. A4
Unclog-QTS. 140.................. L1
Zymeonella 210 .... L2

State Chemical
Klean-Mo Non Butyl Degrease A1

State Chemical & Supply Company
State-Dri............................ J1

State Chemical Company
Big 10............................... C2,B1
BBC.................................. G6
C.F.C. White....................... A4
C.F.C. Yellow...................... C1
Descaler No. 27 .... A3
Dish Brite A1
Dish-L-Powder A1
Dish-O-Det Powder A1
Disol C2
Dry Bleach.......................... B1
Go..................................... L1
GPC................................... A1
Heavy Duty Steam Cleaner A2
Hot Tank Cleaner 202................ A2
LDC................................... A1
Neutral Beads...................... A1
No. 50 FPC Food Plant
    Cleaner......................... A1
Nomal L1
One Wipe A4
Opal Olive Grade A............... E1

Opal Olive Grade AA................E1
Pink Pearl.............................A1
Power Brite............................A1
QS 50...................................D1
Safe.....................................A3
Sanifene...............................A4
Sewer Solv............................L1
Stachlor..............................C2,B1
Staco M.P. No. 1....................B1
Staco Suds............................A1
Staco 3101............................A2
Staco 3102............................A1
Staco 3114...........................A2
Staco 3161............................A1
Staco 3191............................A1
Staco 4000 Chlorinated CIP &
  Spray Clnr...........................A1
Staco 4102............................A1
Staco 4103............................A1
Staco 5100............................A3
Stavac.................................A3
Stazene................................C1
Steam-All..............................A1
Super Cleaner.........................A2
Tri-X-Liquid..........................A1
Tri-X-Powdered.......................A1
Vermononox Aerosol
  Insecticide..........................F1
Vermononox Liquid Insecticide. F1
Vermonox 50% Malathion.........F2
Xodet Concentrate...................A4
1% Vapona.............................F2

State Chemical Manufacturing Co., The

"311 Conveyor Lube"................H2
Acidine................................A3
All Purpose Spray Cleaner..........A4
Away...................................K1
Blue Diamond Cleanser..............A1
Cap.....................................P1
Cherri-D..............................C1
Chloro-Solv Formula.................K1
Cool-Ade...............................G7
Diamond Cleanser....................A1
ELC....................................K2
Fix.....................................C2
Formula C.............................A1
Formula S-800 Drain & Sewer
  Solvent..............................L1
Formula 137...........................A2
Formula 15-7 Oxygen
  Corrosion Control...................G6
Formula 174...........................A1
Formula 185 Liquid Ice-Chek
  Activator............................C1
Formula 222...........................A1
Formula 224 FD-Odor Control . C1
Formula 231...........................A3
Formula 250 De-Foam...............L1,Q5
Formula 273 FFC Terg-O-San
  Cleaner..............................A1
Formula 279 Steam-Off..............A1
Formula 296 State Roach &
  Ant Kil..............................F2
Formula 297 Algaecide..............G7
Formula 298 RAS.....................F2
Formula 312...........................A4
Formula 318 All Purpose
  Cleaner..............................A1
Formula 336 State Gear Oil.......H2
Formula 348 State Stop-Scale ... G6
Formula 350 Liquid
  Chlorinated Cleaner................A1
Formula 351 State Clean-Line .. G6
Formula 358 Mint Kontrol.........C1
Formula 359 Surface Gleam.......C1
Formula 360 Sta-Gel
  Concentrate.........................K1

Formula 362 No Rinse Cleaner/
  Sanitizer............................D2
Formula 555 Rocket Action
  State Ice Cure.......................C1
Formula-251...........................A3
Freez-Kleen...........................A4
G.S.O...................................C1
Gangway...............................A1
GT Cleen-Rite.........................K1
Hands On..............................E1
Hot Soak Tank Cleaner.............A4
Hvy Dty Form. No. 133 Terg-
  O-Cide...............................D1
Instant................................C2
Jonade.................................C2
Kleen All..............................A4
Klenz A................................A1
Klenz-Kream..........................E4
Kontrol................................C1
Lather-Up.............................E1
Lift....................................A4
Lime Solvent..........................A3
Microcide..............................F1
Mud-Out...............................L1
New Form. 160 Car and Truck
  Wash.................................A1
No. MK-50.............................A1
No. 1515-A.............................E1
No. 1516-B.............................E1
No. 1522 Sanico Surgieal Hand
  Soap.................................E1
No. 1532 Sanico Surgical Hand
  Soap.................................E1
No. 1540..............................E1
No. 55 Fogging Mist.................F2
No. 6-55 Fogging Concentrate.. F2
OGL....................................H2
Prestige................................E1
PFC Formula 223.....................A4
PW All Purpose Concentrate.... A4
PW Heavy Duty Degreasant A1
Reserve Cleaner.......................A6
RAS....................................F2
RMK-308..............................F3
RP-3...................................P1
Scat-400 Industrial Stripper......C3
Scram..................................L1
Scrub..................................E4
Sectocide..............................F1
Spec....................................C2
Sta-Zorb...............................J1
Sta-Zyme..............................L2
Stapine................................C1
State Bet..............................P1
State Correct..........................G6
State Dek..............................F1
State Deo..............................C1
State Formula 301 NCNP.........G7
State Formula 319 WAS Wasp
  Killer................................F1
State Formula 345 Mildew
  Stain Remover........................P1
State Formula 52 G.F. Cleaner.. A4
State Formula 64 A Vaporcide
  Brand................................F1
State Lub..............................H1
State MNT.............................C1
State Pen Penetrating Oil H2
State PGR..............................C3
State Sewer Solvent #7.............L1
State SSD..............................K1
Super Eight 80-D.....................A4
Surge..................................C1
SCM-40 Skin Cleaner...............E4
SOK....................................F2
SSC Stainless Steel Cleaner &
  Polish................................A7
SSD Aerosol..........................K1
SSD-55.................................K1
Tef Fluorocarbon.....................H2

Terg-O-Cide...........................A4
Terg-O-Cide Concentrate 315 .... A4,C1
Tile 'N Grout Cleaner...............C2
Touchdown............................A4
U-Ten Pot, Pan And Utensil
  Cleaner..............................A1
Urinal Buoy...........................C2
Urinal-Ade............................C2
Vaporcide..............................F1
X-O Odor Away.......................C1
179....................................C1
180....................................C1
68-A Industrial Cleaner.............A1
999.....................................A4

State Manufacturing Co., Inc., The
Addamine Steam Line
  Treatment 200A......................G7
Addamine Stm Cond. Line
  Treat 200.............................G6
Formula 10-10 Dispersant...........G6
Formula 10-13 Dispersant...........G7
Formula 10-4 Dispersant............G6
Formula 10-5 Dispersant............G6
Formula 10-6 Dispersant............G6
Formula 10-8 Dispersant............G6
Formula 15-2 Oxygen
  Corrosion Contl......................G6
Formula 15-5 Oxygen
  Corrosion Contl......................G6
Formula 17-10 Steam Line
  Treatment............................G6
Formula 17-14 Steam Line
  Treatment............................G6
Formula 17-15 Steam Line
  Treatment............................G6
Formula 17-17 Steam Line
  Treatment............................G6
Formula 17-2 Steam Line
  Treatment............................G6
Formula 17-25 Steam Line
  Treatment............................G7
Formula 17-7 Steam Line
  Treatment............................G6
Formula 17-8 Steam Line
  Treatment............................G6
Formula 19-1 Water Supply
  Treatment............................G7
Formula 19-12 Potable Water
  Supply Treat.........................G7
Formula 19-2 Potable Water
  Supply Treat.........................G7
Formula 4-17 Boiler Water
  Treatment............................G6
Formula 4-2 Boiler Water
  Treatment............................G6
Formula 4-28 Boiler Water
  Treatment............................G6
Formula 4-3 Boiler Water
  Treatment............................G7
Formula 4-30 Boiler Water
  Treatment............................G6
Formula 4-32 Boiler Water
  Treatment............................G6
Formula 4-34 Boiler Water
  Treatment............................G6
Formula 4-35 Boiler Water
  Treatment............................G6
Formula 4-37 Boiler Water
  Treatment............................G6
Formula 4-38 Boiler Water
  Treatment............................G6
Formula 4-40 Boiler Water
  Treatment............................G6
Formula 4-45 Boiler Water
  Treatment............................G6
Formula 4-46 Boiler Water
  Treatment............................G6
Formula 4-47 Boiler Water
  Treatment............................G6

Hose-Off C3,P1
Hy-Foam A1
Hydrosol A3
Hypo-Chlor D2
Hyprochlor X-12 .... D2
Laundry Detergent B1
Manno-Clean A1
Mark 11 .... D2
Mark 13 .... D2
Mechanics Hand Soap C2
MCA Detergent Sanitizer D1
No. 66 Insecticide F1
Nu-Tron A1
Nutra Liquid Hand Soap E1
Odor-Out C1
Pam Activator A3
Pam Detergent A1
Penawet Soft E1
Percleen C2
Pyrosol Super Grade A3
Pyrotex A2
Quik Car Wash A1
Quik Kill Insect Bomb F1
Rack Rat F3
Residual Insecticide Diazinon .. F2
Sanitergent D1
Smokehouse Cleaner-Penawet ... A2
Spray-Det D2
Stannosol Formula M A3
Stannotex A2
Stearns JEB A4
Steramine D1
Stone-Rex A3
Tee-Pak Activator A3
Tee-Pak Detergent A1
Thruton A4
Tripe Cleaner N3
7-11 General Cleaner A1

Steelco Industrial Lubricants, Inc.

Fleet Engine Oil SAE 30 .... H2
6110 Industrial Hydraulic Oil
  SAE 10 .... H2
6120 Industrial Hydraulic Oil
  SAE 20 .... H2
6201 Industrial Air Compressor
  Oil SAE10 .... H2
6202 Industrial Air Compressor
  Oil SAE20 .... H2
6203 Industrial Air Compressor
  Oil SAE30 .... H2
6205 Industrial Air Compressor
  Oil SAE50 .... H2

Steiner Company

Deluxe Skin Cleansing Lotion ... E1
Hygenic Skin Cleansing Lotion . E1
Steiner 200 Creme Hand
  Cleaner E4

Steiner Corporation

Clean All Non Butyl A1

Stepan Chemical Company

Nacconol 40DB A1
Nacconol 40DBX A1
Nacconol 40LB A1
Nacconol 90F A1

Stephenson Chemical Co., Inc.

Ban-Vap 25 Emulsifiable
  Concentrate F2
Diazinon-4E Insecticide F2
Drop Tox Prem. Grade 50%
  Malathion F2
Drop-Tox Aero-Bomb F2
Drop-Tox Potency Spray F1
Durshan 2E Insecticide F2
DDVP Fog Spray Concentrate . F2

DDVP-2 Emulsifiable
  Concentrate F2
DDVP-4 Emulsifiable
  Concentrate F2
DK-11 Emulsifiable
  Concentrate F2
Fly Bait F2
Fog And Mill Spray F1
Formula 382 .... F1
Fumarin Rat Bait Meal F3
Fumarin Rat Bait Pellets F3
Insect Killer F2
Pival Rat Bait Pellets F3
Pyrenone Roach Spray
  Concentrate F1
Pyrethrum and Piperonyl
  Butoxide F1
Pyrethrum Powder F2
Special Fog Spray F1
Stephenson Chem Rozol Redi
  Mix Meal F3
SBP 1382 Fog Compound F1
SBP 1382 O.S. Concentrate F1
Unicron II Fogging Solution ... F1
Warfarin Rat Bait Meal F3
Warfarin Rat Bait Pellets F3

Sterling Chemical Products, Inc.

"131" Liq. Hand Dishwashing
  Detergnt A1
Superior 1000 .... A1
Superior 157 .... A4
Superior 183 Alum. Trailer
  Brightner A3
Superior 23 .... A4
Superior 26 .... Q1,Q4
Superior 269 Sterl-Clean D1
Superior 350 .... L1
25 APC A1
270 Kitchen Multi Purpose
  Degreaser A4
277 Wax Remover A4

Sterling Sanitary Supply Corp.

Eager Beaver Conc. Pressure
  Washer Clnr. A1
Long Life A7
Protect-Aire C2

Sterling Soap and Chemical Co., Inc.

Kleen-Quat D1
Mint Disinfectant C2
Pine Odor Disinfectant C1
Pow'r Plus A1
Start A1
3D Disinfectant, Deodorant,
  Detergnt A4

Sterling Supply

Medi-Sept Hand Soap E1

Stern Chemical Company, Inc.

Assist A1
Big Janitor A4,A8
Freeze Kleen Cleaner A5
Stern's Break-Through A1

Stern Chemical Corporation

#120 Pyrethrins F1
A.C. Cleaner A3
Action D1,Q3
Action Disinfecting Germicidal
  Det. D1
Action Germicidal Spray and
  Wipe Cl. C1
Alltreat 300 .... G7
AP-90 .... A1
Boilerite 100 .... G6
Chlor-N D1

Chlorosan-16 .... D1,Q3
Cleen-Brite A3
Coil-ite P1
CBF-55 .... A2
CIP-250 .... A1
End Con P1
Fantabulus A1
Foamad A1
FAT-10 .... A2
GP-66 .... A1
Hand Soap 20% E1
Hand Soap 40% E1
Jet Det All Purpose Cleaner ... C1
JEC-11 .... A1
JEL-1 .... A2
JEL-7 .... A1
L.D. Cleaner A1
M.P. Concentrate A1
Mark-X K1
Micro-Bac L2
No-Tox Safety Solvent K1
Pine Oil Disinfectant (Phenol
  Coef. 5) C1
Pineaphene C1
PW-50 .... A1
Quixide F1
Red Magic Sewer Solvent L1
S.C. 101 .... A2
Saf-Suds A1
Skat A2
Smoke-X A2
Sparkle Metal Polish A7
Steamrite 200 .... G6
Stern HA Concentrate A4
Stern-O-Lath A1
Stern-O-Lube H2
Sterndex Glass Cleaner C1
SAF-100 .... A1
SD-111 .... A2
T & C Contact Cleaner K2
Tef-Tape Pipe Thread Sealer ... H2
Tower All NC G7
W.T.C. Algaecide And Algal
  Slimicide G7

Stern-Chemtech Corporation

Electro-Ease Safety Solvent ... K1
Koil Kleen Plus P1
Lubezz H1

Stero Chemical Company

Activated Liquid Detergent A1
Amazé Soilgone A2
Electro Charge A2
No. 300 Machine Dishwashing
  Compound A1
Pan Glow A2
Silverware Tarnish Remover ... A3
Stero Chlor A2

Stero Clor Chemical Company

Activated Liquid Detergent A1
Amaze A2
Pan Glow A1
Stero Clor A2

Stero Products

All-Ways-Cleans A1
Dear John Bowl Cleaner C2
Eeze Klean Concentrate A1
Elite Creme E1
Glass & Dish Wash A1
Grease Away A1
Jet Power Cleaner A4
Lemon Aid All Purpose
  Cleaner A4
Lemon Power 10 .... C1
Linen-Brite B1
Liquid Live Micro-Organism ... L2
Pine Concentrate C1

Purple Power Cleaner.................. A3
Stero Glo.................................... C1

Stero-Brite Chemicals

All-Purpose Polish...................... A1
All-Surface Cleaner..................... A1
Concentrated Floor Conditioner C1
Hand Soap Liquid A1
Hand Soap Lotion E1
Heavy Duty Bowl Cleaner........ C2
Heavy Duty Steam Cleaner
Liquid A2
Liquid Drain-Opener And
Maintainer L1
Oven And Grill Cleaner........... K1
Pink Lotion A1
Power Cleaner A4
Surface Scum Remover A4
Wind-O-Shine C1

Stetson, M. D. Company

Stetco Antiseptic Lotion Soap... E1
Stetco Final Disinfectant D1

Stevens Chemical Company

Bowl Glo.................................... C2
Bright Clear............................... C1
Clean Off Concentrate A1
DLT-100.................................... A3
Neutra Scent C1
Quatrocide Germ. Detergent D1
Sta-Gleem A4
Stev-O-Rod L1
Super Drain Eze........................ L1
Superb....................................... A4
Terecide.................................... A4
Terrene A1

Stevenson, Brothers & Co., Inc.

#85 Pure White Mineral Oil H1

Stewart-Hall Chemical Corp.

Acid Cleaner-G A3
Acid Cleaner-H.......................... A3
Acid Cleaner-P A3
Acid Cleaner-S A3
Actabs....................................... G7
Actabs XX:................................. G7
Actflo.. G7
Anti-Foam AF............................ G7
BWT-A...................................... G6
BWT-BC-12 .... G7
BWT-BC-36 .... G7
BWT-BC-6.................................. G7
BWT-BD-11................................ G6
BWT-BD-4.................................. G7
BWT-BD-4F................................ G6
BWT-BD-8.................................. G7
BWT-BD-8F................................ G6
BWT-D....................................... G6
BWT-FG-12................................ G6
BWT-FG-6.................................. G6
BWT-G....................................... G6
BWT-H....................................... G6
BWT-HL..................................... G6
BWT-HM.................................... G6
BWT-HP..................................... G6
BWT-M...................................... G6
BWT-OLA................................... G7
BWT-OLB................................... G7
BWT-OLC................................... G7
BWT-SLF.................................... G6
Corroclean................................. G3
Corroclean F.............................. G3
CT-NF.. G7
CWT-BB-12 .... G7
CWT-BB-25 .... G7
CWT-BC-18 .... G7
CWT-BC-36 .... G7

CWT-BC-6.................................. G7
CWT-BCD-18.............................. G7
CWT-BCD-36.............................. G7
CWT-BCD-6................................ G7
CWT-S....................................... G7
Drainmaster-P............................ L1
ET-BD-14................................... G6
Fortifier F G6
Greasemaster.............................. C1
Grime-Solv................................. A1
Lets-Go H2
Liquid Descalit........................... A3
Outstrip..................................... A3
Scale Strip A3
Scaleclean-F............................... G2
Scalomatic................................. G7
Septi-Solv C1
Silent Run.................................. H2
Steamaster Hi............................. G6
Steamaster Type 1 .... G7
Steamaster-F G6
Sulfite BWT-SF........................... G6
Super Grime-Solv....................... C1
WT-CP....................................... G2
Zipp K2

Stewart-Warner Corporation/ Alemite Marketing Dept.

Alemite All Purpose Spray
Kleen.................................... A1
Alemite Food Machinery H1
Alemite Hi-Temp No. 1............. H2
Alemite Industrial No. 37......... H2
Alemite Industrial No. 38......... H2
Alemite Regular No. 30............. H2
Alemite Regular No. 32............. H2
Alemite Regular No. 33............. H2
Alemite Spray Kleen No. 1........ A1
Alemite Spray Kleen No. 2........ A1
Alemite Spray Kleen No. 3........ A2
Alemite Temprite-White............. H2
Alemite Viscous "H" H2
Alemite Water Pump H2

Stiles-Kem Corporation

Aquadene G1

Stokes Sanitary Supply

S.S.S. 1000 Cleaner A4

Stone Soap Company, Inc.

Ace Laundry Liquid Builder...... B1
Crystal Laundry Liquid Bleach
14%...................................... A1
Great... A4
Heavy Duty Steam Cleaner A1
Kemade General Purpose
Cleaner (H.D.) A1
Kemade Soaper Liquid A4
Medi-Soap E1
Neutral Cleaner A1
Pride Concentrated Powdered
Bleach................................... B1
Pride Concrete Floor Cleaner
Heavy Duty A2
Pride Concrete Floor Cleaner
Reg. Duty.............................. A1
Pride Detergent-Sanitizer D2
Pride Laundry Liquid Fabric
Softener B2
Pride Pressure Wash
Compound............................. A1
Pride-Lo-Dine............................ D2,E3
Star Disinfectant........................ D2
Star Liquid Hand Soap E1
Star Quat D2
Tune Controlled Suds
Detergent.............................. A1,B1
Tune Laundry Liquid
Detergent.............................. B1

Stop Chemical Corporation

Stop 48 Drain Cleaner Enzyme
Type..................................... L2

Stoughton Chemical Sales Co., Inc.

Boiler Water Treatment
Compound............................. G6
Kwick Foam A1
SCS Allbac D1

Stratton Chemicals, Inc.

Dura Solv.................................. K1
Dura 530................................... A4
Dura-Terge II A1

Straub, W. F. & Company

W-T Ten.................................... G2

Strauss Paper Co., Inc.

Power Clean Degreaser
Concentrate A1

Strike Products Division/ Zoecon Industries, Inc.

Strike Insect Strip...................... F2
Strike Rat & Mouse Killer.......... F3

Stucke Chemical Services

Boiler-Treat 100......................... G6
Boiler-Treat 300......................... G7
Boiler-Treat 400......................... G7
Degreaser A1
Degreaser Heavy Duty
Concentrate A1
Lime Remover-Safe Heavy
Duty A3
Lime-Gone Heavy Duty
Concentrate A3
Lube-Jell.................................... H2
Tower-Treat................................ G7

Stutton Corp.

Big K A4
Big R A1
Electra Shield H2
Slipper H1
Straight's................................... H1
Super K A4
Super Solv.................................. K1

Sudbury Laboratory, Inc.

Aqua-Clear (Liquid).................... G3
Aqua-Clear Crystals.................... G3

Suds-Up Company

Floor Cleaner............................. A4
LC-7... A1
S40 Liquid Hand Soap............... E1

Suffolk Chemical Company, Inc.

Sodium Hypochlorite Solution... D2
SB-417....................................... P1

Sugar Beet Products Company

SBS-11 Heavy Duty Veg.-Base
Hand Soap............................ C2
SBS-17....................................... C2
SBS-18....................................... C2
SBS-221..................................... C2
SBS-227..................................... C2
SBS-228..................................... C2
SBS-30....................................... E4
SBS-33....................................... E4
SBS-40....................................... E4
SBS-41....................................... E4
SBS-44....................................... E4
SBS-46....................................... E4

251

SBS-52 Clnr-Disinf. Sanitizer-
   Deod. C1
SBS-60 .... E4
SBS-61 .... E1
SBS-62 Lotion Deodorant Soap E1
SBS-63 .... E4
SBS-65 .... E1
SBS-71 .... E1
SBS-72 .... E1
SBS-76 Sanitizing Lotion Clnsr
   w/Iodine.................................. E2

Suhm Laboratories, Inc.
After Treatment No. 2................ G6
After Treatment No. 2CP.......... G6
After Treatment No. 2CP-1 ...... G6
Anti-Foam No. B4...................... G7
Boiler Water Treat. 21M-P........ G6
Boiler Water Treatment No. 26. G6
BY620 .... G6
Cooling System Conditioner
   Dry...................................... G7
Cooling System Conditioner
   Liq...................................... G7
Cooling System Treatment........ G7
De-Ion Boiler Water Treatment G6
Ex-C(4) G7
EX-A (1) G7
EX-B (2) After Treatment.......... G6
EX-C1 Cooling System
   Treatment............................ G7
EX-C3.._. G6
F-10 Domestic Water
   Treatment............................ G1
F-10-S Domestic Water
   Treatment............................ G2
Floc Conditioner G6
Hot Water Boiler Treatment...... G6
Hydro-Ox Dry Non-Volatile..... G7
NO-ALG.. G7
Poly-Floc...................................... G6
Poly-Treat G6
Res-Kleen P1
S-9-C-20 Condensate Treatment G6
S-9-D-10 Condensate Treatment G6
S-9-D-20 Condensate Treatment G6
S-9-D-30 Condensate Treatment G6
S-9-M-20 Condensate Treatment G6
Sequion G6
Suhm BCY221 .... G6
Suhm BC116................................ G6
Suhm BC211 .... G6
Suhm BC2114 .... G6
Suhm BC213 .... G6
Suhm BC2134 .... G6
Suhm BC216................................ G6
Suhm BC2164 .... G6
Suhm BC316................................ G6
Suhm BC531 .... G6
Suhm BC5314 .... G6
Suhm BE300 .... G6
Suhm BF600 .... G6
Suhm BF650 .... G6
Suhm BPC289.............................. G6
Suhm BPC389.............................. G6
Suhm BPY222.............................. G6
Suhm BP21P4 .... G6
Suhm BP278 .... G6
Suhm BS444 .... G6
Suhm DWP610 .... G2
Suhm HBP010 .... G6
Suhm KR604................................ P1
Suhm RC920................................ G6
Suhm RD910................................ G6
Suhm RD920................................ G6
Suhm RD930................................ G6
Suhm RM920................................ G6
Suhm TA800 .... G7
Suhm TP888 .... G7
Suhm TY730 .... G6

Suhm TY740 .... G7
Supreme Dry G6
Supreme Liquid G6
21M-1 Dry Boiler Water
   Treatment............................ G6
21M-1 Liq' Boiler Water Treat.. G6
21M-3 Dry Boiler Water
   Treatment............................ G6
21M-3 Liq. Boiler Water Treat.. G6
21M-6 Dry Boiler Water
   Treatment............................ G6
21M-6 Liq. Boiler Water Treat.. G6

Sullbrook Service, Incorporated
Formula 101 .... G6
Formula 105-Pulv....................... G6
Formula 109.................................. G6
Formula 156 Pellets.................... G1
Formula 55.................................... G7
Formula 68-Pulv....................... G6,G1
Formula 88.................................... G6
Formula 89.................................... G6
Formula 91.................................... G6
Formula 92.................................... G6
Formula 95.................................... G6

Sumico Lubricant Co. Ltd.
White Alcom Grease.................... H1

Summit Chemical Company
Canner's Special F1
Fogging Insecticide F2
Insect Fogger F1
Mistocide...................................... F1
Mistocide-D F1
Mosquitocide................................ F2
Permacide F2
Permacide Plus F2
Pyrethrins Fogging Concentrate
   II.. F1
Pyrethrins ULV Fogging
   Concentrate F1
Summacide.................................... F1
Super Permacide.......................... F2
SD-5 Detergent Concentrate..... A1

Summit Laboratories, Inc.
Acid Descaler A3
Chlor-Al...................................... A1,N4
Dry Chlorine Bleach.................... B1
H.D. Drain Opener L1
H.D.S. A2
H.D.S. + ... A2
Heavy Duty Cleaner-S A2
High Alkaline Liquid Cleaner.... A1
Laundry Break.............................. B1
Laundry Detergent...................... B1
Liquid S.A.C................................ A1
M.D.P. A1,B1
Q-Foam.. A2
Sum-Brite A3
Sum-J.A.. A1
Summit Foam Additive A1
Summit 295 .... A3,G7
Summit 422 .... G7
Summit 430 .... G6
Summit 455 .... G6
Summit 466 .... G6
Summitt 288 .... G7
Super Laundry Soft/Sour.......... B2

Sun Chem Industries
Enzymes L2

Sun Core Inc.
Sun-Safe.. A1

Sun Oil Company
Sun-Occident Cylinder 11200..... H2

Suniso 3GS.................................... H2
Suniso 4GS.................................... H2
Sunoco Food Mach.Grease
   0240-00/138-625 .... H1
Sunoco Food Machinery
   Grease #2.............................. H1
Sunvis 706 .... H2
Sunvis 747 .... H2
Sunvis 754 .... H2
Sunvis 790 .... H2
Sunvis 9112 .... H2
Sunvis 999 .... H2

Sun Petroleum Products Co.
Crystosol NF 70 .... H1
Crystosol NF 85 .... H1
Crystosol TW 200 .... H1
Crystosol TW110........................ H1
Crystosol TW70........................... H1
Crystosol TW85........................... H1
Crystosol U.S.P. 200 .... H1

Sun Ray Chemical Company
Liquid Steam Cleaner #100........ A1
Liquid Steam Cleaner #50 .... A1

Sun Sanitary Supplies, Inc.
Sun's Excel Conc. Soil &
   Grease Emulsf...................... A1

Sunbeam Products, Inc.
Sunbeam Conquer C1,A8

Sunburst Chemicals/ A Div. of H. V.
   Smith Company
Chlorinated Liq Auto. Dshwsh
   Stn Control.......................... A2
Heavy Duty Kitchen Degreaser A4
Liquid Auto. Dshwsh #202
   Stain Control........................ A2
Skoop Multi-Purpose Cleaner A4
2020 Automatic Bar Glass
   Wash A1
303 Stain Release........................ A1

Sungro Chemicals, Inc.
Numb Bug F1
Sanitiz-It D2
Sun-Bugger #2 .... F1
Sun-Bugger #4 .... F1

Sunnyside Products, Inc.
All-Weather Motor Oil SAE
   10W-40.................................. H2
Compressor Oil Sunex 158 .... H2
Compressor Oil Sunex 159 .... H2
Dripless Oil Sunex 122............... H2
Dripless Oil Sunex 123............... H2
Gear Oil Sunex 160 .... H2
Gear Oil Sunex 161 .... H2
Gear Oil Sunex 162 .... H2
Gear Oil Sunex 163 .... H2
Gear Oil Sunex 164 .... H2
Gear Oil Sunex 165 .... H2
Hydraulic Oil Dynol 363 .... H2
Hydraulic Oil Dynol 364............ H2
Hydraulic Oil Dynol 365............ H2
Hydraulic Oil Dynol 366............ H2
Industrial Lubri. Oil Sunex 143.. H2
Industrial Lubri. Oil Sunex 144.. H2
Industrial Lubri. Oil Sunex 145.. H2
Industrial Lubri. Oil Sunex 146.. H2
Industrial Lubri. Oil Sunex 147.. H2
Industrial Lubri. Oil Sunex 148.. H2
Industrial Lubricating Oil
   Symco 300............................ H2
Industrial Lubricating Oil
   Symco 301............................ H2
Industrial Lubricating Oil
   Symco 302............................ H2

Industrial Lubricating Oil
  Symco 303 .... H2
Industrial Lubricating Oil
  Symco 304 .... H2
Industrial Lubricating Oil
  Symco 305 .... H2
Industrial Lubricating Oil
  Symco 306 .... H2
Industrial Lubricating Oil
  Symco 308 .... H2
Industrial Lubricating Oil
  Symco 309 .... H2
Industrial Lubricating Oil
  Symco 314 .... H2
Parafine Oil H2
Premium Motor Oil SAE 10W .. H2
Premium Motor Oil SAE 20W .. H2
Premium Motor Oil SAE 30 ...... H2
Premium Motor Oil SAE 40 ..... H2
Premium Motor Oil SAE 50 ...... H2
Spindle Oil Sunex 138 .... H2
Spindle Oil Sunex 139 .... H2
Spindle Oil Sunex 140 .... H2
Vanishing Oil Sunex.................... H2
Way Oil Sunex 156...................... H2
Way Oil Sunex 157...................... H2
Way Oil Sunex 174...................... H2
Way Oil Sunex 175...................... H2

Sunrise Supermarket Services, Inc.
Anti Bacterial Hand Soap........... E2
Freezer Cleaner A1
Heavy Duty Degreaser Cleaner . A1
P.S. Plus D1
Sani-Clean-Plus A1

Sunset Products Corp.
Medi Creme E1

Sunshine Chemical Specialties, Inc.
Acid Cleaner-G A3
Acid Cleaner-H........................... A3
Acid Cleaner-P A3
Acid Cleaner-S A3
Actabs... G7
Actabs XX................................... G7
Alumna Brite A3
Anti-Foam AF............................. G7
Atomic Power C1
Atomic Power Flake.................... C1
BWT-A.. G6
BWT-BC-12 .... G7
BWT-BC-36 .... G7
BWT-BC-6................................... G7
BWT-BD-11................................. G6
BWT-BD-4................................... G7
BWT-BD-8................................... G7
BWT-G.. G6
BWT-H.. G6
BWT-HL...................................... G6
BWT-HM..................................... G6
BWT-HP...................................... G6
BWT-M.. G6
BWT-OLA................................... G7
BWT-OLB.................................... G7
BWT-OLC.................................... G6
BWT-SLF..................................... G6
Chain & Cable Lubricant............. H2
Clout Liquid Drain Opener L1
Coil Kleer Hygroscopic Coil
  Cleaner.................................... A1
Corroclean.................................. G3
Corroclean F............................... G3
Crazy Clean C1
Crazy Clean Plus......................... C2
CT-NF... G7
CWT-BB-12 .... G7
CWT-BB-25 .... G7
CWT-BCD-18.............................. G7
CWT-BCD-36.............................. G7

CWT-BCD-6................................ G7
CWT-S G7
Dri-Moly Lubricant H2
Far-N-Away Hand
  Dishwashing Compound........ A1
Fathom Surface Renovator A1
Flash Proof.................................. A1
Flush L1
Foamy... A6
Fortifier F G6
Genie Multi-Purpose Liquid
  Cleaner.................................... A1
Genie Multi-Purpose Liquid
  Cleaner.................................... A4
Giant Mist F2
Greasemaster............................... C1
Grime-Solv.................................. A1
Handi-Clean Waterless Hand
  Cleaner.................................... E4
Haunt.. F2
Industrial Hand Cleaner............. E1
King of Steal............................... A7
Knock Out................................... C2
Lets-Go H2
Lift Station Cleaner L1
Liquid Descalit A3
Liquid Hand Soap E1
Look Out..................................... C1
Meat Head A1
Merlin A3
Mr. Drain L1
No-Sweat Pipe Insulation P1
NBC... A1
Odor Free.................................... D1
Odorid Activated Chemical
  Granules.................................. C1
Outstrip...................................... A3
Pipe-Tite Pipe Thread Sealer H2
Power House Concentrate........... A1
Proficient Freezer Cleaner A5
Protect Skin Shield Cream E4
RF 1127 Rinse Free Non
  Alkaline Flr Cln...................... A4
RF 910 Powdered Hand Soap ... C2
S-100 Disinfectant Deodorant ... C2
Sassy Porcelain and Bowl
  Cleaner.................................... C2
Sassy Porcelain Cleaner.............. C2
Scale Strip A3
Scaleclean-F G2
Scalomatic................................... G7
Scrubby Controlled Sudsing
  Floor Cleaner A1
Septi-Solv.................................... C1
Silent Run................................... H2
Silly Kone Silicone Lubricant.... H1
Slymex Conncentrated Grease
  Trap Solvent C1
Snappy.. C1
Spray Cling Multi-Use Cleaner
  Degreaser C1
Sprint Deodorant-Disnf.-All
  Purp. Clnr............................... C2
Steam Go A1
Steamaster Type 1 .... G7
Steamaster-F G6
Sulfite BWT-SF........................... G6
Sun Jet.. K1
Sun Solv Solvent Degreaser....... K1
Sun Zorb J1
Sun-Clean Concentrate A1
Sunzyme-L................................... L2
Super Cleaner A4
Super Grime-Solv....................... C1
Super Wipe Out.......................... A8
Tronic Contact Cleaner K2
TFE Release Fluorocarbon
  Lubricant................................. H2
W.H.C. Waterless Hand
  Cleanr...................................... E4

White Grease Lithium Base
  Lubricant................................. H2
Wipe Out..................................... A8
WT-CP... G2
12 Volts Battery Terminal Clnr.
  & Protc. P1

Super Chem Corp.
Alk-Super A2
AC Super A3
B R Super.................................... A2
Bravo E1
Deli-Brite.................................... A1
Deli Super................................... A1
Deli-Exxtra.................................. A1
Detail A1
Flite-XL H2
Glade-Super A1
LFI Super A3
Proven A4
Super Clor................................... A1
Super Clor X............................... A2
Super Spray A2
Super XXX A3
Super-Brite.................................. A4
747... A3

Superior Chemical Corporation
Disheen A1
F.E.L.. H1
H.D. Liq. Steam Kleen............... A1
Insect-Terror............................... F1
Polysil Silicone Spray H1
Quat-Am 100/50........................ D1
Special Solv-10 .... A4
Special Solv-50 .... A4
Stainless Beauty-Brite Polish...... A7
Steel-Klene Stainless Steel
  Cleaner.................................... A7
Touch-Up All Purpose Cleaner . C1
TW-5.. A1

Superior Chemical Products, Inc.
Banzol... F2
Deofectant................................... D1
Diazinon 4E F2
Diazinon 4S................................. F2
Durshan 2E F2
Food Plant Spray F1
Iodifectant.................................. D2
Omnicide..................................... F2
Omnicide Industrial Aerosol F2
Omnicide Oil Concentrate
  #3610...................................... F1
Omnicide Pet Special Bomb...... F1
Omnicide VP-5............................ F1
Romix Special............................. F3
Ronnel 24E F2
Saniwash..................................... D1
Saniwash Crystals....................... D1
Super Omni-Dust........................ F2
Super Roach Spray Concentrate
  3610.. F1
Superior EC5 Malathion
  Concentrate F2
Termifectant................................ D1
10-1 Concentrate......................... F2
60-6 O.F. Concentrate................. F2

Superior Chemicals/ Div. of K.I.O. Dairy Supply
San-A-Tize D2
Super Clean................................. A1

Superior Chemicals, Inc.
Sure-Shine A7

Superior Cleaning Supply Co.
New Non Butly Emulsifier......... A1

253

Superior Industrial Products Corp.
Drain-Away L1
Sani-Lube Food Equipment
  Lubricant H1

Superior Industrial Sales
Triple 7 .... A1

Superior Industries
Fogging Spray Concentrate F1
Kleen Waterless Hand Cleaner .. E4
Kleen-It C1
OP-20 .... H1
Safety-Solvent K1
Shield E4
TEF-Tape H2
TFE Release Fluorocarbon
  Lubricant H2

Superior Lubricant Co., Inc.
Industrial & Maríne Cleaner
  (Regular) A1
Super Concentrate Cleaner A1

Superior Manufacturing Corp.
#25 APC A1
Superior ."23" A4
Superior "26" Q1,Q4
Superior 157 .... A4
Superior 183 .... A3
Superior 269 Sterl-Clean D1
Superior 350 .... L1
270 Kitchen Multipurpose
  Degreaser A4
277 Wax Remover A4

Superior Oil & Belting Co.
Blandol White Mineral Oil H1
Carnation White Mineral Oil H1
Kaydol White Mineral Oil USP
  Heavy H1
Pure White Oil H1

Superior Sales Co., Inc.
All Purpose Concentrate A4
Big Job G7
Big Red Concentrate A1
Bora-Soap C2
Brite Creme Cleanser C2
Chloro Flo C1
Compactorcide Spray F1
Concentrated Wax Stripper A4
E-Z Kleen E4
General Purpose Aqueous
  Insecticide F1
Germicidal Cleaner C1
Hard Surface Cleaner A4
Heavy Duty Cleaner-
  Degreaser-Stripper A1
Heavy Duty Degreaser A4
Heavy Duty Steam Cleaner A1
Hefty E4
Liquid Hand Soap E1
Liquid Hand Soap-30% E1
Lotion Soft Dish Detergent A1
Lotion Soft Lotionized Liq
  Hand Dishwash A1
Low Pressure Boiler Water
  Compound G6
Luxury Pink Lotion Handsoap .. E1
Mint Disinfectant C1
Minute Clean 450 .... A4
Minute Clean 460 .... C1
Minute Clean 470 .... C1
Minute Clean 480 .... A4
Neutral Cleaner I A4
Neutral Cleaner II A4
Neutral Cleaner 3 .... A4
Odor Control Deodorizer C1

Pine Oil Disinfectant Coef. 5 ..... C1
Power Plus Prilled Drain
  Opener L1
Quat Disinfectant and Cleaner-
  II D1
Residual Insect. with Pyrn-
  Diazinon Liq F2
Residual Roach Liquid F2
Smoke-House Cleaner A8
Speed Kleen A1
Speed Kleen All Purpose
  Cleaner A4
Super Fresh Hospl Disinf Spray
  Deod C2
Super Spray A4
Super Spray Heavy Duty Clnr-
  Dgrsr-Strip. A1
Sure Kill Insect Killer F2
Swish Bowl Cleaner C2
Valor Insect Spray F1
Wax Stripper #2 .... A4
Whisk Away Germicidal
  Cleaner & Deo. C1
Window Cleaner Spray C1

Superior Solvents Corporation
Sup-R-Klean A4
Sup-R-Strip A4
Sup-R-Triple X A4

Superior Supply Co., Inc.
Break-Down A1
Charge A4
L-5 Lotion Hand Cleaner E4

Superior Supply, Inc.
Superior Solvent Cleaner De-
  Greaser K1

Supply/Service Inc.
Coco-Clean Conc40%
  PureCocontOilHnd Soap E1
Derma Sep Q Anti-Bacterial
  Liquid Soap E1
F-Q.A.C. A1
F-Q.A.C. II A1
Q.A.C. A1

Supreme Consultants LTD.
Sup-R-Kleen A4
Sup-R-Trate A1

Supreme Industries
Super 1001 .... A4
711 P A1

Supro Chemical & Supply Co., Inc.
Acid Cleaner A3
All Purpose Liquid Cleaner A1
Begone Residual Insect Spray F2
Big Power C1
H D Degreaser A1
HDSC A2
Spray & Wipe Ammmoniated
  All Purp. Cln. C1
Supro Chlor Clean A1
Supro Det Formula #55 .... A4
Vaporizer No. 375 Insect Spray F1
Vaporizer No. 500 Insect Spray F1

Surco Products, Incorporated
Odomaster-"Kleen" A1

Sure Chemical Corporation
SCC-1001 Super Pipeline
  Cleaner A1
SCC-1002 Deluxe Pipeline
  Cleaner A1
SCC-200F Disinfectant D2

SCC-2000 General Manual
  Cleaner A1
SCC-2002 Non-Chlorinated
  Manual Cleaner A1
SCC-2020 Egg Washing
  Compound Q1
SCC-300 Liquid Pipeline
  Cleaner A1
SCC-400 Acid Cleaner A3
SCC-444 Heavy Duty Acid
  Cleaner A3

Sure Products
Sure Air Freshener Mint C1
Sure All Purpose A1
Sure All Purpose Polish A7
Sure Concentrated Floor
  Conditioner C1
Sure Dish Pan A1
Sure Disinfectant Spray C1
Sure Drain Opener L1
Sure Emulsion Bowl Cleaner C2
Sure Glass Cleaner C1
Sure Hand Cleaner E1
Sure Ice Away C1
Sure Lotion Hand Soap E1
Sure Oven Cleaner K1
Sure Ox A4
Sure Petro Degreaser C1
Sure Renovator A4
Sure Roach & Ant Kill F2
Sure Safety Solvent K2
Sure Silicone H1
Sure Spray & Wipe A4
Sure Stainless Steel Polish A7
Sure Steam Cleaner A2
Sure Tile & Grout C1

Surpass Chemical Company, Inc.
All Purpose Cleaner L-1P A4,A8
All Purpose Cleaner L-1Y A4,A8
All Purpose Cleaner O-5P A1
Surchlor Sodium Hypochlorite
  Solution D2

Surtec, Inc.
HC-150 .... A4
Surfacide Formula HQ-715 .... D2
SC-210 .... A4

Sutter, A. Ltd.
Taski Profi A4

Swamp Fox Associates
Aci-Det C2
Big-Bull A1
Blue Butyl Cleaner A4,A8
Bowl Kleen C2
Foxy Blue Concentrate A4
Golden Gator C1
Golden Pine C1
H-D 321 Concentrate A4
Lift A4
Red Devil Concrete Cleaner
  H.D. A1
Superior Concrete Cleaner A2

Swamp Fox Chemical & Mfg. Co.
"21" Concentrate A1
"747" HD Jumbo Cleaner A1
Aci-Det C2
All Purpose Cleaner A1
All-Kleen HD A1
Applaud A2
Big-Bull A1
Blue Beads A1
Blue Butyl Cleaner A4,A8
Boni-Fide A8
Bowl Kleen C2

Break-Away A4,A8
Butylful Blue A4,A8
Extra Hi Concentrate A1
Foxy Blue Concentrate A4
Golden Gator C1
Golden Pine C1
Grime Scat S/D-G A4
H-D 321 Concentrate A4
Lift A4;A8
Pink Beads A1
Red.Devil Concrete Cleaner A1
Scale & Slime Remover A1
Steam Cleaner "Medium Duty" . A1
Steam-All Steam Concentrate A1
Superior Concrete Cleaner A2
Tile And Grout Cleaner C2
Total Break A4
White Beads A1
X-Stream ...,.............................. A4,A8

Swan-Mor, Inc.
Super Propel A4,A8

Swift & Company
CIP Kleener A2
CS Kleener A1
Detergent 463 .... A1
Diazinon Roach Spray F2
Dry Chlor Sanitizer D1,Q4
Flexo Powder A1,B1
Germicide Concentrate D1
Hercules Concentrate A4
Hy Score Acid Detergent A3
Institutional Area Spray F1
Institutional Mist Spray F1
Kan Klen Acid Detergent A3
Keystone China Kleen A1
Keystone Cleaner 447A A1
Keystone S.S. Acid Cleaner A3
Morgo Kleener A2
Pan Kleener A2
Seoco Cleaner A4
Smokehouse Kleener A1
Solar HD A1,B1
Solar Neutral A1
Solar 25 .... A1
Sunbrite Laundry Bleach...:....... B1
Trolley Kleener A2
10902 Solar Lube A1
5GE Malathion F2

Swift Chemical & Supplies, Inc.
D.G.D. D1
Kleeneze................................... A4
Spray On Wipe Off C1
Swift Cleaner A1
2125... D2

Swift Chemical Co.
C2M... A1
Kleeneze................................. A4,A8
Lime-Sol.................................... A4
Liquid Hand Soap .:.................... E1
Sarge.. A1
Swift Cleaner A4

Syn D Corporation
Dirtz-Off.................................... E1

Syndet Products, Incorporated
All Klean A2
ALD A1
Cleaner and Degreaser................ A1
Cyclone A2
Fleet-Kleen................................. A1
Jet Klean A1
Kwik-Kleen................................ A4
Liquid Acid Cleaner A3
Power Wash................................ A1

Powerhouse................................ A2
Pre-Soak A1
Smokehouse Cleaner A2
Top Gloss Concentrate................ A1

Syntech Products Corp.
Synquat..................................... D1
Titanic...................................... A4

Synthetic Engineered Lubricants Company
SG-800 Food Machinery
  Grease AA ....;.. H1

Synthetic Laboratories, Inc.
Acid Tile & Wall Cleaner............ A4
All Purpose Cleaner 1................. A4
Cleaner/Degreaser 1-A............... A4,A8
Cleaner/Degreaser 1-B A1
Cleaner/Degreaser 1-C............... A1
Cleaner/Degreaser 1-D............... A1
Cleaner/Degreaser 1-E A1
Cleaner/Degreaser 1-F A4,A8
Cleaner/Degreaser 2 .... A4,A8
CD-IES A4,A8
CD-1 High Foam A1
D-X Chemical Cleaner................ A3
Drain-Away L1
FPD A1
General Purpose Cleaner #1...... A4
General Purpose Cleaner #3....... A4
General Purpose Cleaner #65..... A1
General Purpose Cleaner #7...... A1
GC Super 10 .... D2
GC Super 30.............................. D1
GC Super 40.............................. A4
GC 20....................................... D1
Liquid Hand Dish Compound
  1-A A1
Low Temp Auto Scrub.............. A5
LF 1 Low Foam......................... A1
NBC All Purpose Cleaner A4,A8
PW 3... A4,A8
Steam Cleaner...:...................... A1
2-F Cleaner/Degreaser :............. A4,A8

Sysco Corporation
Action Suds Low Suds
  Laundry Detergent B1
Air Freshner & Sanitizer
  Bouquet................................. C1
Air Freshner & Sanitizer Mint... C1
Ant & Roach Killer.................... F2
Bowl and Porcelain Cleaner....... C2
Chlorocide................................ D1
Disinfectant Deodorant............. C2
Germicidal Foam Cleaner C1
Hi-Suds Heavy Duty Detergent
  Beads..................................... A1
Hi-Temp Mechanical Dishwash
  Cmpd Chlor A1
Insect Killer.............................. F1
Iosan Detergent Sanitizer D2,E3
Lime Solv Liquid Delimer A3
Liquid Germicidal Detergent..... C1
Pine Oil Disinfectant C1
Quat Disinfectant Cleaner A4
Satin Sheen Stainless Steel Cln
  & Polish................................ A7
Scour Power A6
Spark-Ling Floor And Wall
  Cleaner.................................. C1
Stainless Steel Cleaner & Polish. A7
Super Hi-Temp Mech Dish.
  Cmpd H.D. Chlor................... A1
Supreme Suds Hand Dish. &
  Gen Clng Beads..................... A1
Ultra Temp Liquid Mech. Dish.
  Cmpd Chlor A1

T & R Chemicals Inc.
T&R-100 All Purpose Cleaner ... A4

T & W Chemical Co., Inc.
Ind-Hold:................................... K1

T and L Associates, Inc.
Dri-Mol H2
Grip Dressing For V-Flat-
  Round Belts............................ P1
Invisible Skin Shield Cream E4
Kleen Waterless Hand Cleaner ..E4
Mold Release H1
Spray Silicone Release H1
Wyte-Ease Lithium Base White
  Grease................................... H2

T A S C Manufacturing Company
TASC 11 .... A1

T E K Chemicals, Inc.
Tek-P 17 Med. Duty Steam
  Clnr. Powder A1
Tek-P-16 Heavy Duty Steam
  Clnr.Compd............................ A2

T Z L Laboratories
Dyna-Kleen Heavy Duty A4
Dyna-Kleen Plus A1

T. C. Supply Co., Inc.
Tom Cat A1
Tom Cat Super Concentrate A1

T. R. F. Chemicals
7283 T.R.F. Floor Cleaner A4
7284 T.R.F. All Purpose
  Cleaner.................................. A1
7297 T.R.F. Meat Hook &
  Trolley Cl............................... A2

T.C. Products
Chelate...................................... G6
Condensate Control.................... G6
Hex-O-Phos............................... G6
Oxcon #1 .... G6
P-M Control Liquid G6
P-M Control Powder G6
Polyhos...................................... G6

Tabco Industrial Products, Inc.
Bowl Cleaner C2
Tabco All Purpose Cleaner A4
Tabco Brite A1
Tabco No. 2............................... D1
Tabeo Safety Solvent................. A1
Tabco Six Delux Hand Soap...... E1

Tad-Co, Incorporated
Rustrol 102L.............................. G6

Tamaras Supply Company
Liquid Soap............................... E1

Tamco Chemical
Formula #139 .... G6
Formula #50.............................. G6
Formula #70.............................. G6

Tanaka Enterprises, Inc.
Bowl Cleaner Cling Type........... C2
Hard Surface Cleaner................. A4,A8
Industrial And Institutional
  Reodorant............................... C1
Liquid Hand Soap J0 .... E1

Pot And Pan Cleaner A1
Soap Film Remover A4,C1
Tile & Grout Cleaner C2

Tanglefoot Company, The
Difuso F1

Target Chemical Company
SBP-1382 Ready to Use F1

Target Chemical Company
Target Safe A1
Target-Cide D1
Target-Foam A4

Target Industries, Inc.
E-Z Kleen A4
Gentle A4

Targosz, E. & Company
Industrial Detergent Disinf.
Concentrate D2

Tari Industries, Limited
Ter-If-Ic Grime Buster A4

Tarpon Chemical & Supply Co., Inc.
Tarpon Brand Egg Cleaner Q1
Tarpon Brand Hydrolux Q1

Tartan Chemicals, Inc.
Compleat A3
Detail A2
Double Duty 'F' A2
Elbow Grease 'F' A4
Encore A1
Hi-Chlor 'F' A2
Hold A1
Lime Gon A3
Oro-Clean A1
Pro-Tex A3
Sentry A1
Tan-O-Chlor A1
Tan-O-Jet A2
Tan-O-Lac A3
Tan-O-Pan A1
Top-Flite A2
Transform A2
Truck Stop A1
Truck Wash A1
Wite-Brite A1

Tartan Supply, Inc.
Freeze-Clean A5
Soil Solv C1
Spray Wash A4,A8
Suds Up A4
Super Strip A4

Taskmaster Products/ Manufacturing Chemist
Break-Thru A1
Combo Metallo-Ceramic
Resurfacer C2

Tastee Freez International
Ammoniated All Purpose
Cleaner A4
Institutional Liq. Drn Pipe
Opnr. & Maint. L1
Liquid Hand Dish Wash A1

Taurus-Bauer & Associates, Inc.
Acid Free Ice Melter C1
Enzymes L2
Foam-Away A1
Heavy Duty Cleaner A4
Liqui-Drain Opener L1

Liquid Scouring Creme C1
Low Temp. Heavy Duty
Cleaner A5
Oven Cleaner A8
Penetryn L1
Safety Solvent Cleaner &
Degreaser K1
Taurus-Bauer No.601 Odor
Control Grnls. C1

Taverner & Fricke
Butchers 500. .... A1
Packinghouse 155B A1
Saniseptic. E1

Tayco Industrial Products
Aqua Solv A4
Aqua Solv A1

Tayloe Paper Company
Cardinal Concentrated Cleaner .. A4
Clear Lemon Disinfectant C2
Green Kleen A1
Liquid Hand Soap Ready for
Use E1
LD-32 .... A1
Ocide 200 .... C2
Pure Pine Oil Disinfectant C1
Solvs-It A1
Stripper Number 4 .... A4
Super Bowl C2
Taylo-Pine C1
Thrifty Mint Disinfectant C2
Thrifty Pine C1
Winda Shine C1
40% Coco Castile Hand Soap ... E1

Taylor & Sledd, Inc.
Pocahontas Antiseptic Liquid
Hand Soap E1
Pocahontas General Purpose
Cleaner Dgrsr A8
Pocahontas Laundry Detergent . B1
Pocahontas Liquid Oven &
Grill Cleaner A8

Taylor Made Products, Inc.
Glass Cleaner TAC-38 .... C1

Taylor Parts & Supply Co., Inc.
Taylor's Misti A1
Taylor's Regie A1
Taylor's Sudzee A1

Team Associates
B.W.T.-1260 .... G6
Dispersant-1800 .... G6
R.L.T.-1350 .... G6

Team Laboratory Chemical Corp.
350 Multiple Purp Boiler &
Feedwtr Treat. G6
501 Chlorinated Machine Dish
Wash Cmpd. A1

Tech Spray
Blue Shower K2
Color Rid Ox K2
Instant FD Zero Residue
Cleaner K2
Instant Kleen-All K2
Instant Stripper K2
Kleen-It K2
Minus 62 Instant Chiller K2
RX. K2
Slic K2
Zap It Light H2

Tech-Lube Corp.
Conveyor Lubricant H2
Conveyor Lubricant (Greasless) H2
T-NT 100-2500 S.U.S. H1
TG-NT-AAl H1

Technical Industrial Products Co., Inc
Tipco MPC A1

Technical Petroleum Company
Technical Lubricant 285 .... H1
Technical Lubricant 401 .... H1
Technical White Oil 75 T H1
Technical White Oil 75-NF H1
Technical White Oil 85-T H1

Technical Service Division Environmental Maintenance, Inc.
Serv-All Circulator A1
Serv-All Degreaser K1
Serv-All HD Steam Cleaner A1
Serv-All Multi-Circulator A2
Serv-All Steam Cleaner A1
Serv-All Steam Cleaner
Concentrate A1
Serv-All Tank Freshener C1
Serv-All Truck Cleaner C1
Serv-All Truck Cleaner-L A1

Technical Specialties Corp.
TSC 801. .... G6
TSC 802. .... G6
TSC 809. .... G6
TSR #1 .... G6

Tecna Corporation
#2 Colloid Supplement G7
Liquid Special Phosphate G6
Return Line Treatment G6

Tee & Putt Company
Big "G" A1
Chemtronic. A1
GPC-24 .... A4
Thrift Detergent A4
6/49 Concentrate A4

Tek Chemical
Tek H-600 Hospital
Disinfectant & Cl D1
Tek 102 All Purpose Cleaner A4
Tek 143 Pressure Wash A1
Tek 191 Liquid Floor
Degreaser A4
Tek 404 Liquid Detergent
Concentrate A1
161 Heavy Duty Steam Cleaner
Liquid A1

Tek-Trol Chemicals
LD-1 Laundry Detergent B1

Telechem Corporation, The
Break Down A1
Break Down Foaming Cleaner .. A1
Mighty-Clean A1
Teleshine A7
21 Guns A1

Temco Chemical Company
Hi-Test K2
Safety Solvent K1
Sani-Lube H1
Seal-Ease H2
Silicone Spray H1
Stainless Steel And Metal
Polish A7

256

Tempo Chemical Company, Inc.

Aquatone Form Liquid AC....... G6
Greasgo A1
Odorless Insecticide Spray F1

Tennant Company

Tennant 330 Germicidal
  Cleaner.. A4
601 Industrial Grease Remover.. A1
602 Heavy Duty Cleaner........... A1
602-A Heavy Duty Cleaner....... A1
622 Detergent Cleaner A4
6240 Detergent Cleaner A4
651 Cleaner & De-Greaser A4
652 Heavy Duty Cleaner A4
657 Cleaning Solution Medium
  Duty A1
658 SRS Cleaning Solution
  Heavy Duty A4
660 Detergent Cleaner A4
670 Cleaner A4

Tenneco Chemicals

Anderol 495 .... H2
Anderol 497 .... H2
Anderol 500 .... H2

Teoc Industries, Inc.

CB-No. 1 Conc All-Purpose
  Liquid Cl A1
Hi-Lo Alkaline Cleaner A1
Melt-Away Industrial Skin
  Cleaner...................................... E1
Multi-Purpose Cleaner A4
Teoc 500 "The Workhorse"
  Conc Cl-Dgrs...................... A4,A8
Teoc 5000 Heavy-Duty Cleaner A4,A8
Teoc 55 General-Purpose
  Cleaner................................ A4,A8
Uniclean The Liquid Total
  Cleaner...................................... A1

Termac Chemical Company

Tackle A1
Termac Quad 11 Detergent
  Cleaner Disinf. D1

Terminate Control Corp.

Freezer-Cleaner A5
Hard Surface Cleaner.................. A4
Synthetic Cleaner A1

Terminix/Div. of Cook Industries, Inc.

Aero-Term Indust. Insecticide ... F1
Bruce Terminix Pro. Rat Kill F3
GP-Mal Concentrate F2
Inside Residual Dust F2
MFG Concentrate F1
Terminix C4 Concentrate F2
Terminix No-Vex Fogicide 300. F1

Terr-Hunt Company

Fast Klean Concentrate A3

Terra Sphere Chemical, Inc.

Ces-Tic 365 .... C1
TSC Crystals................................ G6
Vulcan Boiler Chemical
  Adjunct #200........................... G6
Vulcan Boiler Chemical
  Adjunct #201........................... G6
Vulcan Boiler Chemical
  Adjunct #202........................... G6
Vulcan Boiler Chemical
  Adjunct #205........................... G6
Vulcan Boiler Chemical
  Adjunct #5............................... G6

Vulcan Boiler Chemical
  Adjunct #500........................... G6
Vulcan Boiler Chemical
  Adjunct #707........................... G6
Vulcan Boiler Chemical
  Adjunct #99............................. G6
Vulcan Boiler Chemical LP G6
Vulcan Steam Line Chemical..... G6
XC-S Boiler Chemical G6
XC-S Tower Chemical G7

Tesch Chemical Co., Inc.

Speedee San D2

Texaco, Incorporated

Aircraft Engine Oil 80 .... H2
Aircraft Hydraulic Oil BB......... H2
Alcaid Oil 60............................... H2
Aleph Oil H2
Algol Oil 100 .... H2
Almag Oil.................................... H2
Anser Oil H2
Bearing Oil S H2
Canopus Oil H2
Canopus Oil 46 .... H2
Canopus Oil 68 .... H2
Capella Machine Oil 32.............. H2
Capella Oil AA Waxfree H2
Capella Oil C Waxfree H2
Capella Oil I 32 .... H2
Capella Oil WF 100.................... H2
Capella Oil WF 22 .... H2
Capella Oil WF 32 .... H2
Capella Oil WF 68 .... H2
Capella 6291................................ H2
Castor Machine Oil Heavy......... H2
Cavis Mineral Cylinder Oil H2
Cepheus Oil 68............................ H2
Cetus Oil 46 .... H2
Corvus Oil 13.............................. H2
Crater A H2
Crater O H2
Crater 1 .... H2
Crater 2 .... H2
Crater 2X Fluid H2
Crater 3 .... H2
Crater 5X Fluid H2
Cup Grease 0 .... H2
Cup Grease 1 .... H2
Cup Grease 2 .... H2
Cup Grease 3 .... H2
Cup Grease 4 .... H2
Cup Grease 5 .... H2
Extra Cutting Oil........................ P1
Geotex HD 30 .... H2
Geotex LA-30.............................. H2
Havoline Motor Oil SAE 10W .. H2
Havoline Motor Oil SAE 20-
  20W... H2
Havoline Motor Oil SAE 30 ..... H2
Havoline Motor Oil SAE 40 ..... H2
Havoline Motor Oil SAE 50 ..... H2
Havoline Motor Oil 10W-30...... H2
Havoline Motor Oil 20W-40....... H2
High Temp Grease....................... H2
Home Lubricant H2
Honor Cylinder Oil 680.............. H2
Hydra Oil 19 .... H2
Leader Valve Oil 680.................. H2
Libra Oil...................................... H2
Low Temp Oil............................. H2
Marfak Heavy Duty 2................. H2
Marfak Heavy Duty 3................. H2
Marfak MP2................................. H2
Marfak 0...................................... H2
Marfak 00.................................... H2
Marfak 1...................................... H2
Marfak 3...................................... H2
Marfak 5...................................... H2
Marine Engine Oil Special.......... H2

Molytex Grease 0 .... H2
Molytex Grease 2 .... H2
Mulifak 2 .... H2
Multigear Lubricant EP 75W.... H2
Multigear Lubricant EP 80W.... H2
Multigear Lubricant EP 85W-
  140... H2
Multigear Lubricant EP 85W-
  90... H2
Nabob Oil.................................... H2
Novatex 0..................................... H2
Novatex 1 .... H2
Novatex 2..................................... H2
Outboard Motor Oil SAE 30 ..... H2
Pinnacle Cylinder Oil 680.......... H2
Rando Oil AA H2
Rando Oil G H2
Rando Oil HD 32 .... H2
Rando Oil HD 46 .... H2
Rando Oil HD 68 .... H2
Rando Oil 150.............................. H2
Rando Oil 32................................ H2
Rando Oil 46................................ H2
Rando Oil 68................................ H2
Regal AFB 2................................ H2
Regal Oil A.................................. H2
Regal Oil B.................................. H2
Regal Oil B R&O H2
Regal Oil C R&O H2
Regal Oil E R&O H2
Regal Oil F R&O H2
Regal Oil H.................................. H2
Regal Oil J R&O H2
Regal Oil K R&O......................... H2
Regal Oil PC R&O...................... H2
Regal Oil PE R&O....................... H2
Regal Oil R&O 220 .... H2
Regal Oil R&O 320 .... H2
Regal Oil 320 .... H2
Regal Oil 390 .... H2
Regal Oil 460 .... H2
Regal Starfak 2............................ H2
Rock Drill Lube XH 320............. H2
Rock Drill Lube 100.................... H2
Rock Drill Lube 46...................... H2
Rust Proof Compound H............. P1
Rust Proof Compound L.............. P1
Rust Proof Compound LB P1
Rust Proof Compound Spray...... P1
Soluble Oil C H3
Soluble Oil CX H3
Soluble Oil D P1
Soluble Oil D............................... P1
Soluble Oil HW P1
Spindle Oil B............................... H2
Spintex Oil 60 .... H2
Startex Antifreeze Coolant P1
Startex Antifreeze Coolant P1
Startex Antifreeze Coolant J P1
Startex Antifreeze Coolant PF... P1
Stazon H....................................... H2
Stazon L....................................... H2
Stazon M...................................... H2
Sultex D P1
Sultex 320.................................... P1
Summitlube 1 .... H2
Taurak 200 .... H2
Texaco Antifreeze Coolant......... P1
Texaco Motor Oil SAE 10W H2
Texaco Motor Oil SAE 20-20W H2
Texaco Motor Oil SAE 30 .... H2
Texaco Motor Oil SAE 40 .... H2
Texaco Motor Oil SAE 50 .... H2
Texatherm 46 .... H2
Texclad 1...................................... H2
Texclad 2...................................... H2
Texol E... H2
Texol K H2
Thermatex EP 1 .... H2
Thermatex 000 .... H2

Threadtex H2
Thuban 140 .... H2
Thuban 90 .... H2
Track Roll Lubricant Heavy H2
Track Roll Lubricant Light H2
Transformer Oil 55 .... H2
Transformer Oil 55 Inhibited ي... H2
Transformer Oil 55 Inhibited W H2
Transformer S H2
Transultex 240 .... P1
Universal Gear Lube EP 140 ..... H2
Universal Gear Lube EP 80 .... H2
Universal Gear Lube EP 90 .... H2
Ursa Oil Extra Duty SAE 10W . H2
Ursa Oil Extra Duty SAE 30 .... H2
Ursa Oil Extra Duty SAE 50 .... H2
Ursa Oil Heavy Duty SAE 20-
  20W H2
Ursa Oil Heavy Duty SAE 30 ... H2
Ursa Oil P-100 .... H2
Ursa Oil P-150 .... H2
Ursa Oil P-68 .... H2
Ursa Oil S-3 SAE 20-20W H2
Ursa Oil S-3 SAE 30 ...'.......... H2
Ursa Oil 150 .... H2
Ursa Oil 50 .... H2
Valor Motor Oil SAE 20............ H2
Valor Motor Oil SAE 30............ H2
Valor Motor Oil SAE 40............ H2
Vanguard Cylinder Oil 460 .... H2
Vega Graphite Grease H2
Water Pump Grease H2
Way Lubricant G H2
White Needle Oil 22.................. H2
650 T Cylinder Oil 1000 .... H2
650 T Mineral Cylinder Oil H2

Texas Chemical Engineering Co.
Tex-Chem P-1 .... G6
Tex-Chem P-2............................ G6
Tex-Chem P-3............................ G6
Tex-Chem P-5............................ G6
Tex-Chem P-6............................ G6
Tex-Chem P-7............................ G6

Texas Industrial Chemicals, Inc.
All Brite All Purpose Cleaner.... C1
Foodlube Food Equipment
  Lubricant H1
Kleer Brite Glass Cleaner C1
Metal Brite Stnls Steel&Metal
  Polish A7
Moto Kleen K1
T.I.C. Lemon 13 .... C1
T.I.C. Mint C1

Texas Phenothiazine Company
Avi-Dine.................................... D1

Texas Refinery Corporation
#880 Crown and Chassis
  Grease H2
#890 Vari-Purpose Gear Lube
  SAE 75 .... H2
#890 Vari-Purpose Gear Oil
  SAE 140 .... H2
#890 Vari-Purpose Gear Oil
  SAE 90 .... H2
Big Blue.................................... A1
Pure Mineral Trans. Lube. SAE
  90.. H2
Pure Mineral Trans. Lube. SAE
  140...................................... H2
Pure Mineral Trans. Lube. SAE
  80.. H2
Quadra-Klean............................ A1
Sure Universal Gear Lube. Grd.
  140...................................... H2
Sure Universal Gear Lube. Grd.
  80.. H2

Sure Universal Gear Lube. Grd.
  90.. H2
TRC Big Red............................. A1
TRC Food Machinery Grease ... H1
TRC Food Machinery Grease
  A.W. H1
TRC Special FM Liquid Lube... H1
TRC 6440 Universal Torque Oil H2
TRC-AM-1................................ A2
TRC-BOP A2
TRC-Elbo-Grease...................... A4
TRC-Super Big Red................... A1
TRC-SCR.................................. A3

Texas-April Company/ Division of Stange Co.
Carbon Remover........................ A8
Formula TA-3 Alkali REmover A3
Formula TA-4 Hood Cleaner
  and Degreaser A1
Fryer 'n Grill Cleaner................ A8
Hot Oven Cleaner A8
TAC All-Purpose Liquid
  Detergent.............................. A1

Texize Chemicals Company
Fantastik Disinf. Bathroom
  Cleaner.................................. C2
Fantastik Spray Cleaner............. A4
Glass Plus C1
Grease Relief............................. C1,B1,A8
Janitor In A Drum C1
Pine Household Cleaner C1
Spring Scent Cleaner A4

Texo Corporation
ATB.. A3
ATC.. A3
Car Wash 384............................ A1
Carwash AD A1
Carwash 438.............................. A1
Chlorex..................................... D1,Q1,Q4
Crystal Base A1,E1,H2
Floorstrip 12 .... A1
Fogicide.................................... F1
Fogicide XL............................... F1
Keg Wash 24............................. A2
Kleen All................................... A1
Kleen-All P................................ A1
Klorotex................................... A1,Q1
Liquitex A1
LP 837...................................... A1
LP-107A.................................... A1
LP-149A.................................... A1
LP-149B.................................... A2
LP-238...................................... A1
LP-269...................................... A3
LP-270...................................... N2
LP-306...................................... A1,B1
LP-311...................................... A2
LP-313 Additive A1
LP-330...................................... A1
LP-336B.................................... A1
LP-354...................................... A1,Q5
LP-44.. N3,N2
LP-452...................................... B1
LP-469...................................... A1
LP-472...................................... A1
LP-492...................................... A1,Q1
LP-509...................................... A1
LP-540...................................... A1
LP-561...................................... A1
LP-562...................................... H2
LP-568...................................... G6
LP-572...................................... A1
LP-578...................................... A2
LP-579...................................... A2
LP-622...................................... Q5
LP-93.. A1

Odor-Gone A4
Pantex A1
Pontentex.................................. A2
Raise C1
Raise Odorless A4
Rinse-Tex A1
Sani-Kleen D2,Q6
Sanitex A6
Sealtex 120 .... A3
Sealtex 26 .... A3
Septodor.................................... D1,Q6
Silvatex..................................... A1
Tank Sheen A1
Tex AC...................................... A1
Texel A1
Texo Han Sope.......................... E1
Texo Industrial A.P.C................ A1
Texo Kleen................................ A4
Texo LP 134 C.......................... A3
Texo LP 37 .... A3
Texo LP 574 .... P1
Texo LP 656 .... A1
Texo LP 701 B H2
Texo LP 708 .... A2
Texo LP 736 .... A1
Texo LP 742 .... A1
Texo LP 750 .... G2,G5
Texo LP 768 .... A1
Texo LP 770 .... A1
Texo LP 771 .... A1
Texo LP 779 .... A4
Texo LP 788 .... A2
Texo LP 789 .... A4
Texo LP 809 .... A1
Texo LP816 .... A2
Texo Pink Pearl......................... E1
Texo RO.................................... A1
Texo Solv.................................. K1
Texo 1050.................................. A1
Texo 220.................................... A1
Texo 261.................................... A1
Texo 354.................................... A1
Texo 481.................................... A3
Texo 51...................................... A1
Texo 528 A A1
Texo 608.................................... A1
Texo 631.................................... G5
Texo 643.................................... Q1
Texo 647.................................... N1
Texo 666.................................... A1
Texo 696.................................... A2
Texo 82...................................... A1
Texolite 100............................... A1,Q1
Texolite 100 NC A1,Q1
Texolite 189............................... A1,Q1
Texolite 200............................... A1
Texolite 200 K A2
Texolite 300............................... A1
Texolite 584............................... A1
Texolite 93 .... D1
Texomatic.................................. A1
Texope #3.................................. E1
Texope 218................................ A1,H2
Texope 25.................................. A1
Texope 33.................................. A1
Texope 338................................ H2
Texope 474................................ A1,H2
Texope 5.................................... C1
Texope 75.................................. A4
Texphene A4
Texstrip 12 .... A2
Texstrip 40 .... A2
Texstrip 50 .... A4
Wet-Tex.................................... N1
XBC.. A1
1013.. A2
1055.. A1
1060.. A4
1070.. A4
115.. A2

Gold Tiger C1
Grease Off A8
H.D. Blue A4
Hitemp G7
Hydrosteam A2
Hygea Dual-Synergist Insect
   Spray F1
Hygea Germicidal Spray
   Cleaner...................................... C1
Hygea Insect Spray Synthetic
   with Fogger F2
Hygea MCP-100 .... H2
Hygea Power Pac...................... F2
Hygea Residual Roach and Ant
   Killer F2
Hygea S.W.A.T. Total Release
   Fogger F2
Kill-O-Cide................................... F1
Kill-O-Cide Concentrate:............ F1
Kill-O-Cide Special
   Concentrate F1
Kitchen Power.......................... A4,A8
Lacto All Purpose Cleaner......... A1
Low Phos Laundry Detergent... A1,B1
Mint-O-Green C1
Multicide D1
MCP-100 .... H2
Oven Magic A2
Pefko C2
PTC-85 .... C1
Residual Insecticide.................... F2
San-Air C1
Sewer Solv L1
Shine-O-Glass C1
Silicone Release Lubricant H2
Smoke House Cleaner.................. A1
Speed-Zyme L2
Stainless Steel Cleaner A7
Steam Quick A2
Suds-O-Foam A1
Super Foam............................... A4,A8
Super Lube HW-15 .... H2
Super Safety Clean:...................... K2
Surprise Non-Butyl Heavy
   Duty Clnr-Dgrsr....................... A1
ST-25 .... A1
T-C 505.. C2
T-C 505 Q C1
T-J0 Heavy Duty General
   Purpose Cleaner..................... A4,A8
T-15 Concentrate....................... A4,A8
T-5 Heavy Duty All Purpose
   Cleaner...................................... A4
Theolac A3
Thermotron A2
Vat Dip.. A2
15% Glyco Soap E1
20% Glyco Soap E1

Therm Processes, Incorporated
Beads-O-Suds A1
Blitz Concentrate.......................... A1
Master... A2
Trojan.. A2

Thermionics Inc.
TI-147E.. G6
TI-152F.. G6
TI-601F.. G6
TI-602F.. G6
TI-603F.. G6
TI-604F.. G6
TI-605F.. G6
TI-610F.. G6
TI-630F.. G6
TI-631F.. G6
TI-632F.. G6
TI-640F.. G6
TI-641F.. G6
TI-647F.. G6

TI-661F.. G6
TI-800 .... G7,G2
TI-804 .... G7
TI-808 .... G7
TI-810 .... G7
TI-814 .... G7
TI-842 .... G7
TI-843 .... G7
750A.. G6

Thero-Chem
B.T.-39 Concentrate A1

Thomas Enterprises
RA-125 Manual Cleaner A1
RB-124 Cleaner Booster A2
RC-126 Acid Cleaner.................. A3
RFB-121 Foam Booster A1
RQ-109 Hand Cleaner................. E1

Thomas J. Brady Company
Bra-Co All Purpose Cleaner A1
Bra-Co Floor Prime A1
Bra-Co Heavy Duty Finish &
   Wax Stripper A4
Bra-Co Liquid Hand Soap......... E1
Bra-Co Power Kiln Wax Strp
   Hvy-Dty Cl Dgr...................... A1
Bra-I-Cide #101 Disinfect.
   Deterg. Clnr............................. D1
Bra-I-Cide Disinfectant Cleaner. A4
Bra-I-Cide Mint Disinfectant...... C2

Thomas, F. J. Chemical Company
Extra Hvy. Dty.-Powder Press.
   Wshr. Cln. A2
Floor And Wall Cleaner............. A1
Heavy Duty Concrete Floor
   Cleaner...................................... C1
Heavy Duty Drain Cleaner........ L1
Heavy Duty Machine Pan
   Wash A1
Medium Duty All Purpose
   Cleaner...................................... A1
Medium Duty-Liquid Pressure
   Washer Clnt.............................. A1
Medium Duty-Powder Pressure
   Washer Clnr. A1
Special All Purpose Cleaner...... A1

Thompson-Hayward Chemical Co.
Calcium Hypochlorite................. D2
Malathion E-5 .... F2
Muriatic Acid 20 degrees BE.... A3
Pyrtox Contact Spray F1
Pyrtox Fogging Spray F1
Silver Swan Coco 40,................. E1
Sodium Hypochlorite Solution... D1,Q4,G3,B1
T-Det N-12.................................. A1
T-Det N-9.5................................. A1
T-H Silicone Antifoam AF-
   10FG Q5
T-H Silicone Antifoam AF-
   100FG Q5
T-H Silicone Antifoam AF-
   30FG Q5

Thompson's Supply Company
Liquid Hand Soap Pink Lotion.. E1

Thompsons Sanitary Supply House, Inc.
Blast-Off A1
Concentrated Bowl Cleaner &
   Disinf. C2
Flying Insect Spray F1
Grease Cutter............................... A1
Lo-Tox Spray F1
Mint Disinfectant........................ C2

259

Pine Oil Disinfectant C1
Slugger Drain Opener L1
Special Cleaner A4
Steam Cleaner A2
Super Non-Butyl Blast-Off A1
Vee Pine Disinfectant C1

Thornton Bros. Paper Co. Inc.
Silicone Spray H2

Thoro Products Company
X-O-X D1

Thoroughbred Chemical Corp.
Insecticide (w/Residual Insect
  Control) F2

Three-M-Supply Company
Bio Cid 101 .... A3
Bio Foam A1
Bio Kleen F-10 .... A1
Bio Kleen F-20 .... A1
Bio Kleen F-30 .... A1
Bio Kleen F-5 .... A1
Bio Kleen HDD A1
Bio Kleen 300 .... A1
Bio Kleen 310 .... A1
Bio Kleen 400 .... A1
Bio Kleen 410 .... A1
Bio Kleen 420 .... A1
Bio Solv A1
Bio Solv 230 .... A1
Bio Solv 240 .... A1
Ferro Guard A2
Meri-Spray A1
Mericide D2
Prodrane L1
Steam It A1
Syntrox C2

Tidewater Chemical Corp.
T-1 General A1
T-2 Heavy Duty A1
Toro Tora A1

Tidewater Sanitary Supply Co.
Cavalier Odorless 20% Liq.
  Hand Soap E1
Hydrolytic Enzyme Bacteria
  Complex L2
Pine Oil Coefi 5 .... C1
Power Concentrate C2
Steam Cleaner Liquid A2
6/49 Concentrate A4

Tidewater Supply Co.
da Antiseptic Waterless Skin
  Cleanser E4
da Antiseptic Wipe-All Lotion ... E4
da Elite Lotion Deodorant
  Soap E1
da Empire Powdered Hand
  Cleaner C2
da Execu-Clean Lotion
  Deodorant Skin Cl.................. E1
da Fortune-Plus Powdered
  Hand Cleaner C2
da Natural Powdered Hand
  Cleaner C2
da Regal Lotion Deodorant
  Skin Cleanser E4

Tifco Industries
Dry Film 8390 Moly Lube H2
Industrial Cleaner Concentrate .. A1
Sani-Lube 7656 Food
  Equipment Lubricant H1
Stainless Steel & Metal 7652
  Spray Plsh. A7

Tifco-SCC Super Cleaner
  Concentrate A1

Tiffany Products
Barracuda Heavy Duty Cleaner
  & Degreaser A4
Butyl Cleaner/Degreaser A4
Cleaner A4
Dio Electric Safety Solvent K1
Emerald 450 .... A4,A8
Inhibited Chemical Cleaner A3
Insecticide Fogger F1
Liquid Drain Pipe Opener L1
Stainless Steel Cleaner A7

Tifton Chemical Company
Tifchem D.D.V.P. Five F2

Tiger Products, Inc.
New No. 1040 Tiger-Power A1
No. 0560 Tiger-San Mint Dsnf.-
  Deo.-Clnr C1
No. 0645 Tiger-Scale All Purp
  Acid Clnr A3
No. 0905 Tiger-Suds A1
No. 2230 Tiger-Dis...................... D2
No.0605 Tiger-Paw Porcln Tile
  , &Marb Clnr C2
Tiger Fang A1

Tim Hennigan Engineering Co.
NLO-119-FDA Boiler Water
  Treatment G6
NLO-119-T Boiler Water
  Treatment G6
1140-T Boiler Water/Cool.
  Water Treat.......................... G6,G7
1785 FDA Boiler Water/Cool.
  Water Treat.......................... G6,G7
2221 L Cooling Water
  Treatment G7
710-T Boiler Water Treatment... G6
716-T Boiler Water Treatment... G6

Time Chemical & Janitor Supply
Timechem Sanisept Disinfectant Q3

Time Chemical, Incorporated
#8535 Carbon Remover A1
#8540 Special Kran-O-Magic
  Clnr. A1
Armour Standard Detergent
  No. 100 .... A1
Armour Standard Detergent
  No. 120 .... A1
Armour Standard No. 101 .... A1
Armour Standard No. 103 .... A2
Armour Standard No. 104 .... A2
Armour Standard No. 105 .... A3
Armour Standard No. 109 .... A3
Armour Standard No. 110 .... A2
Armour Standard No. 112 .... A1
Armour Standard No. 114 .... A1
Armour Standard No. 115 .... N3
Chlor-16....................................... B1
Chlor-9... D1
Chlorinated C1P Cleaner A1
Chlorinated Farm Cleaner.......... A1
Kroger Acid Cleaner A3
Kroger Foam Additive A1
Kroger Gen. Pur. Liquid Clnr... A1
Kroger Heavy Duty Cleaner A2
Kroger Mech. Cleaner................. A1
Kroger Mold Wash Cmpd.......... A2
Laundry Carbonated Alkali........ B1
Laundry Complete Synthetic
  Low-Suds B1
Laundry Liquid Bleach 14%....... B1
Laundry Liquid Builder.............. B1

Laundry Liquid Detergent B1
Laundry Liquid Fabric Softner.. B2
Laundry Powder Fabric
  Softner B2
Morton's Time-Saver Alum.
  Cln & Bright. A3
Morton's Time-Saver H.D.All
  Purp Cl&Dgrs A1
Morton's Time-Saver Liquid-
  Drain Opener L1
Morton's Time-Saver
  Utnsl&Dishware Det.............. A1
Morton's Time-Saver
  Wind&LitDtyGenPurpCl A1
Morton's Timesaver Bowl &
  Porcelain Cl. C2
Morton's Timesaver Odorless
  Sanit. Agent D2
Oven & Grill Cleaner.................. A8
Phos-Bowl.................................... C2
Pink High Active Built Beads.... A1
Power Lube H2
T-800 Liquid Hand Soap E1
T-900 Liq. All Purpose Cleaner. A1
Time Saver Ace Lube 8E........ A1,H2
Time Saver No. 110-E A2
Time-Lo-Dine D2,E3
Time-O-San 1000 .... D1
Time-O-San 545 Acid Sanitizer.. A3,D1
Time-Saver #55 Gen. Cl. &
  Egg Wash............................ A1,Q1
Time-Saver #6748 Truck Wash. A1
Time-Saver #6762 .... Q1
Time-Saver Antiseptic Liq.
  Soap E1
Time-Saver AA A1
Time-Saver AA Chlorinated
  Cleaner A1
Time-Saver AB............................ A1
Time-Saver AC............................ A1
Time-Saver AD............................ A1
Time-Saver AG............................ A1
Time-Saver AJ............................. A1
Time-Saver AP............................ A1
Time-Saver AQ............................ A1
Time-Saver AW........................... A1
Time-Saver AW Chlorinated A1
Time-Saver AW Chlorinated
  Pink Gen. Cl. A1
Time-Saver AX A1
Time-Saver AY A1
Time-Saver Bactericide........... D1,Q3
Time-Saver Blue Dr.Pipe&Sewr
  Opnr L1
Time-Saver Blue Pine Hvy.
  Dty Floor Clnr C1
Time-Saver BB A2
Time-Saver BD............................ A2
Time-Saver BE............................. A2
Time-Saver BF............................. A2
Time-Saver BG............................ A2
Time-Saver BK............................ A2
Time-Saver BN............................ A2
Time-Saver BP............................ A2
Time-Saver BQC#10 .... A2
Time-Saver Campbl#3 Scour.
  Pwdr. A6
Time-Saver Chelate #5.............. A2
Time-Saver Chelate Additive
  No. 4 .... A1
Time-Saver Chelate No. 6.......... A2
Time-Saver Chelate No. 7.......... A2
Time-Saver Chlor Kleen.............. A1
Time-Saver Chlor 16................... D1
Time-Saver Chlorinated
  Cleanser A6
Time-Saver Chlorinated Kleen .. Q1
Time-Saver Chlorinated Mech
  Dshwsh Cmpd A1
Time-Saver CA............................ A6

Time-Mist, Incorporated

Timesaver Products

Tiona Petroleum Company

Titan Chemical Corporation

Trans-Chemco, Inc.

Trans-107 .... Q5

Trans-National Development Co.

TN-101-1-Degreasing Solvent C1
TN-102-A-Low Suds Cleaner A4
TN-103-A-TND Special A4
TN-103-TND Concentrate A4
TN-110-A-Automatic Scrubber
  Soap A4
TN-110-1-Hi Pressure Wash....... A4
TN-175-Sewer Solve L1
TN-243-A-Super Primer &
  Etcher A3
TN-244-Concrete Cleaner &
  Etcher A3
TN-331-Steam Go A1
TN-395-Super Solve L1

Trans/World Chemical Corp.

Clean It-All Purpose Cleaner..... A1
Clean-It.................................... A1
Concentrated Inhibited
  Chemical Cleaner A3
Drain-Flo................................. L1
Emulsifiable Cleaner K1
Extra Strenght Hvy Duty Wtr
  Soluble Cln............................ A1
Flo-Go No. 308 .... C1
General Purpose Liquid
  Cleaner.................................. A1
Germicidal Cleaner C1
Germicidal Cleaner D1
Hand Soap............................... E1
Heavy Duty Organic Drain
  Cleaner.................................. L1
Ice Melting Compound............... C1
JT-650 .... A4
Liquid Hand Soap E1
Liquid Live Bacteria L2
Liquid Solvent Non-Flammable
  Cleaner.................................. A4
Neutral Cleaner A1
Odor Control Granules C1
Organic Acid Detergent A3
Pine Odor Disinfectant C1
Quat Disinfectant...................... D1
Safety Solvent.......................... K1
Sewer Solvent.......................... L1
Super-Kleen A4
Synthetic Cleaner A4
Three-Gly................................. C1
Trans/World Agricultural
  Enzyme.................................. L2
Trans/World Butyl Power
  Cleaner.................................. A4
Trans/World Concrete Cleaner . A4
Trans/World Controlled Suds ... B1
Trans/World Lemon Hand
  Soap E1
Trans/World Odor Control
  Granules C1

Transcontinental, Inc.

Transco All Purpose Cleaner..... A1
Transco Chlorinated Cleaner..... A1
Transco Fast Action Cleaner A4
Transco H. D. Caustic Cleaner.. A2
Transco H. D. Cleaner............... C1
Transco H. D. Liquid Cleaner... A1
Transco Lubricant..................... H2
Transco P. E. Cleaner............... A1
Transco Steam Cleaner A1

Traster, S. Co.

Activated P.W.R. A1
Solv Grand Slam A1

Treasure Chemical

T-426 Silicone Spray................. H1

T-431 Waterless Hand Cleaner .. E4
Treasure T-105 Lotionized
  Hand Soap.............................. E1
Treasure T-121 Organic Acid
  Detergent............................... A3
Treasure T-201 MultPurp
  Boil&Fdwtr Treat.................... G6
Treasure T-202 Low Pressure
  Boiler Comp........................... G6
Treasure T-21 Hospital
  Type(Mint)Dsnf&Cl C1
Treasure T-23 Virucide............. D1
Treasure T-329 Drain Cleaner ... L2
Treasure T-330 Sulfuric Acid
  Drain Opnr............................ L1
Treasure T-332 Liquid
  Bacteria/Enzyme..................... L2
Treasure T-50 Liquid
  Dishwashing Conc.................. A1
Treasure T-625 Foaming
  Cleaner.................................. A1
Treasure T-655 Lemon
  Disinfectant C2

Treasure Coast Paper Co.

Big Jim Grime Fighter............... A4

Trenell Services

Jay's 100 Insect Killer Spray...... F1
Jay's 66 Spray Concentrate
  Insect-Killer F1

Tri Ton Manufacturing Corp.

Adios M.F. F1
Albert Enzyme L2
All Surface Germicidal Spray &
  Wipe Cln................................ C1
B-Tite Belt Dressing P1
Basic 20 Concentrated Liquid
  Hand Soap.............................. E1
Basic 40 Liquid Hand Soap........ E1
Big Daddy................................. A4
Corvex...................................... P1
Davy Flockit Flocculent :........... L1
Delux Hand Dishwashing
  Liquid A1
Dermadrex Proteinized Lotion
  Hand Soap.............................. E1
Ditron Diazinon........................ F2
Double Action A1
Down & Out L1
Draculube Lithium Grease H2
Electron.................................... K1
Excel Lanolized Waterless
  Hand Cleaner E4
Excel Waterless Hand Cleaner... E4
Flow L1
Formula 2000 .... A1
Frankensteam A1
General Purpose Low Sudsing
  Cleaner.................................. C1
Glove Protective Cream............. E4
Hi-P Contact Cleaner................ K2
J.H. 500 .... G6
J.H. 600 .... G6
Lectric Safety Solvent K2
Like New Stainless Steel Polish. A7
Liquid Solvent Non-Flammable
  Cleaner.................................. A4
Loren Green Clean Mint
  Disinfectant C2
Nothin' All Duty Chemical
  Renovator............................... A1
Odor Out Deo-Gran.................. C1
Ovenex..................................... A8
Power Steam............................. A4
Reem Out L1
Royal Flush Sewer Solvent........ L2
Sheen A7
Shield Moisture Barrier H2

Sparkle Glass Cleaner C1
Super Triox A3
Tri Bact L2
Tri-Therm Ice Melting
  Compound.............................. C1
Tri-Vex Deodorizer C1
Tribrite Pot'N Pan
  Concentrated Cleaner.............. A1
Trimax A1
Ultimate Organic Debris
  Dispersant.............................. A4

Tri-Chem Corporation

Big Action................................ B1
Drain-Flow............................... L1
Lectric Kleen K1
Omni Clean A1
Saf-Sol K2
Soft'n Clean.............................. E1
Solv-Away C1
Steam Away A1
Surface Strip C3
Tri-Clean A4

Tri-Chem, Ltd.

Ruff Stuff A1
T.K.O. A1
Tri Breakdown L1
Tuff Stuff................................. A1
TC-24...................................... A4

Tri-Co Equipment Corporation

C2D A4
Flash-Strip C1
Foam-Away A1
Freeze Clean A5
M-38 Cleaner A4
Power Kleen A4,A8
Product #19 Liquid Steam
  Cleaner.................................. A1
Product #21 Stripper/
  Degreaser A4
Speedy Kleen Floor Cleaner...... A4
Tri-Co-Cide Odorless................ D1
Tri-Con Neutral Cleaner........... A1

Tri-County Supply Company

Blast Off A4
HPC... A1

Tri-Flon Company, The

Tri-Flon H2

Tri-Kem Inc.

"Super" Degreaser/Cleaner A1

Tri-Kem Industries, Inc.

Oven Cleaner............................ A8
Remarkable A4
Tri-Bac D2
Tri-Dyne.................................. D2,E3
Tri-Klor................................... D1
Trikelate A2
Tritin A1
Triz A1

Tri-State Chemical Company

Tri-Clean A1
Tri-Crete A1
Tri-Smoke A2
Tri-Test................................... A1
Tri-TW A1

Tri-State Equipment Company

TSE-4 .... A1

Tri-State Manufacturing Inc.

A.C. 57 .... A3
Boost.. A4

263

264

True Chemical Company, Inc.
Tri-Melt C1
True All Purpose Concentrates . A1

True Value Manufcturing Co., Inc.
Manufacturers Concentrate
  Cleaner-Dgrsr. A4,C1

Truett Laboratories/Biochemical Division
Biocatalave (Type PJ) L2
Biocatalave (Type RAH) L2
Biocatalave (Type SIC) L2
Biocatalave Type AE L2
Biocatalave Type CE L2
Biocatalave Type HE L2
Biocatalave Type LE L2
Biocatalave Type PE L2

Trust-X
Liquid Draynamite Drain
  Opener L1

Tryton Construction Corp.
C-100 Foam-Aide A1
C-112 Eagel II A4

Turco Products/ A Division Of Purex Corporation
Aktiv C1
Alkaline Rust Remover A2
Alkaline Rust Remover NF C1
Aqua-Fax A1
Aviation A1
ABC Special A2
Blu-Fax A1
Borzin C1
Botl-Glo A2,N3
Brite Clor A1,Q1
Brite-Fax A2
Chempo C1
Co-Ro-Go A3
Duhl-Brite A1
Dubl-Fax A1
Eaze A1
Econo-Flash A3
Econo-NSF A3
Egg Kleen Q1
Eureka Special A2
Eureka-Special-NP A2
Ferrex B A2
Flash A3
Foamgo Q5
Gem A3
Handi-Brite A1
Handisan C2
Hi-Brite A1
HAC A1,Q1
HAC-NP A1
JC-3 .... A1
Kwik-Solv K1
L.C.A. Q1
Lactop A1
Lactop-NP A1
Line Kleener A1
Liquid Alkaline Rust Remover-
  NF A2
Liquid-Sanilube A1,H2
LCC A2
Meteor A1
Oilift A1
Panguard #2 .... A1
Panguard NC A1
Plaudit A4
Powdered Flash A3,Q2
Powr-Steam A2
Powrinse A1
R.R. #3 .... A1

R.R. #5 .... A1
Ra-Brite A2
Rustgon A2
Saniline A2
Sanitrete A1
Shock A2
Solusoil A1
Sparkleen A1,Q1
Sprayeze LT A1
Stain-Gon Q1,Q2
Steam-Aero A1
Steameze A1
Steamzall A1
Super Ferrex A2
Super Flash A3
Surpass A1
T-1685 .... A4
T-3252 .... A1
T-4331 .... A2
T-5009 .... A2
T-5042 .... A1
T-5450 .... A2
T-5765 .... A1,Q1,Q2
T-5787 .... C1
T-5805 .... A1
T-5865 .... A1
T-5875 .... C1
T-5890 .... A1,H2
T-5921 .... Q1
T-5987 .... A1,H2
Thoral D2,Q3
Tile-O A4
Turbrite #1 .... A1
Turclor D1,Q1,Q2,Q4
Turco Case Cleaner A1
Turco De Fo Mer Q5
Turco Descaler A3
Turco 5668 .... K1
Turco 6012 .... A3
Turco 6162 .... A5
Turco-NSF A3
Turco-Pine C1
Turco-Sudz A1
Turcodine D2,Q6
Turcosan-2 .... D1
Turcosheen A1,Q1
Type C A2
Zeal A1
4819 .... A1
5369 .... A1

Tuzona Mfg., Inc.
Degrease-It A4

Tykor Department/ Klenzade Division
Begone A2
C.I.P. Cleaner A1,Q1
Conveyor Lubricant "IND" H2
Conveyor Lubricant 62 .... H2
D.S. Alkali A2
D.S. Alkali-V-W-4XX A2
Dyna-Solve A3
Dyna-Solve Powder A1
Dynogen D2
Floor Cleaner A4
General Cleaner 30 .... A1
General Cleaner 30CL A1
General Cleaner 40 .... A1
Gunsil A1
Heavy Duty In-Place Cleaner .. A2,Q1
Hi Suds 60 .... A1
Hi Suds 60 CL A1
Klor-Eze D1,Q1,Q4
Mecho-Wash 51 .... A1
MBR Liquid A3
Pan Acid A3
Powdex A2
Prod A1
Quik-Klor D1
Skate-Eze H2

Spray Off...................................A1
Super 88.........................A1,Q1,G1
Tykor Caustic Supplement A1
Tykor Floor Scrub....................A4
Tykor Foam Coat........................A1
Tykor Grease Remover.............A4
Tykor Hand Soap Liquid E1
Tykor Kora-Gard.................D2,Q3
Tykor Mecho Klor.....................A1
Tykor Mecho-Wash 31 ....A1
Tykor Mecho-Wash 31 CL A1,Q1
Tykor Mecho-Wash 4X A1,Q1
Tykor Off..............................A4
Tykor Rin-Klor 50 ....D1
Tykor Sano Grease H1
Tykor Sano-Lube H1
Tykor Soaker Alkali No. 105.....A2
Tykor Soklor A2
Tykor Super Tyroc....................A1
VK C.I.P. Cleaner.....................A1
Zylium.............................D1,Q3

Tysol Products Division
No. 117 ....A2

U

U C C O Chemical Company
Formula 844 ....A1

U.S. Aviex Company
Hand Cleaner..........................E4

U.S. Chemical Corporation
Z-4 Cleaning Compound A1
Z-4 Equipment Cleaning
  Compound.............................A1

U.S. Cleaner Corporation
S-Plus................................A1
Scram.................................A1

U.S. Diamond Chemical Division
USD-980.............................A4,A8

U.S. Polychemical Corporation
Polychem U.S. D401.................K1
Polychem 410 ....A1
ST 21 Poly Spray Jet................A4
ST-21 Poly Spray Jet II............A4

U.S. Professional Laboratories
Accutrol..............................A1
Ekonomeze Liquid Concentrate A4
Foton.................................A1
Grimex #33............................A1
Grimex Supreme......................K1
Grimex 100............................K1
Soaperior.............................E1
Soaperior-Olivena...................E1
Usamine Disinfectant C2
Usamine Spray C2
Usaphene Detergent-
  Disinfectant.......................C2
Usapon...............................D1

Ultra Adhesives, Incorporated
Foam-Ban 119.........................Q5

Ultra Laboratories
Exsol-Chlor..........................D1
U-L Concentrate......................A4
Ultra Ammo Strip C1
Ultra Gem............................A4
Ultra Kleen Quick A4
Ultra 020............................A2

Ultra 113............................A2
Ultra 114............................L1
Ultra 115............................A2
Ultra 130............................A2
Ultra 131............................A2
Ultra 140............................A2
Ultra 220............................N2
Ultra 230............................A1
Ultra 255............................A1
Ultra 270............................A1
Ultra 330............................C2
Ultra 451............................A3
Ultra 510............................A2
Ultra 523............................A1
Ultra 525.........................A1,Q1
Ultra 526............................B1
Ultra 530............................A2
Ultra 550............................A1
Ultra 640............................A4
Ultra 641............................A2
Ultra 642............................A1
Ultra 760............................A1
Ultra 761............................A1
Ultra 853............................A1
Ultra 855............................E1
Ultra 858............................A1
Ultra 862............................A1
Ultra 910............................A3

Ultra-Lube Products Company
E P Food Machinery Grease H1
Ultra Lube Alm. Complx Food
  Mach Grs H1
Ultra Lube EP Alm. Comp.
  Mach. Gr............................H1
Ultra Lube Food Machinery
  Grease..............................H1
Ultra Lube White Oil 350,........H1
White Food Machinery Grease.. H1

Ultrachem, Inc.
Ultrachem Non-Tox Lubricant .. H1
Vischem Non-Tox Grease H1
Vischem Non-Tox Grease II......H1

Uncle Sam Chemical Co., Inc.
#5 Vaporizer Insect Spray........F1
Bombard...............................F1
D-Trans No. 5 Vaporizer Insect
  Spray...............................F1
De-Soil Concentrtd All Purp.
  Detergnt............................A1
Double Strngth Odrls Foramba
  Insect Sp...........................F1
Duzitall Germicidal Cleaner......A4
Duzkleen Liquid Detergent........A1
Formula 12 Wonder...................D1
Formula 50 Wonder Odrls
  Disinf. & Santzr...................D1
Haunt Residual Insect Spray.....F2
Odo-Way Deodorant&Insect
  Spray...............................F2
Odorless Foramba Insect Spray.F1
Proficient Freezer Cleaner A5
Sprayzon Magic Cleaner A1
Super-Solv Emulsion-Degreaser
  Conc................................C1
T K O Detergent Sanitizer.........D2
Up & At'em............................F1
Vaporizer No. 2! Insect Spray....F1
Vaporizer No. 375 Insect Spray F1
Wonder Liquid Soap 15%..........E1
Wonder Liquid Soap 40%..........E1

Uni-Chem of Savannah
A.C. Cleaner..........................A1
Bisodyne.............................D1
Gold Tiger...........................C1
Kill-O-Cide..........................F1
Multicide............................D1

PTC-85 ....C1
Septo Soap E1
Sewer Solvent.........................L1
ST-25................................A1
T-5..................................A4

Uni-Chem Corporation
Action...............................A4
Blanco 76.............................A2
Blué Crete Clean A4
Chlorogent...........................A1
Claro 101.............................A1
Clavel 97.............................A3
Colgar...............................A1
Concentrated Coconut Oil
  Hand Soap...........................E1
Escaldar.............................N1
Industrial Soap Powder A1
Light'n..............................A4
Pink Lotion Hand Soap E1
Soff.................................A1
Special Crete Clean..................C1
UC 49................................L1
UC 747...............................Q1
UC 99................................C2
Verde 100............................A1
White Crete Clean....................A4

Uni-Chem Corporation
Chem-Sol.............................A1
D-Wax................................A4
Derma-Soft 2 ....E1
Dri-Kil..............................F1
Flite-Control........................F1
Germ-A-Pine..........................C1
Ic-A-Wai.............................C1
Instant..............................L1
Kleer................................C1
Minus Plus...........................A5
Quick-Flo............................L1
Sani-Suds............................A4
Steam-Eze............................A1
Tri-Gly..............................C1
UC 25................................K1

Uni-Chem Corporation of Florida
Ammoniated Oven Cleaner A8
Electro-Pure Contact Cleaner....K2
Electro-Sol Safety Solvent K1
Lanolin Waterless Hand
  Cleaner.............................E4
Quick-Off All Purpose Cleaner.. C1
Sani-Lube Food Equipment
  Lubricant...........................H1
Skin Shield Cream....................E4
Super-Dri............................P1
Tef-Seal Pipe Thread Sealer.......H2
TFE Release Fluorocarbon
  Lubricant...........................H2
U-C-10-12 Stainless Steel &
  Metal Polish........................A7

Uni-Crete, Inc.
Uni-Clean............................A1

Uni-Kem International, Inc.
Adios................................A1

Uni-Lab Corp.
"Oxi-Clean"..........................A3
"Power-Sol"..........................A1
"Uni-Jell"...........................K1
Aqua Mist............................F2
Big-U................................C2
Big-U Detergent Concentrate K1
Electric and Mechanical Helper H2
Freezer Clean........................A5
Grease Gun...........................H2
Instant..............................H2

266

Unified Sales & Laboratories, Inc.
Trix Cleaner.................................A1
Tru-Touch Cleaner.......................A4

Unijax, Inc.
All Purpose Cleaner.....................A1
All Star All Purpose Cleaner.....C1
All Star Cleaner Degreaser........A1
All Star Foam Disinfectant
   Cleaner...................................C1
All Star Foamer Foaming
   Degreasing Clnr.....................A1
All Star Hospital Disinfectant
   Deo. 11..................................C1
All Star Klean All.......................A4
All Star Lemon Lime Air
   Sanitizer.................................C1
All Star Mint-A-Dis....................C1
All Star Multi-Purpose
   Insecticide..............................F1
All Star Oven & Grill Cleaner...A8
All Star Pine Odor Disinfectant..C1
All Star Power Punch..................A1
All Star Residual Insect Killer...F2
All Star Stainless Steel Cleaner..A7
All Star Super Power Punch......A1
All Star Vandalism Mark
   Remover.................................K1
B-A..G6
B-FF..G6
B-SF..G6
Barcrobe Germicidal Cleaner.....C2
Blitz Liquid Steam Cleaner
   Concentrate............................A1
Blue Concrete Cleaner................C1
Bole Aire......................................C2
Concentrated Liquid Acid
   Cleaner...................................A3
Concrete Cleaner.........................C1
Corroclean F Water Treatment..G1
Corroclean Water Treatment.....G1
Deep Fat Fryer Cleaner
   Powder....................................A2
Dy-Sul Drain Opener..................L1
Electrical Motor Cleaner............K2
Fly & Insect Spray.......................F1
Fresh And Clean..........................C2
Fryer And Grill Cleaner.............A2
General Purpose Insect
   Killer,.3P.................................F1
Germguard....................................C1
Insect Killer,.6P..........................F1
Multi Purpose Insecticide Spray F1
Multi-Purpose Cleaner................A1
Ocean Blue....................................C1
Oven and Grill Cleaner..............A8
Perfumed Deodorant Blox
   No.16 In Hang-Up...................C2
Perfumed Deodorant Blox
   No.24 In Hang-Up...................C2
Perfumed Deodorant Cakes.......C2
Perfumed-Wired Bowl Blox.......C2
Power Plus....................................L1
Power Punch.................................A4
Power Punch Butyl Cleaner.......A4
Residual Insecticide with
   Baygon....................................F2
Residual Insecticide with
   Diazinon.................................F2
Scaleclean F Water Treatment...G1
Strip-All.......................................C1
Uni-Pine.......................................C1
Unijax Descaler...........................A3
Unijax Drain Opener...................L1
Unijax Foamer.............................A1

Union Carbide Corporation
Sag 710 Silicone Antifoam.........Q5
Sag 730 Silicone Antifoam.........Q5

Ucon Hydraulic Fluid FDC-300 H1
UCON Hydraulic Fluid FDC-
   400..H1

Union Oil Co. of California
Union A Grease 0.........................H2
Union A Grease 1.........................H2
Union A Grease 2.........................H2
Union Aircraft Engine Oil HD-
   100..H2
Union ATF Dexron-II.................H2
Union ATF-Type F......................H2
Union C-3 Fluid 10-W................H2
Union Cable Lube........................H2
Union FM Grease.........................H1
Union FM Grease 2......................H1
Union Gas Engine Oil-HD
   SAE 40....................................H2
Union Guardol Motor Oil SAE
   10W...H2
Union Guardol Motor Oil SAE
   20W/20....................................H2
Union Guardol Motor Oil SAE
   30..H2
Union Guardol Motor Oil SAE
   40..H2
Union H. D. Wheel Bearing
   Grease.....................................H2
Union Hvy Dty Motor Oil
   SAE 10W.................................H2
Union Hvy Dty Motor Oil
   SAE 10W-30............................H2
Union Hvy Dty Motor Oil
   SAE 20W-20............................H2
Union Hvy Dty Motor Oil
   SAE 30....................................H2
Union Hvy Dty Motor Oil
   SAE 40....................................H2
Union Hvy Dty Motor Oil
   SAE 50....................................H2
Union Hydraulic Oil AW ISO
   VG 32......................................H2
Union Hydraulic Oil AW ISO
   VG 46......................................H2
Union Hydraulic Oil AW ISO
   VG 68......................................H2
Union Hydraulic/Tractor Fluid..H2
Union Koolkut..............................P1
Union Koolkut S-8.......................P1
Union Koolkut S-9.......................P1
Union Marok ISO VG 150...........H2
Union Marok ISO VG 220...........H2
Union Marok ISO VG 68.............H2
Union Mineral Gear Lube SAE
   90..H2
Union MP Gear Lube LS SAE
   80W/90....................................H2
Union MP Gear Lube LS SAE
   85W/140..................................H2
Union Premium Motor Oil SAE
   10W...H2
Union Premium Motor Oil SAE
   20W20......................................H2
Union Premium Motor Oil SAE
   30..H2
Union Soluble Oil 10...................P1
Union Soluble Oil 12...................P1
Union Steaval A...........................H2
Union Steaval A 140....................H2
Union Steaval B...........................H2
Union Super Motor Oil SAE
   10W40......................................H2
Union Super Motor Oil SAE
   5W30..H2
Union Turbine Oil ISO VG 100 H2
Union Turbine Oil ISO VG 150 H2
Union Turbine Oil ISO VG 22..H2
Union Turbine Oil ISO VG 220 H2
Union Turbine Oil ISO VG 32..H2
Union Turbine Oil ISO VG 46..H2

Union Turbine Oil ISO VG 460 H2
Union Turbine Oil ISO VG 68 .. H2
Union Turmaco ISO VG 22....... H2
Union Turmaco ISO VG 32....... H2
Union Turmaco ISO VG 46....... H2
Union Turmaco ISO VG 68....... H2
Union Turmaco 465 .... H2
Union Unax AW ISO VG 100 ... H2
Union Unax AW ISO VG 150 ... H2
Union Unax AW ISO VG 220 ... H2
Union Unax AW ISO VG 32.... H2
Union Unax AW ISO VG 320 ... H2
Union Unax AW ISO VG 46..... H2
Union Unax AW ISO VG 68...... H2
Union Unax RX ISO VG 100 .... H2
Union Unax RX ISO VG 150 .... H2
Union Unax RX ISO VG 22...... H2
Union Unax RX ISO VG 220 .... H2
Union Unax RX ISO VG 32...... H2
Union Unax RX ISO VG 46...... H2
Union Unax RX ISO VG 68...... H2
Union Unoba EP-1 Grease H2
Union Unoba EP-2 Grease H2
Union Unoba EP-3 Grease H2
Union Unoba F-1 Grease........... H2
Union Unoba F-2 Grease............. H2
Union Worm Gear Lube 140 ..... H2
Union Worm Gear Lube 250 .... H2
Union Worm Gear Lube 90 .... H2

Unisel Manufacturing Company

Black Out A1
Cold Power K1
Coolmatic P1
Drill L1
E-Lect.. K1
Everything A1
Handy... E1
Heavy Weight............................... A4
Magnet.. P1
Naturally L2
Pit Stop... C1
Rusty... A3
Softy.. A1
Touchdown A1
X-Rated A4

Unisource Industries, Inc.

Caustic-Jell A1
Concrete Cleaner.......................... A1
Deodorant Nuggets....................... C1
Essence of Almond Waterless
  Hand Cleaner E1
Freezit Penetrating Oil................. H2
Gard Lotionized Liquid Soap E1
Gard Lotionized Liquid Soap E1
Process 1 .... A1
Scale-Rid Chemical Cleaner &
  Deoxidizer A3
Sewer-Rod Sewer Solvent L1
Silasan Food Grade Silicone
  Spray... H1
Slip Silicone Spray H1
Solvall.. B1
Sta-Clear Prilled Drain Opener.. L1
Sterile Food Plant Cleaner A1
Swipe.. C2
Thru 22... C1
Triple X Liquid Hand Soap E1
Uni Pine....................................... C1
Uni-Gly Air Sanitizer & Surface
  Disinf. C1
Uni-Kill Non Residual Space
  Spray... F1
Uni-Steam Liquid Steam
  Cleaner...................................... A4
Unicon Concentrate A1
Unisolve 550 Heavy Duty
  Cleaner...................................... A4
Unisolve 990 Degreaser.............. K1

Unizyme L2

Unit Chemical Corporation

Kleen-Quat Q3
U-Chem-Co 3-D D1

United Chemical & Products Corp.

Big "O" A1

United Chemical Company, Inc.

DD Disinfectant D1
Olive Liquid Soap E1
Olive Liquid Soap
  Concentrated E1
Puro Pine Disinfectant................. C1
Q-Nary... D1
Surgasept...................................... E1
U-C Germicidal Soap................... E1
U-C Multi Purpose Spray.......... F2
U-C Olive Liquid Soap................. E1
U-C Pine Odor Disinfectant........ C1

United Chemical Corp. of New Mexico

DP-706... D2
DP-707... D1
IC-7000 .... A1
IC-7010 .... A1
IC-7020 .... A1
IC-7030 .... A1
IC-7040 .... A1
IC-7050 .... A2
IC-7100 .... A2
IC-7110 .... A2
IC-7120 .... A2
IC-7150 .... A1
IC-7155 .... A1
IC-7160 .... A1
IC-7165 .... A1
IC-7170 .... A1
IC-7175 .... A1
IC-7180 .... A1
IC-7200 .... C1
IC-7210 .... C1
IC-7300 .... N3
IC-7310 .... N3
IC-7320 .... N3
IC-7330 .... N3
IC-7400 .... B1
IC-7410 .... B1
IC-7500 .... G6
IC-7510 .... G6
IC-7600 .... L1
IC-7700 .... A3
IC-7710 .... A3
IC-7800 .... E1
IC-7900 .... C1

United Chemical Corp. of Norwood

Detertex AS................................... C1

United Chemical Corporation

Green Magic Heavy Duty
  Cleaner...................................... A1

United Dairymen Of Arizona, The

Dairy Equipment Sanitizer......... D2
Dynemate A1
Foamchek 100............................... A3

United Industrial Sales, Inc.

Alpine-22...................................... A3
AL-109 Waterless Hand
  Cleaner...................................... E4
AL-114 Spray & Wipe Disnft.
  & Deo. Clnr. C1
AL-115 Hospital Disinfectant
  Deodorant C2
AL-130 Van-Off.......................... C1

AL-153 Insecticide Fog With
  Cherry Scent............................. F2
AL-210 Residual Spray
  Insecticide................................. F2
AL-312 Commercial Odor
  Killer C1
AL-67 Liquid Hand Soap........... E1
AL-70 Non-Fuming Chemical
  Cleaner...................................... A4
AL-78 Pearlescent Industrial
  Hand Clnr.................................. E1
AL-82 Heavy Duty Liquid
  Organic Drain Cl...................... L1
AL-83 Body & Hair Shampoo .. E1
AL-88 Automatic Dishwash
  Liquid Conc. A1
ML-110 Silicone Spray H1
ML-19 Liquid Steam Cleaning
  Compound A1

United Industries

Silicone Spray............................... H1

United Industries Corp.

Degreaser II.................................. A4,A8
Degreaser II.................................. A1
Liquid Hand Soap E1
Pot And Pan Cleaner................... A1
Safe Clean A4
Safe Clean Plus............................ A4
Synthetic Cleaner Concentrate .. A4
Tile & Grout Cleaner...................... C2
Un-Scent Cleaner C1
United's A.P. No. 59................... A4
United's Formula E.O.................. A8
United's Formula HS E1
United's Formula No. 671 .... A1
United's Formula No. 711 .... A4
United's Formula RC Concrete
  HD Flr Clnr............................... C1
United's Freezer Cleaner A5
United's No. 102 All Purpose..... A1
United's No. 200 Heavy Duty.... A1
United's Smokehouse Cleaner
  887.. A1
United's TMT-P Tile
  Marble&Terrazzo Clnr............ A1

United Laboratories, Inc.

Biatron.. L1
Liquid Hand Soap Concentrate . E1
Ten Strike Safe-Way Cleaner.... C2
Ul-25 .... K1
United 109 Spray & Wipe
  Cleaner...................................... C1
United 117 Stainless Steel
  Polish A7
United 126.................................... C1
United 127 Protective Cream..... E4
United 136 Disinf. Deod.
  Lemon Scented C2
United 162.................................... K1
United 170.................................... H2
United 170.................................... H2
United 171 Dry Moly Lube
  Lubricant H2
United 351 Disinfectant-
  Deodorant C1
United 41...................................... E1
United 58 with True Grit E4
United 63 .... E1
UL-107.. K1
UL-110.. C2
UL-114.. C1
UL-119.. H2
UL-127.. E4
UL-139 Teflon Dry Lubricant ... H2
UL-14 De-Greaser And Rust
  Inhibitor.................................... K1

UL-149 Insecticide Fog With
Cherry Scent F2
UL-15.. A1
UL-18.. A3
UL-200.. A2
UL-202 Rat Bait F3
UL-206 Residual Insecticide....... F2
UL-21.. L1
UL-213.. A1
UL-217 Non Fuming Clnr &
Descáler.................................... A3
UL-22.. K1
UL-250.. C1
UL-251.. A3
UL-265.. A3
UL-276 All Duty Emulsifier A4
UL-306.. C1
UL-308.. C1
UL-310 Boiler Treatment G6
UL-311 Steam & Return Line
Treatment G6
UL-318 Oxygen Scavenger G6
UL-355 Instant Action Odor
Eliminator................................ C1
UL-360 Slime & Grease Cutter.. L1
UL-376.. A1
UL-377 Solvent Drain
Maintainer C1
UL-401 Conveyor Lubricant....... H2
UL-46 Vaporizing Insecticide F1
UL-51.. A8
UL-53.. A1
UL-54.. A1
UL-55.. L1
UL-57 Ice Melter Pellets C1
UL-62 Disinfectant Cleaner....... A4
UL-64.. D2
UL-66.. A4
UL-70.. C1
UL-80 Odor Neutralizer C1
UL-86.. K1
UL-92.. C2
UL-97.. A1

United Oil Company, Inc., The

"T" Pressure Gun Grease........... H2
Duralene #77 White Grease H2
Duralene Pressure Special
Lubricnt H2
Safti-Lube.. H1

United Sanitary Chemicals Corp.

En-Vi-Ro #256.............................. D1
Formula 1901 .... A4
Jamaica Insect Spray.................... F1
Quick Wash..................................... E1
Spearmint C2
Velva Cream.................................... E1

United States Borax & Chemical Corp.

Boraxo Cool Hands...................... E4
Boraxo Liquid Lotion Soap........ E1
Boraxo Powdered Hand Soap.... C2
Boraxo Waterless Hand Cleaner E4
Pink Luron C2
PCB-100 .... C2
Special Heavy Duty Boraxo....... C2

United States Chemical Co, The

Bio-Kleenrite................................... A4
Break-Thru.. A1
Formula 40/80 .... A1
Kleenocide D1
Non-Butyl Degreaser Cleaner &
Deodorant A1
Nox-Soil A4
USCO Cream Hand Soap............ E1
USCO Disinfectant Deodorant .. C1

USCO Jet Liquid Drain Solvent L1
USCO Liquid Drain Opener L1
USCO Liquid Lustre.................... A7
USCO Royal Concentrate A4
USCO Spray'N Wipe................... C1

United States Chemical Corp.

Rustgonna No. 11A........................ G6
Rustgonna No. 600-5.................... G6

United States Chemical Corp.

Acid Concentrate No. 32........... A3
Active E2
De-Soil-It.. C2
Defoamed Acid No. 117............. A3
Defoamed Liquid Chelate U.S.
No. 66-D.................................... A2
Di-Solv Formula No. 11............. A3
Dig. U.S. Formula No. 62-A..... A1
Dislodge Formula No. 35 .... A2
Dry Bleach No. 45 .... B1
Dual-Drive #220 .... A2
Eggwash Recirculation
Detergent................................. A1,Q1
Farm-Kleen A3
Fryer Cleaner Formula No. 148 A2
Gallant Formula #27 .... A3
General Cleaner No. 107 .... A1
Goal #97 .... A1
Grease Gone #55 .... A1
Grenade No. 87 .... A2
Handyman No. 138......................... L1
Heavy Duty Acid Detergent
No. 15 .... A3
High Temperature Alkali No.
35.. A2
Hydro-Klor A1
HD-148... A2
Immense A4,A8
Inflation Kleen Formula 35........ A2
Kick Formula No. 18.................... A1
Laundry Break Compnd
Formula 35 .... A2
Liquaclean A1
Liquid Chelate U.S. No. 66........ A2
Liquimac Formula No. 142......... A2
Liquitrol Acid Formula No. 25.. A3
Liquitrol Alkali Formula No. 23 A2
Liquitrol Conditioner Form.
No. 24 .... G1
Liquitrol Recirculation Cl. No.
201... A2
Lubriclean No. 243........................ H1
Moderate #195 .... B1
Orbit Formula #2........................... A1
Plasti-Klene A1
Poly-Det Formula No. 236.......... A1
Pro-Tect No. 235............................ A3
Proto-Kleen No. 19........................ A1
Punch No. 298 .... A1
Restore No. 238.............................. B1
Rifle No. 17..................................... A3
San-I-King.. D2
Saniphor No. 240........................... D2
Shurguard D2
Stalwart A1
Standby Kleener #20 .... A1
Super B-Kleen Formula No. 49 . A1
Talent.. A1
Target Formula #39 .... A1
Turbo... A1
U.S. #600 Formula No. 28.......... A1
U.S. Alkali No. 54......................... A2
U.S. Alkali No. 54 G..................... A2
U.S. Boiler Compound No. 56... G7
U.S. Caustic #233........................... A2
U.S. Caustic #234........................... A2
U.S. Caustic Formula No. 43G .. A2
U.S. Chelated Caustic #33 .... A2
U.S. Defoamer Formula No.
200... A1

U.S. Iodophore Formula No. 34 D1
U.S. Pink Liquid No. 4............... A1
U.S. Powerlube................................ H2
U.S. Rust Proofer No.300........... P1
U.S. Rust Proofer No.302........... P1
U.S. Rust Remover No.86.......... A3
U.S. Smokehouse Cl. Form.
No. 174 .... A2
Universal Floor Cleaner No.
128... A4
Wand Formula #131...................... A1
Wij Kleen No. 41 .... A3
Winter Lube Formula No. 241... H2
500 Formula #29............................ A1

United States Chemical Corp.

Dynamic-101 .... A4
Likkety Split A4

United States Rust Control Corp.

Safety Grease.................................. H1

United Supply Company

Foaming Agent A1
Liquid Dishwash Concentrate.... A1
U-100 Wax and Grease
Remover A4
U-3 Liquid Steam Cleaner......... A1
U-35 Pink Concentrate................ A4

United Twine & Paper Co., Inc.

Dishtergent...................................... A1

Univeral Cooperatives, Inc.

Unico Pressurized Dairy Fly
Spray-V F1
Unico Pressurized Dairy Fly
Spray-V F1

Universal Chemical

Enz-Stop... L1

Universal Chemicals & Supplies, Inc.

Cas-AA Trolly Lube....................... H2
LC-125.. A1
Shrowd Cleaner B1
Tripe Cleaner N3
Trolly Wash A2
TK-10 .... A1
TK-100 .... A3
TK-105 .... A1
TK-110 .... A1
TK-140 .... A1
TK-150 .... A4
TK-160 .... A2
TK-20 .... A1
TK-25 .... A1
TK-30 .... A1
TK-40 .... A1
TK-45 .... A1
TK-5 .... A2
TK-50 .... A1
TK-55 .... A1
TK-60 .... A1
TK-70 .... A1
TK-75 .... A1
TK-80 .... A3
TK-85 .... A2
TK-95 .... A2

Universal Chemicals, Inc.

All Purpose Cleaner No. 2 .... A4
Break-Away A1
Drain Opener L1
Dual-27 .... D1
Everbrite C2
Fresh.. C1
Grip P1
Grout And Tile Cleaner A3

Hi-Pure Contact Cleaner K2
Legendary A1
Lemon Cleaner Plus C1
Mint Disinfectant C1
Neutral Cleaner A4
Neutral Soap A4
Perform A1
Pine Disinfectant C1
Quick Clean Oven & Grill
  Cleaner A8
Residual Spray-B F2
Safety Solv 1 .... K1
Safety Solvent K1
Sparkle Protective Coating C1
Stainless Steel and Metal Polish. A7
The General A4
Total Break A1
UBD-5 .... C1
UC 1 .... C1

Universal Laboratories/ Div. of Hanson-Loran Chemical Co. Inc.

Heavy Duty Cl. & Disinf. Cat.
  No. UL-530 .... A4
Quaternary Germicidal Cleaner
  UL-709. .... A4

Universal Manufacturing & Supply Company

HD-10 .... A1
Lucky-7 Liquid Steam Cleaning
  Compound A1
Miracle Cleaner A4,A8
Miracle Cleaner +40 .... A4,A8
Softouch E1
Super Steam Cleaning
  Compound A1

Universal Milking Machine Div. National Cooperatives, Inc.

Bulk Milk Tank Cleaner A1
Chlor-O-San D2
Dairy-Kleen A1
Iodo-Kleen D2
Line-Clean 1 .... A1
Steribalm D1
Universal Egg Wash Detergent . Q1
Universal Gen. Purp. Clning
  Conc A1
Universal Hvy Dty Clng
  Conctrte 2 .... A1
Universal Super Kleen A1

Universal Motor Oils Co., Inc.

Dyna Plex 21C H1
Dyna-Plex 21C Dynamo A4
White Food Machinery Grease.. H1
White Oil 350 .... H1

Upright

Upright Floor Anti-Slip J1

Utility Chemical Company

Absorbent Deodorizer Granules C1
Dairy-Kleen Heavy Duty
  Cleaner A4
Fostan G6
I.S.R. C1
Keltan Special G7
Klick 20% Hand Soap E1
Liquid H.D. Steam Cleaner A1
O'Boy! Waterless Hand Cleaner E4
Polyscour C-105 .... A1
Polyscour M-O-G A4
Sani-San D2
Sewer-Solv L1
Special Quadrasoap B1
Swish A4
Utidyne D2

Utikleen D-O C1
Utikleen 77 .... A4

Utility Chemicals,Inc.

BWT-180 .... G6
BWT-181 .... G6
BWT-270-S G6
BWT-360 .... G6
BWT-361 .... G6
RLT-35 .... G6
U-Cide 401 .... G7

V

Val-A Company

Dinox Concentrate Q1
Go Foam Q5
San-De Egg Wash Q1
Trisol Egg Cleaner Q1

Val-Chem Company, Inc.

Bi-Far B1
Bi-Far Phosphate Free B1
Heavy Duty Cleaner B1
One Shot B1
San-O-Val D1
Softex B2
Val-Chlor Bleach D1,B1
Val-Soft B1
Val-Spec Chlorinated A1
Val-Spec K-10 .... A2
Val-Spec K-56 .... A1
Val-Spec K-61 .... A1
Val-Spec K-65 .... A1
Val-Spec K-66 .... A1
Valdol 1201 .... B2
Valkali B1

Valdes, G. Enterprises, Inc.

El Toro Liqui-Drain Opener L1
El Toro Super All Purpose
  Cleaner A1
El Toro Tempered Drain
  Opener L1

Valdosta Industrial Chemical Co.

Blue Magic Cleaner A1

Valiant Chemical Company

Val Chem X-12 .... A4

Valley Chemical Company

Digest-II L2
V-221 All Purpose Cleaner A4
V-369 .... D1
Val-Freez A5

Valley Chemical Specialties Inc.

AMR A3
AMR 1019 .... A3
Cleaner 100 .... A2
Cleaner 200 .... A1
Cleaner 210 .... A1
Cleaner 210-L A1
Cleaner 240 .... A2
Cleaner 300 .... A1
Cleaner 310 .... A1
Cleaner 320 .... A1
Cleaner 400 .... A2
Cleaner 800 .... A2
Cleaner 900 .... A2
CL-54 .... H2
Defoamer A A1
Foam Clean A1
KC A2
Liquid Steam Cleaner A4

Val-Add A1
Valchlor A1

Valley Cities Paper Company

Degreaser Cleaner A1

Valley Fruit & Produce, Inc.

Val-Pro APC All Purpose
  Clearier A1

Valley Industrial Products, Inc.

VIP Battery Cleaner And
  Protector P1
VIP Dry Lubricant And
  Release Agent H2
VIP Instant Foamer C1
VIP Silicone Lubricant Multi-
  Purp Spray H1

Valley Oil & Chemical Co.

L-20 Cleaner A1
Sulfamic Acid Crystal X
  Technical A3

Valley Products Company

Green Bar Soap H2

Valvoline Oil Company/ Div. of Ashland Oil, Inc.

All Climate HD Motor Oil
  SAE 10W-20W-40 .... H2
Anti-Wear Hydraulic Oil No.
  20 .... H2
Automatic Trans. Fluid Type D
  Dexron 11 .... H2
Automatic Transmission Fluid
  Type FA H2
BG-2 Medium H2
Hydro-Lube SAE 80W-85W-90
  Gear Oil H2
HD Super HPO Motor Oil
  SAE 20W H2
HD Super HPO Motor Oil
  SAE 30 .... H2
Non-Detergent Motor Oil SAE
  20W H2
Non-Detergent Motor Oil SAE
  30 .... H2
Super HPO Motor Oil SAE 10.. H2
Unitrac Fluid H2
X-18 Multi-Purp.GearOil
  SAE80W-85W-90 .... H2
X-18 MD Multi-Purp. Gear Oil
  SAE 85W-140 .... H2

Van Straaten Chemical Company

VFC 30 .... A4

Van Waters & Rogers

Dry Chlorine Granular D1
Dry Chlorine Tablets D1
FD-3 .... A1
Grid-100 .... A3
Grid-30 .... A1
Guardsman Bactericide-Disinf. .. D1
Guardsman Boiler
  Treat.Chemical G6
Namco Pyrenone Concentrate ... F1
Namco Pyrenone Fog
  Concentrate F1
Namco Pyrenone Food Storage
  Spray F1
Namco Synthroid Insect
  Fogger F1
VW & R Mineral Oil NF 9 .... H1
VW & R Mineral Oil USP 35 ..... H1
VW & R Mineral Oil 75 T H1
VW & R Mineral Oil 85 T H1
VW & R Mineral Oil 95 T H1

Varley & Sons, Incorporated See:
James Varley & Sons, Inc.

Vasco Brands Inc.
Insta-Kleen A1

Vasco Chemical Company, Inc.
Clecon Concrete Cleaner C1
Dynamic K2
Glo-N-Go K1
Multi-Purpose Concentrate F1
Pure-N-Sweet C1
Swirlaway A4
Vasco Formula 100-G F2
Vasco-Zyme L2

Vasco Industrial Supply
Vasco Action Foam A1

Veckridge Chemical Company
Nu-Dri J1

Vector Chemicals
Surgi-Klean A1

Vel-Tex Chemical Company
Azone D2,Q4,G4

Velsicol Chemical Corp.
Brisk Germicide A4
Gold Crest Apple Flvrd Bait
  Pellets F3
Gold Crest C.I.O.-20 Oil
  Concentrate F2
Gold Crest C-100 Emulsifiable
  Concntrt F2
Gold Crest C-5% Dust F2
Gold Crest C-50 Emulsifiable
  Concntrt F2
Gold Crest Promar Bait Packs ... F3
Promar Fish Flavored Bait
  Pellets F3
Promar Meal F3
Promar Meat Flavored Bait
  Pellets F3
Ramik Bait Packs F3
Ramik Brown F3
Ramik Green F3
Ramik Green Bait Packs F3
Ramik Red F3
Ramik Red Bait Packs F3
Rotac Insecticide Roach Bait F2
XA-100 Odor Control
  Agent(Aerosol) C1
XA-100 Odor Control
  Agent(Wick) C1

Ventron Corporation
Cunilate Quick-Dry Penetrant ... P1
Cunilate Sealer P1
Cunilate Wood Seal P1
Cunilate Wood Seal Resin P1
Cunilate 2174 WP P1
Cunilate 2419 .... P1
Industrial Cunilate Wood Seal ... P1
Socci Cunilate Penetrant P1

Venus Laboratories, Inc.
#1003 Silicone Spray H1
Acid Free Ice Melter C1,A8
Algaecide G7
Aluminum&Chrome Rust
  Remover A3
Ammoniated Heavy Duty
  Stripper A4
Bowlex C2
Break-Thru Powder L1
Cuts Rust Remover A3
Degreaser M-E K2

Desclr&Delmr for Boil.Swmng
  Pools A3
Emulsion Degreaser K1
Enzymes L2
Fashion Bowl & Porcelain
  Cleaner C2
Fogging Insecticide F2
Germicidal Spray & Wipe
  Cleaner C1
Heavy Duty Multi Purpose
  Cleaner A4
Lemon Odor Disinfectant
  Cleaner A4
Lift Station Cleaner L1
Liqui-Drain Opener L1
Liquid Hand Soap E1
Liquid Hand Soap E1
Liquid Residual Spray
  Insecticide F2
Liquid Scouring Creme C1
Low Temp Heavy Duty
  Cleaner A5
Mint Disinfectant C1
Moisture Barrier & Electrical
  Lubricant K2
Neutral PH Cleaner A1
No.601 Odor Control Granules .. C1
Oven Cleaner A8
Pine Odor Disinfectant Cleaner . C1
Safety Solvent Cleaner &
  Degreaser K1
Super Butyl Cleaner A4
Super Foam Oven and Carbon
  Cleaner A8
Super Neutral Cleaner A1
Super Uni-Tabs C2
Tempered Drain Opener L1
Tile Klenz C2
Touch of Mink E4
Venuscide Insecticide
  Concentrate F1
Waterless Hand Cleaner E4

Verax Chemical Company
Blu Mist A4
Chekmate D1
Maximite A1
Radium Powdered Borax Hand
  Soap C2
Sanishine D1
Snap Degreaser C1
Spra Bryte A1
Test Best D2
Triple De A4
Velva Coconut Oil Liquid
  Hand Soap E1
Wizard A1

Veritas Company, Inc., The
A C Inhibited A1
BCS-1 .... A2
BCS-2 .... A1
BCS-2-B A1
BCS-3 .... A1
Chlorinated LMD A1
Drain Opener L1
Jet A4
Lube-Solv H2
M.P.C.-10 .... A1
Niagara Cleaner A2
Oven Kleen A2
OH-163 .... A2
Safe A3
Single B1
Super Jet A1
Super Wash A1
SHD A1
V-12 Powder C2
V-22 .... A3
V-24 .... A2

oll

271

V-40 .... A1
Veri Foam B A1
Veribrite A1
Verifoam RS A1
Verikwik A6
Verismooth 15% E1
Verismooth 40% E1
Veritas Low Foam A1
Veritas P.W. A1
Veritas 4020 Acid Cleaner A3
Veritas 4650 .... Q1
Veritas 4660 .... Q1
Veritas 6010 Boiler Treatment ... G6
Veritone A4
Veriwhite B1
WT-201 .... G6

Vermatek
Vermatek #1 AA L2

Vermillion Chemical Service, Inc.
Blue Glass Cleaner C1
Concrete Cleaner A4
Du-Fine A3
E-Z-Go A1
LDC B1
Scale Clean A3
Sewer Solvent L1
Steam Kleen It A1
Ster-O-Chloro A1
Super Chloro A1
VC 5 Concentrate A1
Wax Stripper C1

Vernon Sanitation Supply Co.
Visco Bleach B1,P1
Visco Liquid Soap E1
Visco 15 .... A4
Visco 30 .... A4

Vero Chemical Distributors, Inc.
Verochem Silicone Spray H1

Versol Chemical Company
D-17 .... A1
S-45 .... A1
Strippet A2
VR Heavy Duty Clnr-
   Degreaser Concentrate A4
23 .... A1
625 Cleaner A1
707 .... A1
74 .... A1

Vertex Chemical Corporation
Vertex CSS-10 .... D1

Vestal Laboratories
"GR-EM" E4
Alcare E3
Dewaxer A4
Double Barrel A1
Egg Cleaner Concentrate Q1
Environ A4
Environ-D A4
Enviroquat D2
Ezaway A4
Industrial Plant Cleaner A1
Lime Soap Remover A1
LPH A4
Millionite A1
MWC Compound A1
Protective Hand Cream E4
Septi-Soft Skin Cream E4
Skleen A1,Q1
Super Skleen A4
Tri-Klean E1
Ty-ion 10 .... G6
Ty-ion 12 .... G6

Ty-ion 16 .... G6
Ty-ion 18 .... G6
Ty-Ion A39 .... G7
Ty-Ion B10 .... G6
Ty-Ion B12 .... G6
Ty-Ion B16 .... G6
Ty-Ion C25 .... G7
Ty-Ion R40 .... A3
V-5 .... A1,Q1
Vesta Power A1
Vesta-Brite A3,Q2
Vesta-Foam A1
Vesta-Klean A2
Vestal Iodine Scrub (V.I.S.) E2
Vestal Liquid Detergent A1,Q1
Vestal Tri-Klean Solution E1
Vestasol E1
Vestasol Solution E1
1-Stroke Environ D1

Vestal Supply/ Div. of Vestal Laboratories
Oxsorb-100 .... G6
VS-101 Boiler Water Treatment G6
VS-102 Boiler Water Treatment G6

Vicene Specialties Company
Charge A4
Viocide D1

Vicjet Incorporated
Vicjet Hvy Dty All Purp.
   Clning Comp A4

Victor Chemical Company
Victory 7 .... A1

Victor Supply Company
007 Concentrate A4

Vigilant Products Co., Inc.
Cess-Aide L2
Drain Safe L2
Grees-Out L2
Septic Aide L2

Viking Chemical Company
APC A1

Viking Chemical Products, Inc.
D464 Cleaner A1
D473 Freezer Cleaner A5

Viking Chemicals
AB-40 .... G7
AC-10 .... G6
AP-20 .... G6
AP-200 .... G6
Ban-F310 .... G6
Ban-1018 .... G6
Ban-1024 .... G6
Ban-1033 .... G6
Ban-1310 .... G6
Ban-1315 .... G6
Ban-1327 .... G6
Ban-1340 .... G6
Ban-3015 .... G6
Ban-3025 .... G6
Ban-3035 .... G6
Ban-4010 .... G6
Ban-4020 .... G6
Ban-4030 .... G6
BAN-A024 .... G6
BCF-2004-D G6
BCF-34000-D G6
BOX-700-D G6
BOX-7000-D G6
BSC-100 .... G6
BSC-150 .... G6

BSC-1500 .... G6
BSC-200 .... G6
BWL-B200 .... G6
BWL-B200-A G6
BWL-B200-AX G6
BWL-B200-X G6
BWL-2000 .... G6
BWL-2000-A G6
BWL-2000-AX G6
BWL-2000-X G6
BWL-2020 .... G6
BWL-2020-X G6
BWL-2021 .... G7
BWP-5050-D G6
BWP-5075-D G6
BWP-5100 .... G6
BWZ-D400 .... G6
BWZ-D400-A G6
BWZ-D400-AX G6
BWZ-D400-X G6
BWZ-0600 .... G6
BWZ-2000-A G6
BWZ-4000 .... G6
BWZ-4000-A G6
BWZ-4000-AX G6
BWZ-4000-X G6
BWZ-4001 .... G6
BWZ-4005-X G6
BWZ-6000 .... G7
BWZ-6007 .... G7
BWZ-6008 .... G6
RT-30 .... P1
RT-40-D P1

Viking Institutional Products
Stainless Steel A7

Viking Laboratories, Inc.
Mity Mite A4

Viking Laboratories, Inc.
Acid 14 .... A3
Can Wash 318 .... A1
Clor 3 .... A2,Q1
VLH-61 .... G6
VLS-92 .... G6
3X Super Solv A4,A8

Viking Sales, Inc.
Power Steam A1
T-5 Butyl Cleaner A4,A8

Vineland Laboratories, Inc.
Hi-Lethol-50 .... D1
Sani-Squad C1
Vinelab Io-Vine D2,Q6
Vinelab Rat Rid F3
Vinelab Warfet F3

Vineland Poultry Laboratories, Inc.
Detergent Sanitizer Q1
Egg Cleaning Detergent Q1
Hi Lethol-10 .... D1
Hi-Lethol-20 .... D1

Virginia Chemicals, Inc.
Ice Machine Cleaner (Liquid) A3
Lethalaire A-20 .... F1
Lethalaire A-40 .... F2
Lethalaire A-41 Aerosol
   Insecticide F2
Lethalaire B-13 .... F1
Lethalaire D-30 (Aerosol) F1
Lethalaire JR-4 .... F1
Lethalaire V-21 .... F1
Lethalaire V-26 Insecticide F1
Lethalaire V-28 Professional
   Insecticide F2
Pro-Cide F1
Pro-Cide 111 .... F1

Boiler Compound Formula 37-S G6
Boiler Compound Formula 48-S G6
Boiler Compound Formula 593-
F G6
Boiler Compound Formula 593-
SF G6
Boiler Compound Formula 93 ... G6
Boiler Compound Formula 934.. G6
Boiler Compound Formula 935.. G6
Boiler Compound Formula 945.. G6
Cooling Tower Treatment
Form. 310 .... G7
Cooling Tower Treatment
Formula 300 .... G7
Inhibited Sulfamic Acid A3
Steam Line Treatment Formula
11-F G6
Steam Line Treatment Formula
153 .... G6
Steam Line Treatment Formula
253 .... G6
Steam Line Treatment Formula
653-F G6

Walters Chemical Company

BW #20 .... A2
Egg Wash A1,Q1,Q2,Q5
HDC #1 .... A1
Liquid Chlorine Q4
Morgo Cleaner A2
Super Shine A1,Q1,Q2,Q5
Trolley Cleaner A2
Wal-Chem Chlorntd Gen
Cleaner #50 .... A1
Wal-Chem Chloro-Shine A2
Wal-Chem General Cleaner
#100 .... A1
Wal-Chem General Cleaner
#200 .... A1
Wal-Chem H.D.C. #3 .... A2
Wal-Chem H.D.C. #4 .... A2
Wal-Chem M.S.R. 300 .... A3
Wal-Chem M.S.R. 50 .... A3

Waltham Chemical Company

Martin's Contact Insecticide
Spray F1
Martin's Industrial Insecticide ... F1
Martin's Mill Spray F1
Martin's New Improved Rokil F2
Profi Grd Insectcd Pyrenone
Aerosol F1
Professional Insect Killer F1
Professional Type Aer.
Insecticide Spray F1

Walton-March, Incorporated

Anti-Pollution Drain Treatment L2
Bowl Cleaner C2
Floor Corps Cl. for Auto.
Scrub. Equip. A4
Floor Corps Cleaner (Liquid) A4
Floor Corps Cleaner/Degreaser A4
Floor Corps Neutralizer/
Conditioner A4
Flourish Bowl Cleaner C2
Ice-Foe C1
Pipe Dream Drain Opener L1
Power Play C1
Prime Time All Purpose
Cleaner A1
Spike A1
Surfacide D1
True Blue Glass Cleaner C1
Walton-March G.P.C. No. 1 ...... A1
Ways & Means A4

Waper, Incorporated

Steam & Pressure Cleaning
Compound A1

W-3 The New Innovative Clnr.-
Strip. Dgrs. A4
W-4 Solvent Cleaner A4
W-47 All Purpose A4
W-880 .... A4
Waper Engine Cleaner K1

Waples-Platter Companies

A-P-C All Purpose Cleaner
Concentrate A4
Automatic Laundry 204 .... B1
BD-6 Detergent Beads A1
CD-200 Cleaner-Disinfectant C2
Degreaser Heavy Duty Solvent
Cleaner A1
Lemon Disinfectant C2
Liquid Bleach And Cleaner B1
Liquid G A1
LD Creme 33 Liquid Hand
Dishwashing Det. A1
Mint Disinfectant C2
Pine Pine Odor Disinfectant C1
PHD-78 Detergent Beads For
Heavy Duty A1
Q Sanitizer D2,E3
RTU Ready-To-Use Hand Soap E1
Stainless Shine A7
Stainless Steel Cleaner A7
Wash-Rins Ware-Washing
Detergent A2
White Swan Liq. Bleach &
Cleaner B1
X-16 Chlorinated Sanitizer D1

Ward Laboratories

Contact Spray F1
Fogging Concentrate F1
Residual Spray Containing
Baygon F2

Warene

Packers Super A1

Warren Chemical Mfg., Inc.

#501 Wallop Heavy Duty All
Purpose Clnr A4
#503 Coco-Castile Liquid Hand
Soap E1
#518 All Purpose Cleaner
(Powder) A1
Wallop A4

Warren E. Conley Corp.

Liquid Gen. Purpose Cleaner
Code 5150 .... A4
W.C. Cleaner Degreaser A4

Warren Oil Company, Inc.

"Gold Bond" Food Machinery
Grease H1
Gold Bond Packers' Tech 85T .. H1

Warren Refining Division

Plastilube White H1

Warren-Douglas Chemical Co., Inc.

Glide E1
Warlasco Acid Cleaner SS101 ... A3
Warlasco Acid Cleaner SS102 ... A3
Warlasco Acid Cleaner SS221 ... A3
Warlasco Acid Cleaner SS202 ... A3
Warlasco Acid Cleaner SS301 ... A3
Warlasco Acid Cleaner SS302 ... A3
Warlasco Acid Cleaner SS401 ... A3
Warlasco Acid Cleaner SS402 ... A3
Warlasco Auto. Dishwshing
Concntrt A1
Warlasco Bulk Tank Cleaner A1
Warlasco Castor B Lubricant..... H1

Warlasco Cooler Grd Ice
Remvr #9093 .... C1
Warlasco Dairy Alkali Cleaner.. A1
Warlasco Dairy Alkali INP
Cleaner A1
Warlasco Egg Shell Cleaner D1
Warlasco Emul Toilet
Bowl&Urinal Cl. C2
Warlasco Floor & Metal
Cleaner A2
Warlasco Food Equipment
Cleaner A1
Warlasco General Cleaner A1
Warlasco Germicide SH-12 .... D2
Warlasco Grease & Oil
Absorbent J1
Warlasco Hvy Dty Soak Tank
Clnr. A2
Warlasco Hvy Duty Soap Tank
Cl. #71 .... A2
Warlasco Ice Remover C1
Warlasco Industrial Spray F1
Warlasco Packers Oil 4470 .... H1
Warlasco Scale Remover
(Liquid) G7
Warlasco Sewer Cleaner #5 .... L1
Warlasco Sure-Foot J1
Warlasco 10% Sanitizing Agent D1

Warsaw Chemical Co., Inc.

pH 7 .... E4
A-P Insecticide F1
All Purpose 8-217-A A1
Aluminum Stainless-Steel
Cleaner A3
AD 2000 .... A4
Before E4
Borax Base Hand Soap C2
Bright A1
Charge A4,A8
Chlorinated Cleanser A6
Chlorine Liquid N4,G4
CT-5 .... Q1
CT-7 .... Q1
Deep Fat Fryer & Grill Cleaner A1
Detergent Disinf. Deod.
Sanitizer D1
Done D1
Dry Chlorinated Bleach B1
DJ Concrete Cleaner C1
Egg Wash Y-100 .... Q1
Fabrasoft B2
Floor Cleaner Wheel Wash
Steam Clnr. A4
G-65 .... A1
Germ-A-Clean E1
Gruff E4
Grym-Go Hand Cleaner C2
Heavy Duty Cement Cleaner A4
Heavy Duty Cleaner Green A4
Hi-Power A2
HD Cleaner Disinfectant D1
I & I Steam Cleaner A1
Institutional Laundry Detergent B1
Lime Solve A3
Liquid Deluxe Coco Castile E1
Liquid Disinfectant & Sanitizer.. D1
Liquid Scour A6
Lo Shun E4
Low Foaming All Purp. Cl. 8-
221-A A1
Low Foaming HD All Purp.
Cl. 8-220-A A2
Low Foaming Stnls Steel Cl. 8-
222-A A3
Moon A1
Nu Blu B1
NAM #1 .... A3
Odorless Disinfectant C2
Once-A-Week Drain Opener L1

Watco Chemicals, Inc.

Big W A2
Big W Odorless Concrete
    Cleaner....................................... A4
General Cleaner............................. A1
GP-33... A1

Watcon, Incorporated

Watcon 101 .... G6
Watcon 102 .... G6
Watcon 103 .... G6
Watcon 105-A G6
Watcon 105-C G6
Watcon 106 .... G6
Watcon 108 .... G6
Watcon 110 .... G6
Watcon 114 .... G6,G2
Watcon 115 .... G2
Wateou 116 .... G6
Watcon 118 .... G6,G2
Watcon 119 .... G2
Watcon 1230 .... G6
Watcon 1232 .... G6
Watcon 1236 .... G6
Watcon 1237 .... G6
Watcon 1246 .... A3
Watcon 126 .... G6
Watcon 1260 .... G6
Watcon 1262 .... G6
Watcon 128 .... G6
Watcon 1285 .... G2
Watcon 1286 .... A3
Watcon 1288 .... G6
Watcon 1290 .... G6,G2
Watcon 1296 .... G6
Watcon 1305 .... G7
Watcon 1308 .... A3
Watcon 1308A A3
Watcon 1327 .... G6
Watcon 1336 .... G6
Watcon 1354 .... G1
Watcon 4106 .... G6
Watcon 4128 .... G6
Watcon 4232 .... G6
Watcon 4237 .... G6
Watcon 4237 .... G6
Watcon 4296 .... G6

Water Chemicals, Incorporated

BT 703 .... G6
BT-303 .... G6
BT-305 .... G6
BT-404 .... G6
BT-503 .... G6
BT-603 .... G6
BT-605 .... G6
CS 701 .... G7
CT 100 S A3
CT 2000 .... P1
CT 2001 .... P1
CW 870... G7
CW-800... G7
CW-820... G7
CW-821... G7
CW-850... G7
CW-860... G7
CW-880... G7
CW-888... G7
LP 207 .... G6
LP-206 .... G7
PW 100... G2
RL-103.. G6
SW-105... G7

Water Chemistry, Inc.

Aqua-Treat Regular G2
Boiler Power Purge G7
Boiler Power 250.......................... G7
Boiler Power 300.......................... G7

Boiler Power 350.......................... G7
Hot & Cold Circulating Power
    600... G7
Hy-Pro XXXV G1
Liquid Boiler Power 500 .... G6
Tower Power 100.......................... G7
Tower Power 200.......................... A3
Volitile Amines 100..................... G7
Volitile Amines 200..................... G7

Water Chemists, Inc.

Boiler Water Treatment BT-121 G6
Boiler Water Treatment BT-122 G6
Boiler Water Treatment BT-132
    U... G6
Boiler Water Treatment BT-134
    U... G6
Boiler Water Treatment BT-142 G6
Boiler Water Treatment BT-
    146C.. G6
Boiler Water Treatment BT-212 G6
Boiler Water Treatment BT-222 G6
Boiler Water Treatment BT-232 G6
Boiler Water Treatment BT-321 G6
Boiler Water Treatment BT-322 G6

Water Chemists, Incorporated

Antifoam 405L.............................. G6
Boilertreat 402FL......................... G6
Boilertreat 439D G6
Boilertreat 440L........................... G6
BT 408L G6
BT 438 .... G6
BT 438L G6
BT 456L G6
BT-415 .... G6
BT-415L G6
BT-419L G6
BT-450L G6
BT-472L G6
BT-481 .... G6
BT-481L G6

Water Conditioning Consultants

F-30 Foamkill Q5
Saf-Clor G1

Water Guard Division/ Southern Industrial Sales Corp.

Water Guard #101...................... G7
Water Guard #110...................... G6
Water Guard #111....................... G6
Water Guard #112...................... G6
Water Guard #113...................... G6
Water Guard #115...................... G6
Water Guard #130...................... G7
Water Guard #131...................... G7
Water Guard #132...................... G7
Water Guard #161 .... G7
Water Guard #190...................... G2
Water Guard #191 .... G6
Water Guard #210...................... G6
Water Guard #211 .... G6
Water Guard #212...................... G6
Water Guard #213 .... G6

Water Management Division/ Clow Corporation

Boiler Antifoam 1........................ G6
Boiler Antifoam 2........................ G6
Boiler Blend 33001 .... G6
Boiler Water Treamtent No.
    102... G6
Chelyte 1271 .... G6
Chelyte 561 .... G6
Chelyte 563 .... G6
Chelyte 761 .... G6
Chelyte 763 .... G6
Clow 2015 .... A1,G6

Clow 2020 .... G6
Clow 2025 .... G6
Clow 2030 .... G6
Clow 2040 .... P1
Clow 2115 .... G6
Clow 2201 .... G6
Clow 2202 .... G6
Clow 2203 .... G6
Clow 2204 .... G6
Clow 2205 .... G6
Clow 2206 .... G6
Clow 2207 .... G6
Clow 2208 .... G6
Clow 2325 .... G6
Clow 2515 .... G6
Clow 2525 .... G6
Clow 2535 .... G6
Clow 2545 .... G6
Clow 2555 .... G6
Clow 2565 .... G6
Clow 2600 .... G6
Clow 2610 .... G6
Clow 2615 .... G6
Clow 2620 .... G7
Clow 2640 .... G7
Clow 2675 .... G6
Clow 3020 .... G2
Clow 3022 .... G2
Clow 8000 .... L1
Clow 8001 .... L1
Clow 8002 .... L1
Clow 8003 .... L1
Clow 8004 .... L1
Clow 8006 .... L1
Clow 8007 .... L1
Clow 8020 .... L1
Clow 8021 .... G6,L1
Clow 8022 .... L1
Clow 8023 .... L1
Clow 8040 .... L1
Clow 8220 .... L1
Clow 8230 .... L1
Corroban G3
Corrofilm G6
Corrofilm No. 2 .... G7
Custom Formula 31001 .... G6
Custom Formula 31002 .... G6
Custom Formula 33001 .... G6
Custom Formula 33214 .... G6
Custom Formula 33601 .... G6
Custom Formula 34015 .... G6
Custom Formula 35041 .... G6
Custom Formula 36221 .... G6
Liquid Rapid Clean 1601 .... A3
Neufilm G7
Organic, 1 .... G6
Organic 1237 .... G6
Organic 2 .... G6
Organic 3 .... G7
Organic 4 .... G7
Organic 5 .... G7
Oxygen Inhibitor F G6
Oxygen Inhibitor 1254 .... G6
Polymerized Phosphate 142 .... G6
Polyphos F G2
Polyphos 1110 .... G6,G2
Polyphos 1115 .... G6
Polytreat B-200 .... G6
Polytreat B-200-S G6
Polytreat B-200-SV G7
Polytreat B-400 .... G6
Polytreat B-400-S G6
Polytreat B-400-SV G7
Polytreat Z-100 .... G6
Polytreat Z-100-S G6
Polytreat Z-100-SV G7
Polytreat Z-300 .... G6
Polytreat Z-300-S G6
Polytreat Z-300-SV G7
Volamine 1 .... G6

Volamine 1425 .... G6
Volamine 2 .... G6
Volamine 3 .... G7
Volamine 4 .... G7
Volamine 5 .... G6
Volamine 6 .... G6

Water Sciences Division/ Economics Laboratory, Inc.

Elamine 180 .... G6
Elamine 181 .... G7
Elamine 182 .... G6
Elamine 187 .... G6
Elclaw 130 .... G6
Elclaw 132 .... G6
Elclaw 133 .... G6
Elclaw 135 .... G6
Elclaw 136 .... G6
Elgicide 1 .... G7
Ellorate 232 .... G7,G1
Eloxy 162 .... G6
Elphos 121 .... G6
Elplex 111 .... G6
Elplex 112 .... G6
Elplex 113 .... G6
Elplex 115 .... G6
Elplex 116 .... G6
Elplex 117 .... G6
Elplus 170 .... G6
Elplus 171 .... G6
Elplus 172 .... G6
Elplus 173 .... G6
Elprep 210 .... G7
Elsperse 240 .... G7
Elvalent 226 .... G7
Elvalent 227 .... G7
Elvalent 228 .... G7

Water Sciences, Inc.

Formula No. 100 .... G6
Formula No. 101 .... G6
Formula No. 103 .... G6
Formula No. 105 .... G6
Formula No. 203 .... G6
Formula No. 310 .... G6
Formula No. 311 .... G6
Formula No. 400 .... G7
Formula No. 401 .... G6
Formula No. 402 .... G6
Formula No. 403 .... G6
Formula No. 555 .... G2

Water Service Associates

Boiler Water Treatment WSA
    10A G6
Boiler Water Treatment WSA
    104C G6
Boiler Water Treatment WSA
    106C G7
Boiler Water Treatment WSA
    12A G6
Boiler Water Treatment WSA
    14A G6
Boiler Water Treatment WSA
    20A G6
Boiler Water Treatment WSA
    22A G6
Boiler Water Treatment WSA
    24A G6
Boiler Water Treatment WSA
    26A G6
Boiler Water Treatment WSA
    28A G6
Boiler Water Treatment WSA
    30A G6
Boiler Water Treatment WSA
    32A G6
Boiler Water Treatment WSA
    34A G6

Boiler Water Treatment WSA
    36A G6
Boiler Water Treatment WSA
    76B G6
Boiler Water Treatment WSA
    80B G6
Boiler Water Treatment WSA
    82B G7
Cooling Water Treatment WSA
    120D G7
Cooling Water Treatment WSA
    122D G6
Resin Cleaner Treatment WSA
    176F P1

Water Services Company

BWT #33 .... G6

Water Services Division/ UOP, Inc.

A-90 Boiler Water Control G6
A-90V Boiler Water Control G6
A-91 Boiler Treatment
    Briquettes G6
A-91G G6
All Met Descaler A3
Antifoam SL G6
B-85 Boiler Water Control G6
B-85C G6
B-85V Boiler Water Control G6
B-86 Boiler Treatment
    Briquettes G6
B-86G G6
Bulldozer Mud Mover G7
Chemicator Refill Tube #J-2 ..... G7
Chemicator Refill Tube #P-O ... G7
Chemicator Refill Tube #101 ... G7
Chemicator Refill Tube #11 ...... G7
Chemicator Refill Tube #51 ...... G7
Chemicator Refill Tube #515 ... G7
Chemicator Refill Tube #534 ... G7
Chemicator Refill Tube P-40 ..... G7
CLT-50 Condensate Line
    Treatment G6
D-81 Deposit Control G7
D-84 Cooling Water Control G7
Deep Purple Algaecide G7
Descaler G A3
Descaler 140 .... A3
Descaler 60 .... A3
DB-31 Boiler Boilout
    Compound G6
DR-20 Rust Removing
    Compound A3
DR-45 .... A3
E-50 .... G6
E-50V G6
E-57 Boiler Water Control G6
E-57V Boiler Water Control G6
Erlen Glisn A3
Erlin Hifome Liquid Soap E1
F-45 .... G7,G3
F-45N G7,G3
F-91 .... G7,G2
F-91P Corrosion Control G7
G-30 Condensate Line
    Treatment G6
G-40C G6
K-24 Phosphate Pearls G2
K-25 Phosphate Pearls G2
K-26 Phosphate Pearls G2
K-72 Corr. Control for Brine
    Tanks P1
K-81 .... G7
Kleen-Koil A1
KN-10 .... G7
KP-20 Cooling Water Control ... G7
KP-20A G7
KP-40 .... G7
KP-40 A G7
KP-50 Cooling Water Control ... G7

277

#3109 D-139 .... A1
#3203 D-03 .... A1
#3204 D-04 .... A1
#3218 Drain Cleaner.................. L1
#3222 Flor Klene...................... A4
#3224 D-24 .... A2
#3226 D-39 .... A1
#3235 Sprayzol........................... A1
#3257 Paunch Perfect................ N3
#3280 Super Cleaner................. A1
#3291 711 .... A1
#3295 Rust Off.......................... A2
#3302 D-33 .... A1
#3322 Gen Chlor A1
#3330 Whistcip........................... A1
#3340 Utopia A1
#3349 .... A3
#5201 Glo-Wash A1
#5202 LO-137............................. A1
#5203 Glo-Clean A1
#5255 Lance A1
#5256 Mace A1
#5259 Saber A1
#5262 Magic Sheen.................... A1
#5270 Whistlphos...................... A3
#5281 Whistl Fe Off.................. A3
#5501 Liquid Hand Soap E1
#5520 Extra Hand Soap E1
#566 Lily Cleaner A4
#903 D-130 .... D1
LD-129 .... A1

White & Bagley Company, The
Oilzum Multi Purpose
  Lubricant H2
Oilzum White Multi Purpose
  Lubrcnt................................... H2

White Castle Company
White Castle Foam Degreaser
  & Cleaner A1

White Cross Chemicals
Iokleen D1
Killmor D1
Sterakleen D1

White King, Incorporated
Liquid Hand Soap 15% E1
Mermaid A1

White Spot Supply, Inc.
Master Superterge A4

Whitehall-Gardner Industries Inc.
W-G 400 Cleaner-Degreaser A1

Whitlock, W.D. & Associates
All Purpose Fly & Mosquito
  Insectcd F1
CL-10.. A4
E-Z Open L1
Heavy Duty Cleaner.................. A4
WD-22 .... D1

Whitmire Research Labs., Inc.
Prescription Treat. No. 110
  Arsl Gen................................. F1
Prescription Treat. No.500
  Insect Fogger.......................... F1
Prescription Treatment No. 125. F1
Prescription Treatment No. 250. F2
Prescription Treatment No. 260. F2
Prescription Treatment No. 270. F2
Prescription Treatment No. 280. F2
Professional Aerosol Insecticide F1
PT 140...................................... F1
PT 150 .... F1
PT 550 .... F1

PT 565 .... F1

Whitmoyer Laboratories, Inc.
DLP-787 (Rat And Mouse Bait) F3
DLP-787 House Mouse
  Tracking Powder.................... F3
Iofec 20..................................... D2
Iofec 80..................................... D2
San-O-Fec D1
San-O-Fec 50 .... D1
Vacor RatKiller F3

Wichita Brush & Chemical Co., Inc.
All-Clean A4,E1
AC-210 Safety-Solv..................... K1
Beta-Phene C1
Bol Wite C2
Bowl Cleaner C2
Coco Han E1
Con-Septic................................. C2
CL-200 Heavy Duty Hard
  Surface Cleaner...................... A4
CL-21 Industrial Detergent
  Concentrate A4
CL-22 Butyl Base Heavy Duty
  Concentrate A4
CL-270 H.D. Cleaner
  Degreaser A1
De-Clog Prilled Drain Opener... L1
Dish Swish A1
Fly Mist.................................... F1
Foaming Grease Go.................... A1
Gamma-Mene.............................. D1
Gamma-Mene 25.......................... D1
Glass Cleaner............................ C1
Glass Mist................................. C1
Glass Mist (Aerosol) C1
Grease-Go A1
Ice Away................................... C1
Jet Stream A4
Jet Stream Special A1
Liqui Klenz C2
Metal Mist A7
Omega-Mint C2
Phi-So-Creme............................. E1
Pi-N-O-Dis C1
Pi-N-O-Dis II............................ C1
Porcelain Cleaner C2
Roach Mist................................ F2
Sani-Turg.................................. D2
Sigma-San.................................. D2
TBDD Mist (Lemon Scented) ... C2
Wich-Brite............................ A1,E1
Wichcraft Lemon-D-D C2

Wiegand Engineering Corp.
PCW Detergent........................... A1

Wil-Kil Pest Control Company
Wil-Kil Fly Bait......................... F2
Wil-Kil Ready to Use Rat &
  Mouse Bait F3

Wilbur-Ellis Company
Red Top Malathion PG-3 Oil
  Solution................................. F2
Red Top Py-Rin 20 Space.......... F1
Red Top Py-Rin 40 Jet.............. F1

Wilco, Inc.
NB-100...................................... A1

Wildcat Chemical Company
HDC-50 .... A4

Willamette Chemical Company
All Purpose A1
Big-N-Bold A1
Laundry Compound................... B1

Tac A1
Will-Chem-Cide D1
WC-Thirty-One A1
WC-Thirty-Three A2
WC-200 .... A1

Willex Products, Incorporated
Willex-Cleaner-Degreaser A1

William Edwards Company, The
Bleach D2

William F. Nye, Incorporated
Delicate Machinery Oil 100 .... H1
NyoGel 721 .... H1

Williams Chemical Co., Inc.
Action A3
Aqua Magic A1
Caress A1
Chlor-O-Brite A1
Genie A1
Rite Laundry Detergent B2
Rite Laundry Detergent B1
S-700 .... A1
XHD Liquid Steam Cleaner A2
XHD Powder Steam Cleaner..... A1

Williams Chemical Corporation
Blue Wonder C1
Defoamer C Q5
Fumeless L1
Gold Tiger C1
Sewer-Solv L1
Smoke House Cleaner A2
Steam Quick A2
Thermotron A2
Vat Dip A2
Wil-Di-Chlor C1
Wilco AP Cleaner A1
Wilco Bowl Cleanse C1
Wilco Drain Opener L1
Wilco F.E.L. H1
Wilco Mineral Stripper C1
Wilco Silicone Spray H1
Wilco 1001 Hvy Dty All Purp.
  Clnr. A4

Williams Lime Manufacturing Co.
Hydrated Lime N3

Williams, B. E.
Shroud Life B1

Williard Incorporated
Wil Chem 565 .... G6
Wil-Chem BT 1001-X G6
Wil-Chem BT 2001 .... G6
Wil-Chem BT 3001-X G6
Wil-Chem BT 4001 X G6
Wil-Chem Phosmer D-I G6
Wil-Chem Scale Off A3
Wil-Chem Sludge SOL-3 .... G7
Wil-Chem Wil-Meen D-10 .... G6
Wil-Chem Wil-Meen D-20 .... G6
Wil-Chem Wil-Meen D-30 .... G6
Wil-Chem 3132 .... G6
Wil-Chem 3136 .... G6
Wil-Chem 5232 .... A2
Wil-Chem 9500 .... G6
Wil-Chem 9523 .... G6

Willis Supply Company
All Purpose Kitchen Cleaner A1
C.G. Liquid A1
Supreme Automatic Dishwash... A2

Willmarch, LTD.
APC-All Purpose Cleaner A4,B1

Wilmar, Incorporated
Bore L1
Degreasol K2
Egg Clean Q1
Fly Die F1
Formula 224 .... F1
Formula 400 .... K2
Get All A1
Hand Dishwashing A1
Kleen-Out A3
Lemon Cleaner A1
Lemon Drip Fluid C2
Liquithaw C1
Mint CDD C2
Mintolene C2
Pine CDD No. 1 .... C1
Quick-Way A1
Sanitide D1
Scrubkleen C1
Sewer Solvent L1
Shackle Cleaner A1
Shackle Cleaner-P A1
Silkeen C2
Steam Flo A2
Steam-Rite C2
Steamco A2
Steamster A2
Super B C. C2
Surgium Surgical Scrub Soap E1
Truck Brite A3
Wil Float A1
Wil-Fog F1
Wildene A1
Wildene Concentrate A1
Willoteen A1
WM8 .... A4
WM8 Concentrate A1

Wilsel, Industries
Wilsel 1 Extra HD Clnr-
  Degreaser A1
Wilsel 2 Concentrated Lemon
  Clnr. A4
Wilsel 23 Electric Motor
  Cleaner K1
Wilsel 3 Window Cleaner C1
Wilsel 81 Laundry Compound ... B1
Wilsel 84 Powdered Hand
  Cleaner C2

Wilson & Company, Incorporated
#372 General Cleaner A1
#561 .... A3
#658 .... B1
#674 .... B1
APC #10 .... A3
Bon-Gleem A3
Bontex B1
DP-509 .... A1
Heavy Duty Cleaner A1
HD-626 .... B1
Laundry Compound 582 .... B1
Laundry Compound 673 .... B1
P.G.C. A1
T.O.T. A3

Wilson Supply Company
General Purpose Degreaser
  Oven&Grill Cl A1
Heavy Duty Floor Cleaner A1
Lime Solvent A3

Windsor Wax Co., Inc.
All-Purpose Detergent Cleaner
  257 .... A4
Ammoniated Wax Stripper 255.. A4
Biodegradable All-Purp.
  Detrgnt. Cl. 258 .... A1
Industrial Cleaner 252 .... A1

Industrial Cleaner 256 .... A1

Winfield Brooks Company, Inc.
Alpha Beta Acmé Special
  Formula A1
ABA Rinse Treatment P1
Defoamer P-31-1 .... A1

Winfield Industries
Chlorinated C.I.P. Cleaner
  Extra A2
Coconut Oil Liquid Soap E1
De-Grease Hvy Dty Clnr Dgrsr
  & Stripper C1
Delimer A3
Foam-All A1
I-O-Teen Adjusted D2
Io-Det Iodine Germicide D1
Kill Odor C1
Power Kleen Heavy Duty
  Stripper/Dgrsr. A4
Roach & Fly Spray F1
Sewer Solvent L1
Sky-Rite Whisper-Clean C1
Sky-Rite 3-30 Cleaner C1
Smokehouse Cleaner A2
Steam Cleaner A1
Vaporizing Spray F1
W-220 .... A2
W-30 Water-Soluble Degreaser.. A1
W-35 .... H1
Winsurg Liq. Hand Soap
  Concentrate E1
1000 Silicone Lubricant H1
1000 Silicone Spray H1
1200 Freezer And Locker
  Cleaner A5
1800 Inhibited Chem. Clnr. &
  Deoxidizer A3
195 Heavy Duty Chlorinated
  C.I.P. Clnr. A1
2110 Power-All A4
2450 Liq. Boiler & Feed Wtr.
  Treatment G6
3111 Conctd Acid Cl & Stnlss
  Steel Brtnr A3
3210 Acid Cleaner & Milkstone
  Remover A3
3500 Chlorinated C.I.P. Cleaner A1
40 Plus Liquid E1
401 Milkstone Cleaner Acid A3
410 All Purpose Cleaner A1
520 Smoke House Cleaner A2
6209 Heavy Duty Cleaner A1
880 Concentrate A1

Winfield Supply Company
"8000" Professional Cleaner A4
Foaming Degreaser A1

Winner Chemicals, Incorporated
M & P 23 .... A1

Winner Sales And Service, Inc.
Winner C-33 .... A1

Winokur Water Systems, Inc. A Div of Culligan Water Treat. Co., Inc
W-913 .... G2
WB-224 .... G6
WB-226 .... G6
WB-269 .... G6

Winston Chemical Company
Formula C G6
Formula CPD G7
Formula M G6
Formula MPD G7
Formula OSA G7

Formula 200 .... A1
Formula 400 .... A1
Formula 600 Detergent-Sanitizer. D1

World Wide Laboratories, Inc.
Concrete Cleaner 3418-FPP A4
W.W. Lab 3300 .... A3
W.W. Lab 3386 .... A1
W.W. Lab 3399 .... Q1
W.W. Lab 3414 .... A1
W.W. Lab 3426 .... A2
W.W. Lab 3434 .... A1
W.W. Lab 3458 .... A2
W.W. Lab 3675 .... A1
W.W. Lab 3995 .... A3
W.W. Lab 3997 .... A1

World Wide Supply Company
Super-X #281 .... A2
Super-X Cleaner A1
Super-X Food Lube Food
  Equip. Lubricant....................... H1

World-Wide Chemicals
Aura-Pine C1
Dis-N-Clean C2
Double-Duty Toilet Bowl
  Cleaner C2
Drain Storm L1
Fortress Concentrated Cleaner .. A4
Lemony Disinfectant C2
Monday Laundry Detergent B1
Pink Lady A1
Redi-Soap E1
Scent-O-Mint Disinfectant C2
Solvs-All A1
Sparkle A1
Together E1
Tropic Concentrated Hand
  Soap E1
True Grit C2

Worne Biochemicals, Inc.
Polysol L1
Solubac L2

Wren Custom Chemicals
Flo-Eze Liquid Drain Opener ... L1
Formula 101 Heavy Duty
  Detergent Dgrsr. A1
Formula 99 Water Soluble
  Safety Cleaner A1
W-100 Sewer Solvent L1

Wright Chemical Corporation
Hytran 106-A G6
Hytran 132-C G6
Hytran 178 .... A2,G6
Hytran 23-A G6
Hytran 25-A G6
Hytran 41-C G6
Puron 152-A G2
Silspend 180 .... G7
Special Cleaner #9-G A3
Wrico BFL G7
Wrico CRA G7
Wrico D-3082-A G7
Wrico D-3385-A G6
Wrico D-6136 .... G6
Wrico D-6540 .... G6
Wrico D-6541 .... G6
Wrico D-8065 .... G6
Wrico DCB G7
Wrico DFB G7
Wrico DMB G7
Wrico H-1015 .... G7
Wrico H-1116 .... G7
Wrico H-3840 .... G7
Wrico H-7654 .... G7

Wrico NDR G7
Wrico STR G7
Wrico TQC G7
Wrico 12 .... G6
Wrico 42 .... G6
Wrico 42-C G6
Wrico 84 .... G3
Wrico-DCF G7

Wright Custodial Supply
PCK-60 Foam-Up Degreaser A1
PCK60-Heavy Duty Cleaner-
  Degreaser A1

Wright Industrial, Inc.
Blue Concentrate Detergent
  Clnr A4
Bug Trol Insect Spray F1
Bug Trol II F1
Cream & Clean II......................... E1
Foam-Rite A1
Green Giant Plus Cleaner &
  Degreaser A1
Hi-Low C1
Lemon Hand Soap E1
Nu Concentrate A4
Pro-Pink Concentrate A1
Safe-Cide Brand Neutral
  Fragrance D1
Steam-Wright A1

Wright Oil Company
Aero White Food Mach.
  Grease AA H1

Wright Rodent & Pest Control
  Laboratories
Ryte-Rat 25 .... F3

Wright, Inc.
Bold Clean Toilet Bowl Cleaner C1
Chlorinated CIP Cleaner A1
Conveyor Lubricant..................... H2
Dab B1
Dish Clean Chlor. Mach. Dish
  Wash. Cmpd.............................. A1
Drain Clean Sewer Pipe
  Cleaner L1
Du-Rite B1
Egg Wash Non-Foam Egg Clng
  & Destain Cmp Q1
Gusto Oven & Grill Cleaner A8
H.D. 50 Extra Heavy Duty Liq
  Laun. Det. B1
Ideal Bleach D1,B1
Ideal Fabric Softener B2
Ideal Lemon Lotion Hand
  DishwashDetrgnt...................... A1
Ido-Rite Hand Wash Sanitizer .. E2
Liquid Chlorinated CIP Cleaner A1
M.P. Concentrate A4
Muscle Pot & Pan Cleaner A1
Pink-Luster Pink Lotion
  Detergent A1
Punch Chlorinated Ware-
  Washing Detergent A1
Rid Detergent A4,B1
Rite Chlor Dry Chlorine
  Bleach B1
Rite Kleen A1
Rite Off Concrete Cleaner.......... A1
Rite Quick Foam Foaming
  Cleaner-Dgrsr. A4
Rite-Hand Powdered Hand
  Cleaner w/Borax C2
Rite-Soft Extra Conc. Fabric
  Softener B2
Rose Magic Hand Dishwashing
  Beads A1

Special "A" B1
Super Bold Toilet Bowl &
  Porcelain Cln. C1
Super Rite-Off Concrete
  Cleaner A4
Super Safe All Fabric Bleach..... B1
Super-Chlor D2,G4
SC-66 All Purpose Steam
  Cleaner Compound.................. A1
SC-76 Steam Cleaner
  Compound A1
SC-86 Steam Cleaner
  Compound A2
Tru-Rite Bleach D1,B1
VT-84 .... D2
W-100 Industrial Detergent B1
Wright Bleach D2
Wright 20% Dry Bleach B1

Wrighteo Chemical Company
Micro-Non D2
Wrightco Egg Washing Powder Q1
Wrightco Utensil Cleaner
  Powder A1

Wynn Oil Company
Multi-Purpose Grease H2
Safety Solvent.............................. K1
Silicone Lubricant H2
Variplex Moly Grease.................. H2
Viscotene (Aerosol)..................... H2
Viscotene (Bulk) H2
Wynn's 951-1 Synthetic H2

WD-40 Company
WD-40 Formula A H2

X

X-Ergon/ A Div. of Partsmaster, Inc.
Gear-Guard 510............................ H2
Gladiator H2
Handi-Helper E1
Penefree #401 Nonflammable P1
Re-Lube 509 .... H2
Slalom H1
Slick H1

Y

Yale Chemical Division/ See Bio Lab, Inc.

York Chemical Co., Inc.
Certox Insect Killer..................... F2
Certox Insect Spray Code A0·5. F1
Certox Rozol Rat And Mouse
  Bait F3
Food Processing Emulsifiable
  Concentrate F1
Food Processing Fog.................. F1
Oil Concentrate #3610................ F1
York VAP-5................................... F2

Your Brand Products
#602 .... A1

M-50 Multi-Purpose
  Concentrate A1
M-60 Multi-Purpose
  Concentrate A1
SL-20 Multi-Purpose
  Concentrate A1
SL-30 Multi-Purpose
  Concentrate A1
SP-15 Multi-Purpose
  Concentrate A1

Zervis Chemical Company
Brute A1
D-Solv A1
Excel A2
JZ-2... A1
Supreme 5000............................... A1

Zesty Systems, Incorporated
Zestismoke Cleaner 50 .... A1

Ziff, S.N. Paper Company, Inc.
System One Bowl/Tile Prcln &
  Grout Clnr C2
System One Creme Cleanser A6
System One Emulsion Toilet
  Bowl Cleaner C2
System One Heavy Duty
  Power Dgrsr. 750 .... A1
System One Meat Room
  Degreaser 700 .... A1
System One Quat 220................... D2
Ziff System One 128
  Disinfectant D1

Zimmerman, Inc.
Barrier Skin Protectant E4
Complex Z66 .... A1
Flush L1
Hi-Zene....................................... D1
Praize... A1

Zimmite Corporation
ZC-222.. G2,G5

Zip Aerosol Products
D-5440 Fluorocarbon Dry
  Lubricant H2
Min-Oil Mold Lubricant M-304 . H1
MC Cleaner D-5675 .... K1
Pure Liq. Petrolatum USPD-
  5303 .... H1
Pure Silicone Anti-Stick FS-175 H1
Pure Silicone Lubricant D-5100. H1
Q.C. Chemicals Corrsn.
  Prevent. No.2499 .... H2
Q.C. Chemicals MC Cleaner
  No.2044..................................... K2
Zip Chem ZC-010 .... H2
Zip Slip H1
Zip-Aerosol D-5010 .... H2
ZC-100 Pure Silicone Lubricant H1

Zobrist, J.C. Company, Inc.
Brute 4 .... A4
Bu-Solv A4
Deodorizing Cleaner A4
Engine Cleaner K1
Extra Hvy Dty Steam Clnng
  Compound A1
Pink Concentrate Cleaner A4
Power Clean A4
Steam & Pressure Liquid
  Cleaning Cmpd A1
Steam Cleaner A1
Strip-Clean A4

Zodiac Chemical Company, Inc.
Formula 101 .... A4

Gemini 220................................... A3
Gemini 221................................... L1
Gemini 222................................... A3
Leo 35.. A1
Leo 45.. A1
Leo 5.. A1
Leo 65.. N3
Leo 75.. N2
Libra 20 .... A1
Libra 6 .... A1
Roar L1
Taurus 10..................................... A2
Taurus 100................................... A2
Taurus 20..................................... A2
Taurus 30..................................... A2
Taurus 40..................................... A2
Taurus 70..................................... A1
Vesta... A1

Zoecon Industries
Blue Sugar Bait........................... F2
Institutional Insect Strip F2
Institutional Vaporaire Odor
  Neut Strp C1
Starbar Golden Malrin Liquid ... F2
Starbar Trax M F3
Starbar Vapona 20w/DDVP F2
Thuron Fogasect Pressurized
  Fogger F2
Vaporette Insect Strip................. F2
VKP Mist..................................... F2
Zoecon Vaporette Spray F2

Zorbite Corporation
Zorbite.. J1

Zurn Industries, Incorporated
Polyzyme II L2

2

2-Cleen Chemical Laboratories
Degrease-It.................................. A1
Super Degrease-It........................ A4,A8
2-Cleen Lime & Rust Remover.. A3

3

3 D Chemical & Supply, Inc.
Medi Soap E1

3 W's Research Company
Spike L1

3-D Supply Company
3-D A1

3M Company
Foam Scrub No. 202 .... A4
Lithium Spray Lubricant Part
  No. 8915 .... H2
Scotch Brand Contact Cleaner
  1607.. K2
Scotch Brand Silicone 1609........ H1
Scotch Brand 1605
  Demoisturant............................ H2,K2
Scotch Brand 1606 Degreaser K1
Scotch-Grip Industrial
  Adhesive 1870.......................... P1
Sectrol Microencap. Pyrethrins
  Insect....................................... F2
Silicone Spray Lubricant H1
Spray Cleaner No. A-101 .... A4

7

7 Oaks Supply Co.

7HC Corporation

4

4 Share Incorporated

4-D Products, Incorporated

4Tek Industries Incorporated

UNITED STATES DEPARTMENT OF AGRICULTURE
FOOD SAFETY AND QUALITY SERVICE
Washington, D.C. 20250

Official Business
Penalty for Private Use, $300